T0291696

CAMBRIDGE LIBRARY COLLECTION

Books of enduring scholarly value

Darwin

Two hundred years after his birth and 150 years after the publication of 'On the Origin of Species', Charles Darwin and his theories are still the focus of worldwide attention. This series offers not only works by Darwin, but also the writings of his mentors in Cambridge and elsewhere, and a survey of the impassioned scientific, philosophical and theological debates sparked by his 'dangerous idea'.

John Ray, Naturalist

John Ray (1625-1705) was a clergyman and botanist who was dismissed from his post at Cambridge in 1662 for political and sectarian reasons and thereafter funded his research by private teaching and sponsorship. He was admitted to the Royal Society, and was a pioneer of the systematic classifcation of plants; his Historia Plantarum was the first textbook of modern botany. About a century after his death, Sir J. E. Smith (1759-1828), founder of the Linnean Society, praised 'our immortal naturalist, John Ray, the most accurate in observation, the most philosophical in contemplation, and the most faithful in description, of all the botanists of his own, or perhaps any other time.' This biography of Ray was first published in 1942 at the height of the Second World War. It was written by Charles Raven, an eminent theologian who shared Ray's deep respect for intellectual integrity, honest exploration of the natural world, and the value of both theology and scientific endeavour. More than a superb history, this offers an opportunity to reassess the pivotal contributions of a brilliant but often undervalued scientist. Ray's major publications were written in Latin; Raven's linguistic skills – coupled with his passion for natural history – made him ideally and uniquely suited to interpret Ray's scientific legacy. Raven reviews Ray's academic and scientific careers in the context of the dramatic social upheavals of his time. He evaluates the remarkable long-term and widespread influence of Ray's work on the development of science, alongside the significance of the tolerant philosophy in his final book, The Wisdom of God.

Cambridge University Press has long been a pioneer in the reissuing of out-of-print titles from its own backlist, producing digital reprints of books that are still sought after by scholars and students but could not be reprinted economically using traditional technology. The Cambridge Library Collection extends this activity to a wider range of books which are still of importance to researchers and professionals, either for the source material they contain, or as landmarks in the history of their academic discipline.

Drawing from the world-renowned collections in the Cambridge University Library, and guided by the advice of experts in each subject area, Cambridge University Press is using state-of-the-art scanning machines in its own Printing House to capture the content of each book selected for inclusion. The files are processed to give a consistently clear, crisp image, and the books finished to the high quality standard for which the Press is recognised around the world. The latest print-on-demand technology ensures that the books will remain available indefinitely, and that orders for single or multiple copies can quickly be supplied.

The Cambridge Library Collection will bring back to life books of enduring scholarly value across a wide range of disciplines in the humanities and social sciences and in science and technology.

John Ray, Naturalist

His Life and Works

CHARLES RAVEN

CAMBRIDGE UNIVERSITY PRESS

Cambridge New York Melbourne Madrid Cape Town Singapore São Paolo Delhi

Published in the United States of America by Cambridge University Press, New York

www.cambridge.org
Information on this title: www.cambridge.org/9781108004664

This edition first published 1942
This digitally printed version 2009

ISBN 978-1-108-00466-4

JOHN RAY

Ioannes Rajus).
Societatis Regiæ Socius.

W. Faithorne pinx.   G. Vertue Sculp.
                                                              1713

JOHN RAY
NATURALIST

HIS LIFE AND WORKS

by

CHARLES E. RAVEN, D.D.

Master of Christ's College and
Regius Professor of Divinity in the
University of Cambridge

CAMBRIDGE

AT THE UNIVERSITY PRESS

1942

PRINTED IN GREAT BRITAIN

TO ALL
WHO LIKE JOHN RAY
HAVE SACRIFICED SECURITY & CAREERS
FOR CONSCIENCE' SAKE

Contents

Preface

'A good book needs no preface; a bad book deserves none.' That is no doubt true. But when a student of theology turns aside (as it would seem) from his proper concern, when a normally active citizen in the middle of a great war fills much of his time with the life and work of a naturalist of the seventeenth century, it is reasonable that he should give some account of his eccentricity. Hence this personal explanation.

The history of science, with every respect for Mr Crowther and even Dr Hogben, has not yet been written. Nor in these days, when the use and abuse of scientific achievements are so significant, can the subject be regarded as unimportant. But my concern is not with the general record of man's discovery of the scientific method or of his application of it to the service of his needs and ambitions, so much as with one consequence of those events. As a theologian my primary task long ago convinced me of the importance of the change in man's aesthetic, moral and religious outlook which had accompanied and in large measure inspired the scientific movement. It was plain to me as a parson that the mixed folk whom I met as an entomologist and a bird-watcher had found an interest in nature which was singularly rich in educative and recreational value. Observation convinced me that this new resource was almost peculiar to the countries of Protestant Europe and North America; and enquiry disclosed that it was hardly compatible with the traditional devaluation of nature by the Churches. Experience affirmed that though the great poets and artists had seldom lacked it, its widespread influence was comparatively modern. Life in a great city suggested that as an antidote to the effects of urban and suburban environment here was a discovery which the sociologist and the reformer could not ignore.

Of its theological and religious value I have written at length elsewhere. In so doing the question of its origin inevitably arose; and my first answer —Linnaeus—was obviously inadequate. The great Swede had grace of character, diligence, vigour of intellect, a high opinion of his own importance, and the good fortune to hit upon a convenient system of nomenclature. He gave his name to an epoch: he was assuredly not a pioneer.

As I pursued the subject it became clear that the change from the old world of superstition, the world in which there was no settled frame of reference except that fashioned by deduction from the Bible and Aristotle, the world of alchemy and magic, took place not in the eighteenth century but in the seventeenth; that the transition was made by the simultaneous labours of the 'new philosophers' in Italy, France, Holland and Britain; and that in the realm of biology, or at least of botany and zoology, there was one man of outstanding genius, 'our countryman, the excellent Mr Ray'.

It is proof of the neglect into which Ray has fallen that whereas a century ago my discovery was a commonplace, it should come to me as unforeseen and surprising. But the reasons for this neglect are not obscure. No adequate biography of Ray exists, thanks to the default of his literary executor; his books were written in Latin and their contents had supposedly been taken over by his successors; his only writings in English were full of an old-fashioned religion and an unfashionable teleology. Why bother to preserve the memory of a man whose work may have made an era but has now passed out of mind?

The quality of the man himself prevented such an attitude. Behind the vast and varied achievement of one who had laid the foundation of the modern outlook upon nature lay a fascinating and heroic personality. The blacksmith's son who in an age of almost feudal class-consciousness won for himself reverence and friendship and that without sycophancy or aggressiveness; the fellow of Trinity who when consciences were elastic gave up everything sooner than declare that the Covenant which he had not subscribed was no oath; the explorer who 'trusting in Providence and good friends' toured Western Europe for its plants and birds, fishes and fossils, and travelled Britain so thoroughly that he could supply at first hand a list of the rarer plants for every English county; the writer who in the last twenty years of his life, straitened in means and racked with pain, produced some fifteen new books, including the 3000 folio pages of the *History of Plants*, the *Synopses* of animals, reptiles, birds and fishes, the first serious treatises on science and religion, and in addition some nine revised editions of these and earlier works; the naturalist who in the last decade of the century with the help of his four little daughters bred and described the life cycle of nearly three hundred local lepidoptera; could anyone with powers of admiration and interests in natural history resist so intriguing a subject?

To do him justice requires qualifications beyond my range. A deeper knowledge of science and of the seventeenth century would have saved me from blunders; a training in the history and practice of biology would have given me criteria for the estimation of his work and acquaintance with his background and contemporaries; ease of travel would have enabled me to trace his records more exactly. But at least I live in Cambridge, and the influence of the Cambridge Platonists who shaped him has also shaped me; I know enough Latin to read and appreciate his masterly language and descriptions; I have collected nearly all the plants, birds and insects that he records, and often in the same localities; and the thing that he tried to do, to reinterpret the faith of a Christian in the light of a sound knowledge of nature, has been my continuous and chief concern.

The form and length of the book require a word of explanation. It would have been easier and more popular to treat the subject in broad outline and subjectively; to pay my tribute and give my impression without bothering to survey a mass of detail. A sketch of the seventeenth century, that age of vitality and contrasts; a simplified rendering of Ray's character and achievement; a selection from his more memorable writings —the book could have been finished in three months. But if the work was worth doing, it was worth doing thoroughly and objectively. Only by meticulously collating every record of his travel and research could the blunders of his biographers be corrected and the details of his career be disclosed; only by setting out these details could the greatness of his work be established. It may be that others will not share my delight at the first discovery of the Sulphur Clover or the Alpine Bartsia, the Manx Shearwater or the Purple Emperor. For them perhaps it strikes no chord that he too enjoyed the abundance of Jacob's Ladder at Malham Cove or of Cornish Heath at Goonhilly, or that he marvelled at the loop of the trachea in the Whooper Swan, and the sanitary habits of newly hatched hawks, and the presence of Smelts in Rostherne Mere, and the Ammophila dealing with its caterpillar, and the assembling of male moths round a freshly emerged female. When he was confused by the defects of the books on which he relied, when he wrestled with the problems of local variation and plumage change, when he found himself compelled to question the orthodoxies of his day, when his friends reported to him impossible new species and fantastic new speculations, what naturalist will not remember his own experiences and feel a thrill of human sympathy? These things to me are of the stuff of life, and to omit them as trivial would have been to condone a false valuation. Ray must speak for himself, must tell us how he sought and what he found. It is surely good to praise famous men and our fathers that begat us—and never more good than when our dreams of the future are black and blank and when the story of a man who strove to bring order out of chaos and initiated a great epoch of human growth may be an encouragement and an example.

In view of what has been said it would be disingenuous not to add that in days when the first-hand study of nature which has for many years given me health and ability for hard work is impossible, it has been a refreshment to follow such pursuits vicariously and in the setting of an earlier time.

It remains for me to explain that in the case of a man pursuing so many different studies simultaneously a strictly chronological treatment was impossible. To trace his life as a botanist and then to add chapters on his work in other fields was the only practicable arrangement. The dated summary of his career sets out the sequence of events.

In the matter of orthography I have modernised the spelling of his English letters; and have written his own name as Ray even though until

1670 he spelled it Wray. I have translated his Latin letters; and have generally replaced his long and often difficult names for plants by their modern equivalents, printing these in italics. In the controversial matter of the modern nomenclature of flora and fauna (in which at present there seems little prospect of a permanent agreement) I have followed the best text-books rather than the latest alterations. I have adopted the 'new style' for dates.

To express gratitude by name to all those who have helped me would be to compile a long list of friends and acquaintances, and of experts in many fields from whom I have sought and received assistance. I would especially thank the officials of the Libraries of Cambridge University, Trinity College and St Catharine's College, of the Botanical School and of the Balfour Library, and of the Linnean Society of London. The Vice-Master of Trinity, Mrs and Miss Arber, Dr C. F. A. Pantin, Dr C. P. Snow and Dr F. H. A. Marshall have read and commented upon parts of the script; they and many others have given invaluable suggestions. The Rev. J. R. Southern, Rector of Black Notley, Mrs Turner of Dewlands, Mrs Moss of Ray Cottage, and in particular Mr Alfred Hills of The Buck, have helped me on my visits to Ray's home, and by much subsequent information. Lord Middleton gave me news of his discovery of the relics of Francis Willughby's collections, of the dried plants named by him and Ray and of the originals of the pictures illustrating the *Ornithology* and the *History of Fishes*; and his hospitality made it possible for me to see and study them—an unforgettable experience. To Mr A. E. Gunther and to Dr John Johnson, Printer to Oxford University, I am indebted for the loan of proofs of the late Dr R. W. T. Gunther's book on Ray's friend, Edward Lhwyd. Acknowledgments of help received from readers of the University Press are by tradition anonymous. But it happens that the press reader of this book was my friend of old days, Mr W. E. C. Browne, and I wish to thank him for his generous and expert help. Finally I must record my deep gratitude to the Council of the Royal Society for a grant towards the cost of publication, and must apologise to my family, to my long-suffering colleagues in Christ's College, and 'to all others whom it may concern' for the shamelessness with which I have bored them with my enthusiasms.

C. E. R.

1941

Introduction

The need for a fresh study of the life of Ray has long been recognised. Not only are the existing biographical notices (they cannot be called 'lives') admittedly defective, but thanks largely to the energy of G. S. Boulger and R. W. T. Gunther we have now available all the material that is likely to be recovered. It may be that the manuscript of his 'Catalogue of plants grown in the Cambridge gardens', which he seems to have written in or before 1662 and used in his *Historia Plantarum*, may yet be found: it will not add much to our knowledge of him or its subject. There may possibly be a few additional letters, perhaps even the letters to Robinson which Derham used and epitomised, but it is unlikely; and under present conditions search for them is impossible.

Apart from his own books the main sources are as follows:

1. The Life by 'a worthy friend', certainly Samuel Dale of Braintree, printed in *A Compleat History of Europe for the year 1706* under the heading 'Additions to the Remarkables of the year 1705': in 1705 the editor announcing Ray's death had complained that he had failed to obtain a worthy notice of him. This is the Life printed by R. W. T. Gunther, *Further Correspondence of John Ray*, London, 1928, from a MS. in the Bodleian: Dr Gunther was apparently not aware that it had been printed before.

2. *Philosophical Letters*, London, 1718, a volume of extracts from letters to and from Ray edited by Derham and arranged chronologically (with some errors). Derham omitted all personal details and selected what he thought scientifically interesting: of this he was not always a good judge.

3. The notice in Edmund Calamy, *A Continuation of the Account of the Ministers ejected*, London, 1727, I, pp. 120–2. Calamy had mentioned Ray in his *Abridgment of Mr Baxter's History*, ed. I, London, 1702, p. 239, and had given a note and a list of his books in ed. II, London, 1713, p. 87. This expanded notice is mainly concerned with his religious views: it cites I and comments on it.

4. The Life in *A General Dictionary Historical and Critical*, London, 1739, VIII, pp. 692–5: the author acknowledges help from Samuel Dale and his work is a slightly abbreviated and annotated version of I.

5. A similar Life in *Biographia Britannica*, London, 1760, V, pp. 3494–9: this contains little new except a criticism of Calamy's notice.

6. The Life preserved in a small notebook written in 1711–12 by James Petiver and now among the Sloane MSS., no. 3338: this is printed by Gunther, *Further Correspondence*, pp. 9–12: it shows acquaintance with I but is probably by William Derham, a sketch in 1710 for the following.

7. *Select Remains*, the Life written by Derham, the Itineraries, Prayers, etc. published London, 1760, by Derham's nephew, George Scott. This was reprinted as

8. *Memorials of John Ray*, published by the Ray Society, London, 1846, with added notices from 16 and 17 below.

9. *Correspondence of John Ray*, Ray Society, London, 1848, a reprint of 2 with the addition of Ray's letters to Hans Sloane in the Sloane MSS. The editing, by Edwin Lankester, is very defective, the sequence of the letters being distorted by his ignorance that they were dated in the old style, and the footnotes, unless supplied by C. C. Babington, being usually and often grotesquely misleading.

10. *Further Correspondence*, edited by R. W. T. Gunther for the Ray Society, London, 1928, containing series of letters from Ray to Courthope, Lister, Aubrey and Lhwyd, together with 1 and 6 above, with Ray's contributions to the Royal Society, and with Derham's epitomes of his letters to Robinson: a useful volume marred by some carelessness in editing.

11. Certain further letters of Ray to Courthope obtained by Gunther and printed in *Journal of Botany*, August 1934, and in his *Early Science in Cambridge*, 1937.

12. Two letters of Ray to T. Burrell in facsimile in *Early Science in Cambridge*, pp. 351, 354.

13. A letter of Ray to Aubrey printed in Aubrey's *Natural History of Surrey*, London, 1719, v, pp. 408–11, and in his *Natural History of Wiltshire* edited by J. Britton, London, 1847.

Apart from allusions in the letters and books of his contemporaries, and these are very few, this appears to be all the original or early material except his own books. Both as correspondent and as author Ray is singularly reticent about his own life, his feelings and actions. His notes of the localities of flora and fauna give a clue to his journeys: very occasionally there is a reference to his neighbourhood or, in the unfinished *Historia Insectorum*, to his family: and on a few occasions, hardly a dozen in all, he mentions some fact not directly connected with his work. His diary, alluded to by Dale in 1 above, seems to have perished without trace.

More recent records are without exception based upon 2 and 7, Derham's two books: several of the earlier ones show an appreciative knowledge of his writings. The chief of them are the following:

14. Albrecht von Haller in his *Bibliotheca Botanica*, Zürich, 1771, I, pp. 500–6, a very generous tribute to 'the greatest botanist of all time'.

15. Richard Pulteney in his *Sketches of the Progress of Botany*, London, 1790, I, pp. 189–281, probably still the best account of Ray and his works.

16. James E. Smith in *Rees' Cyclopaedia*, London, 1819, a condensed version of 15 with the author's own estimate of Ray's botanical studies.

17. Georges Cuvier and Albert Dupetit-Thouars in *Biographie Universelle*, Paris, 1824, XXXVII, pp. 155–63 or 2nd edition, Paris, 1843, XXXV, pp. 252–6, a careful and appreciative account by two distinguished French savants.

18. Edward Stanley, 'Philosophical Correspondence of Ray and Willughby', in the *Retrospective Review*, London, 1826, XIV, pp. 1–31, a survey of the various scientific subjects dealt with in 2.

19. John Lindley in *Penny Cyclopaedia*, London, 1841, XIX, pp. 317–19, short résumé of 7 with some useful comment on Ray's systematic work.

20. James Duncan, 'Memoir of Ray', in *Jardine's Naturalist's Library*, Edinburgh, 1843, XXXV, pp. 17–70, pleasant but entirely secondary.

21. Mrs Jacob Howell Pattisson[1] in the *Englishwoman's Magazine* for May 1847, II, pp. 257–75, an exceptionally careful and well-informed article, based upon 7, but also upon a MS. life by the Rev. W. L. P. Garnons, and upon local knowledge, her husband, a solicitor of Witham, being at this time owner of Ray's house, Dewlands; she gives an admirable list of sources.

22. G. S. Boulger in *Dictionary of National Biography*, vol. XLVII, in *Transactions of Essex Field Club*, *Essex Naturalist* and *Essex Review*, interesting but not free from errors and guess-work.

23. Albert C. Seward, *John Ray*, Cambridge, 1937, a sketch of his career and character, slight but vivid, often mistaken through reliance on Lankester and Boulger.

In addition there are a number of brief accounts in recent histories of botany: e.g. S. H. Vines in *Makers of Modern Botany*; J. Reynolds Green, *History of Botany in the United Kingdom*; R. J. Harvey-Gibson, *Outlines of the History of Botany*; Ellison Hawks, *Pioneers of Plant Study*. These are all slight and show no signs of serious or independent study.

Besides these printed sources a thesis for the degree of M.Sc. in the University of London entitled 'Studies in the Biological Works of John Ray' was presented in 1933 by D. C. Gunawardena. A typed copy of this in the Library of the Linnean Society of London was lent to me by the kindness of Mr S. Savage. It is a useful survey of Ray's botanical works and traces with considerable insight the development of his ideas of structure and classification. The author is handicapped by incomplete knowledge of Ray's works, of the seventeenth-century background and of the British Flora. A résumé of its findings is in *Proceedings of the Linnean Society*, 27 March 1936, p. 71.

EPITOME OF RAY'S LIFE

1627	29 Nov.	Born at the smithy, Black Notley.
	6 Dec.	Christened, Black Notley church.
1638	13 Aug.	Joseph Plume became Rector.
	?	Ray went to Braintree school.
1644	12 May	Admitted Trinity College, Cambridge.
	28 June	Entered Catharine Hall, pupil of Daniel Duckfield.
1646	21 Nov.	Transferred to Trinity, pupil of James Duport.
1647/8		Graduated B.A.
1649	8 Sept.	Elected Minor Fellow.
1650		Illness: began study of botany.
1651	1 Oct.	Appointed Greek Lecturer.
		M.A. degree.

1 This is stated by the *Gazetteer of Essex*, published 1848, and is borne out by internal evidence.

1653	1 Oct.	Appointed Mathematical Lecturer and Tutor.
1655	2 Oct.	Appointed Humanities Lecturer.
1656	31 Aug.	Death of his father: he built Dewlands for his mother.
		? Again appointed Greek Lecturer.
1657	1 Oct.	Appointed Praelector.
1658	3 Jan.	First letter to Courthope.
		Refused living of Cheadle.
	9 Aug.–	Journey to Derbyshire and North Wales, alone.
	18 Sept.	
	2 Oct.	Appointed Junior Dean.
1659	? March	Death of his friend John Nidd.
	26 Dec.	Appointed Steward.
1660		Published *Catalogus Cantabrigiam*.
	June–	Journey to North England and Isle of Man with Willughby.
	July	
	Aug.	Visited Thomas Browne at Norwich.
	25 Sept.	To Cambridge from Black Notley.
	16 Dec.	Appointed Steward, second year.
	23 Dec.	Ordained in London.
1661	26 July–	Journey to York, Edinburgh, Glasgow, Carlisle with Skippon.
	7 Sept.	
	Oct.	Refused living of Kirkby Lonsdale.
1662	Jan.–	In Sussex with Courthope and Burrell.
	April	
	April	Visited London, Morison's and Morgan's gardens.
	April.	Cambridge, last botanising there.
	8 May–	Journey round Wales with Willughby and Skippon: sea-birds
	16 June	studied, Prestholm, Bardsey, Caldey.
	16 June–	Journey continued to Land's End with Skippon.
	24 July	
	24 July–	At Black Notley, explored Essex.
	30 Aug.	
	24 Aug.	Forfeited Fellowship under Act of Uniformity.
	31 Aug.–	Visited Cambridge.
	10 Sept.	
	11 Sept.	Visited Barnham at Boughton.
	Oct.	In London.
	? 13 Oct.	At Friston as tutor with Thomas Bacon.
1663	19 March	Left Friston for Black Notley.
		Published Appendix to *Catalogus Cantabrigiam*.
	1 April	Met Skippon in Kent.
	18 April	Left Dover for Calais.
		Journey through Low Countries, up the Rhine to Vienna and Venice.
1664	Winter	At Padua, studying anatomy.
	Spring	Genoa, Leghorn, Naples, Willughby leaving for Spain.
	Summer	Went with Skippon to Sicily, Malta, Florence and Rome.
	1 Sept.–	At Rome studying birds and fishes in markets.
	24 Jan.	

1665 24 Jan.– To Rimini, Venice, Bolzano, and across Switzerland.
        June
        Summer At Geneva botanising.
        Autumn At Montpellier with Skippon and Lister.
1666 Spring   From Montpellier to Paris, Calais and Essex.
        ? June   Visited Cambridge and Sussex.
        Winter   At Middleton with Willughby.
                 Composed Tables for Wilkins's *Real Character*.
1667 June   Cambridge (passing) and Black Notley.
        18 June First letter to Lister—from Middleton.
        25 June– Journey to Worcester, Gloucester, Cornwall, Dorset, Hants,
        13 Sept.   with Willughby.
        13 Sept.– At Black Notley seriously ill.
         7 Nov. Admitted Fellow, Royal Society, London.
     ? 24 Nov. In Sussex with Courthope.
1668 April   At Boughton with the Barnhams.
        May–June Much travel, London, Essex, Haslingfield.
        July   Fortnight's journey in Yorkshire and Westmorland alone.
        26 July– At Broomhall with Jessop.
        Sept.
        17 Sept. At Middleton.
        29 Sept. At Black Notley.
        Nov.–Dec. At Middleton with Willughby.
1669 Jan.   At Chester with Wilkins.
        Feb.–March At Middleton, experiments with sap.
        April   At Chester, dissects porpoise.
        May   At Middleton, visit to Dorking, Oxford, Dartford.
        14 Oct. Journey to Wharton, Salop.
1670 28 April– At Wollaton with Willughby.
        29 June
        July   At Middleton.
        22 Aug. Changed spelling of name, Wray to Ray.
                 Published *Catalogus Angliae*.
                 And *Collection of English Proverbs*.
1671 Spring   Jaundice at Middleton.
     ? 21 June Visit to Cambridge.
         3 July Journey to Settle, Berwick, Brignall with Willisel.
        Autumn At Middleton.
         9 Nov.– In London at Royal Society.
         7 Dec.
        Dec.–Feb. At Chester with Wilkins.
1672 Feb.–7 March At Middleton.
        March   Visited Black Notley.
        March–Nov. At Middleton.
         3 July Death of Willughby.
        18 Nov. In London.
        19 Nov. Death of Wilkins: returned to Middleton.
1673 ? Feb.   Published *Observations and Catalogus Exteris*.
         5 June Married Margaret Oakeley at Middleton.

1673 ? Nov.   Published *Collection of English Words*.
1675 ? March Published *Dictionariolum*.
    15 March In London with Hooke and Sir John Cope.
             Left Middleton for Coleshill.
1676 ? 4 April Moved to Sutton Coldfield.
    Summer Visited Essex.
    ? Jan. Published Willughby's *Ornithologia*.
1677 28 Sept. Refused Secretaryship, Royal Society.
    Nov. Left Sutton Coldfield for Faulkbourne Hall near Black Notley.
           Catalogus Angliae, second edition.
1678 Published English version of *Ornithology*.
           Collection of Proverbs, second edition.
1679 15 March Death of his mother.
    24 June Moved to Dewlands, Black Notley.
1682 Published *Methodus Plantarum*.
1683 27 July First letter to Robinson.
1684 12 Aug. Birth of twin daughters, Margaret and Mary.
    11 Nov. First letter to Sloane.
1685 Second Appendix to *Catalogus Cantabrigiam*.
1686 March Published Willughby's *Historia Piscium*.
    June *Historia Plantarum*, vol. I.
1687 3 April Birth of daughter Catharine.
    Sept. Visit to London.
1688 Published *Historia Plantarum*, vol. II.
           And *Fasciculus Britannicarum*.
1689 10 Feb. Birth of daughter Jane.
    21 June First letter to Lhwyd.
           Nomenclator Classicus, second edition.
1690 March Attack of pneumonia.
    May Published *Synopsis Britannicarum*.
           Began collecting insects.
1691 Visited Bishop Compton at Fulham.
           Published *Wisdom of God*.
           Collection of English Words, second edition.
1692 ? Aug. Visit of John Aubrey.
    Feb. Published *Miscellaneous Discourses*.
           Wisdom of God, second edition.
1693 Published *Synopsis Quadrupedum*.
           And *Collection of Curious Travels*.
           Three Physico-Theological Discourses, second edition.
1694 ? Dec. Published *Sylloge Europeanarum*.
1695 11 July Visit of J. Morton.
    ? Oct. Visit of Robinson and W. Moyle.
    ? Nov. Visit of Vernon.
           Published County Lists in Camden.
1696 Published *Dissertatio de Methodis*.
           Synopsis Britannicarum, second edition.
           Nomenclator Classicus, third edition.
1697 July Visit of Krieg.

1698 ? 29 Jan. Death of daughter Mary.
     July   Illness of wife and Margaret.
1699 July   Visit of Petiver and Buddle.
1700 ? Sept.   Published *Persuasive to a Holy Life.*
1701   Visit of Sir T. Millington.
            Wisdom of God, third edition.
1703 April   Visit of Bishop Compton, and of J. Breyne.
     Jan.   Published *Methodus Emendata.*
            Nomenclator Classicus, fourth edition.
1704 March   Seriously ill.
     May   Visit of Derham.
     Aug.   Visit of Sloane.
            Published *Historia Plantarum*, vol. III.
            And *Methodus Insectorum.*
            And *Wisdom of God*, fourth edition.
1705 17 Jan.   Died at Dewlands.
1710 Spring   Publication of *Historia Insectorum.*
1713   Publication of *Synopsis Avium et Piscium.*
            And *Physico-Theological Discourses*, third edition.
1718   *Philosophical Letters.*
1760   *Select Remains.*

ABBREVIATIONS

C.A.   *Catalogus Angliae.* Ray, 1670.
C.C.   *Catalogus Cantabrigiam.* Ray, 1660.
C.E.   *Catalogus Exteris.* Ray, 1673.
Corr.   *Correspondence of J. R.* Ray Society, 1848.
F.C.   *Further Correspondence.* Ray Society, 1928.
H.I.   *Historia Insectorum.* Ray, 1710.
H.P.   *Historia Plantarum.* Ray, 1686.
H. Pisc. *Historia Piscium.* Ray, 1685.
Mem.   *Memorials.* Ray Society, 1846.
Obs.   *Observations.* Ray, 1673.
Orn.   *Ornithology.* Ray, 1678.
S.A.   *Synopsis Avium*, Ray, 1713.
S.B.   *Synopsis Britannicarum.* Ray, 1690.
S.P.   *Synopsis Piscium.* Ray, 1713.
S.Q.   *Synopsis Quadrupedum.* Ray, 1693.

CHAPTER I

BOYHOOD AND YOUTH

He is a person of great worth; and yet humble, and far from conceitedness and self-admiring...a conscientious Christian; and that's much said in little.

JOHN WORTHINGTON TO SAMUEL HARTLIB, *Diary*, I, p. 333.

In these days when we all realise the importance of heredity and early environment in determining and in interpreting character, the student of John Ray will deplore more strongly than did any of his biographers our almost total ignorance of his parents and childhood. To Derham or to Dale the fact that he was the son of a village blacksmith had to be stated but should then be forgotten. It was unconventional if not indecent. They knew it; but neither they nor their successors thought it necessary to amplify it. For more than a century[1] the year of his birth, though correctly given in at least one of his books, was wrongly stated as 1628; until 1847 no one had taken the trouble to search the parish register, and even then he was sometimes identified with the wrong John Ray. W. H. Mullens, who discovered that he had been baptised twelve months before the traditional time,[2] did not prosecute his researches further or spend the few hours needed to run through the long vellum pages that form the scanty annals of Black Notley in the first half of the seventeenth century. Of Ray's birthplace, family, school, and circumstances we are told almost nothing; and to-day it is impossible to recover more than a few fragments of what we have lost. We can only record those fragments, and estimate their importance in the light of the later life of their subject.

The hamlet, for even now it is little more, lies due south of the market town of Braintree, about a mile and a half away, and between the two roads that lead from it, the western to Chelmsford, the eastern above the right bank of the river Brain through White Notley and Faulkbourne to Witham. On this road lie most of the houses, a substantial rectory,[3] an

---

1 E.g. by *General Dictionary*; by Derham, *Mem.* p. 7; by J. E. Smith in *Rees' Cyclopaedia* (*Mem.* p. 65); and by Dale, *F.C.* p. 4; and as late as 1906 when J. Vaughan, *Wild-flowers of Selborne*, p. 127, says: 'The entry of his baptism...runs in almost illegible writing "John son of Roger and Eliz. Wray bapt. June 29, 1628"'—an error all the worse because it claims to be a transcript. It is given as 1627 on the frontispiece of his *Methodus Emendata* and as 29 Nov. 1627 in the *Englishwoman's Magazine* for 1847 (II, p. 257) with a copy of the entry in the Baptismal Register certified by the Rector, and in the *Cottage Gardener* for 1851 (V, p. 221), presumably by G. W. Johnson, the editor.

2 The entry of his baptism was reproduced in *British Birds*, II, p. 296: cf. *F.C.* p. 3.

3 Not the rectory of Ray's time but 'new built by the late incumbent Geoffrey Barton' (Morant, *History of Essex*, II, p. 125). Barton died in 1734.

inn, a small mill and a few cottages where a stream running from the west crosses the road. This is 'the little brook that runs near my dwelling'[1] from which Ray worked out his famous theory of springs. On the rising ground beyond it are the outhouses which still mark the site called the Dewlands on which in 1655 Ray built a house[2] for his widowed mother, the house into which he moved on her death in 1679 and in which he died in 1705. Leez Lane (or as we should spell it Leighs Lane), where he found Field Garlic (*Allium oleraceum*), a plant then new to science,[3] and which led to 'Leez House, the seat of the Earl of Warwick', where he found both species of Reedmace growing together in the rivulet nearby,[4] leaves the Witham road beyond the modern sanatorium, some six hundred yards south of Dewlands. The road across the Brain to Cressing lies closer to the hamlet.

From the rectory a road, now called Baker's Lane, branches off to the south-west and in a quarter of a mile passes to the north of the church and the Hall.[5] These stand on a rounded knoll looking over the valley of the stream. The Hall is a simple fifteenth-century building much reconstructed but with some fine timbered rooms upstairs; opposite to it is a magnificent barn of the same period, a barn larger than the church and plainly of the same age and tiling as its roof. Going along the road one comes to a fork where another road runs north to Braintree, passing the house of Ray's friend James Coker, then called Plumtrees, but now 'The Buck'.[6] Beyond the fork and on the south side of the road stands the forge, now disused, but still containing its two brick fire-places and chimneys, bellows and anvil. Beside and behind the forge is the six-roomed cottage of timber and plaster in which the blacksmith and his family lived, a two-storeyed building running north and south with wash-house, kitchen and parlour on the ground floor and three bedrooms, one with a dormer window, above. Tradition declares this to be Ray's birthplace; and in such a place tradition, supported as it is by the unquestionable age of forge and cottage, is sufficient evidence.[7]

The church, in which on 6 December 1627 he was baptised, is small and faced with flint-work, with a wooden belfry and steeple on the west and two brick buttresses dated 1684 and marked I. P. (that is, Joseph Plume, rector from 1638 to 1645 and from 1662 to 1686) at the corners of the chancel. Its interior has been tragically restored in the nineteenth century. The pews have gone: the floor has been tiled: the rood-screen, which may have been already demolished in Ray's time, is only marked by the sealed doorway to it. The building is said to have been erected by Sir Geoffrey

1 *Miscellaneous Discourses*, p. 74.   2 Destroyed by fire 19 Sept. 1900.
3 *Syn. Brit.* ed. II, p. 229.   4 *Cat. Angl.* p. 308.
  5 For these cf. *Historical Monuments in Essex*, II, pp. 18–20.
  6 Plumtrees probably 'Plumptre's' from the owner's name: *History of Essex*, l.c. speaks of it in 1770 as 'Plumtrees alias The Buck'.
  7 It is accepted as such in 1851: *Cottage Gardener*, V, p. 221.

de Mandeville and is manifestly Norman: the rounded windows and main door still remain in the nave: but later centuries have added windows of varying periods and an attractive porch. To the east of this stands the monument erected originally by Henry Compton, Bishop of London, and other subscribers in Ray's honour and bearing the elaborate Latin inscription composed by the Rev. William Coyte[1] and translated into hardly less elaborate English couplets in the *General Dictionary*.[2] Near it stands the flat-topped tomb of his friend Benjamin Allen, the Braintree doctor. In the vestry of the church is a record of the bequest to the poor of the parish by another friend already mentioned, James Coker 'of the house called Plumtrees',[3] a retired grocer and general trader of Braintree,[4] in whose labyrinth one of Ray's best insects was caught.[5]

The registers of the parish for this period consist only of a single volume containing the baptisms, marriages and burials from 1570 to 1669. The first fifty years are well kept, being apparently copied into this book from earlier records: the writing is good and the ink well preserved. After that the entries are very variable, sometimes defective and often hard to decipher. From 1670 till 1735 the registers have been lost.

The story is neither long nor full. In May 1592 John Ray, subsequently the grandfather of the naturalist, lost his first wife, Mary, perhaps at the birth of a daughter of the same name whose wedding to Richard Nichols is registered in 1610. John must have married his second wife, Elizabeth, shortly afterwards; for Roger, the father of the great John, was baptised on 28 March 1594. There were eight other children; Elizabeth born in 1597, who perhaps married John Calloge at Bocking in September 1636;[6] William, who must have died in infancy,[7] in 1599; Thomas, afterwards father of a family registered at Black Notley, in 1601; another William, who was buried in June 1621, in 1602; Sarah, who seems to have married Thomas Barker, a widower of Braintree, in 1647, in 1605; Ellen, in 1608; Katherine, who died in 1631, in 1611; and John, in 1614. Their mother, Elizabeth Ray senior, died in May 1623.

In 1624 Roger Ray, whose marriage to his wife Elizabeth must have taken place elsewhere, had his eldest son baptised Roger on 16 July: he died in childhood and his burial is registered on 26 December 1632. In 1625 a daughter Elizabeth was baptised on 17 November. In 1627 John, the third child, was baptised on 6 December. There is no further record of Roger's family except that of his own death on 31 August and burial on 1 September 1655.

---

1 Of Balliol College, Master of Woodbridge School, 1703–8; father of William Coyte, the botanist.   2 Vol. VIII, p. 695, published in 1739.

3 Cf. *Syn. Brit.* p. 95.   4 *New History of Essex*, I, p. 421.

5 *Hist. Insect.* p. 219.   6 Cf. Bocking Register.

7 The register of burials 1599–1602 is very ill-kept: there is one entry of a William, parentage and details completely illegible, under 1602.

The other entries refer only to the family of Roger's brother Thomas, whose wife Dorothy had five children: Elizabeth, baptised 26 October 1626; John, 29 June 1628; Thomas, 11 December 1629; Roger, 28 August 1630; and Sarah, 22 June 1637. Of these, Elizabeth died in January 1639 and Sarah in October 1639.

It seems probable that the other children of the older John moved away from Black Notley: but two Rays[1] were buried and a third[2] married there during the period of the second register, 1735–50, these being probably grandchildren of Thomas. The name was not uncommon at that time in Bocking where a Henry Ray, clothier, had a son, also Henry, in 1685 who on his death in 1760 left property in Black Notley.[3]

Two points in these records deserve comment. The first concerns the spelling of the family name.[4] John Ray, as all know, spelled his name Wray on every occasion at Cambridge and until the year 1670 when he published his *Catalogus Plantarum Angliae* as Joannes Raius and told his friend Martin Lister that he had dropped the 'W' because it had been added 'without any adequate reason and altering the old and his father's spelling'.[5] In the register there is no evidence of consistency: Ray, Raye, and Wray all occur apparently according to the taste of the writer. This is not solely due to the illiteracy of the curates or parish clerks of Black Notley: at this time and throughout the seventeenth century orthography in the spelling of names was freely disregarded even by people of education.[6]

The second point is the existence of an elder brother and sister. Roger certainly died in childhood, possibly of the smallpox, which John also had;[7] and to Elizabeth there is no certain allusion later.[8] But the fact that he was not an only child and had companions in his babyhood is not unimportant. It may help to explain the dread of loneliness, the devotion to his family, the genius for friendship which are marked characteristics of his life. If he had been always solitary it would be hard to account for his sociability: if he had not known loss when he was beginning to be impressionable, the fear of failing, or of being estranged from, his friends would be less explicable. He was a man naturally reserved: the few occasions on which he reveals his feelings are all associated with the threat of solitude.

1 John, 26 Nov. 1737; Sarah, a widow, 7 April 1743. This John represented the naturalist on the jury of the Manor Court on 8 Oct. 1695 and sat on it till 1735.
2 William Ray and Lydia Clay, both of Braintree, 23 May 1738.
3 I owe this information to Mr Alfred Hills of the Old House, Bocking.
4 For details cf. transcript of the register in note at end of chapter.
5 *Corr.* p. 65. Lister replied: 'I was pleased with the derivation of your name whilst U was at it.... You well know what Vray in French means': *Corr.* p. 66.
6 Thus of Hooke's Diary the editors write: 'the spelling of a person's name sometimes differs as much as four times on one page', *Diary of Robert Hooke*, p. 2.
7 So Dale, *F.C.* p. 5: 'he had had the smallpox, yet it was in his younger years'.
8 Withering's story drawn from William Atkinson that Ray's sister collected plants for him (*Arrangement*, 3rd edition, II, p. 286) is quite unsupported.

The circumstances of his childhood explain several other features of his character and work. Though near enough to Braintree to secure for him a small circle of congenial spirits, the neighbourhood was not lively and was certainly very rural in its outlook. In 1695, when John Aubrey asked him to procure subscriptions for his *Monumenta Britannica*, he could only reply: 'To tell you the truth this country wherein I live is barren of wits; here being but few either of the gentry or clergy who mind anything that is ingenious'.[1] Though Braintree was 'a great thoroughfare from London into Suffolk and Norfolk', and though a carrier's wagon came weekly through Witham[2] and the posts do not seem to have been very irregular,[3] there is no sign of any newsletter or of the *London Gazette*, and apart from private correspondence isolation was almost complete, even in the days when Ray had become eminent. For the blacksmith's son the world must have been limited to a few miles; and within them nothing of national importance could have touched the boy's life or made him aware of the momentous events that were in fact preparing. Probably the only change that affected Black Notley was the removal of Joseph Plume from the benefice and the intrusion of Edward Sparhauke in 1645; and by then Ray had gone to Cambridge.

It has often been remarked that in all his books and letters there is in fact hardly a single allusion to national or political history. He gives indeed a lively account of the changes in the Colleges at Cambridge after the Restoration in his letters to Courthope:[4] he mentions the Protector's death in his first Itinerary:[5] he deplores in a single sentence 'the Dutch insolency' in 1667 when Ruyter and De Witt sailed up the Medway, but says that 'the particulars of our loss are not certainly known to me':[6] he breaks into fervent thanksgiving for the Revolution and for William and Mary in the Preface of his *Synopsis Britannicarum* in 1690, and alludes to the unsettlement and hope of long life for the king in a letter[7] of the same year. Otherwise there is literally not a word to indicate that he was living in one of the most exciting periods of English history or that he knew or cared what was happening. His silence is not due to lack of intelligence or interest: his *Observations* on his continental travels shows a singular alertness to the condition of culture in the places visited, to their government, education and manufactures. Nor was he personally unaffected by the nation's struggles: his whole career was changed by the Act of Uniformity, and he

1 *F.C.* p. 181: yet in the seventeenth century East Anglia produced a cluster of men famous in science: cf. Pledge, *Science since 1500*, p. 53.

2 Cf. *F.C.* pp. 168, 290: 'the Braintree carrier inns at the Pewter Pot in Leadenhall Street and goes out of town on Friday morning weekly.'

3 *F.C.* pp. 147–8: the post-boy called at Dewlands to deliver and collect.

4 *F.C.* pp. 17–18.   5 *Mem.* p. 130.

6 To Lister, 18 June: *F.C.* p. 113; the raid had been on 10–13 June.

7 *F.C.* p. 204.

was keenly interested in religious and social movements. But his up-bringing had given him no share in the world of government and enabled him to find full satisfaction outside the realm of politics or business. When he realised that if he was to be true to his conscience both fields were closed to him, he accepted the verdict without regret. To live a recluse and a student in his own village and family was his deliberate choice.[1] He was no misanthrope nor pedant: he had indeed a genius for friendship: but he welcomed the world of personal relationships rather than that of ambition, of causes and activities; and was happy.

That he should have found his outlet not in philology or archaeology, subjects then attracting attention, but in the comparatively neglected sphere of botany and zoology, is perhaps also due to his early environment.[2] Black Notley and its neighbourhood are not rich in historic or antiquarian remains. But Stock Doves still nest in the trees round its church; Purple Emperor butterflies were once not uncommon in its woods;[3] the Gladdon Iris still grows 'ad sepes sed rarius where I now live at Black Notley, not far from the Parsonage towards Braintree';[4] and there are Brimstone butterflies and an abundance of marsh plants (even if not now 'the Black Currans or Squinancy-berries'[5]) 'by Braintree river side near the bridge called Hoppet-bridge'[6] which he crossed daily on his way to school. Gardens, though he complained of Dewlands that its soil was cold and its aspect ill-situated, fascinated him, and the little plot behind the smithy was no doubt then as now bright with flowers.[7] Even if he did no scientific work until after his election to a fellowship in 1649 he must have been profoundly interested in nature; for his breadth of range, his power of

1 So his friend, Benjamin Allen (cf. *Essex Naturalist*, XVII, p. 11).

2 It is significant that of the only two reminiscences of his youth in all the mass of his writings one should refer to a plant: 'I remember that when I was a boy I saw the flower of a Buttercup exactly like that which J. Bauhin describes. It was then frequent in gardens near my home': *Hist. Plant.* I, p. 583, dealing with *Ranunculus repens flore pleno*. The other is of a proverb recalling schoolboy antipathies:

'Braintree for the pure
And Bocking for the poor,
Cogshall for the jeering town
And Kelvedon for the whore':

cf. *Collection of Proverbs*, pp. 227–8.

3 He described it when it was new to science: *Hist. Insect.* pp. 126–7.

4 *Syn. Brit.* II, p. 234: for its abundance to-day cf. J. Vaughan, *Wild-flowers of Selborne*, p. 130—an attractive account of the neighbourhood.

5 Cf. Gibson's edition of Camden's *Britannia*, c. 364: reported lost by G. S. Gibson, *Flora of Essex*, p. 124 (1862).

6 Still called Hoppit Bridge on the road from Black Notley; rebuilt 1925.

7 Among his earliest work was a Catalogus Plantarum non domesticarum quae aluntur Cantabrigiae in Hortis Academicorum et Oppidanorum (*General Dictionary*, VIII, p. 693). This was adumbrated in a letter of 1660 to Willughby (*Corr.* p. 2), but was never printed and now seems to be lost. See below pp. 108–10.

acute and accurate observation, his flair for discriminating the vital from the superficial, bespeak not only natural gifts but early habit. He had the authentic love of living things, animals, birds, insects and plants, of the countryside and its denizens, which marks the real naturalist. He enjoyed seeing their growth, watching the development of seeds and the change from caterpillar to chrysalis; and never lost his sense of the continuity and wholeness of the process. He was eager to see them alive and in their natural setting; and though much of his work later in life had to be done from books and dried specimens he always deplored the lack of first-hand field-work and was ready to travel anywhere in order to observe for himself.[1] Only a boyhood spent in daily contact with wild nature could have given him this vital part of his equipment. His greatness as a scientist is constituted by this double capacity: the refusal to concentrate upon one phase or part of an organism to the exclusion of everything else, and the refusal to supplement verified knowledge by guess-work or speculation. If much of his achievement was in the task of description and classification, of discovering synonyms and collating records, yet he belongs to the company of Gilbert White and Richard Jefferies, of Hudson and Selous, rather than to that of the academic and 'museum' scientists.

We can go further and ascribe to the direct influence of his parents some of the outstanding features of his work.

His father was a craftsman. If in 1851 the smithy had fallen upon evil days, it still bore signs of former prosperity;[2] and in Ray's time, when all the transport of the country was by horses and roads were still dangerously bad, the blacksmith was a man of importance. Few crafts require a more exacting artistry; few, as some of us are old enough to remember, are so fascinating to the young. John Ray must have spent many hours watching the delicacy and strength of his father's work and drawing from it his enthusiasm for methods of manufacture. When he visited and described the silver mills at Machynlleth or the alum works at Whitby, the smelting of Sussex iron or the refining of Cornish tin,[3] he revealed an interest native to him, an interest which had a profound effect upon his work as a scientist. It was this desire to see how things are made and how they function that gave him his conviction of the importance of anatomy, his skill in dissection, his insistence that specific distinctions must be based upon structural characteristics, not upon colour or size or habit. It is very unusual to find the poet's sense of wholeness and life combined with the craftsman's

1 Thus in 1671, in spite of ill-health, he travelled with Willisel to the north to see for himself plants of which Willisel had sent him dried specimens.

2 Cf. *Cottage Gardener*, V, p. 221: in *Essex Review*, IX, pp. 243–5 J. W. Kenworthy suggests that Roger Ray was a man of property, a master smith who also farmed his own freehold: this is not impossible, but is hard to reconcile with John Ray's statement to Aubrey, see below, p. 17.

3 These are described in detail in the first edition of his *Collection of English Words*.

concern for details of construction and process. If Ray always refused to
isolate one set of characters from the total make-up of the organism, he was
equally insistent upon the exact exploration of each particular organ. His
birthright on his father's side qualified him for a serious attempt to base
taxonomy upon a study of comparative anatomy and physiology. His field
differed from that of Grew or Swammerdam, Malpighi or Leeuwenhoek:
he had no laboratory nor even a proper microscope:[1] but he was prepared
to accept Grew's discovery of the function of pollen and Leeuwenhoek's
of spermatozoa and to anticipate Malpighi in the study of cotyledons.

His mother's gifts to him were at once more profound and more specific.
We know nothing of Roger Ray save his trade and the fact that he gave his
son the best education in his power. Of Elizabeth we have a single sentence
in Derham's Life, and the note written by her son on the day of her death
'March 15th 1678' (1679 new style)

being Saturday, departed this life my most dear and honoured mother in her
house on Dewlands in the hall-chamber[2] about three of the clock in the afternoon, aged, as I suppose, seventy-eight; whose death for some considerations
was a great wound to me. Yet have I good hope that her soul is received to the
mercy of God and her sins pardoned through the merits and mediation of Jesus
Christ in whom she trusted and whose servant she has been from her youth up,
sticking constantly to her profession and never leaving the church in these times
of giddiness and distraction.[3]

From her he got his religion, the motive evident and avowed in all his
work; the faith that could see life steadily and see it whole, interpreting it
not piecemeal or mechanically but in terms of the many-coloured wisdom
which created and controlled it; the conviction which could surrender success and security for conscience sake, and so gain unforeseeable opportunities for scientific work; the principles which made him indifferent to
money[4] or a career, sensitive to the worth of others, 'charitable to the poor
according to his ability, sober, frugal, studious and religious, allotting the
greatest part of his time to the service of God and his studies'.[5] No doubt
much of his thought on the subject was expressed in the terms of the time,
in the familiar language that his mother would have used; certainly he

1 *F.C.* p. 193: Ray comments on the work of Grew and others: 'I confess for want of
a good microscope I have not observed them for myself': in 1692 Aubrey sent his wife
and daughters a 'glass microscope' (*F.C.* p. 175), probably the sort of simple lens
which he had always used.

2 I.e. the bedroom above the central room of the three into which the house was
divided—Kitchen, Hall, Parlour.

3 *Mem.* p. 37 note. Ray may have inherited from his mother the land at Hockley
mentioned in his will: cf. below p. 481.

4 Thus of the *Synopsis Britannicarum* he wrote to Robinson (epitomised *F.C.*
p. 292): 'would publish for profit not want if had his due, but not genius to be importunate'. In fact he received £5 for it; and it became the standard British Flora for
a century!   5 So his friend Samuel Dale: *F.C.* p. 7.

belonged to an age of transition; and we must not read into him, as some
of his eulogisers have done, ideas of religion or of science that belong to
a later century: but when Benjamin Allen reports him as saying that 'a
spoyle or smile of grass showed a Deity as much as anything'[1] he expresses an outlook radically different from that of almost all contemporary
Christians, an outlook which aligns him with Whichcote or Cudworth and
far apart from the protestants and the traditionalists of his day.

Derham's comment upon Ray's mother draws attention to another trait
that he derived from her. 'She was a very religious and good woman' he
writes, 'and of great use in her neighbourhood, particularly to her neighbours that were lame or sick, among whom she did great good especially
in chirurgical matters.'[2] When her son declares that he 'was by natural
instinct devoted to the study of "res herbaria" from his earliest years',[3]
we can see the origin of his interest. The herbalist and herb-woman held
then a respected position not only among the ignorant but with the higher
ranks of the medical profession. The 'uses and virtues' of plants were not
only an essential part of botany; they were the chief incentive to its study.
Dioscorides among the ancients and many workers in the previous century
had developed a vast and curious lore; and if by some of its students the
subject was mixed up with superstition and magic, by others it was pursued in a worthily scientific spirit. Ray gives explicit evidence that his own
approach to botany was aesthetic and intellectual: he enjoyed plants and
wanted to know them.[4] But even in his first book pharmacology has its
place, and in the *Catalogus Angliae* there is a huge collection of prescriptions and a long list of diseases with the herbs appropriate to their treatment. Though as he insists he is not a medical man his knowledge of herbal
practice is manifestly wide and enlightened. He will have nothing to do
with the doctrine of signatures, still less with astrology and alchemy, and
constantly pleads for observation and experiment. He may well have found
his mother's example an incentive both to the appreciation of traditional
practice and to the investigation of its improvement. For one who stood,
as he did, between the old world and the new such a course was plainly the
way of wisdom.

This matter of his attitude towards the problems of a time of transition
is of primary importance for a right understanding of his work; and for us
living in a similar period it has contemporary interest. His birthright, if it
relieved him of direct concern for the world of affairs, qualified him for a
place among those who were turning from the turmoil of politics to the
exploration of nature; gave him a singular capacity for its enjoyment and
investigation; and enabled him to grasp the general principles and employ

1 Miller Christy in *Essex Naturalist*, XVII, p. 11.
2 So Derham, *Mem.* p. 37.   3 Preface to *Catalogus Angliae*.
4 Cf. Preface to *Catalogus Cantab.*

the skilled technique of the scientist. But in his day there was no such body
of organised knowledge as we now possess. On the contrary, there was a
chaos of legend and superstition; of traditional lore gathered from
Aristotle, Pliny and a dozen other ancient writers into the vast pandects of
Aldrovandi; of fantastic hagiology and demonology preserved in medieval
bestiaries and fostered by folk-tales and magic; and of scarcely more credible stories, 'The anthropophagi and men whose heads Do grow beneath
their shoulders', of strange beasts and birds, sea-monsters and upas-trees,
brought back by travellers from America and the East, and sufficiently
often proved correct to forbid their summary rejection. If the unicorn was
a fiction, what was to be thought of the narwhal or the hippopotamus?
Was he to refuse the roc and yet accept the condor? to challenge the use
of lungwort for tuberculosis and act upon the power of Jesuits' bark [1]
against malaria? What was a student to believe out of the mass of claims
old and new? How was he to strike a mean between credulity and
scepticism? In the very year, 1682, in which Ray published his first
Methodus or classification of plants with its momentous recognition of the
systematic groups of dicotyledons and monocotyledons three women from
Somersetshire were executed for witchcraft after a trial in which solemnly
attested charges of bird familiars and criminal intercourse with the 'black
boy' were argued, recorded and found valid.[2] Which world was real?

Only those who fail to appreciate the confusion of the time will criticise
Ray's attitude towards it. He is ruthless in his demand for evidence and
his rejection of the legendary and the irrational. He denounces witchcraft,
but unlike his friends Sir Thomas Browne [3] or Andrew Paschall [4] never
suggests that he believes in it.[5] He sweeps away the litter of mythology
and fable; treats Aristotle with respect, but accepts or more often denies
his conclusions in accordance with the evidence; sets his face against
travellers' tales and weighs his authorities without prejudice; and always
insists upon accuracy of observation and description and the testing of
every new discovery. There were many who clung to tradition and de-
nounced the new knowledge as impertinent and blasphemous; for them he
has his answer and gives it unequivocally: nature is of God; its study is His

1 Cinchona, quinine, whose discovery and sensational effects were stirring the whole
medical profession.

2 Howell's *State Trials*, VIII, pp. 1017–40: in 1661, when Ray visited Edinburgh,
he saw the heads of Argyll and Guthrie on the city gate and the Tolbooth and noted
that 'women were burnt for witches to the number of 120': *Mem.* p. 157.

3 He affirmed his belief in witches (*Religio Medici*, p. 43) and gave evidence against
two at Bury in 1664: cf. *Works* (ed. Wilkin), I, p. lxxxii.

4 Four letters from him to Ray (*Corr.* pp. 271, 279, 280, 282) deal with tides and
manna: but he contributed a poltergeist story to *Saducismus Triumphatus*, pt. II,
pp. 281–8.

5 As did Jeremy Taylor and Richard Baxter, Henry More and Joseph Glanvill,
all men of liberal and enlightened outlook.

service, its truth His wisdom. There were others, and some of his best friends among them, who were so thrilled with the novelty and value of science that they demanded a repudiation of all previous knowledge, exalted contemporary achievement as final, and made claims for it almost as fantastic as those of the tradition. For them too he has his warning: they must verify before they affirm; must reserve judgment until proof is compelling; must recognise that the new knowledge is in its infancy; must admit its limitations and defects. We may complain that he is over-cautious; that he ought to have adopted Grew's discovery of sex in plants more speedily or followed Jung's conviction that there was no clear difference between trees and herbs or his own that whales were not fishes more consistently; that when he insisted that fossils were actual relics of previous organisms—a belief in which he stood opposed to most popular and much expert opinion—he ought to have done more than hint at the impossibility of Ussher's chronology[1] and grasped at once the concept of geological time. Cautious he certainly was: to be otherwise, till new criteria were firmly established, would have been to betray his loyalty to observation and experiment and to plunge into the speculations which deluded his friend Edward Lhwyd or the exegesis which occupied the leisure of Sir Isaac Newton. Ray has no such vagaries. He sticks faithfully to his chosen field and refuses to be drawn outside it either into theological controversy or into metaphysical speculation or even on to the border-land between science and superstition. Thus we know on his own evidence[2] that in 1667 after his return from abroad he studied 'the business about Greatrakes':[3] but in not one of his writings does he ever mention it or the similar practice of touching for scrofula which busied Charles II and fascinated John Evelyn.[4] His caution was the fruit of wisdom not of prejudice, of the courage which could admit ignorance, of the patience which kept him at work when his body was racked with pain, of the conviction which refused to rest even when he knew that his task could never be finished.

We shall mistake the quality of his achievement if we misunderstand him here. He was not an anticipator of Darwin, even though he admits doubt of the fixity of species and records evidence of their transformation. He is not possessed of a clear and consistent philosophy foreshadowing the modern outlook, though there are abundant signs of a power to grasp the unity and interrelationships of experience which lifts him high above most of his contemporaries. He is not a philosophic theologian, though his *Wisdom of God* supplied the basis for Butler's *Analogy* and is a greater

1 As he does very plainly to Lhwyd: *F.C.* p. 260.
2 *Mem.* p. 17, quoting his letter to Lister.
3 I.e. Valentine Greatrakes, the Irish stroker, who had visited England in 1666.
4 Diary for 6 July 1660 and 28 April 1684.

book than Paley's *Natural Theology*. His greatness is that in a time of transition and universal turmoil he saw the need for precise and ordered knowledge, set himself to test the old and explore the new, and by dint of immense labour in the field and in the study laid the foundations of modern science in many branches of zoology and botany. He studied, corrected and collated the existing literature; he collected, identified, investigated, described and classified mammals, birds, reptiles, fishes and insects, cryptogams and all known plants; he contributed richly to the advance of geology and made observations in astronomy and physics; he was a pioneer in the study of language and first revealed the importance of dialect and folk-speech; he did as much as any man of his time to develop a new understanding and interpretation of religion; more perhaps than any man he enabled the transition from the medieval to the modern outlook. That he could do so is due not only to his own genius and opportunities, but to the character of his inheritance and the circumstances of his upbringing.

Yet that the blacksmith's son, however brilliantly endowed, should have developed intellectual activity for such achievement, compels us to look for some influence in his boyhood other than that of the village. It is hard to believe that the forge and the local school, of which he had a low opinion, were solely responsible for the training of a mind which is conspicuously scholarly, for the first steps in Latin of which he became an outstanding master, and for the habit of exact description and a nice weighing of evidence which characterises all his work. His record demands that in his early years he should have had the guidance of trained and cultured teachers. Any suggestion as to these teachers must be matter of conjecture: but in fact there are obvious probabilities which can properly be mentioned.

Black Notley during Ray's boyhood had two rectors, Thomas Goad, who died in 1638, and Joseph Plume, who succeeded him and, though ejected in 1645, came back in 1662. Both happened to be men of some eminence.

Goad, second of the ten sons of Roger Goad, Provost of King's College, was educated at Eton, admitted to a fellowship at King's before taking his degree, and resident at Cambridge until 1611. He then became chaplain to Archbishop Abbot, was presented in 1618 to the rectory of Hadleigh and in 1627 made his home there. In 1625 he acquired the rectory of Black Notley and held it in plurality;[1] and even when he became Dean of Bocking his interest was centred in Hadleigh, where he was buried in the chancel in 1638. Fuller describes him as 'a great and general scholar, exact critic, historian, poet (delighting in making of verses till the day of his death), schoolman, divine' and adds 'a commanding presence, an uncontrollable

[1] He succeeded Richard Crakanthorpe, the champion of puritanism, rector 1605–24.

spirit, impatient to be opposed and loving to steer the discourse'.[1] He was obviously a man of versatile mind, intellectual interests and dominating personality; and that he should have given up the great world in order to spend his days in country parishes, fostering as he did at Hadleigh a local painter and local activities, marks him out as at least unusual. We know nothing of his contact with Black Notley. He can hardly have been acquainted with the small sons of the village. But in his last years it is not impossible that he discovered a bright boy among them and encouraged his parents to give him schooling.

More important is Joseph Plume, his successor, who according to Newcourt[2] must have bought the right of presentation from the family who owned it and the manor. He came from Suffolk, matriculated as a pensioner at Queens' College in 1622, graduated in 1625–6, was made a fellow in 1629 and ordained in the next year.[3] He took his B.D. in 1636, and became rector on 13 August 1638. By 1640 manor and advowson were sold by Sir Richard Leveson to Thomas Keightly,[4] citizen and merchant of London, who had been M.P. for Beeralston in the Parliament of 1620–1, and had purchased Hertingfordbury Park on the death of Sir William Harrington[5] before 1643 when John Evelyn, his cousin by marriage,[6] stayed with him there. In 1643 Plume's name was included in John White's *First Century of Scandalous Malignant Priests*, a book published by order of Parliament. Number 33 reads:

The benefice of Black Novelty alias Notly in the county of Essex is sequestered from Joseph Plumm [*sic*] Parson thereof, for that he is a common ale-house and tavern-haunter, and hath been divers times drunk, and not only used superstitious bowing himself at the name Jesus, but hath presented the churchwardens for not bowing, and threatened his parishioners because they refused it, commanding his churchwardens to look to them, and hath absented himself from his said cure for the space of eighteen weeks last past and is reported to have betaken himself to the army of the Cavaliers and hath otherwise expressed great malignity against Parliament.[7]

A study of White's list makes clear that the first clause in the charge against Plume is almost stereotyped—parson after parson being entered as 'a tavern-haunter and divers times drunk', generally with the addition 'a swearer by great and bloody oaths' or (as in the case of one of Plume's neighbours) that 'on the monthly fast he suffered football playing in his

---

1 *Worthies*, I, p. 165 (ed. 1811).   2 *Repertorium*, II, pp. 442–3.

3 Venn, *Alumni Cantab.* III, p. 372.

4 Newcourt's version of his name as Knightly is corrected by Morant, *History of Essex*, II, p. 124; both Morant and Wright, *History of Essex*, I, p. 236, give the date of the sale as 1634.

5 Cf. Chauncey, *Hertfordshire*, I, p. 535.

6 He married Rose, daughter of Thomas Evelyn of Ditton.

7 P. 16: epitomised in Walker, *Sufferings of the Clergy*, II, p. 330, where his name appears as Plumin.

own ground'.[1] Laxity in secular matters and strictness in ecclesiastical are the general complaints; and by comparison with most of his colleagues Plume comes off lightly. The case against him would seem to be that he was not a puritan, was on friendly terms with his flock, and had had a tiff with his wardens over his habit of bowing in the creed.[2] Essex was strongly parliamentarian, and Laud had been its bishop: Plume may well have thought it wise to absent himself from his parish rather than promote discord or compromise his convictions. On 9 October 1643, according to the *Journals of the House of Commons*,[3] the living was officially sequestered and Edward Sparhauke or Sparrowhawk, who had graduated at Emmanuel College in 1621–2 and apparently held a lectureship at St Mary Woolnoth in the City of which he had been deprived by Laud,[4] was intruded into the benefice. He was certainly in occupation of the living by September 1645 when his daughter Sarah was baptised there, and in 1650 was reported by the Parochial Inquisition as 'a godly preaching minister'.

When allowance is made for the influence of his parents—Roger and Elizabeth Ray must surely have been superior to the ordinary villagers of the place and period—it is probably right to ascribe to the two rectors of Black Notley a large share in the discovery and training of their son's genius. They were both highly qualified men, and Plume at least lived within a few hundred yards of the smithy. Their influence may not only have attached him to the church but been responsible for sending him to the grammar school in Braintree and encouraging him in his studies there.

The school was then about a hundred years old. In 1535 the south chapel of the fine parish church, now called the Jesus Chapel, had been built as a chantry. At the Reformation Sir John Holmested, priest to the church, set up a school there in 1548, turning the chapel into a two-storeyed building for that purpose: on the chancel pillar adjacent to it can still be traced the marks scraped upon it by the sharpening of slate pencils; and the newel-staircase to the rood-screen no doubt supplied access to the upper storey. In 1626 Martin Holbeach, who moved in the following year to Felsted and did good work there, was master; and in 1646 a Mr Adamson was in charge.[5] But according to Dale, who lived for many years in

1 L.c. p. 36, of Clement Vincent of Danbury.
2 Without wishing to justify the Centurist or Sequestrations we must not forget that in the previous decade thirty clergy had been deprived of livings in the Norwich diocese for refusing to read the Book of Sports and that Laud in his visitation had punished those who did not bow.
3 Vol. III, p. 270.
4 Davids, *Annals of Evangelical Nonconformity in Essex*, p. 432.
5 Cf. *Victoria History, Essex*, II, p. 510: in 1653 he sent a boy to St John's College, Cambridge, who had been under him for seven years: the writer assumes that Adamson taught Ray.

Braintree and must have known the facts,[1] and to Derham,[2] Ray was under a Mr Love, who must have been master between these two. Ray himself, according to Derham, 'used sometimes to lament as a great misfortune to his younger years that it was at that time no good school': but the teaching cannot have been very bad; and the pupil certainly profited by it. He became himself so excellent a teacher that he may have judged Mr Love by a very high standard; and perhaps he compared him and the school with what he knew of the neighbouring and much better school of Felsted, where at this very time Isaac Barrow, his friend at Trinity, was being taught. At least Braintree must have given him a good grounding in Latin—he could not otherwise have acquired so distinguished a style or so large and plastic a vocabulary. It must have trained his memory which, considering that nearly all his later work was done without a library and consisted in recalling and defining minute points of plant and animal structure, must have been remarkably accurate. It must have given him an orderly mind, a mastery of method and a delight in study. It certainly gave him a beautiful and very legible handwriting,[3] which must have immensely simplified the work of his printers and proof-reading.

If Mr Love was a person of small importance, the Vicar of Braintree, Samuel Collins, is in a different category. He is confused both by Walker[4] and by Newcourt[5] with his famous namesake, the Regius Professor of Divinity and Provost of King's College. But the records of their lives and the dates of their deaths show beyond all doubt that they were distinct. Yet though not widely celebrated the vicar was highly regarded in his own neighbourhood. Dr Harold Smith's description of him as 'a diligent parochial minister friendly both with Laud and with puritan leaders'[6] is a fair summary of the evidence; and it is anticipated in his funeral sermon preached by his friend Matthew Newcomen,[7] which testifies to the 'service he had done for this poor town'—to the deplorable state in which he found it and to 'the degree of eminency and outward propriety he advanced it to'.

Collins had graduated at Trinity College, Cambridge, in 1599–1600, been ordained at Norwich in 1601, and been instituted to Braintree on 15 February 1611 on the presentation of Robert Lord Rich, afterwards Earl of Warwick.[8] We first get knowledge of him in the matter of Thomas Hooker, the lecturer at Chelmsford who was deprived by Laud, then

1 *F.C.* p. 4.   2 *Mem.* p. 7.
3 Cf. T. Hearne, *Collections*, VIII, p. 106 'the famous Mr John Raye, tho' he writ so much, writ a fair hand and very slow'.
4 *Sufferings of the Clergy*, II, p. 218.   5 *Repertorium*, II, p. 89 note.
6 *The Ecclesiastical History of Essex under the Commonwealth*, p. 148.
7 Published in London 1658 and quoted by Davids, *Annals*, p. 150 note.
8 Created 1618: his son, who succeeded in 1619, joined the puritans and was the friend of Sibbs, Master of Catharine Hall.

Bishop of London, in 1628. In 1629 Hooker opened a school at Little
Baddow and was again threatened with persecution. Collins wrote on his
behalf to his friend Sir Arthur Duck, Laud's Chancellor, urging that if
Hooker were allowed to depart out of the diocese quietly all would be
well, but that if the bishop punished him there would be upheavals, and
if he suspended him he would be maintained by his supporters in Essex
and continue his ministry in private. Collins writes temperately and tactfully as a man loving peace and forming a truer judgment of public opinion
than the bishop.[1] For a time it seemed that his advice would be taken; but
Laud insisted on Hooker's appearing before him: the Earl of Warwick gave
him concealment; and in 1633 he sailed with John Cotton to Newtown or,
as it was called later, Cambridge in Massachusetts.[2] The case is typical of
the obstinacy which brought Laud eventually to the scaffold and reveals
Collins as a man who while loyal to the church saw the folly and peril of
the policy of 'thorough'.

Laud not only rejected prudent counsel but did not forgive the coun-
sellor. In 1631 Collins fell under his displeasure. The vicar's protest to
Duck is a fine expression of a wise man's dilemma:

It is no easy matter to reduce a numerous congregation into order that has
been disorderly this fifty years.... If I had suddenly and hastily fallen upon the
whole part of uniformity, I had undone myself.... Upon the first notice of alteration, many were resolving to go to New England[3] and others to remove elsewhere.... By my moderate and slow proceedings I have made the stay of some
and do hope to settle their abode with us.... My Lord of London needs not to
implore the power of the High Commission to rule me; the least finger of his
own hand shall suffice. If what I have said and done will not satisfy, I must
submit to his honour's censure.

But with Laud such wisdom was useless. He could not or would not see
the futility of violent measures or realise the effect of his floggings and
nose-splittings upon decent Englishmen.

It is not surprising that Collins despaired of further influencing the
course of events, and threw himself into the affairs of his own town,
winning the affection of his people and being accepted by them through
all the turmoil of the next twenty years. He was maintained in his benefice
when the Presbyterians came into power and was reported as 'an able
godly preaching minister, presented by the Earl of Warwick' by the
Parochial Inquisition of 1650.[4] His refusal to take the Engagement did not

1 A similar wisdom is shown in a letter of 1632 to Laud concerning the treatment
of a young clergyman (Davids, *Annals*, pp. 346–7).

2 For this see Mullinger, *University of Cambridge*, III, pp. 148–202.

3 Some twenty thousand emigrated to New England between 1629 and 1640,
largely from East Anglia. Braintree, as the rhyme 'Braintree for the pure, And Bocking
for the poor' shows, was strongly puritan.

4 So Smith, *Ecclesiastical History of Essex*, p. 293.

affect his influence; and when he died on 16 October 1657[1] he had earned the eulogy already quoted.

We have dealt in some detail with the character of Collins because it not only illustrates the religious situation in which Ray grew up but is typical of an attitude not too common in those days of violent partisanship and exactly similar to that which Ray himself adopted. Men of character and intelligence could not but be repelled by the extravagances both of Laud and of the puritan extremists, shocked by their cruelties and cantings, and constrained to withdraw from close concern with political and ecclesiastical affairs. We do not know whether Ray was directly influenced by Collins; but it was precisely into this shape that his own religious outlook was moulded. He had a deep regard for the realities of religion, an almost complete indifference to its externals, an abhorrence of controversies about it. It is not mere guesswork to conclude that the Vicar of Braintree had a large influence upon the lad who came daily to school in his church.

To that school and his years at it there is in all Ray's published letters only one reference and that in general terms. It is contained in a letter to John Aubrey dated from Black Notley on 27 October 1691, not included in the Aubrey letters printed by Gunther, but appended to the fifth and final volume of Aubrey's *Natural History and Antiquities of Surrey* published posthumously in 1719. The letter is so little known and so revealing that it deserves to be quoted at length.[2] Aubrey had sent him the manuscript of his *History of Wiltshire* on 15 September. Returning it Ray sent a letter and enclosed certain detailed criticisms.[3] In the letter is the following sentence:

Neither is your observation universally true that the sons of labourers and rustics are more dull and indocile than those of gentlemen and tradesmen; for though I do not pretend to have been of the first magnitude for wit or docility, yet I think I may without arrogance say that in our paltry country school here at Braintree ego meis me minoribus condiscipulis ingenio praeluxi; but perchance the advantage I had of my contemporaries may rather be owing to my industry than natural parts; so that I should rather say studio or industria excellui.[4]

At the age of sixteen and a half Ray proceeded to Cambridge. In the early evidence there is no answer to the question how he was enabled to do so: and the generally accepted explanation is manifestly mistaken.

1 He was succeeded by John Argor, ejected in 1662, and then by Robert Carr, fellow of Trinity, Cambridge, 1647, who held the living till 1676 and whom Ray mentions, *F.C.* p. 26.

2 For the rest of this letter cf. below, p. 70: it was known to and quoted by J. Britton in his *Memoir of John Aubrey*, London, 1845, where on p. 80 note it is stated to have been annexed in the Ashmolean to Aubrey's *History of Wilts*.

3 These are in Gunther, *F.C.* pp. 171–3, copied from J. Britton's edition of *History of Wilts*, London, 1847.

4 L.c. pp. 409–10.

Later writers, influenced by the *Dictionary of National Biography*,[1] have ascribed it to the benevolence of Squire Wyvill,[2] who is presumed to have supplied the necessary funds. This statement, traced to its origin, is due to an anonymous article, already quoted, in the *Cottage Gardener* for 1851.[3] The exact words of the article, after mentioning the squire's generosity, are 'a Mr Wyvill if we remember correctly': and there is good reason to challenge the memory. In the church there is an inscribed slab in the belfry to a William Wyvill, rector of the parish in 1830; and shortly before 1770 Sir Marmaduke Asty Wyvill had come into possession of the Hall and patronage by inheritance from his mother and had put his kinsman Christopher Wyvill into the benefice. These facts no doubt suggested the name to the writer. But the Wyvills were a Yorkshire family and in Ray's time heavily involved in the civil war in their own county. There is not the slightest sign or possibility of their having any interest in Essex until a century later. For the list of owners of the Hall as given in Morant's *History of Essex*[4] is the family of Legate in 1401 who were succeeded by those of Spice, Fortescue, Bradbury, Leveson, Keightly, Thorowgood, Pate, Asty. As we have seen, the transfer from Leveson to Keightly took place in 1634; and in 1650 the Parochial Inquisition stated that the right of presentation to Black Notley belonged to 'Thomas Keightly Esquire'.[5] Squire Wyvill is a myth; and so may well be the squire's benefaction.

There is in fact clear proof that when Ray was sent to the University Collins had the largest share in his going. It has always been a mystery why Ray was admitted at Trinity College on 12 May 1644, but entered Catharine Hall on 28 June.[6] Collins, though his own son, also Samuel, had gone up to Corpus Christi College, Cambridge,[7] had himself held a sizarship at Trinity and would naturally recommend to his old College the brilliant pupil in his church school, and hope for similar help for him there. In fact this seems to have been promised: for the Admissions Book[8] contains the entry 'Ray, John, Sizar, May 12 1644. Tutor, Mr Babington'.[9] But it was not forthcoming; and other plans had to be formed.

Fortunately a way out of the difficulty was available. In 1631 Thomas Hobbes of Gray's Inn, not to be confused with his famous namesake of Malmesbury, had bequeathed by will dated 21 February cottages and lands

---

1 Vol. XLVII, p. 339.   2 So e.g. Seward, *John Ray*, p. 7.

3 Vol. V, p. 221: see above, pp. 1, 2, 7. The article is a 'chatty' description of a walk from Braintree to Witham.

4 Vol. II, p. 123.   5 Smith, *Ecclesiastical History of Essex*, p. 307.

6 Venn, *Alumni Cantab.* III, p. 427.

7 Afterwards M.D. of Padua and author of a book on Russia published posthumously, London, 1671: the Samuel Collins, fellow of Trinity, 1640, was son of John Collins and apparently not a relation.

8 *Admissions, Trinity College*, II, p. 379.

9 Humfrey Babington, expelled in 1650, but restored later.

in Braintree in trust for the payment of £5 and after the death of the lessee £6 yearly 'to the Vicar of Braintree and his successors, being painful in the ministry for their better maintenance'; £5 yearly for a Catechising Lecture at Catharine Hall; £3 yearly to augment a scholarship at Pembroke Hall; and 'all the rest of the said rents and profits towards the maintenance to two or three hopeful poor scholars, students in the University of Cambridge, namely in Catharine Hall and Emmanuel College, or one of them, being also of sober and Christian conversation'. He gave preference to the sons of 'godly poor ministers' and to Catharine Hall. He adds: 'my will is that during the said lease the said scholars shall from time to time be chosen by them that have the interest of that lease and by and with the consent of the Master of the said College or Hall where the election shall be made'[1]—a clause which would give to the Vicar of Braintree a controlling voice in the selection.

Here was Collins's opportunity. He was in close touch with the authorities at Catharine Hall:[2] he had a boy of exceptional promise: he would transfer him from Trinity and recommend him for a scholarship at the Hall. In 1644 the College audit contains a payment for the admission of 'Wray a scholar'.[3]

NOTE. *The Rays in the Register at Black Notley*

By the kindness of my friend the Rev. J. R. Southern, Rector of Black Notley, I have been able to search and transcribe the records relating to the Ray family during the period 1570–1669 covered by the earliest volume. They are

Roger Ray the sonne of John Ray was baptized the xxviii day of Marche 1594.
Elizabeth Ray the daughter of John Ray was baptized the xiii day of October 1597.
Willm̃ Ray the sonne of John Ray was baptized the xxvii day of November 1599.
Thomas Ray the sonne of John Ray was baptized the xi day of Ffebruarie 1600.
Willm̃ Raye the sonne of John Raye was baptized the xxviith daye of December in the yeare of our Lord God 1602.
Sarah Raye the daughter of John Raye was baptized the xxvth day of Aprill Ano Dni 1605.
Elen Ray the daughter of John Ray was baptized the 20 of Aprill 1608.
Katheryne Ray the daughter of John Ray was baptized the viiith day of Decembre 1611.

1 The will is printed in Philpott, *Documents relating to St Catharine's College* (published 1861), pp. 112–15.

2 Richard Sibbs, Master till 1635, and specially mentioned in the Hobbes trust had been intimate with his patron, the second Earl of Warwick. Trust business must have strengthened the connection.

3 I am indebted for this to Dr W. H. S. Jones, President of St Catharine's.

John Raye the sonne of John Raye was baptized the 13 of October 1614.

1624 Roger Wray son of Roger and Elizabeth his wife bapt. July 16.

1625 Elizabeth the daughter of Roger and Elizabeth Wray Nov. 17.

1626 Elizabeth of Thomas and Dorothy Wray Octob. 26.

1627 John of Roger and Elizabeth Ray Decemb. 6.

1628 John son of Thomas and Dorothie Wray bapt. June 29.

Thomas the sonne of Thomas and Dorothy Wray was baptised the xith of December Ano Domi 1629.

Roger the son of Thomas and Dorothie Wray was baptized August the 28th Ano Domi 1630.

Sarah the daughter of Thomas Ray and Dorothy his wife was baptized June 22° 1637.

Richard Nichols and Mary Ray were married the 2 of Novemb. 1610.

1646 Thomas Barker of Braintree widower and Sarah Ray single woman of this parish were maried Janu, 14.

Mary the wife of John Ray was buried the vi day of May 1592.

Willm̃ Ray the son of John Ray buried June 24 1621.

Elizabeth Ray the wife of John Ray was buried May 23 1623.

Katherine the daughter of John Wray was buryed Febru. 27 ano dom 1630.

Roger Wray was buryed December 26 1632.

Elizabeth Ray the daughter of Tho: Ray was buried Jan. 21 1638.

Sarah the daughter of Thomas Ray was buryed Octob. 18 1639.

1655 Ray. Roger Ray dyed Aug. 31 was buryed Sept. 1.

CHAPTER II

AT CAMBRIDGE UNIVERSITY

Ad Joannem Raium
Quondam discipulus Rai mihi docte fuisti:
Nunc cedo ac herbam porrigo, amice, tibi.
Multa Dioscorides, Theophrastus, Plinius ingens,
(Dicet posteritas) omnia Raius habet.

JAMES DUPORT, *Musae Subsecivae*, p. 395.

When Ray came into residence Cambridge, or at least the centre of the town, was externally not unlike its modern self. The river had been brought into its present course; the island opposite Trinity had been removed; the sites of the chief bridges had been fixed; the main streets had taken their present form. The Colleges, save for the four foundations of the nineteenth century, were all established. The great period of building which created the library of St John's, the first court of Clare and the fellows' building of Christ's, had just closed. Apart from the third court of St John's and the work of Wren, the chapels of Pembroke and Emmanuel and the library of Trinity, the town was the same as it is in Loggan's map, published in 1688, and almost as it remained for more than a century. There was no Senate-house: Ray's friend, Isaac Barrow, failed to persuade the University to emulate Fell's Sheldonian at Oxford; and until 1722 dwellings, the Devil's Tavern and the Regent's Walk occupied its present site and most of the west side of Trumpington Street. Exercises were kept and degrees conferred in the University Church, where even in Puritan days the broad witticisms of Praevaricator or Tripos were not often taken amiss.[1] There were no laboratories until Bentley amid much quarrelling with his colleagues founded one for Vigani in Trinity in 1707. There were no playing-fields: each College had its tennis-court, that at Trinity being at the west end of the north wing of Neville's Court:[2] several of them had bowling-greens, that at Trinity being laid down while Ray was an undergraduate:[3] St John's had two groups of fish-ponds. The gardens were differently arranged;[4] for the great storm of 1703[5] worked havoc with them and

1 Seth Ward, afterwards Bishop of Salisbury, praevaricator in 1640, took so many liberties that the Vice-Chancellor suspended him from his degree—for one day: Pope, *Life of Ward*, pp. 11–12.

2 Willis and Clark, *Architectural History of Cambridge*, II, p. 518.

3 Ray alludes to it as a locality for *Sagina procumbens: Cat. Cant.* p. 151.

4 Much was being done in Ray's time: the lime-walk of Trinity was planted in 1671–2 and the gardens were being laid out as in Loggan.

5 Ray refers to this storm of 27 November till 1 December when the wind is thought to have reached a steady 100 miles an hour: *Corr.* p. 438.

compelled a period of planting and reconstruction. But in spite of recent expansions the place has changed comparatively little; most of the localities that Ray mentions in his *Cambridge Catalogue* are easily identifiable and many of them can still be visited.

Internally the condition of the University was vastly different from what it is to-day, and at few periods of its history has it been more chaotic. In 1644 it was suffering from its third spoliation. Two years previously, in June 1642, the king's demand for a loan had bereft many of the Colleges of their plate: next year Cromwell's fortification of the castle had destroyed six of their bridges and Dowsing's visitation had wrought havoc in their chapels:[1] at the very time of Ray's admission the Earl of Manchester was enforcing subscription to the Solemn League and Covenant upon all who held office and had already deprived many fellows and declared their possessions forfeit. Cambridge had in the main favoured the Parliament, and ever since the outbreak of war been a centre of military activity: but the University, if its theology was generally Puritan, had in it many who were loyal to their oaths to the Crown and unwilling to forswear themselves. The wholesale expulsions threw many Colleges into confusion: the religious controversy dominated all other interests: scholarship, education, discipline came momentarily to a standstill.

It must not be supposed that these changes, after the first upheaval, involved an intellectual loss: their result, as soon as the newcomers settled in, was definite gain. The policy of 'thorough' interpreted by Laud had disgusted the best minds of England, alienated them from the Church, and because ordination was obligatory upon most fellows of Colleges[2] debarred them from the University. John Milton would obviously have remained at Christ's and perhaps spent his life in Cambridge if he could have done so without accepting Laudian doctrines and orders in the Established Church;[3] and he was one of very many, including the cream of English scholarship. No doubt the evictions were drastic and in some cases unjust: certainly the introduction of men of less experience was unsettling: but it had been recognised for many years and by both parties that the Universities badly needed reform, and the men 'intruded' were in the main intellectually and spiritually more vigorous than those whom they displaced. Milton's charges that the students were fed 'with nothing else but the scragged and thorny lectures of monkish and miserable sophistry' and sent home 'with such a scholastic burr in their throats as hath stopped and hindered all true and generous philosophy from entering'[4]

1 F. J. Varley, *Cambridge during the Civil War*, contends that his ruthlessness has been much exaggerated.

2 Conditions varied in different Colleges.

3 Cf. Masson, *Life of John Milton*, I, pp. 323–33: 'Church-outed by the Prelates.'

4 *The Reason of Church-government urged against Prelaty* (published 1641): cf. Mullinger, *University of Cambridge*, III, p. 204.

may well be biased: but at Oxford Laud, for all his insistence upon dialectic and Aristotle, declaimed equally violently against the decay of learning. Change was plainly overdue; and although the evidence as to its effects is naturally conflicting and the criterion was rather religious and political than academic, the results were generally salutary. Certainly Manchester made a real effort to secure a high standard, in many cases preferring merit to partisanship and in some leaving good men in office in spite of their views. When Ray came into residence, the standard of education and of religion was unquestionably higher than it had been ten years before.[1] The period which produced the Cambridge Platonists[2] is not one of which the University need be ashamed.

With the change in the personnel came also an opportunity for a measure of reform in the ancient curriculum. In Milton's time, 1625–32,[3] attention was still directed almost exclusively to the studies of the traditional 'trivium', Grammar, Logic and Rhetoric. To these the undergraduate devoted his four years, attending chapel, hall and lectures, and proving his proficiency by disputations in his own College and by two Acts and two 'opponencies' in the Schools of the University before taking his degree. Milton's *Prolusiones Oratoriae*, published by Aylmer in 1674, the first held in College on the subject 'Day is more excellent than Night' and the second in the Schools on 'The Music of the Spheres',[4] are typical examples of the themes and treatment required. Such education was calculated to promote knowledge of Latin and felicity of style; for to embellish a hackneyed subject—appropriately called a Commonplace, though the term had hardly its modern meaning—demanded verbal ingenuity and a measure of inventiveness. It resulted in the elaborate tropes, metaphors and classical allusions which congest the sermons and poetry of the time. It did not stimulate or encourage thought; and in a world yeasting with new knowledge and struggling to adapt itself to revolutionary changes was pitifully irrelevant. Mathematics, science, even philosophy and theology in any profound sense, were wholly lacking. Old men and vested interests loved to have it so. No wonder youths of intelligence and vigour spurned it. No wonder the future poet aligned himself with the Ramists, Baconians, and Platonists who were already in revolt.

The new men whom Manchester intruded were on the whole representative of those who had refused to conform to the old ways, men not

1 There is a brief account of an undergraduate's experience at Trinity in Ray's time in Oliver Heywood's autobiography; cf. R. Slate, *Memoirs of Oliver Heywood*, pp. 11–14: he came up in 1647 and gives an attractive picture of the College, of its Master and his tutor.

2 For a valuable account of them and of the general thought of the time, cf. B. Willey, *The Seventeenth Century Background*.

3 Cf. especially Masson's account of Cambridge in *Life of Milton*, I, pp. 111–320.

4 Printed with a translation in *Works of John Milton*, Columbia, 1936, XII, pp. 628–30.

content to spend their lives easily ploughing the exhausted fields of the 'totum scibile' and blind to the 'fresh woods and pastures new' that were opening before them. Henry More, who came up to Christ's in Milton's last year there and who exercised a large influence upon Ray, has left a brief record[1] of his own discontent with the tradition and of his unsatisfied search for 'natural' knowledge. Ten years later Bacon's protests and prophecies were beginning to gain acceptance; Descartes had a substantial following; mathematics, in the wider sense, were being seriously studied; and young men were exploring, at first surreptitiously and 'out of school', subjects outside the old curriculum. Progress was slow; indeed when Ray came up had hardly begun: Barrow, his friend and contemporary, whose *Prolusiones* as a fellow of Trinity have been preserved,[2] deals with themes less 'trivial' than those of Milton; but subjects like 'that the world is not and cannot be eternal' or 'the creature cannot create' do not suggest a very radical reform.[3] Progress was slow; but it was in the air: and if, under the Protectorate and by the energy of John Wilkins, Oxford far outstripped Cambridge, yet many of the Oxford pioneers were alumni of the other University. Progress was made; and two of the greatest scientists of the century, John Ray and Isaac Newton, were its first-fruits. As we shall see, Ray and his circle at Trinity were able to initiate and foster the necessary changes.

Progress was slow because the opposition, though by no means so conscious or intelligent as has been supposed, was very weighty. The conflict was not between religion and science: that, so far as England is concerned, is a suggestion due to reading back into the seventeenth century the story of the nineteenth. Few if any of the rivals saw the full significance of what was happening or the real issues at stake. They were only aware of a struggle between the old philosophy and its educational curriculum on the one hand and the new methods of observation and experiment on the other; between scholars like Duport loath to admit that Aristotle had not said the last word, ignoring Descartes and explicitly rejecting Copernicus,[4] men who, as Ray afterwards put it, could not bring themselves 'to lose in age the lessons of their youth',[5] and the new men, living in the thrill of a renaissance, witnesses to the success of attacks upon the established order in politics, and impatient of the conservatism and classical pedantries of

1 In the Preface to the Latin edition of his works published in 1679.
2 Cf. Napier's edition of his *Works*, vol. IX.
3 Ray himself in the Commonplaces delivered about 1660 complains: 'I am sorry to see so little account made of real Experimental Philosophy in this University and that those ingenious sciences of the Mathematics are so much neglected by us; and therefore do earnestly exhort those that are young, especially Gentlemen, to set upon these studies and take some pains in them': *Wisdom of God*, pp. 125–6.
4 Cf. Duport, *Musae Subsecivae*, p. 65.
5 *Syn. Brit.* Preface, p. 8: see below, p. 251.

their teachers. The struggle overlapped, even if it did not coincide with, the political issue of the time. The Royalists, sharing Laud's devotion to the Aristotelianism and dignities of the old régime, regarded the new philosophy as symptomatic of the unrest and indiscipline of the rebellion. The Puritans, revolting against authority in the interests of liberty and independence, recognised that the Universities were strongholds of tradition and privilege: extremists among them argued for suppression; the majority, like Milton, was content with reform. It is not an accident that the Invisible College and its successor, the Royal Society, were largely Puritan in membership, and only secured public recognition because their experiments amused the king and his family. The general attitude is perhaps best expressed by Samuel Butler, who both satirised Puritanism in *Hudibras* and ridiculed the Royal Society in his *Elephant in the Moon*.

At Cambridge in Ray's time the opponents of science, though until 1660 passive in their resistance, were yet numerous and influential. Henry More, who in 1648 had written enthusiastic letters to Descartes, was obliged to apologise at length for his ardour and to explain that he disagreed with much of the Cartesian position. Barrow, though he abandoned his Greek professorship with thankfulness that he could now escape 'e grammatico pistrino in mathematicam palaestram',[1] 'from philology to philosophy', had been content to defend the classical tradition and to argue in prose and verse that the Cartesian hypothesis was contrary to nature.[2] Even in 1669, when the University entertained Cosimo de' Medici to a Latin disputation, the subject was 'an examination of the experimental philosophy and a condemnation of the Copernican system'.[3] On the one side was the old culture with its protestant scholasticism, its imitations of Horace and Martial, its control of the Schools and dominance in the teaching posts; on the other were the advocates of the exploration and control of new worlds, offering fascinating possibilities of the expansion of human power, and the hope of escape from the scandals of controversy and the learned trifling that so easily lapsed into obscurantism or obscenity. As a multitude of pamphlets shows, the issue was being fought out in Ray's undergraduate years. He was qualified by them to play a large, if at the time inconspicuous, part in the struggle.

Fortunately for Ray Catharine Hall had not suffered as much as more royalist societies; for it gave him two years of steady tuition and time to acclimatise himself. Under Richard Sibbs, who had been Master from 1626 till 1635, its numbers and endowment were greatly increased. He was a strong Puritan, a man of piety and influence, who had according to the report to Laud in 1636 ordered the College after the manner of Emmanuel.[4]

1 *Barrow's Works*, IX, p. 202. 2 L.C. IX, pp. 79–104, 441–3.
3 Cooper, *Annals*, III, p. 536.
4 Cf. Browne, *St Catharine's College*, p. 101.

Ralph Brownrig, his successor, though loyal to Charles, was on excellent terms with the Presbyterians and in theology a strong Calvinist. He had secured for the College the possession of the house and stables of the famous carrier Hobson[1] in 1637, thus opening up the site to Trumpington Street. He had acquired a number of ecclesiastical preferments, culminating in 1641 in appointment as Bishop of Exeter. In 1643 he was Vice-Chancellor and succeeded in saving the College from the first outbreak of Puritan severity. Indeed, though he was stripped of his positions in the Church, he was allowed to continue as Master and made a member of the Assembly of Divines. He seems to have kept to the use of the Prayer Book and to the doctrine and order of the Church of England; but his preaching was highly appreciated by the leaders of the parliamentary party and in its length, wealth of detail and attention to the text of Scripture was typical of their sermons. Under him the fortunes of the College prospered; for he was deeply concerned with its welfare, with the supervision of its students and the personal examination of the younger scholars. His influence upon Ray, of whom he certainly formed a high opinion,[2] may well have been considerable.

Nevertheless Ray, according to Derham,[3] put it on record that he found Catharine Hall overmuch concerned with disputations by comparison with the more liberal culture, 'the politer arts and sciences' of Trinity. This no doubt means that the teaching in the Hall was more rigidly disciplined, that the traditional exercises, compositions and essays, as we should call them, were strictly enforced, and perhaps that Brownrig's supervision was a trifle exacting. It was no longer a small society, having profited in this respect by the Puritan ascendancy: but it did not include many men of intellectual eminence; and the brilliant young scholar probably suffered from lack of companions of his own calibre and perhaps from too much concentrated attention. He had come up from his village to the wonder-land of the University and may well have resented having to devote his days to conventional discourses upon hackneyed themes. His outspoken condemnation of the old learning in his later life[4] is partly due to his experiences of the 'Commonplaces' of his freshman's year.

On his arrival he was put under the tuition of Daniel Duckfield, an Essex man[5] and probably the son of the Vicar of Childerditch near Brentwood. Duckfield had graduated at Catharine Hall in 1635–6 and gained his fellowship in the next year. His younger brother John came up in 1642

1 'Hobson's choice' is included in the later editions of Ray's *Collection of English Proverbs*.

2 In 1661 John Gauden, Brownrig's friend, eulogist and successor at Exeter, wished to send his son to Ray at Trinity: cf. Ray's letter to Courthope (Gunther, *Early Science in Cambridge*, p. 467).

3 *Mem.* p. 7.   4 E.g. in the Preface to his *Synopsis Britannicarum*.

5 Venn, *Alumni Cantab.* II, p. 71.

and also gained a fellowship, though he was not admitted to it until after the purge that followed upon refusal to subscribe to the Engagement in 1650. Daniel, of whom nothing else seems to be known, died on 3 May 1645 while Ray was still his pupil; and his death may have been one of the reasons that led Ray to migrate to Trinity. Brownrig's expulsion for a royalist sermon in 1645 may have influenced him in the same direction.

It was less his discontent than a combination of the evidence of his brilliance and of the desire of Collins to send him to Trinity that produced the change. On 21 November 1646 he was transferred;[1] a sizarship was awarded to him, and he was placed under James Duport who, although he had become Regius Professor of Greek in 1639, was allowed to continue as a tutor by the College. He had been born in Cambridge—his father was Master of Jesus—and been at Trinity since 1622, gaining a great reputation as a scholar and teacher. Though a decided royalist he had escaped from the Earl of Manchester's visitation with the loss of his prebendal stall at Lincoln, and in 1646 was elected to the Lady Margaret's preachership, which he held along with his professorship and tutorship.

In 1646 along with Ray he had received as his pupil Isaac Barrow, who had gone up two years earlier from Felsted School to Peterhouse, where his uncle, also Isaac Barrow and afterwards Bishop of St Asaph, was a fellow, and had transferred to Trinity on his uncle's expulsion. Barrow and Ray, educated within a few miles of one another, were thus living together and reading under the same tutor; and Duport is said to have regarded them as the two most brilliant pupils of his whole career.[2] Certainly they became close friends; Barrow, when he left Cambridge in 1658, sent to Ray the earliest of the letters printed by Derham[3] and giving an account of his travels in the East; and his Oration of the year 1654 plainly contains a playful allusion to his colleague's interest in science.[4]

This speech, condensed and commented on by Whewell in his essay on 'Barrow and his Academical Times',[5] gives a vivid account of education in the University at the time. Delivered on a formal occasion and in the tone of eulogy appropriate to the beginning of the year's work it no doubt represents too rosy a picture. Superlatives are in fashion at such functions, and Barrow's rhetoric and humour express themselves in exaggeration. Nevertheless the survey shows that in spite of the dislocation of the times learning was not neglected and a wider range of subjects than is usually recognised was being studied. Latin was of course the language of all

1 'Wray, John. Of Essex. From St Catharine's Hall. Subsizar, November 21, 1646. Tutor, Mr Duport' (*Admissions, Trinity College*, II, p. 395).

2 So Brokesby, who was a fellow of Trinity with Ray, to Derham: *Mem.* p. 8.

3 *Phil. Letters*, pp. 1–8: it is addressed to the fellows of Trinity College from Constantinople.

4 Cf. below and Orat. ad Acad. in *Barrow's Works* (ed. Napier), IX, p. 46.

5 *Barrow's Works*, IX, pp. i–liii.

educated men. To be able to write it with distinction, to understand it in
lecture and conversation, to turn a neat epigram or compose a congratulatory ode was a necessary part of University life. Greek, Hebrew,
Arabic, French, Italian, Spanish, all according to Barrow had their
votaries. Mathematics with Geometry and Algebra, Optics and Mechanics;
Natural Philosophy with Anatomy and Botany; Chemistry; Moral Philosophy; these are hardly less in evidence. It all sounds very modern and
impressive; and Whewell apparently regarded it as such. But it must not
be forgotten that there was not then a single laboratory of any kind in the
College or the University;[1] that though by the Elizabethan statutes,
chs. xv and xvii, two dissections were required for the M.B. and three or
at least two for the M.D. Francis Glisson, the Regius Professor of Physic,
only came to Cambridge to supervise the Acts and did no regular teaching
there; that Ray on his own confession searched through the University in
vain for a mentor to help him in the study of plants; that chemistry and
anatomy were practised by him and his friends in the privacy of their own
rooms; and that as late as 1660 he deplored[2] the lack of interest in 'experimental philosophy and the ingenious sciences of the mathematics'.[3]
Moreover, as Barrow himself constantly asserts, Orations of this kind were
always expected to be facetious. It is difficult not to conclude that in much
of his speech, as certainly in his allusions to zoology and botany, he is
being deliberately exaggerative and ironical.

Under Duport's guidance Ray may well have been encouraged to widen
his range; for his tutor seems to have had some interest in the study of
nature.[4] But languages were certainly his chief subject. He must have
already acquired a sound knowledge of the classics; for though he never
burdens his writings with the recondite references and elaborate allego-
rising of his contemporaries he has the rich vocabulary, the dignity and
clarity of style, the ease of expression and the sense of rhythm which only
come with long and intimate knowledge. His Latin even in letters to his
friends is never slipshod and hardly ever ungrammatical. The language is
not that of the Augustan age: he delights in rare words and writes not an
archaeological parody but a living speech, easy to translate except where

1 Vigani only began to be recognised in 1692, when De la Pryme 'went a course
in chymistry' and described him as ' very learned and a great traveller, but a drunken
fellow' (*Diary*, p. 25).   2 Cf. *Wisdom of God*, pp. 125–6.

3 For the almost complete neglect of mathematics cf. John Wallis's autobiographical
letter in *Works of Thomas Hearne*, iii, p. cxlvii: he was at Felsted under Holbeach
(see above p. 14) 1630–2, and then at Emmanuel College.

4 Duport bequeathed to Trinity some books on scientific subjects, including
Clusius's *Curae Posteriores* and Schonevelde's *Ichthyology*. He was himself however a
strict Aristotelian, opposing the new philosophy, scorning Descartes and even rejecting
the Copernican astronomy: cf. his Latin poems and J. H. Monk, *Memoir*, p. 26. That
in spite of this he had a real regard for Ray is evident from the verses addressed to him
after the publication of his *Catalogus Angliae* in 1670 (*Musae Subsecivae*, p. 395).

some unfamiliar term or technicality discloses the seventeenth century, and capable of doing justice either to the niceties of a botanical description or to the solicitudes of friendship. In some of his more studied work, in the dedications and prefaces of his books, he reveals a real eloquence and beauty of construction and a freedom from affectation in pleasant contrast to the pomp and pretentiousness of the period. That until the end of his life he wrote Latin as easily as English is recorded by one of his biographers[1] and attested by all his work. There is in fact a distinction about his Latin style which his English rarely achieves. He had obviously been carefully disciplined in the language of culture, but had not been encouraged to gain a similar craftsmanship in his native tongue. With the exception of the English version of his *Ornithology* all his scientific writings are in Latin, and this points to an instinctive conviction that the older language was the medium in which he was best able to express himself as well as that whereby he could best reach the world of science. That the great Dr Gale, High Master of St Paul's School and reputed the best Latinist of the day, should have thought highly of his vocabulary,[2] and that Dr Wilkins of Wadham and Trinity, Cambridge, should have chosen him as the best possible person to translate his *Real Character*,[3] suggests that his instinct was correct.

In the documents relating to Cambridge preserved in the libraries of Trinity College and the University it has only been possible to find two specimens of Ray's Latin compositions during his residence.[4]

The first of these, mentioned by J. E. B. Mayor in his notes on Uffenbach's record of his visit to Cambridge in 1710,[5] is a poem of twenty-seven Alcaic stanzas in the volume presented to Oliver Cromwell in 1654 and entitled *Oliva Pacis*. It is a congratulation to the Protector on his undertaking the charge of the State and concluding peace with the Dutch; is written in the style which Duport and the seventeenth-century scholars encouraged, a style less strictly classical in its language and prosody than the rigorists of the next century approved; and is a typical ode of the kind then in fashion, neatly phrased, eulogistic and to our taste fulsome in its flattery. Its interest lies in the fact that it is the earliest specimen of Ray's work known to us, and that it was produced in distinguished company. Among his fellow contributors were the following, whose names will occur hereafter in connection with his life: John Arrowsmith, Master of Trinity; Benjamin Whichcote, Provost of King's; Antony Tuckney, Master of St John's, Whichcote's former tutor and recent critic; Ralph Cudworth, then Master of Clare; John Worthington, Master of Jesus; James Duport, Ray's

1 So Petiver, *Life of Ray*: F.C. p. 11.
2 Miller Christy quoting Allen in *Essex Naturalist*, XVII, p. 13: Ray sent greetings to Gale in a letter to Aubrey in 1692: F.C. p. 180.
3 *Mem.* p. 22.
4 A copy of them is in the note appended to this chapter.
5 In *Cambridge under Queen Anne*, p. 272.

tutor at Trinity; William Lynnet, his friend there; and William Croone, then an Inceptor in Arts of Emmanuel and afterwards the distinguished doctor. The second is the congratulation to Charles II on his return mentioned by Calamy, and is printed in the large collection of such pieces entitled 'Sostra sive ad Carolum II reducem Gratulatio' and offered to the king by the University in 1660. It is a poem of thirty-six lines written in the metre of Horace, Epode xv, called Pythiambicum i, that is a hexameter followed not by a pentameter but by an iambic dimeter. Its contents are in striking contrast to most of the collection; for it is a quiet and dignified prayer that out of the past calamities king and people may learn a new appreciation of one another, a new respect for law and order, and a concept of kingship based rather on merit than on birth. There is not a fulsome or insincere word in it; its honesty stands out against the denunciations, exculpations, sycophancy and affectation of almost all the others. Compositions of this kind suited Barrow better than his colleague.[1] Ray had in fact been absent from Cambridge when Cromwell died and did not contribute to the smaller collection offered to his son Richard in 1658. It is hardly surprising that he was not included among the contributors to the 'Dirges' in 1661 on the deaths of Henry Duke of Gloucester and Mary of Orange or to the Epithalamia on Charles's marriage in 1662. He was an admirable Latinist: he could never be a courtier.

His choice of Latin as the language of his books is a matter of importance in considering his influence. He lived, as we have seen, in a time of transition. When he went to Cambridge, the old tradition which regarded Latin as essential for all serious and academic writing, as the hall-mark of the educated man, and the indispensable means of international communication,[2] was still unchallenged. The Roman liturgy and the Latin Bible had been banished; but for the gentleman and the student Latinity was *de rigueur*. 'He is so ignorant that he cannot even write Latin without solecisms' is Ray's severest criticism of an opponent.[3] His own mastery of the language gave a world-wide currency to his work. Gerard and Parkinson, writing in English, were unknown outside Britain: Ray was read by Tournefort and Hermann and Malpighi as readily as by his own countrymen or in the next generation by Linnaeus. His Latin gave him a great reputation as not only the most eminent of contemporary naturalists but in the eyes of Cuvier and Haller a principal founder of scientific

1 Barrow, both in verse and prose, is always verbose and fulsome, and usually heavily facetious: cf. his eulogy of Duport, a series of dreary variations on the theme of 'a great mind in a small body'.

2 Though even then Englishmen pronounced it in a way almost unintelligible to foreigners: cf. Calamy, *History of my Life*, p. 219; Cooper, *Annals*, III, p. 536.

3 *Corr.* p. 42.

zoology, ornithology, ichthyology and entomology[1] and 'the greatest botanist in the memory of man'[2]—testimonials the more remarkable as coming from French and German sources.

But in his own country and for his influence here it must be recognised that there is another side to the picture. The growth of the new men and new families, indeed the whole trend of the time, was against him on this issue. Latin might remain in the Universities: it was ceasing to be used even by the aristocracy and clergy. By the end of the century Ray was himself complaining that there was only one printer in London, Benjamin Mott, who could be trusted to set up a Latin book;[3] and his publishers, who sold many editions of his *Wisdom of God*, found it impossible to continue producing his *Synopses*;[4] such books simply did not sell. The Universities, entangled in Classics and Mathematics, failed to develop any scientific interests except of the most desultory kind: the educated came to regard Latin as a dead language, of interest only to the Schools: the people in general had to wait for help in natural history until White of Selborne: in several departments, notably in entomology, Ray's lead was never effectively followed. No doubt the slackening of intellectual interests, so notable with the second decade of the eighteenth century, is largely to blame. But it is hard to believe that if Ray's books had appeared in both tongues they would not have exerted a vastly greater influence. It is at least significant that the branch of study which attracted the largest amount of attention and is still probably the favourite with Englishmen— ornithology—is precisely the one in which Ray produced an English version of his work.[5]

Nevertheless it is to his Latinity that his peculiar excellence as a naturalist is largely due. When Gilbert White of Selborne wrote:

Foreign systematics are, I observe, much too vague in their specific differences; which are almost universally constituted by one or two particular marks, the rest of the description running in general terms. But our countryman, the excellent Ray, is the only describer who conveys some precise idea in every term or word, maintaining his superiority over his followers and imitators in spite of the advantage of fresh discoveries and modern information,[6]

he spoke truth. Ray is supreme in description, in the concise accounts that he gives of the characteristics of plants, birds, fishes and insects. In most

1 Cf. Cuvier in *Mem.* pp. 104–6: 'His works are the basis of all modern zoology.'

2 So Haller, *Mem.* p. 65.

3 *F.C.* p. 239: 'Truly I know no other printer in London that one may trust with the printing of a Latin book.'

4 His final *Methodus* had to be printed in Holland; and the *Synopsis Avium et Piscium* was held up for more than ten years.

5 Thomas Pennant, the most considerable naturalist of the next century in England, declared that he owed his interest in the subject to the gift of the *Ornithology* to him by Richard Salisbury at the age of twelve.

6 To Barrington: letter of 1 Aug. 1771.

of his books it was impossible to afford the expense of plates; and he had therefore to set before him the ideal of enabling any careful reader to identify accurately each species that he is describing. This, as anyone will know who has tried it, is an exceedingly difficult art. It is one for which Latin is a peculiarly suitable medium; and the learning of Latin is a discipline best fitted to promote accuracy in the use of words, brevity of style, and a habit of fastening upon essentials. No man could have produced the two thousand folio pages of the *Historia Plantarum* packed with many thousand recognisable descriptions of species, or the hundreds of concisely accurate definitions of moths and their larvae in the *Historia Insectorum*,[1] if he had not had the mental training that Duport and his other classical teachers had given him.

Latin was a language of necessity. Greek and Hebrew, the other subjects in which he won academic distinction, were less obligatory. Unfortunately we have little evidence by which to judge his proficiency in them. His books contain many fragments of quotation and a few of discussion. He slips into an appropriate Greek phrase if there is no suitable Latinism: he had obviously read widely and acquired a rich vocabulary. He knows enough Hebrew to discuss the language of the Old Testament or to speculate upon the influence of Hebraic on Greek or Latin words. But though he held the Greek lectureship as his first office at Trinity and had some reputation as a Hebraist, nothing of his studies in these fields has been preserved.

Of other languages he is credited by Petiver[2] with French and Italian: of the former he writes to Lister, with reference to a request that he would print the French names of plants in a new edition of his *Catalogue*, that he is 'but a smatterer in that language'[3]—in which however he was able to read Swammerdam's book on Insects; of Italian he knew enough to quote its proverbs and probably to speak a little, and this he had picked up in his two winters in Italy. He had unquestionably a gift for languages, a remarkable memory and an absorbing interest in words, in their derivations and meanings,[4] in varieties of dialect, in etymology and philology; and this did more than equip him with a clear style and mastery of exposition.

Among his earliest, and not least important, books are a *Collection of English Proverbs* and a *Collection of English Words*, the former embodying

1 It is perhaps worth recording as proof of this that I worked through all Ray's descriptions of Lepidoptera before consulting Bodenheimer, identified nearly all of them, and found that in most cases my results were the same as his.

2 Cf. *F.C.* p. 11.   3 *Corr.* p. 123.

4 His greatest work, the *Historia Plantarum*, is full of etymological notes often admirable: cf. e.g. II, p. 1513 on the Apricot: 'Mala Armeniaca Romanis, autore Dioscoride, Praecocia dicuntur; nominantur et Praecoqua...Βερικόκκια vox quâ recentioribus Graecis hi fructus dicti sunt proculdubio à Latino Praecocia corrupta est: a quo fonte nostrum Apricock et Gallicum Abricot profluxit'—which is virtually a Latin version of Murray's account of the word.

the homely wisdom of the common people, the latter preserving the dialects of northern and southern speech. Both had a considerable vogue; and the second is the earliest attempt to gather, interpret and discuss the remnants of older and local language as a contribution not only to the better understanding of the provincial communities, but to those antiquarian studies which were leading men to a knowledge of the Anglo-Saxon and Celtic tongues. Ray's work in this field was, as Skeat testifies,[1] most important, and in this as in others that of a pioneer. He never regarded it as a major interest, and was inclined to speak lightly of it as a personal and peculiar hobby excusable on utilitarian grounds. But to an understanding of his character and circumstances it is highly significant; and we cannot but see in it an indication of his own early difficulties. He had come up from the forge at Black Notley and the little church school at Braintree to the brilliance and culture of a great University. His ability and fineness of quality gave him a place among the intellectuals, a place which he held with distinction through a long life of comparative poverty. We know him as 'the excellent Mr Ray', the intimate friend of Aubrey and Sloane and Compton, respected and visited by the world of fashion. It must have taken a struggle for the lad with his provincialisms of speech and village background to acquire the language and manners of the aristocracy, and gain the poise, the dignity and modesty, which enabled him to hold an honoured place in a society in which class distinctions were still almost feudally rigid.[2] We know from a mass of evidence, contemporary and more recent, how cruel a College could be to the poor boys who came up to live on charity and sizarships; and though Cromwell's ascendancy had compelled the gentry to recognise that greatness was not determined solely by quarterings, the country squire was the favourite butt of the London wits and the lot of the humbly born was often bitterly humiliating. Ray's interest in local dialects illuminates the secret of his struggles and his success. He made a study of what others regarded with shame, and avowed without apology or concealment his belief that the thought and speech of common folk were worthy of the attention of the learned.

Thus he accepted his own position in society without aggressiveness or subservience. Punctilious in using the correct form of address to men of high station, in the nice distinction between Mister and Esquire, he does so without a trace of self-consciousness. They are, by the grace of God, what

1 Cf. his edition of Ray's book published 1874, Introd. p. v.
2 An illustration is found in his letter of June 1703 to Thoresby, who was collecting letters of famous men and had asked Ray for specimens: he replied: 'Those great men of the Royal Society took but little notice of so mean and inconsiderable a person as I must needs own myself to be. I had indeed some acquaintance with Bishop Wilkins, yet never held any correspondence with him by letter': *Correspondence of R.T.* II, p. 23.

they are; it is discourteous and insensitive to ignore it: and he is, by the same grace, what he is; and sees no reason to hide it. Among the most moving of his letters are those in which he tells John Aubrey of his pleasure at having him as his guest at Dewlands and his regret that the entertainment could not express that pleasure more lavishly,[1] or apologises to Hans Sloane for the fact that in that crowded household some precious books lent by the great doctor have been stained with soot from the fire and spattered with ink by one of the children.[2] Aubrey was a wit and a gossip, critical and sometimes salacious; but for Ray he has an invariable friendliness. Sloane was an Irishman, genial, warm-hearted, impulsive, but a busybody and something of a snob: yet he sends his presents of sweetmeats to the family and advice to his patient without a shade of patronage or embarrassment. Only a very great man could have won such a relationship or been able to be himself, natural and at ease in the world of the Restoration. It says much for the society of his day that such relationships were possible: it says more for the man that he maintained them. But such freedom can only be obtained at a price; and the price must have been paid in those first years at Trinity.

Here Barrow's friendship may well have been invaluable. Pope's rather spiteful humour has described[3] for us Barrow's slovenliness of attire and indifference to the conventions—how he arrived with his collar buttonless and his bands hanging loose, and had to be put into Pope's nightshirt while the tailor sewed on his buttons and clothed him with decency; how a great congregation seeing his deplorable appearance stampeded out of the church, climbing over the pews in their haste to escape the sermon of such a guy; how the vergers at Westminster Abbey when Barrow was beginning his second hour of discourse prevailed on the organist to blow him out of the pulpit. Such a man exposing himself at College to the ridicule of his contemporaries and offering grounds of more serious attack by his royalist opinions, yet won for himself the respect and affection of his opponents: 'let him be; he is a better man than any of us', said the staunchly puritan Master:[4] and his partner under Duport may well have profited by the tolerance.

1 *F.C.* p. 174: 'Your reception and entertainment here was such as my meanness could afford, not such as your merit did exact; and were you not a philosopher would rather need excuse than deserve thanks. The best of it was a hearty welcome.'

2 *Corr.* p. 365: he adds: 'I intend to have the books myself' (they were a volume of Hermann and two of Boccone) 'and if you please to send me back these again, I will order Mr Smith to get them bound for you in the same manner that these are.'

3 *Life of Ward*, ch. 20: Napier's warning that 'absolute veracity seems not to have adorned Pope's character' must be remembered.

4 Thomas Hill, Arrowsmith's predecessor, cf. *Barrow's Works*, I, p. xl.

Barrow too was Ray's 'socius studiorum' in mathematics;[1] and this also gives evidence of his ability. For Barrow, though an excellent classic and theologian, was primarily a mathematician. When in 1654 his youth and political views prevented him from election as Duport's successor in the Regius Professorship of Greek, he recognised that his real endowments lay elsewhere, and left Cambridge to travel in the East. Returning before the Restoration he received the Greek chair in 1660; succeeded Rooke as Geometry professor at Gresham College in 1662; returned to Cambridge as first Lucasian Professor of Mathematics in 1663; and resigned in favour of Isaac Newton in 1669. He was a man of great and varied learning and did good work as a pioneer in the development of the Differential Calculus. Though his popular reputation rests upon his preaching—he was 'the exhaustive preacher' of the Caroline age—and upon his Mastership at Trinity, mathematics[2] was the field in which he did his best academic service.

If Ray was fit to study alongside Barrow he must have had mathematical gifts of a high order. In his letters this only appears in connection with his visit to Francis Jessop at Broomhall[3] near Sheffield in 1668, when he writes of the progress that Jessop is making in a way which suggests that he has been guiding an amateur along the right lines of research.[4] Probably his bent was towards applied rather than pure mathematics; for he was always keenly interested in machinery, in the water-wheels and canal-locks of his foreign tour,[5] not less than in processes of manufacture[6] or in Hooke's experiments with watch-springs and pendulums.[7] But his mind was essentially concrete; and there is no evidence of interest in the researches of Newton and his contemporaries. It is not true to say that he rejected abstract thought and confined himself to observation and description; but he plainly distrusted the theorisings of men like his friend Bishop Wilkins and believed that what was needed was primarily a closer attention to scientifically verifiable data. He may well have been confirmed in his opinion by the readiness of even the greatest of his contemporaries to plunge into fantasies as soon as they left their proper fields of study. Ray was very far from being a materialist, but he distrusted speculations which could not be checked by exact knowledge.

If Catharine Hall had enslaved him to disputations, he confesses to finding the atmosphere of Trinity more favourable to liberal culture. He had probably realised that his abilities had marked him out for a fellowship, and was able to see beyond the necessity for satisfying the examiners.

1 So Abraham Hill in *Barrow's Works*, I, p. xlii.
2 Cf. two letters of his to Willughby in Derham, *Phil. Letters*, pp. 360, 362.
3 Built in Henry VIII's time: now a district of the city.
4 *Corr.* p. 25. 5 *Observations*, pp. 3, 4.
6 Cf. his account of industrial processes in *Collection of English Words*.
7 *Corr.* pp. 23–4.

Inevitably much of his attention must have been centred in the political and religious controversies of the day: for these created a series of crises in the life of the University and of every one of its members.[1] The imposition of the Solemn League and Covenant had been enforced before he came into residence, though the consequent expulsions continued for several years. But almost before these upheavals had subsided, others were brewing. Men who had reconciled themselves to Presbyterianism were now seeing the triumph of the Independents for whom 'new presbyter was but old priest writ large', and were confronted with the menace of signature to the Engagement. If in a University where men professed to honour truth vicars of Bray were less numerous than elsewhere, there were many, and some of the best divines among them, who accepted what was not far from perjury and like Cudworth and More[2] remained in possession through all the changes; and there were others who did not scruple to plot against their colleagues and intrigue without shame for the right to replace them; and because the niceties of theological and ecclesiastical dogmas provided the avowed criteria, both truth and religion were brought into disrepute. It is a sorry tale: men of honest opinions were almost inevitably driven into retirement; the prizes of office were available for those who had few qualifications but party spirit or elasticity of conscience; and as Cromwell was to discover the way was opened for every crank and careerist to push his claims under the pretext of piety.

Cambridge was in fact fortunate. There were scandals and schemings—some of them notorious: there was insecurity; for at any time another purge might be administered: but on the whole the University stood for a genuine tolerance, and even at the worst there were men who cared for learning rather than for politics, for educational ability more than for dogma. On the whole its atmosphere enabled Roundhead and Cavalier to live together and appreciate one another; and if debate was often acrimonious, much of the sound and fury was due to the conventions of the time and not incompatible with mutual appreciation. Ray and Barrow, the one temperamentally of the puritans, the other a royalist and churchman, would find their common interest in religion and scholarship a bond that such differences would not too seriously strain.

In religion indeed the great school of which Joseph Mead of Christ's had been the forerunner and Benjamin Whichcote[3] was the leader and

---

1 Cf. the account in Cooper, *Annals*, III, p. 423 of the battle of June 1648 arising out of 'some disgraceful expressions in the Schools against the Parliament'... 'in the fight divers were wounded, the number slain was not brought in; the scholars of Trinity did gallantly'. For this cf. G. B. Tatham, *The Puritans in Power*, pp. 129–30.

2 More declares that he never signed the Covenant: he is silent about the Engagement. Barrow signed it, but speedily recanted (*Works*, I, p. xl).

3 Westcott's essay in *Essays on Religious Thought in the West*, pp. 362–97, is still the best appreciation of his thought and influence.

champion was already establishing its position against both Calvinists and Laudians. Its Platonism is generally reckoned a third and characteristically English element in the religious life of the period, an alternative to the choice between protestantism and catholicism, a theology more profound than the *via media* of the Elizabethans or the latitudinarianism of Bishop Hoadly. The appeal to reason as 'res illuminata illuminans', the insistence upon morality as manifesting the present energy of the Spirit of God, the refusal to set nature and grace in antithesis, these were accompanied in it by a conviction of the grandeur and scope of the divine activity, of the gravity and responsibility of man's task, and of the need for loyalty to truth as against both tradition and enthusiasm. Whichcote, who had grown out of the puritanism which he had learned from Anthony Tuckney[1] at Emmanuel College, had come back to Cambridge as Provost of King's[2] in the year of Ray's entry at Catharine Hall; and had then resumed the famous sermons in Holy Trinity Church which mark an epoch in the history of Anglicanism and which Ray certainly heard.[3] He was a thinker rather than a scholar, a teacher not by books but by personal influence; and the University and especially its younger members were deeply impressed. Ray himself in all his religious writings reveals his debt to Whichcote,[4] to Henry More of Christ's, whose *Antidote against Atheism* supplied the theme of the Commonplaces which afterwards became his *Wisdom of God*, and to Ralph Cudworth, Master of the same College, whose *True Intellectual System* underlay his own philosophy of creation. These three, with his later friend John Wilkins, gave him a theology in which reason and science could find full exercise and the highest kind of mysticism goes harmoniously with observation and exact knowledge.

Whether he had any personal contact with the three great Cambridge Platonists is unknown; but John Worthington, Master of Jesus from 1650 to 1660, who married Whichcote's niece and edited Joseph Mead's *Works* and John Smith's *Discourses*, was certainly his friend. It is from Worthington that we get our only description of Ray in his Cambridge days. Writing to Samuel Hartlib,[5] who had supplied him with the manuscript

1 His correspondence with Tuckney began in 1651.

2 He succeeded Collins, but never took the Covenant and supported the fellows of his College in refusing it.

3 Cf. *Miscellaneous Discourses*, p. 210: 'I have heard it often from a great divine, Dr Whichcote, in his sermons, "there is but a thought's distance between a wicked man and hell"': on p. 208 he had quoted the most famous of Whichcote's aphorisms: 'Some do think heaven to be rather a state than a place' (cf. *Aphorisms*, No. 464).

4 Thus e.g. his saying recorded by his friend B. Allen, 'a spoyle of grass showed a Deity as much as anything' (*Essex Naturalist*, XVII, p. 11), reproduces Whichcote: 'Every grass in the field declares God', Discourse LVIII (*Works*, ed. 1751, III, p. 176).

5 To whom Milton had dedicated his tract, *Of Education*, in 1644: 'everybody knew Hartlib' (Masson, *Life of Milton*, III, p. 193).

of Jung's work[1] which he had lent to Ray, and sending to him a letter from Ray and a copy of the *Cambridge Catalogue*, he says:

He [J.R.] told me that he had thoughts heretofore to have sent you his book; and nothing hindered but modesty, he being a stranger to you. He is a person of great worth; and yet humble, and far from conceitedness and self-admiring. He is a conscientious Christian; and that's much said in little....I think there are not many that have attained to so great a knowledge in this part of natural philosophy; which he is still adding to. He hath a little garden by his chamber which is as full of choice things as it can hold: that it were twenty times as big I could wish for his sake.[2]

This reference is the only clue that we possess as to the rooms which Ray occupied in Trinity. Probably as an undergraduate subsizar he shared a room with Barrow and perhaps two others of Duport's pupils and may have had as his study one of the closets attached to it. But as a fellow he would have a 'chamber' of his own; and the account of his garden, 'a little spot of ground belonging to his chamber',[3] fixes its position. Only two rooms could be so described, those on the ground floor to the south and on the first floor to the north of the Great Gate, which each possessed a small garden on the east between the court and the street. In that on the north of the gate lived afterwards Isaac Newton; for in the College accounts for 1683 is the item 'to mending the wall between Mr Newton's garden and St John's College'. It is therefore possible that the two greatest scientists of Cambridge in the seventeenth century occupied the same 'chamber'.[4]

But though he may well have moved into these quarters on his election in 1649,[5] the first dated reference to his occupation of them is in 1658, when he writes to Courthope[6] of planting out the roots sent by him from the Sussex downs. In 1660 he confesses that his garden has been neglected.[7] In the *Catalogus Angliae*, speaking of the specific difference between the British Tree Mallow (*Lavatera arborea*) and that described by

1 Cf. Ray, *C.C.* Pt. II, p. 87 and *Hist. Plant.* I, Preface, p. 6. Hartlib wrote frequently of Jung to Worthington: *Diary*, pp. 82, 123, 175.

2 Twenty-four letters from Worthington to Hartlib written between Nov. 1660 and Feb. 1662 were printed in *Miscellanies by J. Worthington*, edited by Edward Fowler (London, 1704): letters X–XVII contain references to Ray. This passage is *Misc.* pp. 254–6: reprinted in *Diary and Correspondence of J.W.* (ed. J. Crossley), I, pp. 332–4. In this letter Worthington describes the *Cambridge Catalogue* in terms of strong praise.

3 Worthington to Hartlib, *Misc.* p. 262, *Diary*, p. 345: he adds 'it hath at least 700 plants in it'. See below pp. 108–10.

4 I owe this paragraph to the kindness of the present Vice-Master, Mr D. A. Winstanley.

5 In *C.C.* Preface he mentions growing wild plants in his 'little garden' at the outset of his botanical studies, i.e. certainly before 1654 and probably in 1650–1.

6 Letter printed by Gunther in *Journal of Botany*, LXXII, p. 218.

7 L.c. p. 220.

the Bauhins, he says: 'I cultivated both of them for several years in my little garden at Cambridge';[1] and of the hybrid Avens (*Geum intermedium*): 'we found this in our gardens at Cambridge which we had brought out of the neighbour-fields'.[2] As this plant is not in the *Cambridge Catalogue* but in the Appendix published in 1663, it must have been noticed in 1661, the last summer that he spent there.

These records and the occupancy of the chamber belong to the last years of his life at Cambridge, when he had devoted himself to natural history. When he took his B.A. degree in 1647–8 there is no evidence of any special bent in that direction. We must return to his history at that point.

NOTE. *Ray's Latin poems*

I. From Oliva Pacis, Cantabrigiae 1654.

> Ad serenissimum OLIVERUM,
>   Angliae, Scotiae, & Hyberniae
>   Protectorem gratulatio,
>   Tam de suscepta Reipub. cura, quàm de pace
>   cum Belgis firmata.

MAgnus Magister militiae, Ducum
Flos purus, almae fortis amasius,
   Victoriae felix alumnus,
      Imperii morientis haeres,
Tandem cacumen pertigit arduum
Throni eminentis, non temerario
   Saltu, at verecundè per omnes
      Militiae meritíque scalas.
Divina dudum tempora laurea
(Celsi laboris laurea praemium
   Exile) cinxit; siqua Hybernis,
      Siqua viret Scoticísque campis.
Mutantur auro ferra, gravi modò
Majora nutant pondera vertice;
   Victríxque dignam fortitudo
      Fronte sua recipit coronam.
Calcata sprevit saepius atria,
Visásque vulgi lumine porticus
   Honoris, ignotum cubile
      Atq; adyta ingrediens tremenda.
Vidi merentes quando humeros Tyros
Amplexa primùm est, purpura Principe
   Se sensit indignam, rubénsque,
      Ipsa magìs decorabor, inquit.

Regale sceptrum contremuit, stupens
Se sustinentis robora dexterae,
   Fascésque visi sunt recentis
     Justitiam Domini timere.
Clamore festivo omnia perstrepunt;
Stupore discusso ora Quiritium
   Aures terunt, raucísque sudum
     Aëra comminuunt susurris.
Campi virentes fertilis Angliae,
Montésque praerupti Caledoniae,
   Lacúsque Hyberni voce faustâ
     Tergeminâ resonant, sed unâ.
Aulae canorus moenia perstrepens
Clangor resurgentísque palatii
   Grato optimatum consopitas
     Murmure conciliavit aures:
Dudum profundi nocte silentii
Ignava stertens nobilitas, suae
   Oblita vitae, excita tandem
     Exoriens veneratur Astrum.
En qualis ornat regna sui Ducis
Armis refulgens turma micantibus:
   Alumna tantae disciplinae,
     Fronte minax & onusta palmis!
Miscere gestit castra palatiis,
Sagúmque Martis tingere purpurâ,
   Gaudétque tot victoriarum
     Praemia tot retulisse regna.
Quin gratulantur jura ruentia
Tantum Statorem; Curia Principem
   Plausu recepit laeta, dextrae
     Tuta sub auspiciis potentis.
Quondam Gradivum blanda Cupidinum
Nutrix amavit, nunc quoque suavibus
   Excepit ulnis, atque ridens
     Alloquitur placidè Tonantis
Astraea magni nata severior;
Depone multo sanguine civium
   Ensem rubentem, prima laudis
     Principia ac elementa vestrae;
En sanctiorem, Dextera vindicem
Quam vestra juris vibret & aureae
   Pacis, Ducem dedisce, Regem
     Indue justitiâ verendum.
Compesce Martem Marte tuo, artibus
Pacem serenam protege bellicis,
   Mavortias fac serpat inter
     Et vireat tua Oliva lauros.

Hanc ambientes supplicibus feri
Votis Mosellae divitis accolae,
   Vestri favoris munus almum
     Vertice prono humiles fatentur.
Insigne Regni principium tui
Pacâsse terras, oceanum, polum;
   Pacâsse Hibernos & Britannos,
     Et Batavos trepidósq; Jutas
Cum pace donas omnia; copiam,
Pulchrum decorem, jura, fidem, ordinem,
   Pontúsque quas merces apertus,
     Terráque quas dat amica fruges.
Agmen jocosum claudit ovantium
Non ultimus voto Aonidum chorus;
   Et fundit immortale carmen
     Principis acta vetans perire:
Illi innocentis praesidium togae,
Coelestis illi carminis otium,
   Suúmque Parnassum, suúmque
     Pierides Helicona debent.
Debere pergant: gloria te manet
Haec sola postquam viceris omnia,
   Ut supplices vulgi à furore
     Protegeres clypeo Camoenas.
Hanc, quam reservârunt tibi sydera
Victoriarum impone coronidem:
   Postrema vestros ut superba
     Barbaries timeat triumphos.
Sic te perenni semideûm sacra
Reponet albo musa: tuebere
   Te fati iniquitate, saevíque
     Invidiam superabis aevi.
Praeclara Martis progenies vale,
Quin & Minervae, & jam Jovis ultimò
   Annum illa currentem beatè
     Numina dent tibi, túque nobis.

JOHAN. WRAY, A.M.
Coll. Trin. Soc.

II. From Sostra, Cantabrigiae 1660.

REX REDUX

Nos ludos fortuna facit, mirámque faceta
   Exercet histrioniam;
Per Tragicos luctus, & acerbae tristia scenae
   Deducit in comoediam.
Concutit, ut possit meliùs stabilire, premítque
   Attollat ut mox altiús.
Obvolvit tenebris, & moestâ nube recondit,
   Ut clarior lux emicet.

Diruit imperium penitus, durabile firmi
    Fundamen imperî ut struat.
Inducit miserae genti civilia bella,
    Clades, rapinas, funera;
Evertit leges, confundit jura, Senatus
    Pellit, trucidat Principes,
Templa Deûm stabulis aequat, pallatia Regum,
    Et curias Senatuum,
Ut pax certior emergat, reverentia legum,
    Ordo, nitórque illustrior:
Ut redeat tecum diuturnior, Optime Regum,
    Et gratior felicitas.
Praeteritis olim memores ut mitia sceptra
    Magis aestimemus ex malis.
Charius ut regnum Tibi sit, tu Charior illi,
    Ex mutuis absentiis,
Aptior ut celso officio Tua casibus esset
    Probata virtus asperis.
Ut meritis sis Rex, plusquam natalibus altis,
    Votísque nostris, quàm Tuis.
Ut reliquâ Regum turbâ felicior, atque
    Videare Diis acceptior;
Queis Regale decus tribuit natura, sed uni
    Tibi dedit miraculum.
Denique majorem quò promereare triumphum,
    Ineasque regnum laetiús:
Quae regnaturum comitantur gaudia, semper
    Regnanti ut adsint, comprecor.

                  Jo. WRAY, Coll. Trin. Soc.

CHAPTER III

FIRST STUDIES IN SCIENCE

I am very glad of the florid pursuits of that useful scholar Mr Wray.
SAMUEL HARTLIB to JOHN WORTHINGTON,
Worthington's *Diary*, 1, p. 342.

The twelve years which Ray spent at Trinity after securing his fellowship
were a period of profound importance for the intellectual life of Britain.
Macaulay, summarising the condition of England in 1685, declared that
'the English genius was effecting in science a revolution which will to the
end of time be reckoned among the highest achievements of the human
intellect...the civil troubles had stimulated the faculties of the educated
classes and had called forth a restless activity and an insatiable curiosity...
the torrent which had been dammed up in one channel rushed violently
into another'.[1] The suggestion that England was a pioneer in this field is
not of course correct: in Italy, France, Germany and the Low Countries
scientific studies had been attracting attention for at least a century. But
Green's record—'From the vexed problems, religious and political, with
which it had so long wrestled in vain, England turned at last to the physical
world around it...and its method of research by observation, comparison
and experiment, transformed the older methods of enquiry in matters
without its pale'[2]—emphasises more judiciously the same characteristic of
the period. If in Britain the start was delayed, if until the middle of the
seventeenth century Bacon, Gilbert and Harvey are the only names of
international importance, the country made up by the energy and range
of its activities for its lateness in beginning, and in the next fifty years produced work of epoch-making value in almost every department. In the
Royal Society, whose earliest history[3] written by Thomas Sprat, afterwards Bishop of Rochester, was translated into French and widely circu-
lated, was gathered a group of men who attained eminence in mathematics,
physics, chemistry, astronomy, physiology, botany, zoology, engineering
and political science.

It is customary to ascribe to the Royal Society and the savants who had
gathered in London and Oxford in the decade before its official foundation
the credit for the change; and to them the influence of the movement, both
at home and abroad, is unquestionably due. But in fact, as the passages
quoted indicate, the origin of the 'new philosophy' was widespread; and
the career of Ray at Cambridge reveals a contributory element in it.

1 *History of England* (ed. 1886), 1, p. 198.
2 *History of the English People* (ed. 1898), p. 609.   3 Published in 1667.

In the year in which he came into residence John Milton had indicted the curriculum with a vigour due not only to his dislike of Laudian theology and ritual but to his conviction that the whole character of academic study was effete, conventional, deadening. He was himself disgusted by the interminable debates which supplied him with the starting-point of his epic but seemed fit occupation only for fiends in hell; and he did not hesitate to urge men to turn from the study of 'man's first disobedience' and God's 'foreknowledge absolute' to the world of nature.[1] But his complaints bore little fruit until the end of the Civil War. Then, as has been truly observed,[2] religious controversies, political convulsions and patriotic fervour combined to send many scholars to the study of archaeology and the antiquities of Britain. They were not less potent in arousing an enthusiasm for chemistry, physics and biology. Those like Ray who hated controversy and were not deeply interested in history found a wonderland of almost limitless attraction in natural philosophy. To the thoughtful among them it was obvious that the old outlook was vitiated by its ignorance of accurate data and that, if a new interpretation of the universe was to replace an outworn tradition, observation and experiment must precede its formulation. To the adventurous the exploration of East and West, the invention of telescope and microscope, the discoveries of Harvey and the speculations of Descartes opened up a field of practical achievement which was beginning to appear intelligible and remunerative. To the religious, shocked as they were by the bitterness and trivialities of sectarianism, the *vis medicatrix naturae* offered at once a refuge and a starting-point, a realm of beauty and order and a field of wholesome activity. It is not surprising that the best of them turned to it with eagerness.

Ray had obviously been born with a bent in that direction. He was fortunate in finding among his contemporaries those who shared and fostered his interest; and among them his first and closest friend.

John Nidd, or Nid as he and Ray spell the name,[3] had been admitted a sizar at Trinity in 1640 and graduated in the year of Ray's arrival. It is probable that he was the son of Gervase Nidd,[4] also of Trinity, a Doctor of Divinity and Rector of Sundridge in Kent from 1615 until his death in 1629. He was admitted to his fellowship in 1647, became a tutor in 1650, Junior Dean in 1652–3 and Senior Dean in 1657–8, and was elected a Senior Fellow shortly before his death.

1 The almost total lack and the great value of 'natural knowledge' were emphasised by Milton, *Of Education*, pp. 4, 5, in 1644, and again by John Hall, *The Advancement of Learning*, in 1650.

2 Cf. Douglas, *English Scholars, a study of historical and linguistic Research*, 1660–1730, pp. 20–4.

3 The signature 'Jo. Nid' in his books may be due to the Latin form Nidus.

4 So Venn, *Alumni Cantab.* III, p. 258.

Thomas Pockley, a subsizar from Yorkshire just junior to Ray, became a fellow in 1650 and apparently devoted himself to medicine. He is nowhere mentioned in Ray's books and apparently did not share the enthusiasm for botany or ever become so close a friend as Nidd. But in letters to Courthope we hear of him as physician to the British troops in Dunkirk[1] in 1660 and as dying there of a 'squinancy which dispatched him in less than two days'[2] in the autumn of 1661. Whether he had been deprived of his fellowship or had merely left Cambridge during the upheavals of the Restoration is not clear.

William Lynnet, the other contemporary who is mentioned in Ray's early letters, had been elected to a fellowship at the same time as Nidd. His father had been Vicar of Sompting in Sussex and had died in the year of his election. He was primarily interested in divinity, taking his B.D. in 1662 and his D.D. by Royal Letters in 1671,[3] and though very friendly to Ray[4] seems never to have had any interest in science.

Thomas Millington, who had come up from Westminster in 1646, taken his degree three years later and then migrated to a fellowship at All Souls, was certainly known to Ray in later years and may well have been one of his earliest scientific friends.[5] He had a distinguished career as a doctor, was one of the original members of the Royal Society, was knighted in 1680 and died as President of the Royal College of Physicians in 1704.

John Mapletoft,[6] member of a family which played a considerable part in the intellectual and religious life of the time and brought up on his father's death in 1635 by his godfather Nicholas Ferrar of Little Gidding, came up as a pensioner in 1648, graduated in 1651–2 and was elected a fellow in 1653. He was in residence till 1658, but then went as tutor to the son of the last Earl of Northumberland. He travelled for a time abroad,[7] returned in 1663 and took his M.D. in 1667; became Professor of Physic[8]

1 Gunther, *Early Science in Cambridge*, p. 465.

2 *F.C.* p. 27; Dunkirk had been won from Spain by Cromwell and was soon afterwards sold to France by Charles II.

3 He became Vicar of Chesterton in 1665 and of Trumpington in 1673 and died in 1700: Venn, *Alumni Cantab.* III, p. 88.

4 Ray says that 'on his advice and direction Sir Edward Goring admitted his son my pupil': Gunther, l.c. p. 467.

5 He visited and prescribed for Ray at Black Notley in 1701: *Corr.* p. 397. He is credited by Grew with the discovery of the male nature of stamens.

6 There is a brief account of him and some letters to him from John Locke and others in the *European Magazine*, Nov. and Dec. 1788, XIV, pp. 321–3, 401–4, and Jan. to Sept. 1789, XV and XVI. When Barrow died in 1677, Mapletoft wrote his epitaph (Pope, *Life of Ward*, p. 168).

7 So Ray in Gunther, *Early Science in Cambridge*, p. 467. He went in order to study medicine and was in Rome with Algernon Sydney.

8 There is a reference to him in Ray's *Hist. Plant.* I, p. 586, as supplying a note of the medicinal use of *Ranunculus hederaceus*.

at Gresham College in 1675 and a Fellow of the Royal Society in 1676; and later took orders and in 1686 became Vicar of St Lawrence Jewry. In the one letter that mentions him Ray speaks of him with warmth and admiration.

Walter Needham came up from Westminster in 1650 and became a fellow of Queens' College in 1655, his work being in medicine and anatomy. After the Restoration he lived in Shrewsbury and Oxford and at Robert Boyle's request wrote his *Disquisitio anatomica de Formato Foetu*, published in London in 1667 and reviewed in *Philosophical Transactions*,[1] work for which had been done at Cambridge thirteen years before. He was elected to the Royal Society in the same year. In 1672 he moved to London as physician to the Charterhouse, and Ray came again into touch with him. They were always close friends.[2] Ray constantly acknowledges his help,[3] particularly concerning the medical properties of plants in the *Catalogue* of 1670. He supplied the localities of one or two species;[4] is mentioned among Ray's special helpers in the *Historia Plantarum*; and is quoted as an authority on anatomy in the *Synopsis Quadrupedum* and elsewhere.[5]

Another friend, less intimate but probably a member of the group, was Thomas Allen, who came up to Trinity in 1648 and after graduating became a fellow of Caius College, 1651–60. He then moved to London, became physician to the Bethlehem Hospital, and was elected to the Royal Society in the same year (1667) as Ray. In spite of this he refused the request of the Society to perform experiments in the blood-transfusion, then fashionable, upon one of his mad patients.[6] Ray seems only to mention him in connection with an introduction to a citizen of London who had a large number of foreign birds caged in his house.[7]

Though Ray's friendship with the first three of these men certainly goes back to his undergraduate days it is unlikely that science played much part in his activities till later. Even if disputations were not so rigidly enforced as at Catharine Hall, he must have had the normal exercises and acts to keep before he could proceed to his B.A. It was in 1649, when he and Barrow were elected together into minor fellowships on 8 September, that his freedom to study at his own choice began. Even then it was limited by College duties. It was customary to lay upon the younger fellows in turn the annually awarded lectureships in particular subjects. Ray was appointed Greek lecturer on 1 October 1651 and again in 1656, Mathematical

1 No. 28, pp. 509–16.   2 Cf. the tribute to him in *Cat. Angl.* p. 2.
3 *Corr.* p. 53.   4 *C.A.* pp. 2, 7.
5 An account of his work in embryology is in J. Needham, *Chemical Embryology*, I, pp. 162–5.
6 Munk, *Roll of the Royal College of Physicians*, I, p. 361.
7 *Ornithology*, p. 266.

lecturer on 1 October 1653 and Humanities lecturer on 2 October 1655.[1] This last post Barrow had held in the previous year, and from his inaugural lecture printed among his Latin Orations[2] we learn that its duties consisted in delivering a course on one of the great figures in Latin literature, Barrow choosing Ovid on the ground that he was less 'austere and morose' than Vergil and less salacious than Horace!

By this time the activities of the little circle had certainly begun to attract attention, though until recently little was known of them. Nidd was apparently their leader. A reference in the *Wisdom of God*[3] showed that he kept a vivarium in his room and studied the breeding of frogs:[4] the letter to Courthope in January 1659[5] warning him that Nidd is suffering from a malarial fever and that his health is seriously affected: an affectionate reminiscence in the *Historia Plantarum*:[6] the information that his funeral sermon was preached by Ray in the College Chapel on what seems a curiously ill-chosen text:[7] the glowing tribute to his qualities and help in the Preface to the *Cambridge Catalogue* and a single reference to him in its text[8]—these and the suggestion that if 'an iron retort like to Glauber's in the second part of his Philosophical Furnaces'[9] can be procured for Pockley their experiments, presumably in chemistry, can go forward, were all that we possessed. Even the discovery that some of Nidd's books passed on his death to the Trinity College Library and that among these, signed with his name in ink on their title-pages, were the 1619 edition of Lyte's *New Herbal*, the *Icones Stirpium*, an album of woodcuts arranged according to De l'Obel's system and published in 1591 by the house of Plantin, and the identical volume of Glauber, *A Description of New Philosophical Furnaces*, does not tell us much.

A more illuminating reference was disclosed when Gunther secured the full text of Ray's letters to Courthope and printed the new material in his *Early Science in Cambridge*.[10] The letter from Friston dated 20 January 1663, as printed in *Further Correspondence*,[11] had read as a description of Ray's collecting birds on the Alde. The excised sentence transformed the meaning. 'I remember', Ray had written, 'some years ago at Mr Nid's

1 Particulars as to these appointments are derived from the Trinity College Registers.
2 Cf. *Works* (ed. Napier), IX, p. 132.   3 Pt. II, p. 84 (2nd ed.).
4 Contrast with this 'This year, 1645, a woman was hanged at Cambridge for keeping a tame frog, and it was sworn to be her imp': Cooper, *Annals*, III, p. 398.
5 *F.C.* p. 16.
6 'My intimate friend Mr Nid, a notable botanist, used to say "Lovage smells like a drug-store"': *H.P.* I, p. 437.
7 *Mem.* p. 11: the text is Psalm xxxix, 5.
8 *C.C.* pp. 6, 7 and 57: cf. also a reference in the Appendix published in 1663.
9 Published in Amsterdam, 1648, London, 1651. Courthope, who like his cousins the Burrells was interested in Sussex ironworks, was to get this retort made: cf. *F.C.* p. 17.   10 Pp. 346–7.   11 P. 35.

chamber in Trinity College I was present at the dissection of four birds. The names of three of them were a Bittern, a Curlew and a Yarwelp; the fourth was like a duck...whose cases hung up in the cupboards over Mr Nid's portal.'[1]

From this evidence, which is in keeping with a reference to his own dissection of a Bustard in the *Cambridge Catalogue*, with Needham's record of his Cambridge dissections[2] and with many indications in other books, it seems that the primary interest of the group was in comparative anatomy. This was in fact a natural consequence of the scientific situation. Harvey's discovery of the circulation of the blood had been the first achievement of an Englishman in the field which Vesalius had begun to explore at Padua a century before. It had thrilled his fellow-countrymen and made them eager for the practice of dissection. His subsequent book, *De Generatione Animalium*, published in 1651, deepened and directed their enthusiasm: it may well have led to the studies of foetal life which Needham and Ray carried out at this time.[3] Nor was the interest confined to students of surgery: it was extended to wider circles by the example and influence of René Descartes. The *Discourse of Method* in its Latin version and the *Principles of Philosophy* appeared at Amsterdam in the very year of Ray's arrival in Cambridge and had an immediate effect upon the men by whom he and the University were most powerfully affected. Whichcote, Cudworth and especially More show evidence of deep interest: and this was probably strengthened by personal contact through John Alsop.[4] If they did not realise the full significance of the Cartesian mechanistic theories, they were stimulated by his elaborate investigations of anatomy and physiology and provoked by his particular treatment of the lower animals as mere automata.[5] The body was after all an intricate but recognisable mechanism; and whether or no Descartes was right in locating the soul in the 'conarion' or pineal gland, it was clear that his claims could not be correctly gauged until the subject had been investigated. While the mathematicians discussed his doctrine of vortices, the anatomists explored his evidence for automatism. So what was proper to surgery became also vital for philosophy and religion.

1 In the manner common in museums and shown on the title-page of Olaus Worm's *Museum* and seen in the Elk's foot from Tradescant still preserved at Oxford.

2 Cf. *De Formato Foetu*, Preface, in which he states that most of the observations were made in 1654 at Trinity College: these included dissections of horse, cow, pig, dogs, cats, rabbits, birds, frogs, fishes and a viper.

3 Ray alludes to his dissections of embryos in *Wisdom of God*, p. 57 and quotes Harvey in l.c. ed. II, p. 67: Needham also quotes Harvey, l.c. p. 161.

4 So Peile, *Biographical Register of Christ's College*, I, p. 315: Alsop was a fellow of Christ's 1623–39, who after Laud's death fled to France, became intimate with Descartes, and introduced him and his work to More.

5 Cf. More's correspondence with Descartes in 1648–9—four letters and two replies.

This last-named element must not be ignored. There was good warrant in Scripture, quite apart from the theology of the Christian Platonists, for regarding the study of the works of the Lord as a true part of the religious life. Solomon, Job and the Psalmists were agreed that the wisdom revealed in creation was of God, and even St Paul testified that divinity was manifested by the things that were made. Nothing is more evident in all Ray's writings than his conviction that the study of nature is essentially a religious duty, appropriate to mankind as a means to God's greater glory and a deeper understanding of his ways. Aesthetic satisfaction and intellectual interest played a large part: nature supplied him with a medium of worship which he could not find in the churches and with a stimulus to thought and research which was lacking in the tradition of the schools: but beyond these was the conviction that here was something reasonable and reliable, an antidote to superstition and an instrument of enlightenment. A man could find in the investigation of nature a truer concept and a better use of life than in the intrigues of politicians or the sophistries of the sects. In 'Mr Nid's chamber' Ray began the education for his life's work and developed that knowledge of structure and of comparative anatomy which was the basis of his contribution to biological science. It is evident from all his books, and especially the *Ornithology*, the *Historia Piscium*, the *Synopsis Quadrupedum* and the *Wisdom of God*,[1] from his dissections at Padua and contacts with Malpighi and Tyson, and from the testimony of his successors that he had in fact at some time in his early career worked hard at the subject. The fragment to Courthope if it stood alone might be dismissed: in fact it supplies an indispensable clue to his history.

It must have been soon after his election that he began the study of botany. In the *Cambridge Catalogue* he has left a record of his first enthusiasm during a period of ill-health which compelled leisure and exercise, his efforts to find a teacher, his solitary struggles and increasing absorption in the subject. Certainly his expeditions into the fens and woods of the neighbourhood had become familiar to his friends by the autumn of 1654, if Whewell is right in assigning Barrow's Oration, already mentioned, to that year.[2] For it is plainly to Ray and his circle that Barrow is alluding, and his words spoken to raise a laugh at their expense give a vivid idea of their activities.

At what time, I ask you, since the foundation of the University has a murderous curiosity wrought savageries of death and dismemberment against so many

---

1 If the dissections mentioned in this book could be certainly assigned to his Cambridge days, the extent of his work there would be demonstrated: cf. below, pp. 372–3.

2 Napier, *Barrow's Works*, IX, pp. i–iv: the date is significant in view of Needham's activities: see note 2 on previous page.

dogs, fishes, birds, in order to notify you of the structure and functions of the parts of animals? A most innocent cruelty! An easily excusable ferocity! What am I to say of the history of plants, a subject eagerly explored by your freshmen? Men have hardly struck their roots in the seed-plots of the Muses before they are able to recognise and name all the plants that grow wild in the fields or cultivated in gardens—plants most of which Dioscorides himself if he were still alive would fail to identify.

We can see the representative of tradition shaking his finger at the little group of his colleagues that conducted its researches in anatomy in John Nidd's chamber and singling out for special banter the young botanist whose plant-hunting seemed an amiable eccentricity.

To their other studies, and especially to the chemistry for which Glauber's retort was needed, there is hardly a single allusion except that in the *Wisdom of God*. There, after a paragraph dealing with the immutability of metals, Ray adds: 'Note. This was written above thirty years since, when I thought I had reason to distrust whatever had then been reported or written to affirm the transmutation of metals one into another.'[1] This is very significant. Glauber, though he may claim to be one of the first true chemists, yet like Libavius and van Helmont and indeed all the alchemists firmly believed in transmutation;[2] they had devoted most of their experiments to the making of gold: Robert Boyle and Isaac Newton believed in it not less firmly. If Ray in 1660 or earlier had broken away from this tradition, he was a man far ahead of his time. The note also suggests another point. By heritage and temper he was deeply interested in the working of metals:[3] the age was very favourable to the chemical experiments and metallurgical studies which Boyle and Hooke were developing: Ray's genius might well have taken that road. It is probable that his dislike and distrust of the alchemists deterred him and swung his interest over from inanimate to animate nature. In any case there is no evidence after his days at Trinity that he ever took part in chemical or physical research.

In 1653 he received his first pupil as a College tutor, and two terms earlier there came up the man with whom his name has always been most intimately connected, Francis Willughby. He was the only son of Sir Francis[4] Willughby of Middleton Hall in Warwickshire, some five miles south of Tamworth, and of Wollaton in Nottinghamshire, and of the

1 P. 71.

2 Cf. J. Read, *Prelude to Chemistry*, p. 31, etc.

3 Cf. the account of various industries, silver, tin, iron, etc. in his *Collection of English Words*.

4 Son of Sir Percival, one of the Willughbys of Eresby, Lincolnshire, who had married Brigid, eldest daughter and heiress of Sir Francis Willughby of Middleton: for the pedigree cf. Dugdale, *Antiquities of Warwickshire*, p. 757 and Thoroton, *Antiquities of Nottinghamshire*, p. 221.

Lady Cassandra, daughter of Thomas Ridgeway, first Earl of London-derry; and came up to Trinity as a fellow-commoner and pupil of Duport.[1] He was a man of remarkable beauty of feature and charm of expression, of delicate physique, of ardent and restless temperament, and of great ability and industry. At Cambridge he had a distinguished career, especially in mathematics. On this subject two letters to him from Barrow are printed in Derham's *Philosophical Letters*.[2] Unfortunately these and letters from Skippon, Bishop Wilkins and Jessop,[3] a few letters from him to his cousin Peter Courthope,[4] a letter to Ray describing his tour in Spain in 1664[5] and his travel-diary printed in Ray's *Observations* represent practically all that we possess of him except what is found in Ray's books; and none of them gives us much insight into his character or scientific knowledge. The paragraph acknowledging his help in the *Cambridge Catalogue* and the obituary notice prefixed to the *Ornithology* testify to Ray's affection for him and admiration of his gifts. But there can be no doubt that these tributes have misled their readers into attaching a very exaggerated importance to his work as a naturalist. When they were to-gether the older man fired his friend, as he did so many others, with his own enthusiasm and profited much from his ingenuity and industry: when they were apart, there is little sign that Willughby's interest was main-tained except in entomology;[6] and there the results hardly indicate any large achievement. His vitality was an invaluable inspiration to the more cautious and conservative Ray. His generosity in financing the tours and giving a home during his life and an annuity in his will gave Ray the courage and the means to devote himself to science. His vision of a Systema Naturae fired Ray's ambition and laid on him a sense of obliga-tion. When he burnt himself out, the fulfilment of the task which he had planned became to Ray a sacred charge. He must always claim the credit as the short-lived father of their joint child: the child was formed and brought to birth by the life-long labour of his partner.

1 Shortly before his death Willughby received a Latin poem from Duport (*Musae Subsecivae*, pp. 315–17), expostulating on his devotion to science, begging him to look after his health which was never robust, and reminding him that at Trinity he had always been a glutton for books who needed the reins rather than the spur. Another poem (l.c. p. 495) after his death laments that the advice was disregarded.

2 Pp. 360, 362–5.

3 *Phil. Letters*, pp. 361, 367–9.

4 In Gunther, *Early Science in Cambridge*, pp. 343–5.

5 *Corr.* pp. 7–9. Boulger in *D.N.B.* dates this tour a year too late: Willughby was in England before the end of 1664.

6 His only contribution to *C.C.* (p. 136) and the only record from his Spanish tour (*H.I.* p. 62) are on this subject; and Ray, *F.C.* p. 69, writing after his death, says: 'the History of Insects is that wherein Mr W. did chiefly labour and most considerably advance'.

In 1655 two fellow-commoners, one of them his pupil and both his firm friends, came into residence, Peter Courthope of Danny near Hurst-pierpoint in Sussex, whose name is joined with Willughby's in the Preface to the *Cambridge Catalogue*, and Philip Skippon, only son of Sir Philip Skippon, Cromwell's Major-general who had fought at Naseby and written *A Salve for every Sore*.[1]

Courthope was a cousin of Willughby and his mother was a sister of Walter Burrell of Cuckfield, 'one of the chief iron-masters in Sussex',[2] whose fifth son Timothy, born in January 1642, came up to Trinity in 1659 and was a pupil and firm friend of Ray, and of Thomas Burrell, who was responsible for the publication of his two small books, the *Collection of English Words* and the *Dictionariolum Trilingue*. Ray's letters to Court-hope,[3] now available in full, are the earliest and most intimate of all his correspondence. They and the constant visits to Danny and Cuckfield testify to his affection for the family and to the part that it played in his life at its most difficult period. The dedication of one of his last books, the *Synopsis Quadrupedum*, to 'his honoured friends Peter Courthope and Timothy Burrell, the last survivors of his early associates', and the two letters to Burrell preserved in the Library of Trinity College, prove the endurance of his attachment. Courthope is mentioned as a helper in the *Cambridge Catalogue* and was elected to the Royal Society in 1668, but neither he[4] nor Burrell showed much interest in natural history: they were country gentlemen occupied with their estates and ironworks.

Skippon, although Ray's pupil, was in the early days less intimate with him. Unlike Courthope, whose inheritance of Danny in 1657 prevented him from taking a degree, he graduated in 1660, though without special distinction. In spite of his strong political connections he was unmolested at the Restoration, and accompanied Ray on his journeys in 1661, 1662 and 1663–6. Edward Browne,[5] son of the famous author of *Religio Medici*, met the two in Rome, wrote to his father of the collections they were making and told him how the first instalment of these, gathered in Germany and North Italy, had been sent home, presumably from Sicily, in charge of a servant of Skippon's and how the servant had been captured

1 Cf. *D.N.B.* for details of his career: from 1642 till his death in 1660 he was a mainstay of the Parliamentary forces.

2 So Ray in his *Collection of English Words*, p. 129, to which W.B. supplied details about iron-working and farming. He is described as 'Esquire deceased' in 1674. He died in 1671. For the Burrell family cf. J. Comber, *Sussex Genealogies*, II, Ardingly, pp. 48–51.

3 The version of them in Gunther, *F.C.* pp. 13–39, is imperfect and corrected in his *Early Science in Cambridge*.

4 He was one of many Fellows of the Royal Society (John Locke among them) listed for arrears of subscription in 1675 and 1685: Birch, *History of the Royal Society*, III, p. 328 and IV, p. 421.

5 *Sir T. Browne's Works* (ed. Wilkin), I, p. 86.

and his freight destroyed by Barbary pirates—a disaster to which Ray refers in connection with his *Historia Piscium*.[1] Skippon's own record of their journey, long, careful and rich in archaeological details, was not published until 1732:[2] it is a valuable supplement to the *Observations* and gives a favourable impression of its author's character and ability. He was elected to the Royal Society on the proposal of Wilkins in 1667 and sent several papers to it. A few fragments of his letters to Ray, most of them dealing with experiments[3] at meetings of the Society soon after Ray's election, are included in Derham's *Philosophical Letters*. In 1669 he married[4] Amy, the elder daughter and heiress of Francis Brewster of Wrentham Hall in Suffolk.[5] In 1671 he settled at Wrentham on his father-in-law's death, and wrote to Ray about William the five-year old son of the Rev. Henry Wotton, an infant prodigy who could already read Latin, Greek and Hebrew as well as English[6] and whom he presented to the Royal Society in July 1679 when he had already graduated at Cambridge at the age of $12\frac{1}{2}$ years.[7] He was knighted in 1674 and became Member of Parliament for Dunwich. After his marriage he seems to have done little in the way of natural history, but his friendship with Ray remained: in 1684 he supplied some pictures and an account of the Herring-fishery at Yarmouth for the *Historia Piscium*,[8] and in 1687 stood as godfather to Ray's third daughter Catharine.[9] He died in 1692.[10]

Of his other pupils we know little more than the names, and these of no special significance. But in his letters to Courthope, and especially in one that describes the illness of a freshman, the son of Sir Edward Goring,[11] there is evidence of the care with which he directed their studies, watched over their health (Cambridge being visited at this time with the plague), and sacrificed his plans in their interest. Few tutors would write so tactfully in urging an undergraduate to pay a bill as he does when he reminds Courthope that though of course 'Mr Brian'[12] has been assured of the safety

1 *Corr.* p. 166.

2 In *A Collection of Voyages and Travels*, VI, pp. 359–736, London.

3 Two of them (*Corr.* pp. 22–3) with transfusion of sheep's blood into a man, Arthur Coga, in Dec. 1667 (Birch, l.c. III, p. 225), two others with inventions of Hooke's (l.c. pp. 240, 251); one mentions More's *Enchiridion Ethicum* as newly published.   4 Cf. Ray to Lister, 7 May: *F.C.* p. 31.

5 She died in 1676: cf. Norfolk Record Society, V, p. 253.   6 *Corr.* pp. 87–8.

7 Cf. Evelyn, *Diary* (ed. 1854), II, pp. 130–1: he was given his B.A. from Catharine Hall after some discussion in the University. He became a fellow of St John's 1682, and in 1693 was described by De la Pryme as "a most excellent preacher but a drunken, whoring soul" (*Diary*, p. 29).

8 *F.C.* p. 148.   9 *F.C.* p. 291.   10 Le Neve, *Pedigrees of Knights*, p. 299.

11 Gunther, *Early Science in Cambridge*, p. 269.

12 Gunther, l.c. p. 375. Presumably Alderman William Bryan, Mayor in 1650–1, one of Cromwell's commissioners in 1654 and a commissioner for raising taxes in 1657: cf. Cooper, *Annals*, III, passim: according to the *Diary of Samuel Newton* (published by Cambridge Antiquarian Society, 1890), p. 109, he was a confectioner.

of his money yet 'Gratia ab officio quod mora tardat abest'; or would show such distress over Goring's death[1] or Burrell's 'feverish distemper'.[2]

One other junior contemporary, though not a pupil, deserves mention, Francis Brokesby, from whom Derham derived information about Trinity. He came up as a sizar in 1652, gained a fellowship in 1658 and was Rector of Rowley in Yorkshire from 1668 till his ejection presumably as a non-juror in 1690. Ray received from him a list of north-country words which he included in the second edition of the *Collection*, but there is no allusion to him in the letters.

In the town he had one friend with whom he was on intimate terms and from whom he received much help, Peter Dent the apothecary. His father, also Peter, had lived in Cambridge but apparently was unable to send his son to the University. His name first appears in the earliest of Ray's letters to Lister, that of 18 June 1667, where he speaks of having visited Dent for a few hours and sends respects to him and 'his consort'.[3] But he contributed to the *Catalogus Angliae* in 1670, and carrying on Ray's exploration of Cambridgeshire produced the second Appendix to the *Cambridge Catalogue* in 1686. He helped in the study of fishes, obtaining and dissecting Skates through James Mayfield, the Cambridge fishmonger; and sent a consignment of wildfowl to Ray in 1674. He seems to have lived in St Sepulchre's parish in a house whose garden ran down to the King's ditch.[4] He obtained his M.B. from Lambeth in 1678 and was incorporated at Trinity in 1680, but died in 1689. Little seems to be known of his professional skill or scientific attainments except what Ray has recorded: but Robert Tabor, the eminent doctor who first succeeded in managing the dosage of quinine and is described on his epitaph as 'febrium malleus', started his career as apprentice to Dent's father, which suggests that there was science in the family.

It must not be supposed that the influence of Ray and his circle was confined to Trinity. As Worthington's letters prove,[5] interest was aroused throughout the University. At least two of Ray's most valued helpers, Ralph Johnson, who entered St John's in 1648 and was afterwards Vicar of Brignall from 1662 till his death in 1689, and Edward Hulse, at Emmanuel from 1653 to 1662 and afterwards a famous doctor, came into touch with him at Cambridge: and although the evidence is slight it seems clear that his example and work had some effect upon the other great botanist,

1 *F.C.* p. 19.
2 *F.C.* p. 20: Ray postponed his Scottish tour rather than leave the invalid.
3 *F.C.* p. 113.   4 Cf. *Diary of Samuel Newton*, pp. 103, 39.
5 Cf. his letter to Hartlib quoted by Hartlib in a letter of 2 Feb. 1658 to Boyle (*Boyle's Works*, ed. 1772, VI, p. 101): 'We have divers fellows of Colleges who have made excellent progresses in anatomy.... They have also much travelled in botanics, and have got together many hundreds of plants in several gardens here.... They intend to publish a Phytologia of such herbs as are within ten miles of Cambridge.'

Nehemiah Grew, who matriculated at Pembroke in 1659. The University never again relapsed into the state of indifference to science in which Ray found it: indeed from his time the succession of eminent scientists has continued unbroken.

After 1657 Ray seems to have been free from official lectureships, but to have discharged a number of College offices. Derham states that he was Praelector Primarius in 1657–8, Junior Dean in 1658–9, and Steward twice in 1660 and 1661,[1] appointment to the former offices being made in October but to the stewardship in December.[2] That he had served his College well and was qualified for a high place in academic life is clear not only from the leniency with which he was treated at the Restoration, but by his selection to deliver the funeral sermons of his friend Nidd and of the Master of the College, John Arrowsmith, in the spring of 1659.[3] It seems impossible to trace copies of either of these discourses, but Derham apparently possessed them and speaks of them in terms of eulogy. The selection of Ray to commemorate the Master implies that he was highly esteemed by his colleagues; for the task is one that could only be laid upon a junior man if he had outstanding ability and general affection—more particularly as at this time he was not in holy orders and the question of his status had been raised.

Arrowsmith's death gave to him and to Trinity a few months of John Wilkins, who was transferred to the Mastership on 17 August 1659[4] from the post of Warden of Wadham, which he had filled since 1648. Wilkins was one of those rather rare men who without specialised academic eminence by the range of their interests and quality of their personality exercise a creative influence upon history. He was not a great thinker or scholar, scientist or divine: he was not even a great organiser or administrator or leader: but judged by his effect upon his contemporaries and his achievement under circumstances of exceptional difficulty he was certainly a great man. With no help from birth or wealth, the son of Walter Wilkins, a goldsmith of Oxford, he derived from his father a love of intricate mechanism and a practical ingenuity which in that first age of applied science attracted to him men like Boyle and Hooke. Winning distinction early in life and being chaplain to various noblemen he found himself in a world wherein the choicer spirits were turning away from tradition and faction to the exploration of nature and in a position in

1 These statements are borne out by the College Registers.
2 Thus Ray speaks of giving up his accounts as Steward in Nov. 1661: *F.C.* p. 27.
3 Samuel Salter, in the Preface to his edition of the correspondence between Whichcote and Tuckney, says that Ray spoke along with Widdrington and Templar at the funeral of Arrowsmith's predecessor, Thomas Hill, in 1653: this is probably a mistake.
4 Cooper, *Annals*, III, p. 474.

London and at Oxford to offer them hospitality and encouragement. He gathered round him at Wadham a remarkable group,[1] Robert Boyle, revered of all men for his intellectual acumen and integrity of character; Ralph Bathurst, afterwards President of Trinity College; Jonathan Goddard, Warden of Merton and Professor of Physic at Gresham College; Laurence Rooke, of King's College, Cambridge and Wadham, Professor of Astronomy and Geometry at Gresham; Thomas Millington, Ray's contemporary; John Wallis, of Emmanuel and Queens' Colleges, Cambridge, and then Savilian Professor of Geometry; Seth Ward, of Sidney Sussex, then at Wadham and Savilian Professor of Astronomy; Thomas Willis, doctor and anatomist; and Matthew Wren, son of the Bishop of Ely and cousin of Christopher.

Wilkins had imagination of a high order and courage both speculative and moral; energy to accomplish and to stimulate hard work; and charm which enabled him to survive the cataclysms of the time without compromising his opinions or ruining his career. If in his discussion of a voyage to the moon he was a forerunner of Jules Verne,[2] in his universal language he anticipated the inventor of Volapuk. When the manuscript of his *Real Character* perished in the Great Fire, he rewrote it on a grander scale in the next two years. Having married Robina French,[3] Cromwell's sister, and being a convinced supporter of the people's cause, he was inevitably a victim of the Restoration: but he accepted his expulsion without complaint, moved to London, assisted at the foundation of the Royal Society and in 1668 became Bishop of Chester. It is easy to sneer at him as an amateur, a busybody, a time-server: the first may be a fair criticism, but it applies to almost every one of his colleagues; the second and third are manifestly unjust.

Upon Ray he had an influence of which we hear little but can gather abundant evidence. Their acquaintance began at Trinity: it was strengthened by Wilkins's friendship with Willughby. It led not only to an invitation to compile tables of animals, birds, fishes,

1 Cf. Pope, *Life of Ward*, p. 29; a similar list with the addition of Sir William Petty and Christopher Wren is given by Sprat, *History of the Royal Society*, p. 55.

2 *Discovery of a New World and Probability of a Passage to the World in the Moon* published in 1638. Dryden's lines in *Annus Mirabilis* 164:

'Then we upon our globe's last verge shall go,
And view the ocean leaning on the sky,
From thence our rolling neighbours we shall know
And on the lunar world securely pry'

are not meant to be sarcastic—unlike Butler's *Elephant in the Moon*.

3 The widow of Peter French, Canon of Christchurch. Pope, with his usual malice, after describing the struggle between puritans and 'moral men' at Oxford, says that Wilkins protected himself by matrimony (*Life of Ward*, p. 44). Tillotson married French's daughter and so became Wilkins's stepson.

reptiles and plants for the enlarged *Real Character* and to translate it into Latin,[1] but to a series of lengthy visits[2] and on Ray's part to a lifelong admiration.

In the summer of 1660, when the Restoration produced its purge of Cambridge, Ray was absent, travelling the north of England with Willughby. He returned to find Dr Ferne and 'the old gang' in possession, many of his colleagues ejected, and his own case reserved for further consideration. In a letter to Courthope he gave a racy description of the upheavals, revealed his own inclination—'no promise of conformity and no orders, rebus sic stantibus', recognised that this would make forfeiture of his fellowship inevitable, and showed how distressed he was by the changes and how difficult he found it to see his own proper course. In spite of the scanty evidence it is easy to understand his dilemma. He had accepted the return of Charles: his congratulatory verses make this plain. He hoped for a generous settlement in accordance with the king's pronouncement at Breda, and would have gladly co-operated in it. But the summary expulsion of his friends and the intrigues which were disgracing the University disgusted him; and had it not been that the College urged him to stay he would doubtless have followed his impulse. When the first shock was over he decided that it was his duty to remain. This at once raised the question of ordination; and the matter is so important for our judgment of his character and has been so confusedly discussed by his biographers that it requires somewhat full consideration.

The points that arise are two. How came it that at a time when in most Colleges fellows were bound by statute to proceed to ordination within a fixed period after election,[3] Ray in spite of his sermons to the College and University was still a layman? What influenced him at the Restoration to seek what he had till then refused?

The first question is easy to answer: indeed, it can only be asked by those who like Derham[4] seem to ignore the facts of the time. Whewell in his essay on Barrow has given and documented an analogous case, that of William Croyden, a fellow of Trinity slightly senior to Ray. In 1650, the very year when Ray should have been ordained, the College investigated the position of its lay fellows in view of the requirements of the Statutes and agreed to a minute, dated 17 June, in the 'Conclusion Book', that

---

1 Cf. Skippon's letter, *Corr.* p. 22: Ray undertook the task: see below, p. 147.

2 Calamy, *Account of Ministers Ejected*, I, p. 121, says that Ray lived sometimes at Chester, sometimes at other places, but for the most part at Middleton: Allen, the Braintree doctor, records that 'Ray lived seven years with Wilkins': Miller Christy, *Essex Naturalist*, XVII, p. 12.

3 The requirement at Trinity was that with the exception of the Professors of Civil Law and Medicine all fellows must be in priests' orders after completing seven years from their M.A. (*Statutes*, ch. xix).

4 *Mem.* p. 13.

under existing conditions 'since bishops are abolished and the power of Classical Presbyteries has expired, no person can now be legally ordained a presbyter' and therefore the penalty of expulsion is void. Croyden then made a formal statement that he was no Refuser; 'seeing the formality of the Statute cannot be obeyed I will obey the purport of the Statute, that is I will use my gifts for the propagation of the Gospel in Trinity College; and for the Outward Call I shall be made to seek and embrace it as it shall be held out by Parliament'.[1] Certain of the body seem to have revived the question later: for in 1658 the 'Conclusion Book' shows a decision on 2 October that notwithstanding the decision of 17 June 1650 the statute imposing ordination 'is now in force'.[2] This certainly set Ray thinking about his position, as is clear from his first letter to Courthope:[3] but he then decided not to be ordained, and no action against him was taken. His own attitude was that of Croyden, strengthened by his strong general dislike of the imposition of oaths under threat of poverty and as a condition of employment: this, as a letter to Lister[4] makes plain, he always regarded as immoral. He had never taken the Covenant or the Engagement: he was never either a Presbyterian or an Independent. But like many hundreds of the clergy he had a deep dislike of ritual and a large sympathy with puritanism: he was profoundly religious in life and thought, but his religion like that of Whichcote or Wilkins was based upon experience interpreted by reason and expressed in morality; and for such the right to minister must be a matter of conviction and fitness rather than of external appointment. There is in all his writings not a sign that he cared for the points at issue in ecclesiastical disputes. He hated controversy in any field of endeavour and would have echoed Chillingworth's aphorism 'Protestants are inexcusable if they do offer violence to other men's consciences'[5] or Whichcote's 'Nothing is more unnatural to religion than contentions about it'.[6] So long as he was allowed to say his own prayers and preach in his College Chapel[7] he was content to stay with Croyden as one not unwilling to be ordained but not at present obliged or indeed able to comply with the statutory obligation.

The second question is hardly more difficult. He held his fellowship on condition that he should be ordained whenever practicable, and had never formally separated himself from the Church. When on his return from his tour he found that 'they have brought all these things here as they were in 1641, viz. services morning and evening, surplice Sundays and holy days

---

1 *Barrow's Works*, IX, p. xii.   2 L.c. p. xiii.
3 *F.C.* p. 16.   4 *F.C.* p. 120.
5 Quoted by Green, *History of the English People*, p. 613.
6 *Aphorisms*, no. 756.
7 Tenison's saying, 'Ray was much celebrated in Cambridge for preaching solid and useful Divinity instead of that enthusiastic stuff', is quoted by Derham, *Mem.* p. 10.

and their eves, organs, bowing, going bare, fasting nights',[1] he assumed
that he would be rejected and deliberately stayed away from Cambridge
till his fate was settled. To his surprise his place was kept open and he was
urged to come back to it. His colleague, Stephen Scandrett,[2] has recorded
that 'the College was peculiarly desirous to keep him in':[3] and in the
'Conclusion Book' is a minute of 2 October 'That Mr Wray have time
till 16th Oct. for the making up his accounts of the Stewardship and
giving in his final resolution as to conformity.'[4] They could hardly have
behaved more considerately. He was always very sensitive to kindness;
as he wrote eighteen months later: 'I hope I may say without ostentation
I am deeply sensible of and most kindly affected with every courtesy done
me, every civility shown me; this is the best quality that ever I perceived
myself to have.'[5] He was deeply attached to his mother, wholly dependent
on him since his father's death, and could not lightly sacrifice her. He
loved his work and must have known that he was doing it with marked
success; to leave his pupils and his research would be a wrench if not a
betrayal. He had been brought up a churchman and, though he disliked
the way of Laud, the Church was largely served by men of Puritan
sympathies like his friends Worthington and Wilkins; the Convention
was hopeful of a settlement along the lines proposed by Archbishop Ussher
and of a reformed and united Church such as he would gladly have served.
These things must have weighed with him. But if his conscience had not
been clear he would certainly have done in 1660 what he did two years
later when the royal policy was declared. In fact there was then no ground
of principle on which if he could overcome his dislike of the ritual changes
he need hesitate. He decided that the invitation of the College must be
accepted and that his life's work was in the University. He was therefore
ordained in London by Bishop Sanderson of Lincoln in his chapel in
Barbican to the diaconate and the priesthood on 23 December 1660.[6] He
could not foresee either the temper of the Cavalier Parliament or the
instability and profligacy of the king. That he could say, many years later,
to John Aubrey, 'Divinity is my profession',[7] and refuse on that ground
the Secretaryship of the Royal Society,[8] must clear him of any suspicion of
having regarded his orders lightly or sought them for the sake of his career.

So he returned to Cambridge, settled down to his work as tutor and
steward, took new pupils and carried on with his studies. He was not left
long in peace. Charles, when he took refuge among the Scots, had himself

1 *F.C.* p. 18. The wearing of surplices had been dropped by order of the House of
Commons in Jan. 1643: Cooper, *Annals*, III, p. 336.
2 Chaplain 1659; expelled by Ferne for refusing to read the Prayer Book services.
3 Cf. Calamy, *Account of Ministers Ejected*, I, p. 122.
4 Quoted by Matthews, *Calamy revised*, p. 405.   5 *F.C.* p. 29.
6 *Mem.* p. 13. Sanderson had preached in London on 18 Dec.: cf. Evelyn, *Diary*.
7 *F.C.* p. 163.   8 *F.C.* p. 159.

sworn allegiance to the Covenant. No Stuart ever troubled himself about perjury. His father had taken literally the adage that the king can do no wrong: he himself cared little for right or wrong, provided his personal advantage was secured; and an oath taken from political necessity could be broken for a similar reason. But others also had pledged themselves and might be less able to forget it. There must be a general dispensation and a legitimate revenge upon recusants. If the clergy declared that 'There lies no obligation upon me, or on any other person, from the oath...and that the same was in itself an unlawful oath', they would be united with the Crown if only in a common falsehood. So it was enacted by Sections VIII and IX of the Act that before St Bartholomew's Day 1662 all clergymen, and in the Universities all who bore any office, must make a solemn declaration[1] to this effect on pain of immediate forfeiture. The issue now was plain; and Ray could not but see it plainly. He had not himself taken the Covenant: but an oath was an oath, whatever kings and parliaments might pronounce: to accept the Act was to subscribe to a lie:[2] a man of honour, a teacher concerned with truth, could not thus sacrifice truth to expediency. He could not retain his fellowship or hold any position in the Church on those terms.[3]

Many of his colleagues, a larger number than we, or Ray,[4] might have expected, thought otherwise. Derham, who has no sympathy with his action, ascribes it partly 'to prejudice of education in unhappy times' and blames him that 'he was absent from his College where he might have met with satisfaction to his scruples and was among zealous Nonconformists'[5] —he was in fact touring Cornwall with Skippon, and on his return went quietly to his mother at Black Notley in order to avoid a useless and painful controversy. No doubt in the changes of the past twenty years pledges had been enforced so frequently that men had lost all sense of their seriousness. No doubt there were many who honestly believed that the Covenant accepted under duress could be legitimately disavowed. Ray had enough of the puritan in him to dislike the half-tones of casuistry and to see the issue in black and white. It cost him a bitter struggle with himself, though he has no word of bitterness for others. 'I cannot do it' is his sole assertion: 'whatever happens I cannot do it'. He had already refused

1 For the exact form of declaration cf. Cooper, *Annals*, III, p. 499.
2 This seems to be the real reason for his refusal. Writing to Courthope in July he professed his 'disgust' with 'oaths and subscriptions': cf. Gunther, *Early Science in Cambridge*, p. 376. The matter is argued by Calamy as against Dale's Life; and in *Biographia Britannica* against Calamy: but the arguments ascribe over-elaborate and meticulous scruples to him.
3 As late as 1691 he refused an offer of preferment from the Archbishop.
4 *F.C.* p. 32: 'not many in this University have refused to subscribe, in all twelve fellows'; the names given in a note by Ray are in *Select Remains of J. R.*, pp. 15–16 note.   5 *Mem.* p. 16.

the livings of Cheadle in 1659 and of Kirkby Lonsdale in 1661:[1] he now resigned his offices in the College and cast himself, as he said, 'upon Providence and good friends'.[2]

So on 24 August he found himself free and unemployed, a teacher without pupils, a cleric without a charge, debarred by his profession from secular employment, debarred by the law from his profession. The blacksmith's son had returned to his village, a failure by his own act.

In writing to his friends[3] he breaks through his habitual self-effacement and reveals how much the decision has cost him—his relief that the crisis is over, his fears and anxieties, his sense of all that has been surrendered, his hope that they will stand by him. The loss to himself and the world was indeed very great. He was cut off from the resources, the scholars and books, of a University, from financial security and official status. He must work in poverty and dependence; work as opportunity offered. What he would have accomplished if he had been able to retain the advantages of his life at Trinity, it is idle to speculate—though we can see from his books something of what he missed. But the loss was plainly not absolute, was probably more than balanced by the gain.[4] To sacrifice his integrity would have been to damage the whole character of his work. To accept an academic career might have meant absorption in tuition or removal to a bishopric. He had in fact found a vocation exactly suited to his gifts and inclinations; and if he pursued it under a handicap, at least he pursued it gloriously. His fears were not fulfilled: his friends did not fail him: he became John Ray, the naturalist.

Before we consider in detail his claim to that title it will be appropriate to ask what manner of man he was. At thirty-five and after eighteen years at the University his character was mature, his mind disciplined, his tastes fixed, his principles tested. Although his life's work was hardly begun, the man himself was fullgrown. It ought to be possible to give some picture less vague than those that have come down to us.

Physically the first point to be discussed must be his health. For ever since Boulger stated in the *Dictionary of National Biography*[5] on the authority of Calamy that he was consumptive, this has become a commonplace[6] and is even said to be obvious from his portraits. Calamy is in fact a contemporary and reliable witness: but no reference is quoted; the state-

1 *F.C.* pp. 16, 22.   2 *F.C.* p. 25.
3 Cf. especially *F.C.* pp. 25, 28, 30–1. In the most intimate passage he cloaks the strength of his feelings by breaking into Latin.
4 Green's summary: 'If the issues of St Bartholomew's Day have been harmful to the spiritual life of the English Church, they have been in the highest degree advantageous to the cause of religious liberty' (*History of the English People*, p. 623), is true of Ray's own life.   5 *D.N.B.* XLVII, p. 343.
6 E.g. Hawks, *Pioneers of Plant Study*, p. 204, states that he undoubtedly suffered from tuberculosis in later life.

ment cannot be traced; and it is probable that here as elsewhere Boulger is in error. There is in one of Ray's letters to Lister a statement that his lungs had once been affected and that the cure may not be complete.[1] But all that this involves is that as a young man, and perhaps when he began botany, he was attacked, but that the infection was checked and his health not seriously impaired. For in the same letter he states that almost all his illnesses have been due to digestive trouble.[2] It is hard to believe that if there had been permanent tuberculosis he could have carried through the long days on horseback in the Scottish tour, the stormy and interrupted voyage to Malta, and the ride through the Engadine and across Switzerland in March. It is harder still to suppose that in his old age, when he was constantly ill and was treated by Millington and Sloane, pulmonary weakness could have been ignored by them and by Allen, his Braintree doctor, and by Dale, his intimate friend and biographer. Dale, who was also a medical man, states explicitly that he knew of no illnesses except smallpox in his childhood, measles in 1668, and ulcers in his legs in his last years.[3] Allen, who treated him for pneumonia in 1690 and has left an account of the attack, makes no mention of consumption.[4] On his own confession he was always affected by cold weather—'when I was young I was every winter much troubled with itching tumours on my feet which in this country we call bloudy fals'[5]—and this increased the pain of his sores in later years. According to Lister he was far too thin and ate far too little;[6] and this, with the facts that he suffered from jaundice in 1671[7] and from attacks of violent diarrhoea in old age suggests that his digestion was much weaker than his lungs; and in view of the medicines of the period a defect there must have been given every opportunity of becoming serious. Considering the hardships of his years of travel and the vast amount of sedentary work that followed, it is ridiculous to suppose that he was either physically delicate or constitutionally an invalid.

Of his appearance and features there is little to be said. A noble bust by Roubiliac in the Library of Trinity College;[8] two portraits in oils, both by Mrs Mary Beale, one in the Zoological Department of the British Museum[9] the other in the National Portrait Gallery;[10] and a crayon drawing by William Faithorne, engraved first as Ray tells us[11] in June 1690 by

---

1 *Corr.* p. 18, letter of 1 Oct. 1667.   2 *F.C.* p. 114.
3 *F.C.* p. 5.   4 *F.C.* pp. 205–6 and *Essex Naturalist*, vol. XVII.
5 *Corr.* p. 361.   6 *Corr.* p. 17.   7 *Corr.* p. 86.
8 Presented by Edmund Garforth in 1751: a very poor engraving is the frontispiece of *Correspondence* and plates of it and of the terracotta model for it are in K. A. Esdaile, *Roubiliac's Work at Cambridge*, pp. 12, 14.
9 Reproduced in A. C. Seward's *John Ray*.
10 Reproduced in J. Duncan, 'Memoir of Ray', in *Jardine's Naturalists' Library*, vol. XXXV; and in Gunther, *Further Correspondence*.
11 Cf. the epitome of a letter to Robinson, 'Faithorne chased': *F.C.* p. 293.

W. Elder and published as a frontispiece to the *Wisdom of God* and others
of his later books, then by A. de Blois for the Dutch printing of the
Methodus Emendata, and then with some difference of brow, hair and
gown by G. Vertue in 1713 for the third edition of the *Three Physico-
Theological Discourses*; these are all the genuine evidence that we possess.[1]
Whether it be the failure of the artists or the character of their subject the
results are hardly satisfactory. Roubiliac, whose statue of Isaac Newton
has been called the greatest portrait-sculpture since the Greeks, did his best
to create a similar image of an inspired visionary. But in 1745 he had little
material to work upon—presumably only the pictures already mentioned.
He produced a work of art and imagination rather than a likeness. The
portraits are at least contemporary: but the paintings are poor specimens
of a bad period, giving no impression of character or ability, typical of the
hack-work turned out all too freely in the days before photography.
Mrs Beale's first portrait represents him without a wig and in early middle
life, perhaps at the time of his election to the Royal Society. Though
Seward claims that it 'conveys a much clearer impression of the man than
is given by the bewigged portraits of a later date', others may find its
composition dull, its expression wooden, its features unimpressive. The
second is a better painting, and is with Faithorne the source of Roubiliac's
bust. But like most of her work it is weak and characterless, a study of a
bored and rather querulous gentleman, not unhandsome but certainly neither
great nor distinguished, the sort of 'family portrait' with which we are all
painfully familiar. The engraving has much more life about it. Faithorne
was a respectable artist, and some of his portraits are among the best work
of the period. The fact that Ray chose this picture for insertion in his books
proves at least that he regarded it as recognisable. It should certainly be
accepted as the best extant likeness, preferably in Vertue's engraving: for
Vertue was a skilled worker and conscientious in his efforts to depict his
subjects faithfully. It is a strong face, grave and enquiring with a touch of
humour and much sensitiveness in the large mouth, a trace of rigour in the
eyes, and plenty of intelligence in brows and forehead. But the features
are not handsome or specially distinguished; full of character but in themselves ordinary: it is the sort of face that can still be found in East Anglian
parish councils where men listen patiently, come slowly to a decision, and
are often uncannily wise in their judgments and inflexible in their resolves.

When we turn from the pictures to his writings the case is not much
better. No man who has left so large a mass of books and letters has told

1 The portrait in St Catharine's College is almost certainly not of Ray. There is an
engraved portrait copied from Vertue in the *Select Remains of J. R.* with the names
W. Hibbart and Bathon sculp. beneath it and another in the *Englishwoman's Magazine,*
II, p. 257 with 'Lely pinx*t*, Meyer sculp.' beneath it, which are plainly redrawings of
the Faithorne picture.

so little of his own feelings and private affairs. He was always reticent; and Derham deliberately excised from the volume of correspondence at his disposal everything that has any human interest. Fortunately Gunther's researches have disclosed many more intimate letters and enable us to get a clearer impression of the man behind the philosopher. But, even so, it is seldom that his self-repression is relaxed or that personal information is given. He used all the customary means of self-revelation, diaries, observations, records of research, private and public letters, books, sermons, prayers; and in none of them designedly, in few even accidentally, revealed himself.

The explanation lies largely in the character of the bulk of his correspondence. Letters such as he wrote regularly and for many years to Robinson or Sloane, Lister or Aubrey or Lhwyd are not to be judged as the utterances of private friends; and were frequently exchanged between men who had never met and knew nothing of one another outside the field of their common interest. Thus Ray often expresses the wish to correspond with young naturalists whose work has been reported to him; and after ten years of contact writes to Lhwyd 'you may possibly have heard, though I do not remember I ever told you, that I have four daughters'.[1] In these busier days letter-writing of this kind has entirely disappeared: in the seventeenth century it was a recognised and universal practice, a literary form with its own technique and tradition. Oldenburg, the first Secretary of the Royal Society, had developed it to an almost unbelievable degree:[2] the published volumes of Ray's correspondence show how massive was his own output. In the next century Gilbert White proves how vigorously the tradition was maintained. Indeed it only ceased when periodicals and meetings of learned societies superseded it and when means of communication made travel and talk easy.

Nevertheless, even in the purely informal correspondence with Courthope and a few of the letters to Lister where friendship breaks down convention, there is little subjectivity or expression of personal tastes and temperament. We see a man of strong human sympathies, quickly responsive to affection, generous in his judgments, shy of avowing emotion, sensitive to the feelings of others. There is one instance of strong, almost contemptuous, indignation when he had been wantonly insulted by Robert Morison;[3] there is one sharp criticism—of Plukenet[4]—when he had to admit his own over-estimate of his knowledge. Otherwise there is not a

1 *F.C.* p. 267.
2 It was stated that he spent many hours daily over his correspondence, and after his death seven volumes of letters copied by him were in the possession of the Royal Society: cf. Birch, *History of the Royal Society*, IV, p. 260.
3 *Corr.* pp. 41–2: he makes it clear that his outburst is confidential.
4 *Corr.* pp. 307, 371.

word of condemnation, still less any flavour of malice or jealousy or self-satisfaction. The man is genuinely humble in the true and Christian sense[1]—one of the meek who inherit, and enjoy, the earth. He is never thinking about himself, even when he is facing the calamity of his expulsion from Cambridge or the death of his daughter; but always about his friends, their common work, the tasks that life has laid upon him. In an age of gossip, affectation and intrigue his letters to Aubrey or Sloane are as wholesome and straightforward as they are to Courthope or Robinson. He is no prude: he can enquire after his friends' wives in childbirth[2] with the frankness of his day; and although he apologises gravely and delicately for the appearance of an indecent proverb in his collection, he includes and defends 'slovenly and dirty words' which the *Oxford Dictionary* suppresses.[3] He is no ascetic: when Lister criticised his leanness of body and abstemiousness of diet, he replied[4] that on the contrary he was a regular glutton. Certainly he liked his sweetmeats[5] and beer[6] and tobacco,[7] and had acquired at Trinity an interest in food and a cultured taste in wine. But he is not dependent upon luxuries and endured fifteen years of pain without self-pity. As he grew older the essential simplicity of his temperament led him to an almost austere condemnation of any artificiality in speech or living: 'oratory which is the best of these arts', he writes to Robinson, 'is but a kind of voluptuary one, like cookery, which sophisticates meats and cheats the palate, spoiling wholesome viands and helping unwholesome'.[8]

But if he was always something of a puritan, this did not affect his splendid capacity for friendship, or set bounds to his power of appreciating the worth of his fellow-men. His friends were not limited by creed or party or class. Skippon, who left the University for ever at the Restoration; Lister, who was given a fellowship at St John's by Royal Mandate at the same time; Wilkins, the son-in-law of Cromwell; Sir Robert Barnham, the Cavalier M.P.; Samuel Dale, a staunch nonconformist, his

1 Cf. e.g. *Corr.* p. 346, the close of a letter to Sloane: 'I cannot take leave without telling you that I dare not own anything of worth in myself meriting respect from any man, but the less I deserve it, the more I am obliged to them that give it, but especially you, sir, who must needs see through me and discern how mean my skill is in anything, and must therefore be partial to your most affectionate friend.'

2 E.g. to Lister: *F.C.* pp. 125–7.

3 *Collection of English Proverbs*, Preface to 2nd edition.

4 *Corr.* p. 18: he advocated a natural diet by reference to the health of animals.

5 Sloane regularly sent 'sugar' to the household: cf. *Corr.* pp. 294 ('the sugar you design me I cannot without some violation of modesty receive') and 303 ('we often taste of your kindness'), etc.

6 *Observations*, p. 51 ('thick beer they call it, and well they may').

7 Seward, *John Ray*, p. 33, infers from a burn on one of his letters that he was a smoker. Tobacco is an item in a household groceries bill scribbled on the back of a letter from Lhwyd; *F.C.* p. 151—'tobacco ¼...2d.'

8 *Corr.* p. 229.

neighbour at Black Notley; Henry Compton,[1] the Bishop of London, who visited him there;[2] Charles Hatton, the 'incomparable'[3] aristocrat; Peter Dent and James Petiver the apothecaries, the one a friend of his youth, the other of his old age; these and many others, diverse in all else except their interest in nature, were equally the objects of his affection. After Willughby, Tancred Robinson, with whom he interchanged letters for twenty years, became his 'amicorum alpha': but he shows the same comradeship to Thomas Willisel the tramping plant-collector or Ralph Johnson the country parson, to Thomas Lawson the Quaker or Hans Sloane the Court physician.

Nor was he, so far as can be judged, so absorbed in his men friends and scientific interests as to be indifferent to feminine society or unconcerned with the daily happenings of life. There is something more than convention in his messages to Courthope's mother and sisters,[4] in his visit to Sir Robert and Lady Barnham after the death of his friend their son,[5] in his affection for the Lady Cassandra, Willughby's mother, or his standing as proxy at the christening of his doctor Benjamin Allen's baby.[6] We know very little of his intimate life, of his care for his mother in her widowhood, of his marriage to Margaret Oakeley, of the birth of his four daughters and of the domestic circle at Dewlands. But from a multitude of small hints and allusions[7] it is evident that he was always generous of his gifts, sociable to his kind and devoted to his home; indeed he could not have accomplished the colossal labours of his later years in a small house crowded with books, collections and children unless he had been contented and happy. After many adventures and over almost insuperable obstacles he had found his haven in the village of his childhood. 'He was of most recluse and pious life, and told me that he preferred that with just a sufficiency than to expend time wastefully and enter into temptations which he could not see to avoid in making address for preferment or public business. This was his deliberate choice.'[8]

Benjamin Allen, who put this on record, knew him only in his last years when old age and ill-health had straitened his outlets. In his prime he was an eagerly welcomed guest, living with Wilkins at Chester or Willughby at Middleton or Thomas Bacon at Friston as an honoured member of the

1 *F.C.* p. 294. Compton collected exotic trees at Fulham.

2 *Essex Naturalist*, XVII, p. 159: Black Notley was at this time in his diocese.

3 So North's *Lives* (ed. 1826), III, p. 314; the editor notes that he knows no reason for the epithet.

4 *F.C.* pp. 30–1, 35, 38.   5 *F.C.* p. 117.

6 On 12 Aug. 1697: cf. *Essex Naturalist*, XVII, p. 11; and he did not approve of godparents!

7 Scattered through his letters and particularly in the *Historia Insectorum*: see below, pp. 394–6.

8 *Essex Naturalist*, l.c.

family, competed for by Courthope and the Burrells, and lent Faulkbourne Hall by his 'very kind and constant friend Edward Bullock'.[1] He was at home with the Royal Society at Gresham College or among the virtuosi at Naples; and was held in honour by Malpighi[2] and Boccone[3] and Camelli[4] and Tentzel[5] not less than by Thomas Browne or John Tillotson. He was, as Dale puts it, 'affable, being not puffed by his learning, and always communicative of anything he was master of',[6] always encouraging his friends to publish their results, always scrupulous to a fault in acknowledging indebtedness.[7] Everywhere for all his modesty he was recognised as a man of eminence, a leader who had the power of bringing out the best in other men and enthusing them with his own love of nature.

It is no small testimony to his greatness not only of intellect but of personality that the impression of his contemporaries has been endorsed by all subsequent students who have become intimate with the records. They cannot speak of him without enthusiasm and the use of superlatives; and the more they know the stronger becomes their praise. Pulteney, whose *Sketches* is still the best record of early British botanists and who surveys Ray's work carefully and in detail, writes: 'the character of Mr Ray cannot be contemplated without a high sentiment of respect and gratitude'[8] and 'he became without the patronage of an Alexander the Aristotle of England and the Linnaeus of the time'.[9] Sir James Smith, whose studies fitted him to form a sound opinion and who has written a concise account of Ray's work, describes him as 'our immortal naturalist, the most accurate in observation, the most philosophical in contemplation, and the most faithful in description of all the botanists of his own or perhaps any other time'.[10] Professor Boulger, author of the account of him in the *Dictionary of National Biography*, calls him 'the greatest naturalist with one possible exception that England has ever produced'.[11] Sir Albert Seward, whose essay gives a fresh and vivid picture of his character, concludes the account of his work with the words: 'Ray set truth above tradition and had the courage of his convictions. We do homage to him as one of the founders of modern science; we think of him as a prophet and preacher of the new gospel in an age when the dawn was beginning to

1 *F.C.* p. 182.
2 Cf. Robinson's account of his meeting with Malpighi, of M.'s respect for Ray and criticism of Lister: *Corr.* pp. 142–3.
3 He wrote to Boccone in 1674: cf. *Corr.* pp. 108–9.
4 Cf. *Corr.* pp. 377–9.
5 Cf. his letter about the bones of the mammoth: *Corr.* pp. 322–3.
6 *F.C.* p. 6.
7 He condemned secretiveness (e.g. in Plukenet, *Corr.* p. 307) and was severe to plagiarism.
8 *Sketches*, I, p. 280.
9 L.c. p. 188.   10 *Mem.* p. 87.
11 *Proceedings of Essex Field Club*, IV, p. 171.

break after a long night of comparative darkness. He stands as a beacon set on a hill penetrating the mists of ages with shafts of light, giving warmth to the hearts and stirring the imagination.'[1] These are certainly the four men who have known their subject best; and it is clear that they have all fallen under the spell of his character.

It is this power to fascinate, combined with the sheer mass of his intellectual output, that gives him his unique place in the history of human progress. If in the strict sense he left no school, he created a great following by his example and advice, his friendship and encouragement. Willughby, Lister, Robinson, Sloane, Dale, Lhwyd, Vernon, Petiver, Morton, Moyle, all owed to him their interest in natural history; and in the next century his influence affected all the leaders of science and initiated that love of plants, insects and birds which has become so characteristic of the people of Britain. If to-day it amazes our continental neighbours to find the greatest English newspaper reporting the annual arrival of Cuckoos and Swallows or the capture of a Camberwell Beauty and the spread of the Comma, the development of that trait in our national life is due to John Ray far more than to any other man. And that is an achievement of some importance in our history.

For this achievement he was singularly well endowed. Early in his career as a scientist, when he was invited by the Council of the Royal Society 'to entertain them yearly with one discourse grounded upon experiment', he wrote to Oldenburg, the Secretary, a reply which gives his own estimate of his capabilities:

> For my part [he says] I do not think myself qualified for such an undertaking, for though I am as willing as any to contribute what I can, yet I find not that ability which is requisite to such a performance. To speak the truth I have neither ability nor leisure to make experiments in any kind: ability I mean neither of wit nor purse, having no good projecting or inventive faculty. And therefore if such an exercise the Society expects as is grounded on experiment, I dare not promise anything and must desire to be excused. But if something that I have observed in the history of plants or animals (in which particulars only I can pretend to the knowledge of anything not common) may be acceptable, I shall not be wanting to do my part in what has been propounded.[2]

This verdict, far too modest as it certainly is, reveals the bent of his genius. It was written in 1674 after the publication of the experiments that he and Willughby had performed on the sap of trees, and at the very time when the *Collection of English Words* with its full accounts of manufacturing processes was being printed. He may have been lacking in inventiveness; he was certainly not ill-qualified for experimental work. But Willughby's death had left him poor, dependent and busy. For laboratory studies and the chemical and physical problems which intrigued

1 *John Ray*, p. 30.   2 *F.C.* p. 67.

Hooke and amused the virtuosi he had neither resources nor much inclination.

The sort of observations which he was able to make were in point of fact more important at this stage in the development of science than the experiments which he declined to offer. If alchemy was to be replaced by ordered knowledge, the first step must be an objective and unbiased study of nature. Description and classification must precede interpretation and speculation: we must know what things are and how they work before we can begin to experiment profitably. When men do not know the names and properties of natural objects, and are ready to believe any fanciful superstition about them, they cannot even see and record accurately, and their experiments, even if occasionally successful, are merely empirical and meaningless. Ray recognised his vocation rightly when he gave his superb powers of exact and discriminating observation, of retentive memory and precise definition, to the description of nature. So doing he laid the foundations for systematic study and influenced the whole of zoological and botanical science. When he fulfilled Oldenburg's request by his papers upon the structure of seeds and the determination of specific distinctness he contributed to the advance of knowledge more usefully than if he had devised some ingenious mechanism or pleasant toy.

His quality of mind and outlook can be best illustrated from the one incident in his whole career which raised the question of witchcraft. Early in 1694 his young friend Edward Lhwyd, Keeper of the Ashmolean Museum at Oxford, had been much excited by some unexplained stack fires in Merionethshire.[1] These had been reported to him by his friend the Rev. Maurice Jones, Rector of Dolgelly,[2] who apparently regarded them as due to witches.[3] Lhwyd himself associated them with a plague of locusts that had appeared in Wales during the previous autumn and suggested that the dead insects engendered an inflammable vapour. He submitted the problem to Ray; and the answer was given by return of post.[4] It was, in effect, that there can be no connection with the locusts: witchcraft and miracle are the natural explanation by country people of the facts as reported: no wonder they are alarmed and helpless. He adds: 'This I have written because you required me as seeming most probable according to the phenomena communicated by your friend. Possibly if one were upon the place, one might upon full information of all circumstances and particulars be able to guess better at the causes of these accidents, and those who have a comprehensive knowledge of nature and good inventive heads

1 Cf. his letters to John Lloyd (Gunther, *Life of Lhwyd*, pp. 218, 227–8) and to Lister (l.c. pp. 218–20, 225, 233). The fires took place on three nights before and after Christmas.

2 Gunther, *Life of Lhwyd*, p. 221.

3 L.c. p. 228.   4 *F.C.* p. 243.

might judge whether anything could be done'—which is Ray's tactful way of saying 'don't be alarmed by superstition; accurate observation is the first requisite'. His modesty compels him to add: 'For my part I am none of these, never having been favoured by Providence to invent any new or useful thing in all my life.'

Equally tactful and even more frank in its warning is the letter, already quoted, which he sent to John Aubrey in comment upon his *History of Wiltshire* on 27 October 1691.[1] After urging him to 'speed it to the press' he says:

Whatever you conceive may give offence may by the wording of it be so softened and sweetened as to take off the edge of it as pills are gilded to make them less ungrateful. As for the soil or air altering the nature and influencing the wits of men, if it be modestly delivered, no man need be offended at it, because it accrues not to them by their own fault, and yet in such places as dull men's wits there are some exceptions to be made. You know the poet observes that Democritus was an example

> Summos posse viros et magna exempla daturos
> Vervecum in patria crassoque sub aere nasci.

Then after the passage about his school[2] he continues:

I think (if you can give me leave to be free with you) that you are a little too inclinable to credit strange relations. I have found men that are not skilful in the history of nature very credulous and apt to impose upon themselves and others, and therefore dare not give a firm assent to any thing they report upon their own authority; but am ever suspicious that they may either be deceived themselves or delight to teratologize (pardon the word) and to make show of knowing strange things.

It says much for Aubrey, whose *Miscellanies* prove how exactly the last sentence fits him, that he accepted the letter so readily as to attach it to the script of his History. It says more for Ray that in the very letter in which he acknowledged his humble origin he could express so plainly his condemnation of credulity and of speculation divorced from accurate observation.

But when his critics fasten upon this insistence and urge that he is nothing but a taxonomist or even a maker of catalogues, it must be remembered that his work was not confined to description. No man could have produced the two books in which he set out the interpretation of nature, the *Wisdom of God in the Works of Creation* and the *Three Physico-Theological Discourses*, if he had been a mere observer and systematist. Those books reveal what is indeed apparent in all his work, an acute and inquisitive insight into the problems of form and function and a genius for

---

1 The letter is printed in Aubrey's *Natural History and Antiquities of Surrey*, v, pp. 408–11.
2 Quoted above, p. 17.

fastening upon the essentials in them and for asking the right questions. He knew (and was wise in knowing) that the time for full answers was not yet; that guesswork only darkened counsel; and that patient investigation, perhaps for generations, must precede any satisfactory solutions. But he saw what was worth studying; he was content to make a sound beginning; he refused to let his imagination run away with him or to confuse fancy or tradition with knowledge. It is time to see how he was drawn into his life's work.

NOTE. *Ray's pupils at Trinity as given in the Admissions Register*

Benskin, Richard. Subsizar, 4 May 1653 (B.A. 1656).
Hunt, Thomas. Of Essex. Subsizar, 10 November 1653 (B.A. 1657–8).
Skippon. Fellow-Commoner, 1655.
Burrell, Timothy. Pensioner, 22 June 1659 (Scholar 1661, B.A. 1662–3).
Colvill, Thomas. Pensioner, 27 June 1659.
Coney, Joseph. Subsizar, 5 July 1659 (B.A. 1663–4).
Howard, Richard. Subsizar, 5 March 1660 (B.A. 1663–4).
Bouchert, Matthew. Pensioner, 9 March 1660 (Scholar 1661, B.A. 1663–4, Fellow 1664).
German, Henry. Pensioner, 26 March 1660.
Allen, John. Sizar, 19 April 1660 (Scholar 1662, B.A. 1663–4).
Goring, Edward. Fellow-Commoner, 1 June 1660.
Grace, Robert. Sizar, 13 June 1660 (B.A. 1663–4).
Grace, Job. Sizar, 13 June 1660 (Scholar 1664, B.A. 1664–5).

Of these Goring, as we have seen, died almost immediately after coming into residence; of Colvill and German nothing seems to be known; all the others except Skippon were ordained.

CHAPTER IV

THE *CAMBRIDGE CATALOGUE*

The Catalogus Plantarum will be a florid ornament to Cambridge.
SAMUEL HARTLIB to JOHN WORTHINGTON,
Worthington's *Diary*, 1, p. 174.

Of the condition of science at the time when Ray first devoted himself to it there is abundant evidence in his own writings. The literature in each department of zoology will best be considered when we treat of his contribution to it. But though he had certainly read the works of Gesner and his English friends Caius and Mouffet before he published his first book and had a good knowledge of Aristotle, Pliny, Dioscorides and the other ancients, it was then only in botany[1] that he was a master.

Here so far as Britain is concerned the foundation had been laid by Ray's Cambridge predecessor, William Turner, a student at Pembroke Hall, elected to a fellowship in 1531 and compiling his first essay, *Libellus de re herbaria novus*,[2] in 1538. Turner gives a clear account of the total ignorance of botany in the University at that time: 'I could never learn one Greek neither Latin nor English name, even amongst the physicians, of any herb or tree...and as yet there was no English Herbal but one full of unlearned cacographies and falsely naming of herbs.'[3] Enthusiasm for the Reformation drove Turner from Cambridge into exile: he travelled widely, met Gesner and studied botany at Bologna under Luca Ghini, the first professor of the subject,[4] and published a second tract in 1548.[5] After a short spell in England as Dean of Wells, during which he published the first part of his Herbal in 1551, he was again banished, went to Germany and published the second part at Cologne in 1562. The third part, incorporating the others and containing the figures used by Fuchs in 1545, was published also at Cologne in 1568, the year of Turner's death. It was intended for the use of apothecaries, but showed evidence of his originality, knowledge of plants and interest in their localities. Ray speaks with admiration of his sound learning and judgment.

1 The best account of early botanists is that in Agnes Arber, *Herbals*, 2nd ed. 1938. Pulteney, *Sketches of Botany*, 1790 and Sachs, *History of Botany*, translated by Garnsey and Balfour, are useful.

2 Reprinted with notes and a life of Turner by B. D. Jackson, 1877.

3 Pulteney, *Sketches of Botany*, 1, p. 60: on p. 48 he says that Turner is referring to the *Great Herbal*, printed by Peter Treveris in London 1516.

4 He founded the Physic Garden and taught Cesalpino: Pulteney, l.c. p. 61.

5 *Names of Herbs in Greek, Latin, English, Dutch and French*: an edition by J. Britten, 1881.

During the last quarter of the sixteenth century the study of the 'uses and virtues' of plants was powerfully stimulated in Britain by the work of the great trio of botanists, Rembert Dodoens (1517–85), Jules-Charles de l'Ecluse (1526–1609) and Mathias de l'Obel (1538–1616), who with the help of the famous printer, Plantin of Antwerp, and the University of Leyden, gave lustre to the Low Countries. Clusius, who used his remarkable knowledge of languages in translating many works otherwise inaccessible, whose travels gave him unrivalled acquaintance with the flora of Europe, and whose character and personality commanded general respect, visited England at least three times.[1] William Harrison, writing 'Of Simples',[2] says of him:

The chief workman (or as I may call him the founder of this device) is Carolus Clusius, the noblest herbarist, whose industry hath wonderfully stirred them up unto this good act. For albeit that Matthiolus, Rembert, Lobel and others have travelled very far in this behalf, yet none of them hath come near Clusius....If this man were in England but one seven years, he would reveal a number of herbs whereof our physicians have no knowledge.

Ray, in his County Lists in Camden's *Britannia*,[3] mentions two of the plants found by Clusius in Kent, *Calamintha acinos* and *Gentiana amarella*, and in the *Cambridge Catalogue* gives a full chronological list of his translations and original writings. The chief of these were the *Rariorum Stirpium Historia*, published at Antwerp in 1576, and the ten books of *Exotica* in 1605 which Ray used for birds and fishes as well as for plants.

Lobelius, as Ray calls him, was like Clusius a pupil of Guillaume Rondelet[4] at Montpellier, the greatest school of botany in Europe. At the age of twenty-one he came to England, and in 1570 with Pierre Pena, of whom little is known, published his *Stirpium Adversaria nova* and in it set forth his system of classification. Going to the Low Countries he published his *Observationes* with 2116 figures at Antwerp in 1581. Some ten years later he settled in England, superintending a botanical garden at Hackney and being appointed King's Botanist by James I. He died before his final work, *Stirpium Illustrationes*, was printed. Of this work Parkinson seems to have bought the right to use the manuscript, and was severely criticised by How for his borrowings from it. De l'Obel's long residence in England gave him a good knowledge of its flora and 'he is responsible for more than eighty "first records" of our native plants';[5] but he wrote in Latin, and

1 He was in touch with Sir Philip Sidney and also with Drake and Raleigh.
2 *Elizabethan England* (ed. L.W.), p. 30; *Description of England* (ed. Furnivall), I, p. 329).
3 Gibson's edition, 1696, cc. 224, 226: for the Gentian cf. *Cat. Angl.* p. 129.
4 Rondelet seems to have written nothing on botany except a *Pharmacopoea*, included by De l'Obel in the 1605 edition of his *Adversaria*: but his *De Piscibus Marinis* was one of Ray's chief authorities in the *Historia Piscium*.
5 Arber, *Herbals*, p. 90.

bad Latin at that, and his works did not attain wide influence. Ray's opinion of him is given in the *Historia Plantarum*:

If I remember aright Jean Bauhin [who worked with De l'Obel at Montpellier] called him unreliable and vain: this seems too severe: but I have noticed that in his descriptions he is inexact—*parum curiosum*—and trusting his memory too often made mistakes especially in localities: he gives many plants as natives of England which no one else has observed and which cannot be found in the places specified, or perhaps anywhere at all.

More important for the general development of botanical studies in England was the translation by Henry Lyte of De l'Ecluse's French version of Dodoens's *Cruydeboeck*, originally published in 1554. Lyte's work, though actually printed in Antwerp 'to receive the advantage of the figures',[1] was published in London in 1578 in folio with 870 cuts taken from Plantin's stock, and reprinted without pictures in square octavo in 1586 and 1595 and in an abridged form in 1606. Dodoens was a physician rather than a botanist, and the work became the normal text-book for medical students. It is curious that Ray, who gives an account of the Flemish, French and Latin versions, does not mention the English or its translator.[2] In 1583 Dodoens published a final edition of all his works, *Stirpium Historiae Pemptades*: and of this Gerard obtained Dr Robert Priest's unpublished translation and used it as the main and unacknowledged source of his famous *Herbal*.

Few writers have obtained a great reputation less deservedly than John Gerard. Born at Nantwich in 1545 and trained as an apothecary he had a herb-garden in Holborn of which he issued a catalogue in 1596, supervised Lord Burleigh's gardens and in 1588 offered to create a similar garden in Cambridge.[3] In 1597 he published his famous *Herbal or General History of Plants*, taking almost all the figures from Jacob Dietrich of Bergzabern (Tabernaemontanus), whose *Eicones* had appeared in 1590. The text was certainly derived from Priest, though Gerard explicitly denies this, but it was rearranged after De l'Obel's order and expanded by borrowings from De l'Ecluse. The charm and interest of the book are apt to blind its readers to its defects, the attachment of plates to the wrong descriptions, the reckless multiplication of species, the credulity and errors, the false claims and statements, and the blunders due to ignorance of Latin. Ray, who discovered these faults by experience and duly noted them, used the *Herbal* in the revised and corrected form published by Thomas Johnson[4] in 1633 and reprinted in 1636.

1 Pulteney, *Sketches of Botany*, I, p. 92: the figures are mainly those originally cut for Fuchs.

2 So far as I can discover Ray never mentions Lyte by name: this is the more curious as Nidd certainly possessed a copy of the *New Herbal*, 1619 edition.

3 Cf. letter to Burleigh in Gunther, *Early Science in Cambridge*, p. 371.

4 Cf. H. Wallis Kew and H. E. Powell, *Thomas Johnson Botanist and Royalist*, 1932.

Johnson, who was also a London apothecary, had printed in 1629 an account of a plant-hunting excursion into Kent and a list of plants on Hampstead Heath. In 1634 he published his *Mercurius Botanicus*, recording a journey through Oxford to Bristol and back by the Isle of Wight and Guildford, and in 1641 a second part of the *Mercurius* on a visit through Wales to Snowdon, where he found *Saussurea alpina*, *Saxifraga stellaris* and *Sedum roseum*.[1] The friends who accompanied him on his journeys, and especially his distinguished correspondent, John Goodyer, contributed much to the value of his writings. His death from wounds in the defence of Basing House was a serious loss to British botany. Ray, who always speaks highly of him as *Gerardi emaculator*, declared that his general scholarship and botanical skill richly deserved the doctorate which Oxford conferred on him. But the *Herbal*, although much improved, is still a very faulty volume: the figures and descriptions by no means always agree; many species appear more than once; varieties are multiplied; and it is no easy task to identify plants from it.

In his introductory remarks 'To the Reader' Johnson gives a long historical account of the state of botanical knowledge as he understood it. His list of authors is as follows: Solomon, Theophrastus, Aristotle, Dioscorides, Pliny, Galen, Paulus Aegineta, Aetius, Macer, Apuleius (for whom Ray[2] referred his readers to Johnson), and the Arabians; Bartholomaeus Anglus, whom he identifies wrongly with B. Glanville, author of the *De Proprietatibus Rerum* in 1397;[3] the *Hortus Sanitatis*; Ruel, Brunfels, Bock (Hieronymus Tragus), Fuchs and Gesner; Lonicer, Mattioli, Amatus and Dodoens; Pierre Pena and Mathias de l'Obel, who 'did here in London set forth *Stirpium Adversaria nova*'; de l'Ecluse, Pona, Cesalpino, Camerarius, d'Aléchamps and the book 'vulgarly termed *Historia Lugdunensis*'; 'J. T. Tabernaemontanus', who published 2087 figures—'with these same figures was this work formerly printed'; Alpino, Colonna, Gaspard Bauhin, Besler and three Italians; W. Turner, whose *Herbal* had the figures of Fuchs; Henry Lyte and J. Parkinson's *Paradisus Terrestris*. Of Gerard he adds: 'His chief commendation is that out of a propense goodwill to the public advancement of this knowledge he endeavoured to perform therein more than he could well accomplish.'[4] This bibliography proves Johnson to have been well qualified to emend Gerard, though in fact the merits of his work are due less to books than

---

1 But not, as Pulteney (*Sketches of Botany*, I, p. 133) states, *Meconopsis cambrica*, which he inserts only on Parkinson's authority.

2 *C.C.* Explicatio Autorum.

3 The identification of Bartholomaeus Anglus with Glanville is a mistake due originally to Leland: his book is also dated wrongly; for a copy of it was made in Oxford in 1296. Ray never mentions him.

4 He fully exposes Gerard's plagiarism from Priest.

to his excursions and to the skilled assistance of his friend John Goodyer of Mapledurham and Petersfield.[1]

Gerard and Johnson had chiefly in view the needs of herbalists and apothecaries, men of their own profession. John Parkinson approached botany from the side of its other utilitarian interest and primarily as a gardener. In his first book, written in English despite its punning title, *Paradisi in Sole Paradisus terrestris*—'the Park on earth of the Park in sun'—he limited himself to the plants suited for gardens, and produced a work which alike for the quaint felicity of its style and its range of knowledge has a unique place among lovers of flowers.[2] Published first in 1629 with woodcuts specially prepared for it, but for the most part borrowed from and inferior to those of Clusius and L'Obel, it went into a second edition in 1656. Ray was of course more indebted to his second book, the *Theatrum Botanicum* published in 1640, which professed to include all known plants except those in the *Paradisus* and in fact 'contained more species than any other History yet extant'. These were divided into classes 'on a new method according to their uses and qualities'.[3] In the third of his letters to John Aubrey[4] Ray gives an interesting account of Parkinson's sources:

First he takes in all the plants contained in G. Bauhin's Pinax... to which he adds the Canada plants of Cornutus;[5] some out of Ferrarius his Flora[6] and Tobias Aldinus his Hortus Farnesianus:[7] and some hundreds of new ones which he took partly out of Lobel's papers by him purchased, and partly had the seeds and patterns of plants from Wm. Boel, a Frenchman and a skilful herbarist who travelled Spain, Parkinson saith at his charges,[8] and from Jo. Tradescant[9] who brought home sundry rare exotics out of Virginia:[10] some few perchance might be of his own discovery... to enumerate them exactly would be very difficult by reason of the brevity and obscurity of his descriptions.

In the *Cambridge Catalogue* Ray complains that he multiplies species unnecessarily and adds: 'I do not criticise his industry: I too often deplore his lack of judgment and accuracy.'

1 For Goodyer cf. Gunther, *Early British Botanists*, pp. 1–196.
2 It was reproduced in facsimile in 1904.
3 So Ray, *Hist. Plant.* I. Explicatio Nominum.   4 *F.C.* p. 162.
5 Jacques Cornut of Paris, *Canadensium Plantarum Historia*, 1635.
6 Giovanni Battista Ferrari S.J., *Flora seu De Florum Cultura*, Rome, 1633.
. 7 *Exactissima Descriptio Rariorum Plantarum in Horto Farnesiano*, Rome, 1625. This book was apparently written by Pietro Castelli, Prefect of the garden at Messina: 'falso sibi vindicavit Aldinus' (Séguier quoted by Pritzel, *Thesaurus*, p. 58). I owe this information to Mrs Arber.
8 For Boel cf. Pulteney, *Sketches of Botany*, I, p. 153: he brought some 200 seeds and many plants to Parkinson in 1608.
9 Founder of the famous garden and of the museum at Lambeth, eventually given by his son to Elias Ashmole who presented it to Oxford University.
10 For his Garden Lists cf. Gunther, *Early British Botanists*, pp. 328–46.

Ray's verdicts upon Gerard and Parkinson were the result of long and intimate study of their works. When he set himself to collate the names given by all previous authorities to the plants that he found around Cambridge, he speedily discovered that his two best known predecessors had been wholly uncritical in their use of earlier books, obsessed with a desire to include as many entries as possible, and entirely without any clear idea of specific distinctness. If they found records of plants not obviously already listed or differing in size or flower-colour, they cheerfully inserted them. The problem of identifying their descriptions is still often insoluble except in the case of familiar or striking species: in the more difficult families there is complete chaos. Much as he respected their achievements he learned not to accept their results without caution and, in his later writings, a word of warning. They are pioneers in British herbalism and floriculture: they can hardly be regarded as scientific students of botany.

Of British students the only other that deserves mention is the author of the *Phytologia Britannica*, a small alphabetical catalogue published anonymously in 1650 but ascribed to William How, the London doctor, by Merret in 1666, and mentioned as edited by him in the *Cambridge Catalogue*. This is little more than a reprint of the two parts of Johnson's *Mercurius* with a few added records from Richard Heaton of Irish plants, from Walter Stonehouse of Yorkshire[1] and others. Ray is critical of its accuracy both in identifications and in localities. Indeed the grave defects of the work were responsible for Ray's own determination, confessed to Willughby in the spring of 1660, 'to put forth a complete P.B.'[2]

When How's work went out of print, Christopher Merret, of Gloucester Hall and Oriel, Fellow of the Royal College of Physicians in 1651 and of the Royal Society in 1662, and keeper of Harvey's Library until its destruction in the Great Fire, was asked to produce a book to replace it. He was an able and versatile man, but had no real knowledge of the subject and little time to devote to it. Thomas Willisel,[3] the old soldier who played a large part in the development of field botany and afterwards worked with Ray, was employed by Merret for five years as a collector—presumably between 1661 or 1662 when he was engaged by Robert Morison at St James's Park and 1668 when he was appointed by the Royal Society. Merret also procured 800 plates of plants engraved for Thomas Johnson and still preserved in the British Museum. High hopes were formed of the promised book and Ray with real generosity contributed to it the list of plants obtained on his first visit to Cornwall in 1662. When the *Pinax Rerum Naturalium* appeared in 1666 it was found to be little more than a list of names, even the plants, only 410 in number, being catalogued very carelessly and mostly derived from Gerard and Parkinson. Ray felt that

1 Cf. Pulteney, *Sketches of Botany*, I, p. 172; Kew and Powell, *T. Johnson*, pp. 136–9.
2 *Corr.* p. 2.   3 See below, p. 151, etc.

his own work had been forestalled—'the world is glutted with Dr Merret's bungling Pinax'[1]—but the book was so obviously a failure that it became an incentive to the production of his own *Catalogus Angliae*.

For Ray's work in serious botany the influence of continental scholars is more important than any of his English predecessors. In his second letter to Aubrey,[2] in which he gives a brief but careful summary of botanical studies and literature, Parkinson is included on the ground of the new species that he had described: but the other names are all of foreigners. There is an admirable list of them at the beginning of the *Cambridge Catalogue*; and before considering that work it will be well to say something about the most important of them.

Ray himself leaves us in no doubt as to where his greatest obligation lies. In the *Cambridge Catalogue* the principal authorities named in the Preface are Jean and Gaspard Bauhin with Gerard and Parkinson; in the bibliography the Bauhins are singled out for praise; in the book itself he usually follows them and is often mistaken when he does not do so.[3] The two brothers are so obviously the heroes and guides of his early studies[4] that it is worth reproducing his tribute to them at some length.

Of Jean, the elder brother, born in 1541, trained first at Tübingen under Leonhard Fuchs[5] and then at Montpellier under Rondelet, the friend of Conrad Gesner,[6] the collaborator of Jacques d'Aléchamps,[7] Ray writes:

He was chief physician to the Duke of Wurtemberg, a man of eminent learning, high character, profound scholarship, ripe judgment, expert in all botanical literature ancient and modern, master of every field of humane and serious literature, in a word the Prince of Herbarists. Besides certain small books published in 1591, 1593 and 1598, he began his *Historia Plantarum Universalis*, but died before it was entirely finished. It was completed by Dominic Chabrey of Geneva and is the greatest book yet published on the subject. It contains almost everything that deserves record either among the older or the more recent students together with the necessary synonyms and apparatus. It consists of three volumes. If he had lived to supervise its publication he would certainly have improved its arrangement and made alterations and additions so as to fulfil his motto 'not to seem but to be'.

1 *F.C.* p. 112.   2 *F.C.* pp. 159–60.
3 As e.g. in the Henbit (*Lamium amplexicaule*), *C.C.* p. 8, which both Bauhins class as a Dead-nettle, but Ray following Gerard and Parkinson as a Chickweed.
4 Ray's estimate of the pre-eminence of the Bauhins is fully endorsed by Mrs Arber: summarising her survey of early botany she writes: 'in the works of Gaspard Bauhin classification, nomenclature and description reach their high-water mark': *Herbals*, p. 268.
5 Author of *De Historia Stirpium*, 1542, illustrated by admirable woodcuts.
6 Probably the greatest naturalist of his time, author of the *Historia Animalium* and a great botanist.
7 Joint-author of *Historia Generalis Plantarum*, Lyons, 1587.

Of Gaspard, born in 1560, studying at many universities, collecting a great herbarium and becoming Professor of Medicine at Basle, he gives a list of books:

Phytopinax, containing 2460 names of plants with some synonyms, 164 descriptions of new species, and eight plates: Basle, 1596, quarto.

All the works of Pierandrea Mattioli[1] (these not having been printed before in Germany) he corrected and edited, adding 330 figures and 50 new plants with synonyms: Frankfurt, 1598, folio.

A criticism of the *Historia Generalis* of d'Aléchamps, showing that 400 figures had been twice or thrice repeated: Frankfurt, 1600, octavo.

The *Kreuterbuch* of Jacob Dietrich of Bergzabern (Jacobi Theodori Tabernaemontani Historia)[2] he corrected and enlarged, adding many new figures, describing some of them and giving synonyms in the first part: Frankfurt, 1613, folio.

Prodromus Theatri Botanici, containing some 600 plants first described by him and 140 new figures: Frankfurt, 1620, quarto.

Catalogus Plantarum circa Basileam sponte nascentium,[3] with synonyms and localities: Basle, 1628,[4] octavo.

Pinax Theatri Botanici, the result of forty years' work, containing the names of about 6000 plants with synonyms and characteristics: Basle, 1623, quarto.

In 1658, many years after his death, his son Jean Gaspard Bauhin[5] edited the first volume[6] of his *Theatrum Botanicum* or *Historia Plantarum*: this being a twelfth of the whole work as is clear from the *Pinax*.

In the title of the Explicatio from which these extracts are taken it is stated that the list is mainly drawn from Gaspard Bauhin, and in fact the entries relating to him with the exception of the two last are copied verbatim from the bibliography in his *Pinax*. Ray did not always copy Bauhin: the names in the two lists are not identical: and in many cases Ray has written his own notes. He evidently aimed at giving full information about such authors as contributed anything to the study of British plants, and added books published since 1623. Both lists are alphabetical: but Bauhin takes the first letter of the first name of his authors while Ray starts with the abbreviation used in his *Catalogue—Ad Lob.* is De l'Obel's *Adversaria*; *Lob.* his other works; *C.B.* Gaspard Bauhin; *Ger. emac.* Johnson's edition of the *Herbal.* It must not be assumed that Ray had read

1 Born at Siena, 1501: wrote Commentaries on the six books of Dioscorides, 1544 and many editions.

2 Pupil of Brunfels and Bock (Tragus): the figures in the *Kreuterbuch*, 1588 were published separately, *Eicones Plantarum*, 1590.

3 Further evidence of Ray's knowledge of this book is contained in *Obs.* p. 101: 'if anyone desires a more particular account of plants about Basil, C. Bauhinus his Catalogus will give him full satisfaction.'

4 This is a misprint: the date should be 1622.

5 J. G. Bauhin and his son Jerome were lecturing in the medical faculty at Basle when Ray visited it in 1663: *Obs.* p. 98.

6 The cost of further publication proved prohibitive: cf. *F.C.* p. 160.

every book mentioned. He put on record long afterward that he had sought on his Italian tour for the books of Fabio Colonna and failed to obtain them.[1] There is no evidence in the *Catalogue* that he had as yet seen the *De Plantis* of Andrea Cesalpino, though it figures on his list and was afterwards of great value to him. D'Aléchamps's *Historia* may have been known to him at first hand when he composed the section dealing with the derivations of the names of plants. Jung's *Isagoge*[2] was lent to him in manuscript by Samuel Hartlib through his friend Worthington before his book was finished, though it is not mentioned in the *Explicatio*. Summing up the evidence, we may say that he had certainly studied Turner, Gerard, Johnson, Parkinson and How, the Bauhins, Dodoens, De l'Ecluse, De l'Obel and Gesner, and with these some at least of the classics.

Internal evidence shows that the Bauhins had the largest share in the shaping of his botanical interests. It can hardly be a coincidence that in the Library of Trinity College are to be found J. Bauhin's *Historia Plantarum* and G. Bauhin's *Phytopinax, Prodromus, Pinax* and *Catalogus Plantarum circa Basileam sponte nascentium*. One of these, the *Prodromus*, has a contemporary note in ink on its title-page: 'Trin. Coll. Cant. 1673. Ex dono Mḡri Henr. Dove huius Coll. Socii.' Dove[3] was admitted in 1658, gained a fellowship, became a tutor and resigned in 1673. It is hardly far-fetched to suppose that he got the book under Ray's influence and gave it to the College when he left Cambridge. All these books have the same bookplate, that used in the middle of the seventeenth century. It seems impossible to trace precisely when or from whom the other volumes came into the Library. One of them, the *Catalogus*, is very rare and of no general interest; all are uncommon. Surely Ray must have been directly responsible for their appearance.

Particular significance attaches to the Basle Catalogue;[4] for it is obviously the model on which Ray's first book was planned. The title is clue enough: he drops the *sponte*, for he included cultivated with wild plants: he substitutes *Cantabrigiam* for *Basileam*: otherwise the two are identical. Bauhin's Catalogue is arranged in the sequence of his *Pinax*, according to its author's system of classification; the notes on each species form the bulk of its contents and are followed by a full index: Ray preferred the alphabetical tradition which dispenses with the need for an index.[5] His notes follow Bauhin's pattern—a list of descriptive synonyms with their authors, and then a record of locality, for the commoner species a few words in

1 *F.C.* p. 162.   2 *C.C.* Pt. II, p. 87: see below, p. 106.
3 He became Vicar of St Bride's, Fleet Street, in 1673, Archdeacon of Richmond 1678, and Chaplain to Charles II, James II, William and Mary—a good Church and State man.
4 Mrs Arber has drawn my attention to an article on this Catalogue by H. Christ in the *Basler Zeitschrift*, XII (1913), pp. 1–15.
5 As used e.g. in Johnson, *Mercurius*.

Latin: 'In hedges', 'In meadows and pastures very common', 'Where water lies stagnant in winter'; for the rarer a precise and often very detailed definition, which Ray gives in English. Descriptions of the plant's appearance and structure are seldom given unless it is new to science. The two books are both small octavo volumes with pages of the same size.

'Moles parva, vis magna'[1]—so the learned Dr Pulteney summed up his account of the 'extraordinary production' with which Ray began his career as a scientist. The *Catalogus Plantarum circa Cantabrigiam nascentium*, published at Cambridge by John Field, printer to the University, at the expense of William Nealand, bookseller,[2] in 1660,[3] carrying no author's name on its title-page or in its contents, a small octavo volume suitable for the pocket, is certainly an unpretentious piece of work. Few books of such compass have contained so great a store of information and learning or exerted so great an influence upon the future; no book has so evidently initiated a new era in British botany.[4]

Its Preface, a charming account of the circumstances and purpose of its composition, is the best starting-point for an understanding of its achievement; though an English summary can do no justice to the dignity and the simplicity of its Latin.

I had been ill, physically and mentally [so he begins], and had to rest from more serious study and ride or walk. There was leisure to contemplate by the way what lay constantly before the eyes and were so often trodden thoughtlessly under foot, the various beauty of plants, the cunning craftsmanship of nature. First the rich array of spring-time meadows, then the shape, colour and structure of particular plants fascinated and absorbed me: interest in botany became a passion.

So he records how he searched the University for a 'preceptor and mystagogue'; found no one among all the lights of learning; despaired of ever surmounting the mountainous obstacles without help, and yet thought it shame to let this precious and necessary element of 'natural philosophy

1 *Sketches of Botany*, I, p. 197.

2 Nealand, who apparently had a London shop at the Crown in Duck Lane, seems to have gone out of business in 1662: cf. Plomer, *Dictionary of Booksellers 1641–67*, pp. 134–5. According to Worthington, *Diary*, I, p. 333, 'the most active booksellers, Allestree and his partners at the Bell in S. Paul's Churchyard, bought all the impression of one in Cambridge for whom the book was printed'. Copies in which 'London: from Jo. Martin, Ja. Allestry, Tho. Dicas at the sign of the Bell in S.[t] Pauls Churchyard' is substituted for Cambridge are known: cf. *Catalogue of Cambridge Books*, p. 40: that they are rare is perhaps due to the destruction of premises and stock in the Great Fire; Worthington, *Diary*, II, p. 211, etc.

3 John Goodyer bought his copy on 10 May for 2s. 6d.: cf. Gunther, *Early British Botanists*, p. 222: Willughby's was sent off on 25 Febr. Cf. *Corr.* p. 3.

4 Its pagination is peculiar. Preface and list of authorities are unnumbered: the Catalogue has 182 pages: in the Appendix—English names, Localities, Derivations, Botanical terms and Classification—the pages are numbered afresh 1–103. It is probable that the two sections could be bought separately.

and history' remain totally neglected. He had leisure if not ability: the study of Phytology would be of value to the University: it would be at once a pleasure to himself and perhaps a profit to others. A start could be made by collecting simples, growing them in his garden, and exploring the neighbourhood.

He continues with an account of his efforts, which no one who remembers his own first adventures, even in these days of text-books and museums, will fail to appreciate.

I had first to become familiar with the literature, to compare the plants that I found with the pictures, and when there seemed to be a resemblance to go fully into the descriptions. Gaining skill by experience, I enquired of any unknown plant to what tribe and family it belonged or could be assigned: this taught me to notice points of similarity and saved a vast deal of labour. Then the desire arose to help others in their difficulties. I was eager to make progress myself: I wanted to entice my friends to share my pursuits: so the idea of the *Catalogue* was formed.

Six years after the commencement of his researches he started on his book: nearly three more he devoted to its compilation and improvement, encouraged by his friend and inseparable companion John Nidd, whose death saw the script nearly finished,[1] and by his younger associates Francis Willughby and Peter Courthope, who had no special knowledge of the subject but gave him constant help and support. His object he defines as follows: 'it was in general to illustrate the glory of God in the knowledge of the works of nature or creation; then to enhance the reputation of my Alma Mater, the University of Cambridge, which must suffer abroad if its equipment in this field were defective; and finally to enrich the common life and extend the advantages which such studies can bestow.'

So he turns to a description of the methods that he has followed. The names of the plants are given in alphabetical order: the titles under which they appear in the four principal authors, Jean Bauhin, Gaspard Bauhin, Gerard and Parkinson, are added in every possible case: if other authors give anything characteristic or important, or if a list of synonyms would furnish a description, further titles are included. The type is so chosen as to distinguish the name of the plant from its epithets. Localities are given in English not in Latin. Plants previously undescribed or described obscurely and with confusion are furnished with descriptions. Observations either confirmed by personal experience or drawn from reliable sources or in themselves probable are also appended. To the *Catalogue* itself certain additions have been made, first, and for the benefit of novices, an index of English names with their Latin equivalents; secondly, a list of places round

1 The suggestion that Nidd was responsible for joint authorship is not borne out by this paragraph. Indeed his influence hardly appears except in the note on the rings of trees (*C.C.* p. 57), where Ray rejects his view, and in the insertion of four species of *Veronica* (*C.C.* pp. 10–11), which Ray in his Appendix of 1663 says was due to Nidd.

Cambridge and of the rarer wild plants found in them; thirdly, an explanation of the derivation and meaning of the names of plants; finally, a glossary of botanical terms and an outline of the customary classification.

The Preface closes with a recognition of the inevitability of errors in the work, although he has spent much labour on its revision:

> Pardon for mistakes will encourage further efforts, but I promise to be careful never to publish what has not been the subject of many days' emendation: my present purpose involved a measure of haste in order to revive the almost extinct study of botany....There are larger prospects ahead. This little book may excite others to a similar survey of their own localities and so to a complete *Phytologia Britannica*.[1] There are many plants like Foxglove, Wood-sage, Butcher's broom, Oak-fern, Fine-leaved Heath which we hope to find some day here: others like Walnut, Hornbeam, Hellebore, Wild Radish, Poppy, Bistort etc. which I suspect did not spring up naturally, but were casually introduced: others may well disappear from the places noted by me, some as annuals that fail to establish themselves, some destroyed by cattle, some torn up by root-collectors....We would urge men of University standing to spare a brief interval from other pursuits for the study of nature and of the vast library of creation so that they can gain wisdom in it at first hand and learn to read the leaves of plants and the characters impressed on flowers and seeds. Surely we can admit that even if, as things are, such studies do not greatly conduce to wealth or human favour, there is for a free man no occupation more worthy and delightful than to contemplate the beauteous works of nature and honour the infinite wisdom and goodness of God. We are sure that the pursuit of plants can appeal to the young; for we have seen many sons of Trinity College finding in it both bodily exercise and mental satisfaction. Of course there are people entirely indifferent to the sight of flowers or of meadows in spring, or if not indifferent at least pre-occupied elsewhere. They devote themselves to ball-games, to drinking, gambling, money-making, popularity-hunting. For these our subject is meaningless. We offer a hundred banquets to the Pythagoreans or rather the true philosophers whose concern is not so much to know what authors think as to gaze with their own eyes on the nature of things and to listen with their own ears to her voice; who prefer quality to quantity, and usefulness to pretension: to their use, in accordance with God's glory, we dedicate this little book and all our studies.

We have quoted this Preface at length since it gives a plain picture of Ray's intention. He is not producing a scientific treatise or a rival to the great Herbals of the Bauhins: he is not claiming to be an expert or to be making a contribution to human knowledge. All he desires is to create interest in a neglected subject, to save others from the toil that he has had to undertake, and to enable them to learn the alphabet of the science. Identification must come first. The beginner if he knows the common name of his plant can look it up in the English index; the more advanced

---

1 W. How had published his little book with this title in 1650. It is alphabetical in arrangement, and contains names and localities: but the notes are very slight, the list very incomplete, the identifications often inaccurate.

can go to one of the places specially listed and find there plants peculiar to that locality; the student of Gerard or Parkinson can turn up their names for any species and see at once whether it is found in Cambridgeshire; the expert can learn the meaning of the names, find out something of the technicalities of botany, from the notes discover how wide a field of enquiry is open to him—all this in a book that can be slipped into the pocket as a companion on any country ramble. Even in these days when hand-books and pocket-guides are innumerable no better system could well be devised.

The Explicatio or bibliography which follows the Preface has been already described. It contains some fifty names, ranging from Theophrastus to How's *Phytologia*. The list is 'so accurately and instructively drawn up as not to have lost its utility to this day'—for the historian of botany Pulteney's praise is still true.[1]

That he should have succeeded in the immense task of discovering what species his various authorities really intended, and of bringing together the several synonyms, is little short of amazing. There are indeed occasions on which the task is beyond him. Of the genus *Salix* he writes:

Wonderful is the confusion and obscurity of botanists in describing and distinguishing the Willows. Some divide them into trees and shrubs, others into broad-leaved and narrow-leaved, paying no attention to the texture and quality of the leaves, and so both separating kindred species and joining those widely divergent. Moreover, most descriptions are so brief and obscure as to convey no clear meaning at all. So...we have tried to arrange them more appropriately and to describe those of our species which are mentioned obscurely or not at all more accurately, so that anyone who pays careful attention cannot fail to recognise and identify them.[2]

A similar note follows his careful accounts of the 'Yellow Charlock with a smooth jointed cod', which he recognises as identical except in the colour of its flower with the white *Raphanus raphanistrum*, Linn. and the 'Yellow Charlock with a rough joynted cod', the *Brassica alba*, Linn.[3] Of the sedges, after setting out seven species, he adds: 'We have two others which we think have not yet been described; and we have not been able to distinguish accurately enough those whose synonyms we have given. There is wonderful confusion and discrepancy among authors in their descriptions of this group.'[4] Anyone acquainted with the books to which he refers will realise how temperate are his criticisms.

In other cases, and they are among the most interesting, his treatment of particular species gives us an insight into the thoroughness and the difficulty of his work. On the one hand he is resolutely opposed to the

1 *Sketches of Botany*, I, p. 196.
2 *C.C.* pp. 141–2. It must be admitted that his classification is not a great success.
3 *C.C.* pp. 132–3.   4 *C.C.* p. 67.

multiplication of species which had been so recklessly undertaken by his predecessors: he will not admit that any plant is new until he has ransacked all the authorities and tried to trace a mention of it. On the other their plates and descriptions are so faulty that certainty is sometimes impossible.[1] Typical examples of this dilemma are found in his treatment of the two clovers, *Trifolium ochroleucum* and *striatum*, of *Caucalis daucoides* and *Orchis pyramidalis*. 'Trifolium pratense hirsutum majus flore albo-sulphureo, nondum, quod scimus, descriptum' is fully described, root, stem, leaves, and flowers. He continues:

> As to a picture of it, the nearest approach is the 'larger white-flowered hoary Clover of Clusius' in J. Bauhin: it is rather different from the 'white meadow Clover' depicted by Fuchs or the 'male Clover' of the same author: but each of these, unlike our plant, has crenate leaves, and their root-leaves are more numerous and larger, ours are much smaller.[2]

'Trifolium dilute purpureum glomerulis florum oblongis sine pediculis caulibus adnatis', an admirable characterisation of *T. striatum*, is then described: it is noted as an annual and said to vary in size from barely three inches to more than six according to situation. 'Is this the "Clover flowering at the nodes or next the leaves" of the *Phytologia Britannica?* Or the "Clover whose stalks bear oblong clusters from the joints" of Jean Bauhin?'[3] Considering how closely allied *T. striatum* is to *T. glomeratum*, which at this time Ray had not found, and how confusing the genus is, his hesitation is not surprising. 'Caucalis tenuifolia flosculis subrubentibus' is another plant that he feels it necessary to describe in detail:

> It seems to be the white-flowered Caucalis of Gerard and the common one with white flowers of Parkinson. But G. Bauhin calls Gerard's white-flowered species the 'field Caucalis with prickly seeds and large flowers', whereas our plant has small flowers. There is a slight difference between it and J. Bauhin's 'whitish Lappula with smaller flower or fine-leaved' because, while in his plant the spines along the sides of the seeds are deep purple, ours are green.[4]

'Orchis sive Cynosorchis purpurea spica congesta pyramidali' is also clearly and fully described, with a note that it is the last of the Orchids to come into bloom.

> Perhaps [he surmises] this will prove to be the same as the one that J. Bauhin entitles the 'Orchid with the small or purple flower'.[5] Or is it the 'middle-sized military Orchid' of G. Bauhin? Or Parkinson's 'Red military Orchid'? Or Clusius' 'Orchid with very elegant red flower'?

1 Cf. in addition to the examples cited his careful notes on Chamaecistus (*Helianthemum vulgare*), p. 32; Cirsium Anglicum (*C. pratensis*), p. 35; Cyperus longus (*Cladium mariscus*), p. 43; Hieracium luteum (*Crepis virens*), p. 75; Lithospermum (*L. officinale*), p. 90; Orchis lilifolius (*Liparis loeselii*), p. 106; Sesamoides (*Silene otites*), p. 155; Stellaria aquatica (*Callitriche verna*), pp. 160–1.
2 *C.C.* pp. 167–8. 3 *C.C.* pp. 168–9.
4 *C.C.* pp. 30–1. 5 *C.C.* p. 108.

In these four species it is interesting to note that ten years later when he published his *Catalogue of English Plants* he accepted both Clovers as species first described in his *Cambridge Catalogue*,[1] repeated his question as to J. Bauhin's Caucalis,[2] and identified the Orchid with that of Clusius.[3]

In other cases his researches at the time of his first book yielded no result. 'Saxifraga graminea pusilla flore parvo herbido et muscoso',[4] identified by Babington as *Sagina apetala*,[5] he does not describe at length but says—in English—'If any one desires certainly to know what plant we mean, he may be sure to find it among the stones in the stone-walk in the Fellows garden at Trinity Colledge. We think this plant hath not been yet described.' Ten years later[6] he identifies it with Gerard's Saxifraga anglicana.

'Lactuca sylvestris laciniata minima nondum descripta', the *Lactuca saligna*, Linn., is given no further description except a detailed account of its locality: 'This was found on a bank and in a ditch by the side of a kind of drove or lane leading from London road to the river, just at the water near a quarter of a mile beyond the spittle-house end.'[7] But in the *English Catalogue*[8] he drops his own correct Lactuca and enters it as 'Chondrilla viscosa humilis C.B., Park., Ger.'. If only he had been free to work out plants for himself or had lacked the scholar's reverence for his predecessors, his gifts would have had more scope. But like all true naturalists he was always more interested in the plants themselves than in their nomenclature.

Two further instances will demonstrate the handicap which loyalty to the writings of others imposed upon him. The Linnaean plant *Ceratophyllum demersum*, which he names 'Equisetum palustre ramosum aquis immersum', he describes in detail, and adds, very truly:

> In sluggish waters almost everywhere—which makes it the more surprising that it is mentioned in no other that I have consulted, or at least that after careful investigation I have not been able to discover it. Is it the Hippuris lacustris foliis mansu arenosis of Gerard?[9]

In the *English Catalogue*[10] he identifies it with J. Bauhin's Millefolium cornutum and adds that J.B.'s description fits it well enough, but that G. Bauhin's picture is mistaken and misleading: 'The flowers in it, if this is the plant, derive from conjecture not inspection.' The Linnaean *Polygonum amphibium* he describes in deference to the authority of his big four,

1 *C.A.* pp. 304–5.   2 L.c. pp. 62–3.
3 L.c. p. 227.   4 *C.C.* p. 151.
5 *Flora of Cambs*, p. 34, but see below, p. 90 n. 2.
6 *C.A.* p. 277.
7 *C.C.* p. 83: Spittle-house end is marked in Loggan's map as the part of Trumpington Street on the extreme edge of the town extending from the site of Addenbrooke's Hospital to Lensfield Road.
8 P. 37. In *Syn. Brit.* p. 4, he replaces Lactuca.
9 *C.C.* p. 49.   10 Pp. 62–3.

the two Bauhins, Gerard, and Parkinson, as Potamogeiton angustifolium. But his mind is not happy:

The leaves of this plant when under water, as you can see it on the banks of rivers, are smooth and glossy, but those outside the water are hairy, rough and altogether without gloss. The same is the case when, as very often, the whole plant grows out of water. This must be carefully noted; otherwise error will be very likely, and the careless observer will certainly suppose that the two forms are different plants. Hence I suspect that the plant in the catalogue of indigenous species issued at the garden in Leyden which is entitled the 'Hairy perennial Persicaria' is the same as this Potamogeiton; for when out of water it somewhat resembles a Persicaria in appearance.[1]

In the *English Catalogue*[2] he has gained more confidence in his own judgment and insists definitely that it belongs to the Persicarias.

It must be admitted that in thus accepting the classification of his authorities Ray made himself an accomplice to mistakes which must have strained his botanical conscience. One of the points on which his critics since and including Linnaeus have fastened is that he was rather an observer and collector than a taxonomist, one beyond praise for the extent and accuracy of his knowledge, but deficient in truly scientific abilities. If the charge were true, it might well be replied that in the existing state of confusion, when huge and recklessly inflated lists of plants and animals were being compiled, the primary business of the student was to bring order into the chaos, to sift the available data, to examine and reject spurious or redundant species, and set out accurate descriptions of the material of which a classification could then be made. Without Ray's preliminary work there could have been no Linnaeus. That is certainly true; and as regards his first book a Catalogue is all that its author attempted. But from his own subsequent efforts to reconsider and reclassify the species then listed it is plain that the charge is unjust. If we examine the two groups of species in which his regard for authority produced the most unsatisfactory results, we shall be able, when we deal with his later botanical writings, to see how steadily and systematically he corrected inappropriate or mistaken attachments. These two groups are unquestionably those named Alsine and Millefolium, each of which contains what at first sight appears a miscellaneous collection.

Under Alsine[3] there are twelve species, of which we give Ray's descriptive titles and their modern equivalents as fixed by Babington.

Alsine aquatica major—on this definition all his authorities agree[4]— is *Malachium aquaticum*, Fries.

Alsine foliis trissaginis, L'Obel, Gerard, Parkinson; Chamaedryfolia (which is the same as foliis trissaginis, 'oak-leaved'), G. Bauhin, is *Veronica agrestis*, Linn.

1 *C.C.* p. 124.   2 P. 250.   3 *C.C.* pp. 7–9.
4 The lists of synonyms have been condensed in most of the plants here listed.

Alsine foliis veronicae, Tabernaemontanus, Gerard, so too Parkinson; and G. Bauhin, is *Veronica arvensis*, Linn.

Alsine hederacea, Tabernaemontanus, Gerard, G. Bauhin, Parkinson; Alsines genus Fuchsii folio hederulae hirsuto, J. Bauhin; Morsus gallinae folio hederulae, L'Obel, is *Veronica hederaefolia*, Linn.

Alsine hederula altera, Gerard; Hederulae folio major, Parkinson; Morsus gallinae folio hederulae alter, L'Obel; Galeopsis sive urtica iners folio caulem ambiente, J. Bauhin; Lamium folio caulem ambiente, G. Bauhin, is *Lamium amplexicaule*, Linn. (Here Ray's respect for the Bauhins does not save him from error.)

Alsine hirsuta myosotis, L'Obel; spuria 3, Dodoens; Auricula muris, J. Bauhin, is *Cerastium vulgatum*, Linn.

Alsine longifolia uliginosis proveniens locis, J. Bauhin; aquatica media, G. Bauhin; 'fontana, Tabernaemontanus, Gerard, may belong to this plant', is according to Babington *Stellaria uliginosa*, Murr.—and this is borne out by Ray's later writings.

Alsine major, Dodoens, Gerard; maxima, L'Obel, Parkinson; altissima nemorum, G. Bauhin; 'in hedges. The flowers of this and the following consist of five petals so deeply divided that the careless would think them ten.' This Babington[1] identifies with Ray's Alsine aquatica major and his own *M. aquaticum*. Yet Ray and all his authorities clearly distinguish the characters and localities; and this may well be the large form of *Stellaria media* or the kindred species *S. neglecta*, Weihe. In later writings Ray drops this plant and in his *Historia Plantarum*[2] indicates that it is *M. aquaticum*, and that Gerard and Parkinson have made one species into two.

Alsine media, all authorities, is *Stellaria media*, Vill.

Alsine minor multicaulis, G. Bauhin; minima, J. Bauhin, etc. 'with five undivided petals', is *Arenaria serpyllifolia*, Linn.

Alsine palustris foliis tenuissimis sive saxifraga palustris alsine-folia, Johnson; Sax. palustris Anglica, Parkinson; 'flower very large for the size of the plant', is *Sagina nodosa*, Meyer.

Alsine tenuifolia, J. Bauhin, 'not found in any other botanist'. . . 'when flowering it resembles Linum catharticum'. This is still called by the same name.

Under Millefolium[3] there are eight entries—though the two last are referred to a single species.

Millefolium aquaticum dictum Viola aquatica, J. Bauhin; so G. Bauhin and Parkinson; Viola palustris, Gerard; Foeniculum aquaticum 2, Tabernaemontanus, is *Hottonia palustris*, Linn.

---

1 *Flora of Cambs*, p. 38. There are two or three instances in which Ray enters the same plant twice over as Babington correctly shows.

2 II, p. 1030.   3 *C.C.* pp. 98–100.

Millefolium aquaticum ranunculi flore et capitulo, Parkinson; so Clusius, G. Bauhin, Gerard; Ranunculus aquat. omnino tenuifolius, J. Bauhin; Ranunc. trichophyllon, Colonna, is presumably *Ranunculus trichophyllus*. It is strange that Ray did not follow J. Bauhin here as he lists R. aquatilis, which appears to be *R. heterophyllus* or perhaps *peltatus*. He refers it to Ranunculus not Millefolium in his later writings.

Millefolium aquaticum minus, Parkinson, J. Bauhin; sive flosculis ad foliorum nodos, G. Bauhin; aquat. 6, Johnson, is apparently *Myriophyllum verticillatum*, Linn.

Millefolium aquaticum pennatum spicatum, Parkinson, G. Bauhin, etc., is *Myriophyllum spicatum*, Linn.

Millefolium palustre galericulatum, Johnson; aquat. flore luteo galericulato, J. Bauhin, Parkinson, etc.; Lentibularia et Meon aquaticum, Gesner; 'it is hard to find except when in flower', is *Utricularia vulgaris*, Linn.

Millefolium tenuifolium, Johnson, 'so far as the picture goes; the description does not correspond'. 'This in respect to its seed-spike resembles a Fontalis and is more correctly assigned to that group than to the Milfoils: so far as I have seen there is no description or picture of it except in Gerard' (i.e. Johnson as Gerard's 'emaculator'). This plant seems rightly identified by Babington[1] with *Potamogeton pectinatus*.

Millefolium vulgare album, G. Bauhin, Parkinson; terrestre vulgare, Gerard; stratiotes pennatum terrestre, J. Bauhin; Stratiotes millefolia, Fuchs; Achillea, Dodoens, is *Achillea millefolium*, Linn.

Millefolium vulgare flore diluti ruboris, Bock, Gerard, Parkinson, etc. 'These two do not seem specifically distinct.'

No one studying these lists along with those in Gerard[2] and Parkinson and considering the changes made later in them by Ray[3] will fail to recognise that ease of reference to the most familiar authorities—that is the practical utility of his *Catalogue*—was a primary consideration with him. As we shall see, his own successive books reveal a steady development of systematic accuracy. In this, his first list, he strove to bring order into the chaos of Gerard and Parkinson and to help commencing students to identify their plants. Classification could and in fact did come later. His business at this stage is to collate the most familiar names and to fix the species to which they belonged.

1 *Flora of Cambs*, p. 251.
2 Thus Johnson's edition of the *Herbal* has, chapter 192, of Chickweed, 1, Great C. (i.e. *Malachium aquaticum*); 2, Smaller C. (i.e. *Stellaria media*); 7, Speedwell C. (i.e. *Veronica arvensis*); and others less easily identified; chapter 300, of Water Yarrow, 1, Water Violet (i.e. *Hottonia palustris*); 2, Water Yarrow (probably *Oenanthe fluviatilis*); 3 and 4, Water Crowfoot (i.e. *Ranunculus*); 5, Species 'bearing yellow gaping floures fashioned like a hood or the small Snapdragon' (i.e. *Utricularia*).
3 See below, pp. 260–1.

In this task he was amazingly successful. Of the species contained in Babington's *Flora of Cambridgeshire*, 558 are listed by Ray in the *Catalogue* as having been seen in the county by him personally, and in the vast majority of these there can be no doubt at all as to his accuracy. Babington is certainly justified in his remark that 'the alphabetical arrangement and the obscurity attendant upon the old nomenclature make the book rather difficult to consult':[1] and this is a very modest indication of the learning and skill with which he has disentangled its problems.[2] There is in fact (and this is a notable tribute both to Ray and to his successors) only one species in the *Catalogue* which defies identification,[3] and that is not unnaturally a Sedge: it is followed by the note already quoted.[4] In others, and particularly in the grasses which were then little known, there is occasionally room for doubt; as, for example, in that named by him Gramen alopecuroides minus,[5] which Babington fails to identify and which Ray himself later[6] replaces by the alopecuroides majus with a different list of synonyms—this being evidently the modern *Alopecurus pratensis*. Otherwise every entry, even those with outlandish names like 'Behen album, the Spatling Poppy',[7] or 'Perchpier Anglorum, the Parslypiert',[8] can be given its proper equivalent.

There are indeed, even outside the notoriously subdivided Brambles, Roses and Hawkweeds, a few cases in which he includes two modern species under a single heading. Thus Saxifraga graminea pusilla,[9] identified by Babington[10] with *Sagina apetala*, is said by Ray to be found both in marshy places and also sometimes in much drier ones, upon Newmarket Heath and in the stone-walk of Trinity College. Thus it probably includes *S. procumbens* with *S. apetala*. Of Orobanche,[11] though he only names one species, he says that it is found 'nigh the Church at Cherry-Hinton', still a famous locality for *O. elatior*, and also plentifully in the broom fields at Gamlingay, where it must be *O. rapum*. So too under Tithymalus helioscopius[12] he writes: 'This includes two species, one with crenate leaves, larger, whose stalk almost invariably splits at the same spot into five

1 *Flora of Cambs*, p. vii.
2 I have only found one case of a definite error, where on p. 33 he mistakenly identifies Ray's Saxifraga Anglica facie Seseli pratensis with *Sagina procumbens* (although subsequently on p. 99 and rightly identifying it with *Silaus pratensis*); this may be a slip for Sax. Anglica occidentalium.
3 *C.C.* p. 67, Gramen cyperoides ex monte Ballon spica divulsa: this in *Syn. Brit.* II, p. 268 he admits to having wrongly named and enters as G. c. spica e pluribus spicis brevibus mollibus compacta. From his description there it is perhaps *Carex disticha*. Babington in the MS. notes in his copy of *C.C.* identifies it with *C. divisa*, which is not recorded from Cambridgeshire.
4 *C.C.* p. 67, see above p. 84. 5 *C.C.* p. 64. 6 *C.A.* pp. 137–8.
7 *C.C.* p. 20: in App. p. 43 he declares that the name is Arabian.
8 *C.C.* p. 116: from French perce-pierre. 9 *C.C.* p. 151.
10 L.c. p. 34. 11 *C.C.* p. 110. 12 *C.C.* p. 163.

branches, five crenate leaves at their base; the other with smooth-edged leaves and smaller: these species also differ in other respects, notably in flower and seed: see and compare'—the former manifestly *Euphorbia helioscopia*, and the other hardly less certainly *E. peplus*.[1]

On the other hand he gives two entries in a number of cases where only one species is properly recognised. One of these is merely a slip: *Vicia hirsuta* appears both as Aracus sive Cracca minor and as Vicia parva sive Cracca minor;[2] the synonyms are identical under both headings, as is the English name 'Small wild Tare or Tine-tare'.[3] In some fifteen species the double appearance is due to difference of form. For example, he divides *Ranunculus flammula* into two, with smooth and indented leaves respectively, and quotes a mass of authorities of whom only J. Bauhin urges that the two forms belong to a single species[4] (*R. flammula* varies markedly and has been split up into sub-species by recent botanists). To *Lysimachia salicaria*, or purpurea, as he calls it, he adds trifolia caule hexagono[5] and describes it in detail, though ten years later he admits that it is a variety not a distinct species.[6] So too with *Hydrocharis morsus-ranae*[7] he adds a species 'with double and strongly scented flower' which he found 'by the side of Audrey causey plentifully': this he kept distinct in his *English Catalogue*[8] and listed in the *Synopsis*[9] as 'probably a variety'. *Echium vulgare*[10] has a form with smaller flowers and longer stamens and Babington[11] identifies this with Ray's second species: Ray himself declared later[12] that he had only seen the common species in England and thought the insertion of the other a mistake of L'Obel's. *Pimpinella saxifraga* is a notoriously variable species: it is not surprising that Ray should list two forms of it and hint at a third.[13] *Thymus serpyllum* is almost equally variable: to insert a hairy form as distinct,[14] even if he cannot have meant *T. chamaedrys*, is a very venial fault. And subdivision in the cases of *Ulmus*[15] and *Salix alba* and *purpurea*[16] becomes a positive merit.[17] Finally, there are four

1 Babington notes the two species under *Orobanche*, but does not comment upon Ray's contention here and identifies his entry only with *E. helioscopia*: cf. pp. 161 and 207.

2 *C.C.* pp. 15, 175: corrected in Appendix published in 1663.

3 Babington, l.c. p. 63, indicates a similar error in *V. sativa*: but there Ray's second entry refers only to the cultivated Vetch, *C.C.* p. 175.

4 *C.C.* p. 131. In *C.A.* p. 258 he says that the two are not certainly different.

5 *C.C.* p. 93.   6 *C.A.* p. 203.   7 *C.C.* p. 101.

8 P. 221.   9 P. 207.   10 *C.C.* p. 47.

11 L.c. p. 157.   12 *C.A.* p. 97; cf. also Appendix, 1663.

13 *C.C.* pp. 118–19.   14 *C.C.* p. 154.

15 *C.C.* p. 178.   16 *C.C.* pp. 142, 146.

17 For completeness of record it may be noted that the other species divided by Ray into different forms are *Brassica campestris* (Napus sylvestris and N. rapum), *Hedera helix* (H. helix and H. arborea), *Carduus lanceolatus* (C. lanceatus and C. l. angustifolius, the latter being possibly the modern *C. crispus*); *Veronica anagallis* and *V. beccabunga*

cases in which difference of colour in the flower is the basis of separation—
Viola odorata, purple and white; *Centaurea scabiosa*, purple and white;
Achillea millefolium, white and red; and *Anagallis arvensis*, red and blue:
in this last case Babington argues[1] that Ray's *A. foemina* is probably the
blue form of the common species, and not that now usually recognised as
specifically distinct.

When we turn from the character and structure of his list to its contents,
admiration for his achievement increases. He claims to have worked for
nine years at the study of plants, and to have done so single-handed. The
knowledge that he had gained of the flora of Cambridgeshire would have
been remarkable if he had possessed the books and collections of a modern
student: considering his handicaps, the lack of any reliable authorities, the
pressure of his other work and the difficulty of travel, it is evidence not
only of rare energy and powers of observation, but of a genius for natural
history, a flair for locality combined with a fine sense of the characteristics
of plant life. From the saltings at Wisbech, where he discovered *Althaea
officinalis*[2] and *Spergularia salina*[3] to the undrained bogs and heathland of
Gamlingay with *Hypericum elodes*[4] and *Teesdalia nudicaulis*[5] and *Nardus
stricta*[6]—all these being unknown elsewhere in the county—he explored
and collected and identified. In Kingston Wood on his western boundary
he found *Epipactis helleborine*[7] and *Paris quadrifolia*[8] as well as *Campanula
trachelium*[9] and the true Oxlip, *Primula elatior*.[10] To the east in and around
Newmarket *Asperugo procumbens*,[11] 'lost for many years',[12] *Medicago
minima*[13] and near it *Silene otites*[14] (both 'in the gravel pits as you go to the
nearest windmill on the Northside of the town', and probably outside
Cambridgeshire), *Antennaria dioica*,[15] on the heath 'on the right hand of
the road from Cambridge' and 'in a close near the beacon on the left hand
of the way from Cambridge in great plenty', *Veronica spicata*,[16] the last
a record of peculiar interest because, having been completely lost in this

(each into a major and minor); *Plantago major* (P. latifolia and P. major paniculata,
the latter 'found once by Dr Strachey: we could never find it', an abnormality, but
included in his later works, e.g. *C.A.* p. 246, as found by Willisel at Reculver); *Sagittaria
sagittifolia* (Sagitta major and minor) and *Phleum pratense* (Gramen typhinum maximum and medium). Many of these are corrected in the Appendix of additions and
emendations printed in 1663.

1 *Flora of Cambs*, p. 190.   2 *C.C.* p. 9.   3 *C.C.* p. 159.   4 *C.C.* p. 17.
5 *C.C.* p. 24.   6 *C.C.* p. 159.   7 *C.C.* p. 73.
8 *C.C.* p. 74.   9 *C.C.* p. 164.   10 *C.C.* p. 126.
11 *C.C.* p. 9: in *Syn. Brit.* p. 76, recording it from Newmarket, he adds 'where I hear
it is now lost'.
12 Babington, l.c. p. 155.   13 *C.C.* p. 166.
14 *C.C.* p. 154.   15 *C.C.* p. 64.
16 *C.C.* pp. 174–5: in *C.A.* p. 302 he says 'in closes adjoining Newmarket Heath
beyond Bottisham'. Evans, *Flora of Cambs*, p. 124, appears to be mistaken in saying
that Ray recorded it from Newmarket: his records are from 'closes' not the open heath.

locality ever since Ray's time, it has been rediscovered recently and probably on the self-same spot. On the chalk of the Gogmagogs he found *Anemone pulsatilla*,[1] *Linum perenne*,[2] *Senecio campestris*,[3] *Thesium humifusum*,[4] *Orchis ustulata*,[5] *Ophrys apifera*,[6] and the long lost *Calluna vulgaris*;[7] and on the undrained bogs of 'Hinton moor' beneath them *Parnassia palustris*,[8] *Sagina nodosa*,[9] *Drosera intermedia*,[10] *Lythrum hyssopifolia*[11] and the finest of all his discoveries, *Liparis loeselii*, which he describes fully and names 'the Dwarf Orchies of Zealand'.[12] He found *Melampyrum cristatum*[13] 'almost in all woods of this county plentifully', though it had not then been described by any British author, *Ajuga chamaepitys*[14] 'on the layer about the borders of Triplow heath' and *Teucrium scordium*[15] 'in the osier holts about Ely city'. Moreover, he knew the waterways of the county and made notable observations on their flora: 'about Stretham ferry', where it is still abundant, *Villarsia nymphaeoides*;[16] 'in the river about the same ferry and about Audrey causey' (the causeway at Aldreth), where it is now rarely seen, *Stratiotes aloides*;[17] 'in the rivulet Stour by the little islet which it makes above the Paper mills' *Myriophyllum verticillatum*;[18] and 'in the river Cam in many places' three species then new to science, *Potamogeton zosterifolius*,[19] *P. pusillus*,[20] and *Zannichellia palustris*.[21] He has also recorded several other plants now very rare or lost to the county, *Myrica gale*,[22] 'in the fens in the Isle of Ely in many places', but since destroyed by cultivation except in the sanctuary at Wicken; *Caucalis daucoides*, now very rare,[23] and *latifolia*,[24] now quite unknown; *Sedum telephium*,[25] 'in a lane at Shelford and at Burrough-green in a grove'; *Ribes nigrum*,[26] 'by the river's side at Abington'; *Atropa belladonna*,[27] 'in the lanes about Fulbourn plentifully'; and *Amaranthus blitum*,[28] 'in some osier holts by the river: also in a ditch on the backside of S. John's College in a close on the north of the back-gate': and at least three now extinct in Britain, *Roemeria hybrida*,[29] 'in the cornfields beyond Swaffham as you go to Burwell'; *Senecio paludosus*,[30] 'in many places about the

1 *C.C.* p. 128.	2 *C.C.* p. 89.	3 *C.C.* p. 80.
4 *C.C.* p. 88.	5 *C.C.* p. 108.	6 *C.C.* p. 109.
7 *C.C.* p. 50.	8 *C.C.* p. 70.	9 *C.C.* p. 9.
10 *C.C.* p. 139.	11 *C.C.* p. 71.	12 *C.C.* pp. 105–6.
13 *C.C.* p. 95.	14 *C.C.* p. 32.	15 *C.C.* p. 152.
16 *C.C.* p. 104.	17 *C.C.* p. 98.	

18 *C.C.* p. 99: the Stour is the brook which bounded on the east the field in which Stourbridge Fair was annually held: the paper-mills were near the spot at which it joined the Cam: the islet formed by the mill-leet can still be traced.

19 *C.C.* pp. 124–5. Ray's description suits this species better than *P. compressus*, the obscure form with which Babington, l.c. p. 250, identifies it.

20 *C.C.* p. 125.	21 *C.C.* p. 125.	22 *C.C.* p. 47.
23 *C.C.* p. 30. I found it in 1939 near Linton.		24 *C.C.* p. 29.
25 *C.C.* p. 162.	26 *C.C.* p. 139.	27 *C.C.* p. 157.
28 *C.C.* p. 23.	29 *C.C.* p. 111.	30 *C.C.* p. 37.

Fens, as by a great ditch-side near Stretham ferry' and *S. palustris*,[1] 'in the fen ditches about Marsh and Chatteresse in the Isle of Ely'. On the evidence of the *Phytologia Britannica* he lists *Artemisia campestris*,[2] or rather two species which he afterwards refused to separate, as found on Newmarket heath ('we have searched diligently, but can as yet find neither there'); and on that of Parkinson 'Galega', *G. officinalis*,[3] reported to have been found in meadows near Linton ('we could not find it there and suspect that it is not there to be found') and 'Staphylodendron',[4] *Staphylea pinnata*, said to grow at Milton three miles from Cambridge ('we could not find it out by search nor hear of it by enquiry, howbeit we deny not but possibly it may grow there').

As we chose two of his groups to illustrate the difficulties of his task it may give a better idea of his success if we examine another, the Geraniums. He lists eight species[5] as follows:

Geranium arvense vel minus, Tabernaemontanus; cicutae folio, Gerard, G. Bauhin; Moschatum inodorum, Parkinson and J. Bauhin (obviously *Erodium cicutarium*), 'on the banks in the highway between Cambridge and Barnwell and elsewhere frequently'. Ray also includes it among the plants found on the 'Hill of Health'—in his day an area famous for the smaller plants *Cerastium arvense*,[6] *Draba verna*,[7] *Saxifraga tridactylites*[8] and *S. granulata*,[9] *Erythraea centaurium*[10] and *Echium vulgare*,[11] but now built over and unrecognisable.

Geranium batrachoides, all authorities (*G. pratense*), 'in the hedges about Bigwin closes' (Biggin Abbey between Fen Ditton and Horningsea), 'about Cherry-hinton and Histon and many other places'. It is tolerably common to-day, especially on the west side of Cambridge.

Geranium columbinum, Gerard, Parkinson; folio rotundo multum serrato, J. Bauhin (*G. molle*): no localities are given: it is in fact universally common.

Geranium columbinum majus dissectis foliis, Johnson, etc. (*G. dissectum*), 'in meadows and hedgerows everywhere': it is generally common.

Geranium malacoides sive columbinum minimum, Parkinson, G. Bauhin (*G. pusillum*), 'on the Hill of Health and elsewhere in waste ground: it varies in the flower-colour': a species not too easily distinguished from *G. molle*, but found sparingly round Cambridge.

---

1 *C.C.* p. 37.   2 *C.C.* p. 1.   3 *C.C.* p. 59: cf. *Theatrum Botanicum*, p. 418.
4 *C.C.* p. 160 and *T. B.* p. 1417: later, probably in 1661, he reported this 'in hedges near Pontefract but not so frequent that I would venture to call it wild': *C.A.* p. 293—an instance, as Lees, *Flora of West Yorks*, p. 186, remarks, of his acuteness.
5 *C.C.* pp. 61–2.
6 *C.C.* p. 19.
7 *C.C.* p. 113. Ray couples this and the next under the name Paronychia.
8 *C.C.* p. 113.   9 *C.C.* p. 150.
10 *C.C.* p. 31.   11 *C.C.* p. 47.

Geranium haematodes, Parkinson, etc.; sive sanguineum maximo flore, G. Bauhin (*C. sanguineum*), 'found on Newmarket Heath in the Devil's ditch, also in a wood adjoining to the highway betwixt Stitchworth and Chidley' (i.e. Cheveley). This is still the only locality in the county, a small piece of the ditch ending in the wood and road, in which *G. sanguineum* abounds.

Geranium robertianum, all authorities (*G. robertianum*), 'hedgerows', now as then.

Geranium saxatile, Johnson; lucidum, J. Bauhin; anemones folio rotundo, G. Bauhin (*G. lucidum*), 'on the bank and in the hedge on the right hand of the lane leading from Cambridge to Chesterton plentifully, and elsewhere': a casual, nowhere abundant.

The list thus completed gives a very illuminating picture of the thoroughness and competence of Ray's work. The Crane's-bills are not an easy or very noticeable family. Apart from the two species *G. pratense* and *G. sanguineum* they are wayside weeds of the kind which the casual student easily ignores or finds perplexing. Of *G. pusillum* even Bentham and Hooker remark that it is 'perhaps sometimes mistaken for *G. molle* and thus overlooked';[1] and though like *Erodium cicutarium* it is abundant in the Breck sand district, in Cambridge one may readily pass it over. Ray has noted all the species that occur to-day in the district except one; for *G. phaeum*, *G. columbinum* and *G. rotundifolium*, though included by Babington,[2] are at best only the rarest of casuals. *G. pyrenaicum*, his one striking omission, is now a common plant, to be found not only at Red Cross and beyond Girton on the Huntingdon Road, but on the wayside about Ely: nevertheless it is a recent colonist, commoner now than in Babington's day, and in Ray's time was apparently unknown in Britain. It is interesting to note that ten years later, when he wrote his *Catalogue of English Plants*,[3] though unable to add to the Cambridgeshire species, he could describe *G. sylvaticum*, 'in hill pastures of Westmoreland and Yorkshire abundantly'; *G. columbinum*, 'I have seen it in plenty round Swanley in Kent'; *Erodium moschatum*, 'in Craven-Common and near Bristow on a little Green you pass over going thence to S. Vincents Rock', and *E. maritimum*, 'in sandy places along the Western shore abundantly, as on the islands of Man and Prestholm [Puffin Island off the Menai Straits], near Carnarvon and in Cornwall round Penzance'.

Mention of *G. pyrenaicum* naturally draws attention to the species which Ray omits. Some, very many indeed, of these were such as might easily escape notice and were added in the supplementary lists issued in 1663, in the *English Catalogue* in 1670, and by Peter Dent in 1685. But there remain a few which were apparently unknown in Cambridgeshire at that date. The most obvious of these is the Beech, *Fagus sylvatica*, planted so freely

---

1 *British Flora*, p. 96. 2 *Flora of Cambs*, pp. 44, 46–7. 3 *C.A.* pp. 130–3.

in Cambridge gardens after the great storm in 1703, and now so splendid an ornament of Ray's beloved Gogmagogs and of the road to Newmarket before it crosses the Devil's Ditch. The Lime, *Tilia vulgaris*, almost equally familiar in Ray's own College[1] and throughout the county, was unknown except as a manifest alien in his time. *Cardamine hirsuta*, still curiously uncommon though elsewhere a garden pest, was unknown to him: so, of course, was *Linaria cymbalaria*, which did not begin to invade our walls until a century after his death. *Seseli libanotis* he came to know before his death;[2] but not apparently in his favourite locality at Cherry-hinton, where it is now common. *Carum bulbocastanum*, found to-day though more rarely in the same neighbourhood, he never knew—perhaps because its specific distinctness is not readily recognised. *Peucedanum palustre* is so familiar to visitors to Wicken that we should expect him to have seen it—until we remember that it is strictly confined to a locality which in his time was hardly accessible. Ray was writing before the great drainage of the fens had taken effect, and the whole area in which plants like *Lathyrus palustris* and *Lastrea thelypteris* are now and *Sonchus palustris* was once common, was a wilderness only crossed by the tracks to 'Audrey causey' or Stretham ferry: the Isle of Ely was an island indeed, and the swamps around it, though rich in plant life, were more impassable than the sea. The marvel is not that Ray missed certain species, but that he saw so many. The places which he selected for special study were in fact representative and well-chosen.[3]

The *Catalogue* is very defective in the lower forms of plant life, which had then hardly received any attention. There are six species listed under Equisetum[4] but three of these belong elsewhere, *Hippuris vulgaris*, *Ceratophyllum demersum* and *Chara vulgaris*. In Ferns Cambridgeshire is notably weak; Ray knew *Ophioglossum*,[5] 'in Grantchester meadow abundantly', but only five others. Six Mosses, one the 'Moss from a human skull' of medical fame,[6] two Lichens, four Fungi, Conferva Plinii, 'the Hairie River-weed',[7] and Ustilago,[8] 'burnt, blasted or smutted Corn', complete his record. The progress that he made in these Orders as revealed in his later books is not the least of his contributions to natural history.

On the other hand the *Catalogue* is enlarged by the insertion of the plants grown as crops in the county. These, though to the modern student out of place, were always included by Ray in his plant-books, and throw

---

1 The lime-walk was first planted in 1671–2.   2 *Syn. Brit.* p. 70.

3 Thirteen localities are mentioned with lists of the rarer plants found in them in the Appendix or second part of his book, pp. 30–5.

4 *C.C.* pp. 48–9.   5 *C.C.* p. 105.

6 *C.C.* p. 101—an ingredient in a popular ointment.

7 *C.C.* p. 36.   8 *C.C.* p. 181.

an interesting light upon the agriculture of the period. The list is as follows: Avena, 'Common or manured Oates'; Cannabis, 'Hempe, the male
and female or Winter and Summer Hempe, about Ely it is sown'; Crocus,
'True Saffron'; Faba major, 'Great garden Beans', and Faba minor,
'Common Beans or Horse-Beans'; Glastum sativum, 'Woade planted
about Littleport'; Glycyrrhiza, 'Common Liquorice, planted in good
quantity at Elme in the isle of Ely'; Hordeum distichum, 'Common
Barley'; distichum minus, 'Sprit-Barley, Battledoor-Barley'; poly-
stichum, 'Winter square-Barley'; Lens, 'Lentills, often mixed with
barley'; Linum sativum, 'Manured Flax'; Lupulus, 'Manured Hops but in
few places of this county'; Pastinaca sativa, 'Common Garden Parsnep';
lutea, 'Yellow and white rooted Carrot'; atrorubens, 'Red-rooted Carrot'; Pisum, 'Field Pease of three kinds, White, Gray and Maple'; Rapum,
'Round Turnep'; radice oblonga, 'Long Turnep'; sylvestre, 'Wild Rape
sown in the Isle of Ely'; Secale, 'Rie'; Triticum of seven kinds, 'White
Lammas', 'White-eared red', 'Kentish or Red Lammas', 'Red with a
bearded white ear', 'Bearded Kentish', 'Gray', including Red Pollard and
Gray Pollard, and 'Red-eared bearded'; and finally Vicia, 'Common
Vetch or Tare'. Celery, though he includes the wild *Apium graveolens*[1]
and calls it 'Smallage', was not then cultivated: nor apparently was Beet,
at any rate in Cambridgeshire, nor any of the varieties of Cabbage.
Potatoes, though he mentions 'Battata' in his Appendix[2] as an example
of the Tuberous tribe, had not yet been generally grown, although three
years later the Royal Society urged their introduction. Though he includes
Juglans, 'the Wallnut Tree', without questioning the propriety of so
doing,[3] he does not mention any of the cultivated fruit-trees or shrubs.

If his list gives an insight into the life of the countryside, some of his
localities contain fascinating references to the town and its Colleges.
Allium vineale[4] is found 'on Jesus Colledge wall, nigh the gate which
opens out of the road into Garlick fair;[5] also about a gravill-pit near the
footway leading from Christ's Colledge to Cherry-hinton'. *Chenopodium
olidum*[6] is 'under the wall that joynes to Peter-house Tennis-court'; *Polypodium vulgare*,[7] 'in Trinity-hall wall by the lane's side which leads to
Garrett-hostle-bridge'; *Populus alba*,[8] 'in the hedge of S. John's Colledge
bowling-green'; *Viburnum opulus*,[9] 'by the ditch of one of the closes on
the back-side of Clare-hall'; and *Convolvulus sepium*,[10] 'in the privet-hedge
in the fellows garden at Trinity Colledge'. Farther afield he found

1 *C.C.* p. 14. 2 *C.C.* App. p. 101.
3 *C.C.* p. 80. In *C.A.* pp. 177–8 he writes 'though it does not grow wild among us
or perhaps elsewhere in Europe...we may regard it as naturalised if not indigenous'.
4 *C.C.* p. 6.
5 Then held on a site south of the College Chapel on 16 August: cf. *Diary of
S. Newton*, p. 21, Willis and Clark, *Architectural Hist. of Cambridge*, II, p. 115.
6 *C.C.* p. 17. 7 *C.C.* p. 122. 8 *C.C.* p. 122. 9 *C.C.* p. 147. 10 *C.C.* p. 38.

Ononis campestris,[1] 'with a white flower in Huntington rode not far from Howes bowling-green'; *Pulicaria vulgaris*,[2] 'beyond the Castle in the road to Histon'; *Alopecurus geniculatus*,[3] 'beyond the Castle near the Windmill'; *Ranunculus hederaceus*,[4] 'in a little lane leading out of Coton road up to the bowling-green beyond the Castle'; *Trifolium medium*,[5] 'in an enclosed ground near the river Cam not farre from Newnham by the footway to Grantcester'; *Apium graveolens*,[6] 'in divers ditches about the town as at the Spittle-house end'; *Saponaria officinalis*,[7] 'in a place paled about on the right hand of Barnwell as you go to Sturbridge fair'; *Nepeta cataria*,[8] 'in a lane on the right hand of Barnwell in going thither which leads down to the moor on which stand the pest-houses'; *Medicago maculata*,[9] 'in the lane which leads from Barnwell to the pest-houses or the Common called Coldhams'; *Origanum vulgare*,[10] 'in the yard where the yew trees grow at Barnwell Abbey'; *Adoxa moschatellina*,[11] 'by a grove of elmes on the north side of Chesterton'; *Vicia tetrasperma*,[12] 'as you goe from Hoginton[13] to Huntington rode'; and *Tanacetum vulgare*[14] 'at Histon near the ruines of an old Chappell', that is, the Church of St Etheldreda destroyed in the year 1600 to build Madingley Hall. Such glimpses of bygone scenes give a vivid charm to his *Catalogue*.

The greatness of the book is not only in what it accomplished but in what it foreshadowed. Interpolated in the dry record of patiently discovered synonyms are a number of notes on points which the author thought of interest. Most of them are extracts from previous writers and have to do with the medical uses of herbs—a subject to which he devoted very much space in his next botany book. These are carefully chosen: Ray has no sympathy with astrology or alchemy or with the legends that superstition had fastened upon so many plants.

Pennyroyal when dried is said to flower at midwinter: so Cicero in the *De Divinatione*, bk. II. Costaeus tells the same story and says that there is a similar example in the case of the Black Woodpecker, whose body hung up by a string has been observed to shed its old feathers in the spring and grow new ones. Utrumque hoc incredibile

—both are travellers' tales.[15] He attacks in a very elaborate note under Sambucus[16] 'the foolishness of the Chymists who chatter and boast so loudly of the signatures of plants':

We have paid close attention to the matter and are moved to assert that the signatures are not indications of natural qualities and powers impressed on

1 *C.C.* p. 12.   2 *C.C.* p. 37.   3 *C.C.* p. 65.   4 *C.C.* p. 131.
5 *C.C.* p. 168.   6 *C.C.* p. 14.   7 *C.C.* p. 150.
8 *C.C.* p. 97: the pest-houses had been built in 1655 (Cooper, *Annals*, III, p. 464).
9 *C.C.* p. 166.   10 *C.C.* p. 110.   11 *C.C.* p. 130.   12 *C.C.* p. 175.
13 Now known as Oakington.   14 *C.C.* p. 161.
15 *C.C.* p. 128.   16 *C.C.* pp. 148–50.

plants by nature. Of the plants specifically said to be appropriate to a particular portion of the body or to a disease far the greater number have no signature. Different parts of the same plant have signatures not merely different but contradictory. Many plants resemble natural or artificial objects for which they have no affinity as Orchid flowers look like flies, spiders, frogs, bees or butter-flies. Parts of some plants represent parts of the body with which they violently disagree: the juice of the Spurges is like milk, but no one is so imbecile as to give it to nursing-mothers. There is such a vast number of plants that, even if they had come into existence altogether at haphazard, any ingenious and imaginative person could have found as many signatures as are known to-day.

We have summarised Ray's criticism at some length; for it must not be forgotten that he was dealing not with an exploded fallacy but with a fashionable and widely advocated conviction, and that Nicholas Culpeper's *Herbal* was published in 1653, Robert Lovell's *Pambotanologia* in 1659, and Robert Turner's *Botanologia* in 1664. Credulity in some and dis-honesty in others maintained superstitions disastrous to any scientific pro-gress. We have only to glance at the Appendix on the etymology of plant-names in Ray's *Catalogue*[1] to see how great a majority of them refer to fancied potencies or resemblances in the world of magic or demonology.

The most interesting of his notes are those in which he records his own observations. A few of these are dietetic or practical. Thus under *Cheno-podium bonus-henricus*[2] he writes 'the asparagus or tender sprout of this is put into boiling water, cooked for a quarter of an hour, and eaten with butter and salt—a pleasant dish not unlike ordinary asparagus'; under *Tragopogon pratensis*[3] he commends 'its roots cooked until they are tender and served with butter like parsnips: the Italians make much use of this plant and call it Sassefrica and Sassifica'; under *Artemisia absinthium*,[4] 'those who go about the country looking for plants, if they happen upon acid and nasty-tasting beer (*cerevisia*) can improve it both for the palate and for digestion by an infusion of the Common Wormwood. Bitterness removes acidity even better than sugar'; under *Rhamnus catharticus*,[5] 'if the berries are gathered in autumn, painters get from them a colour called in English Sap-green'; and under *Aspidium filix-mas*,[6] 'I am reliably in-formed that in Shropshire it is often used when dried instead of hops in the brewing of beer'. Others are more strictly botanical. Under *Humulus lupulus*[7] he describes how 'this and probably all other creepers in climbing follow the course of the sun, that is they twist from east through south to west, and never in the reverse direction'; under *Ulmus campestris*[8] how

1 *C.C.* App. pp. 36–83: the list is a very complete and scholarly treatise.
2 *C.C.* p. 23.   3 *C.C.* p. 164.   4 *C.C.* p. 1.
5 *C.C.* p. 138.   6 *C.C.* p. 53.
7 *C.C.* p. 91: he corrects this note in *C.A.* p. 200, where he says 'some twine from right to left like the Hop, but others from left to right like the Large Convolvulus'.
8 *C.C.* p. 180.

the growth of trees in the open reveals the quarter of the prevailing wind: 'thus trees on the shore are bent towards the land: in mid England treetops point to the east because those parts of the country are exposed to fierce westerly gales'; under *Cuscuta europaea*,[1] 'this plant is peculiar: it has neither leaves nor root; it is destitute of leaves always, of root as soon as it is mature'; and under *Fraxinus excelsior*[2] how the rings in the trunks and branches of trees explain their growth and age:

These rings in trees growing in the tropics are equidistant all round and have the heart of the tree at the true centre; in other regions south or north they are expanded towards the equator and contracted towards the pole so that the heart is always eccentric: skilled phytologists warn us that in transplanting trees they should be placed in aspect and position, relative to the sky, similar to that from which they have been moved.

He adds:

We have observed the following points: 1. the age of a tree or branch is disclosed by the number of rings; unless the tree has stopped growing the number of rings equals the number of years: 2. normally the inner rings are closer together owing to pressure; probably in trees of great girth and growing old, the outside rings may be narrower through lack of vigour: 3. the pith is compressed as the tree ages; this is evident in the Elder: 4. the wood is harder and often darker in the inner rings: 5. the tops of trees have fewer rings, and the inner rings of the trunk can be seen drawing to a point as they rise; the pattern thus formed is called in English the 'grain of the tree': 6. it is likely that each tree has an absolute limit of age and rings fixed by nature, or so Mr Nid[3] conjectures: I cannot agree; my opinion is that so long as the tree is alive it adds a new ring even if a narrow one every year: the age of an old tree cannot be determined because its inside decays and the external rings become too narrow to count.

Two further points may perhaps, as Mr D. C. Gunawardena[4] suggests, be foreshadowings of his later work. Under Betula[5] he has a note on the flow of sap if the bark is pierced, this being derived from Bock (Hieronymus Tragus). It seems unlikely that this reference had any influence in promoting the experiments which he and Willughby afterwards carried out: but the subject may already have attracted his interest. The second is a longer note on the germination of seeds.[6] In this he states the traditional view of the ancients, with a reference to Theophrastus, that in corn the root springs from the thick or bottom end of the seed and the shoot from its top, whereas in beans both root and shoot derive from the same opening; against this he sets the evidence of Dr Thomas Browne,[7] with whose

1 *C.C.* p. 42.   2 *C.C.* p. 55.
3 This is the only place, apart from the Preface, in which Nidd's name is mentioned. Yet some of Ray's biographers, following Derham, *Mem.* p. 12, have asserted that Nidd contributed largely to the *Catalogue*.
4 In his dissertation, *Studies in the Biological Works of John Ray* (1933), preserved in the Library of the Linnean Society. The reference is to pp. 22–3.
5 *C.C.* pp. 21–2.   6 *C.C.* pp. 171–2.   7 *Works*, III, p. 408.

son Edward he was acquainted at Trinity, who in *The Garden of Cyrus*, Ch. III, disproves this and argues that it is not true that each part of the plant derives from a distinct part of the seed. Here again his point is drawn from the reading of Browne's book and there is no evidence that he had himself yet begun to study seed-leaves: but his attention had been aroused and the discovery of cotyledons was the sequel.

Other notes, though they have no very close connection with his subject, reveal the range and variety of his interests. In problems of language and etymology he had obviously been occupied: Latin, Greek and Hebrew were his subjects of study in the University, and he was soon to produce books on English dialects and proverbs: the derivations of the names of plants in this *Catalogue* gave him scope for much ingenious discourse. But his enthusiasm becomes almost irrelevant when, in discussing the derivation of hyssop 'not, as scholars absurdly state, from what is poured over the face but as I am convinced from the Hebrew, since the Septuagint translates eyob as hyssop', he fills nearly two pages[1] with a list of Greek and Latin words derived from Hebrew, or 'Punic' as given in the vocabulary of Caninius.[2] But, lest on this score we write him down a pedant, he is equally ready with a note on the attraction of Cat-mint for cats.[3]

This mint when transplanted from the fields into a garden I have often seen bruised and completely destroyed by cats: they tear it with their teeth and roll upon it so that the soil round its battered remains is flattened out by them. No other plant is treated in this way: this is invariably destroyed unless covered up with spikes: I have tried it half a dozen times and in different places. If it is protected until it establishes itself and flowers, it will then be as safe as if it had been grown from seed. Cats are rabid for it when it is withering, they ignore it when it is healthy. Hence the English rhyme about Cat-mint:

> If you set it the cats will eat it;
> If you sow it the cats can't know it.

Here is a pleasant note bearing on another of his pursuits: it is under Cicuta, that is, *Conium maculatum*:[4] 'I dissected the crop of a bustard[5] and found it stuffed with Hemlock seeds: there were only four or five grains of corn mixed with them: so even at harvest-time the bird leaves corn for hemlock.' So under *Atropa belladonna*[6] he observes that snails and slugs eat it freely and adds a note on their hermaphroditism.

But in view of Ray's subsequent achievements it is his notes upon insects that are the best worthy of attention. In botany there had long been

1 *C.C.* App. pp. 59–61: curiously enough in the case of hyssop he is right—the Greek word is derived from the Eastern and Hebrew. Most of his other examples are of course coincidences: see below, p. 171.
2 *Institutiones linguae Syriacae* etc., Paris, 1554.
3 Under Mentha, i.e. *Nepeta cataria*: *C.C.* p. 98.   4 *C.C.* p. 34.
5 'Otidis sive Tardae avis': *Otis tarda* is its Linnean name. In Ray's time the Bustard still nested in Cambridgeshire.   6 *C.C.* pp. 157–8.

a widespread interest: gardeners and herbalists and a few more dis-
interested students had produced a large literature: Ray stands out for the
extent of his learning and the soundness of his judgment, for his independence in rejecting traditional beliefs and demanding proof, for the
thoroughness of his observations and the accuracy of his statements. In
entomology the field was almost unexplored: Thomas Mouffet or Moffet
had left a large book based upon Edward Wotton, Conrad Gesner and his
friend Thomas Penny which had been published by the energy of Sir
Theodore Mayerne in 1634, thirty years after his death: but the *Insectorum
Theatrum*, of which we shall speak in detail later, is a very incomplete, ill-
arranged and unreliable compilation. Ray, whose concern with insects
has been often ascribed to the influence of Willughby, reveals in these
early notes not only a knowledge of all that had been written but the acute
insight, the power of exact objective description and the indefinable flair
for a correct interpretation which are the marks of the great scientist. In
a pre-scientific age, when speculation was limited by no experience of what
could or could not happen, Ray's power of discarding legendary lore and
fanciful explanations and of fastening upon the right line of investigation
establishes for him a strong claim to be one of the fathers of modern science.

Two or three of the notes are brief. Under *Urtica*[1] is a mention of
Mouffet's nettle-caterpillar, its chrysalis and imago, obviously the Small
Tortoiseshell, *Vanessa urticae*. Under *Ligustrum vulgare*,[2] after reference
to Mouffet's 'noblest of the greens',[3] the Privet Hawk (*Sphinx ligustri*),
he describes its irritability when touched, its change to a pupa with a sheath
for the proboscis projecting, and finally the moth that emerges from it:
he adds: 'It will not be improper (for the matter can easily be observed)
to draw attention to the mistake which Mouffet[4] and others make in
asserting that the head of the caterpillar is changed into the tail of the
butterfly: in every caterpillar that I have seen the exact opposite is the
fact'. Under *Corylus avellana*[5] is an account of Mouffet's hazel-caterpillar,
or rather of its construction of a cocoon of hair, of its pupa, and of the
moth, probably the Pale Tussock (*Dasychira pudibunda*), which he compares with the Silk Worm Moth. This last note he follows with one upon
the luminous Myriapod (probably *Geophilus electricus*)[6] called Julus from its
resemblance to Hazel catkins (*juli*), which he declares to be a Scolopendra.
Under *Dipsacus sylvestris*[7] is one upon Mouffet's solitary caterpillar
living inside the Teasel-heads: 'these', Ray says, 'are shut up in a goose-
quill and hung round the necks of quartan-ague patients as amulets or

1 *C.C.* p. 181.   2 *C.C.* p. 87.
3 Mouffet divides caterpillars according to colour.
4 Ray gives the reference *Theatr. Insect.* Bk. 2, c. 1 ad fin.   5 *C.C.* pp. 39–40.
6 Cf. *Hist. Insect.* p. 45 and F. G. Sinclair, *Cambridge Nat. Hist.* v, p. 34.
7 *C.C.* pp. 44–5: cf. *Hist. Insect.* p. 341.

drugs by our country-folk'; he then describes how these caterpillars run backwards when touched,[1] and most carefully how they bore the pith, prepare a chamber, pupate and in the following spring emerge either as a fly (probably *Omorga mutabilis*) or as a moth (*Penthina gentianana*).

This same subject occupies one of his longest notes, that under his Rapum sylvestre[2] (i.e. *Brassica napus*). He begins with a note on the relationship between the Wild Rape and the Cabbage, and states that caterpillars reared on cabbage readily eat rape, though refusing other plants: 'the history of these caterpillars, since I have never read, seen or heard anything like it, let me describe even if it is too prolix to suit the plan of my book'.[3] He then describes the caterpillar—of the Large White (*Pieris brassicae*)—in great detail; and continues:

I shut up ten or so of these in a wooden box at the end of August 1658. They fed for a few days, and fixed themselves to the sides or lid of the box. Seven of them proved to be viviparous or vermiparous: from their backs and sides very many, from thirty to sixty apiece, wormlike animalcules broke out; they were white, glabrous, footless, and under the microscope transparent. As soon as they were born they begin to spin silken cocoons, finished them in a couple of hours, and in early October came out as flies, black all over with reddish legs and long antennae, and about the size of a small ant.[4] The three or four caterpillars which did not produce maggots after a longer interval changed into angular and humped chrysalids which came out in mid-April as white butterflies. The viviparous caterpillars died a few days after the birth without any metamorphosis.

Ray then reports that Mr Francis Willughby has observed in the case of several kinds of caterpillar that they change not into chrysalids but into oblong eggs like ant's eggs but dark-coloured. His comment is:

Are these eggs produced in this form or like those of the Cabbage butterfly do they first emerge as worms? We hope to answer this question next summer: meanwhile I think it likely that just as butterflies come out of chrysalids and chrysalids out of caterpillars, so these flies come out of eggs and the eggs out of maggots. Yet it is not altogether unlikely that the caterpillars are sometimes changed directly into eggs as they regularly are into chrysalids. I must also examine the suggestion that the caterpillars which produce eggs are weaker ones for which nature could not fulfil her first intention of producing a butterfly and so to avoid total loss formed a less perfect creature, the fly.

Further, Willughby notes that these eggs are of two kinds, the one shorter, dark and ringed, the other oblong, almost transparent, and covered by a skin; and that different kinds of fly come from each. Ray says that on consideration and having had no time for experiment he thinks that, just as some caterpillars spin a cocoon and others go through the change

---

1 'Amphisbaenae sunt haec animalcula.'   2 *C.C.* pp. 134–8.

3 This account is condensed in *Hist. Insect.* pp. 113–14, under Papilio alba; p. 254, under Vespa ichneumon parva; and among various notes on p. 260.

4 *Apanteles glomeratus* is the Ichneumon in question.

naked, so these maggots do the same; the naked or brown are the pupae of flies, the others are not pupae but cocoons in which the pupae are concealed. 'We have not yet sufficiently studied the problem and do not venture upon any rash pronouncement.'

From this account Ray's acumen is perfectly clear. He had observed what Mouffet and Willughby had failed to observe, the emergence of the parasites as maggots; and though he does not actually deny the speculation about caterpillars turning into fly pupae, he evidently distrusts it, and is already on the track of the true solution. This he reached and records in his *Historia Insectorum*:[1]

> Whence these maggots arise is a great problem. I think that the ichneumon wasps prick these caterpillars with the hollow tube of their ovipositor and insert eggs into their bodies: the maggots are hatched by the warmth of them, and feed there until they are full grown: then they gnaw through the skin, come out, and spin their cocoons.

Ray's gratitude to and reverence for Willughby has too often led to the belief that except in botany Willughby was the abler of the two. This early note, though it pays tribute to Willughby's keenness and observations, plainly indicates that Ray has a clearer grasp of scientific method and a truer insight into the principles of natural history.

A second note under Rosa sylvestris[2] deals with the rose-gall Bedeguar or Spongiola.

> If you cut open this gall you will find it packed with certain small white maggots—this on the evidence of Verulam, *Nat. Hist.* Cent. vi, Exp. 562; Spigel, *Isag.* i, 10; Mouffet, *Theat. Insect.* ii, 20 and of experiment. Spigel and Mouffet in the passages quoted and Aristotle, *Hist. Anim.* v, 19, say that beetles are born of these maggots. This is incorrect. I saved some of these galls; and the maggots hidden in them through the winter came out in the following May as flies: their shape and proportion are like those of a winged ant, their size is a little smaller.

He then describes them in detail, and concludes: 'some of these flies are armed with sting or spike always protruding from the tail, others altogether lack these: and therein is perhaps a difference between the sexes'.

Having thus corrected his predecessors Ray followed up his own observations and reported more fully in *Historia Insectorum*.[3] There he repeats, with a reference to the *Cambridge Catalogue*, the note already quoted and continues with an account of Mentzel's observations. He now classifies the 'fly' under Smallest Ichneumon-wasps and calls it Musca Cynosbati spongiolae or Vespa rosea. Its modern name and style is *Rhodites rosae*.[4]

---

1 P. 114.   2 *C.C.* pp. 139–40.   3 Pp. 259–60.
4 See P. Cameron, *British Phytophagous Hymenoptera*, iv, pp. 40–1.

Finally, under Papaver spumosum,[1] a plant which he lists elsewhere as Behen album, that is *Silene inflata*, Linn., is a note on 'what is called in English Woodseare and Cuckow-spittle':

Of the cause and origin of this foam the common crowd of philosophers is under a gross illusion; some call it star's spit and believe that it rains down from heaven, like Manna and honey dew—which are found in abundance on very few plants; others that it is an exhalation from the earth; others an exudation from the plant itself. I have discovered that it is vomited from the mouth[2] of an insect, a tiny creature that always lurks in the middle of the spittle. For if you wipe off the froth of foam, you will see for yourself the same foam very soon again poured out from the creature's mouth in such abundance that it will soon enwrap and conceal itself in it: so it can lie hid there in safety from all harm from frogs, small birds and other enemies while it is still feeble and cannot save itself by jumping or flight. This insect is almost like a louse in shape but shorter for the size of its body, yellow-green in colour with large protruding eyes: its hind legs are shaped for jumping, whence we are ready to agree with the learned men who have asserted by experiment that it turns into a locust such as we call Grasshopper. We, to confess the truth, have not yet investigated its origin or its final state.

Here again he pushed forward researches and published in *Historia Insectorum*[3] an emended account under the heading of Locusta-pulex, Swammerdamio: nobis Cicadula dici solita (the modern name is of course *Philaenus spumarius*, Linn.). After condensing the record of the *Catalogue* he continues:

The hind legs are scarcely longer than the rest: rudiments of wings appear on the shoulders: a long proboscis is bent back on the belly between the feet: they often change the skin which is found next the spittle. While they are still en-wrapped in the spittle, they crawl; after leaving it, they always move by jump-ing, the hind feet being now stronger and longer. They resemble cicales more than locusts: they fly higher than locusts: their wings conceal the whole body.

Before closing our account of the *Catalogue* reference must be made to the subject briefly mentioned in its closing pages. In the 'Interpretation of terms' there is one point of great importance—the appearance of the name of Joachim Jung of Lübeck and his book the *Isagoge Phytoscopica*. This Ray describes as 'not yet published but communicated to me by the famous Samuel Hartlib',[4] and this account is amplified in the Preface of his *Methodus Plantarum*,[5] where he writes: 'Jung's *Isagoge* in manuscript and received from Mr Samuel Hartlib was communicated to me by that

1 *C.C.* p. 112.
2 This is not strictly accurate: cf. Imms, *Textbook of Entomology*, p. 374.
3 P. 67.
4 *C.C.* Pt. II, p. 87: the *Isagoge* (Hamburg, 1678) is a pamphlet of 40 pages.
5 Published in 1682; see below, pp. 192–4.

excellent and learned man Dr John Worthington, my close friend, now deceased.'[1] Hartlib was of course the distinguished German to whom in 1644 Milton addressed his little *Tract on Education*,[2] whose works on husbandry attracted great interest at this time[3] and who was in close touch with continental writers on plants and farming; and Worthington was Master of Jesus College and Rector of Fen Ditton. It would seem that Ray only received Jung's book just before the completion of the *Catalogue*; for he is not included in the list of authorities nor elsewhere, and appears here in a form which suggests last-moment additions. Jung is important as a source of Ray's interest in plant-structure and of his ideas of classification; and although some estimates of his influence are gravely exaggerated, the *Historia Plantarum*[4] acknowledges very freely the value of Jung's work upon physiology, bracketing him with Grew, and incorporating much of the *Isagoge*.

The final point is that of Classification.[5] Ray's purpose as defined in his Preface debarred him from any attempt to introduce a system into his work: nor was the time ripe for it. While plants themselves were largely unknown, accurate identification was the first necessity: while botany was mixed up with legendary notions of the virtues and affinities of plants, scientific determination of their relationships could not be undertaken: the ground had first to be explored, data and reliable observations be collected, study of structure and habits, of anatomy and physiology be undertaken. Without far wider knowledge than was then available the characteristics upon which taxonomy could be based could not be settled. As Ray himself stated in the last sentence of his *Catalogue*: 'It would be possible to classify plants in many other ways, for example in respect of their roots or stems or flowers or seeds or leaves, but to pursue the subject further is not at present my intention.'

Nevertheless he gives a rough and ready scheme, styling it 'the more customary classes or divisions of plants' and obviously regarding it as little more than a conventional arrangement. We shall have to consider in detail later his long and important work on the subject; for the search for true principles of classification was one of his chief interests. But as a starting-point and as representing the sort of ideas prevalent at the time this scheme must be briefly summarised. He divides plants 'perfect and properly so called' (i.e. excluding fungi and lichens and perhaps mosses)

1 Preface, p. 6 (unnumbered). The *Isagoge* had by this time been published. In the correspondence between Worthington and Hartlib in *Diary and Correspondence of J. W.*, edited by Crossley, vol. 1, there are several references to Jung both in 1655 (pp. 66, 82) and in 1659 (pp. 123, 175), but no clear allusion to this loan.

2 Cf. Masson, *Milton's Poetical Works*, 1, p. 23.

3 He published many treatises on the subject between 1650 and 1659.

4 In the general essays in vol. 1.

5 *C.C.* App. pp. 100–3.

into trees, shrubs, sub-shrubs and herbs. These, after mentioning that trees and shrubs are divided into evergreen and deciduous, he divides as follows:

Trees into eight classes: 1. Bearing fruits without stones, Apple, Lemon, Fig, etc. 2. With stoned fruits, Plum, Peach, Date-palm, Olive. 3. With nuts, Walnut, Chestnut, etc. 4. With berries, Laurel, Mulberry, Juniper, Elder, etc. 5. With acorns, Oak, Ilex, Beech, etc. 6. With cones, Pine, Fir, Larch, etc. 7. With pods, Laburnum, Judas tree, etc. 8. The rest, some with catkins, Hornbeam, Willow; some with membraned seeds, Ash, Sycamore, Elm, Lime, etc.; some perhaps infertile: these might be called woodland trees.

Shrubs into two: 1. With thorns, Berberis, Buckthorn, Gooseberry, etc. 2. Without thorns, Broom, Jasmine, Privet, etc. 'Shrubs might be divided into flowering, fruit-bearing, climbing, etc.'

Sub-shrubs—one family, not numerous and mostly scented: Hyssop, Lavender, Wormwood, etc.

Herbs are so numerous that it is difficult, indeed almost impossible, to divide them into fixed and self-contained classes so that none can belong to several different classes or be ambiguously placed. He proposes the following: 1. Bulbous, Lily, Daffodil, Crocus, etc. 2. Tuberous, Fig-wort, Potato, Asphodel, etc. 3. Umbelliferous, Fennel, Anise, Parsnip, etc. 4. Whorled, Horehound, Motherwort, some of the Mints, etc. 5. Spiked, Loosestrife, Speedwell, etc.: 'almost all can be placed among the Whorled'. 6. Climbing, either turning, Hop and Bindweed, or with tendrils, divisible into fruit-bearing, Melon, Pumpkin, Cucumber, etc., or pod-bearing, Pea, Vetch, Lentil, etc. 7. Corymbiferous, Aggeratum, Tansy, etc., or following J. Bauhin, Daisy, Marigold, etc. Others interpret corymb differently so as to describe plants with circular flower-heads, Ivy, Guelder-rose, Leek, etc. 8. Pappus-bearing, Dandelion, Groundsel, Goatsbeard, etc. 9. Capitate, Knapweed, Scabious, etc. 10. Bell-shaped, Campanula, Foxglove, Rampion, etc.: 'many of these are climbers'. 11. Garland-makers, 'which are sought for the sake of the flower and used for weaving crowns', Pink, Carnation, etc. 12. Round-leaved, Asarabacca, Cyclamen, etc. 13. Nerve-leaved, Plantain, Helleborine, etc. 14. Starred, 'whose leaves radiate starlike from the stalk', Woodruff, Galium, Madder, etc. 15. Cereal, including grain and vetches. 16. Succulent, Stonecrop, Orpine, etc. 17. Grassy-leaved, Corn, Grasses, Reeds, etc. 18. Vegetable, Endive, Spinach, Beet, Cabbage, etc. 19. Aquatic, Water-soldier, Water-lily, etc. 20. Marine, Sea-weed, etc. 21. Rock-loving, Fern, Maidenhair, etc. Moreover in general Herbs are either: 1. Permanent with a perennial root although stems and leaves perish yearly; 2. Perennial, stems, branches and even leaves live all year round; 3. Annual, living only one year whether sown in spring or autumn.

Of the origin of this classification Ray gives no sign beyond calling the divisions 'usitatiores'.[1] We should expect it to be derived either from De l'Obel's *Adversaria*, which was used in the Plantin album of wood-cuts, or from Gaspard Bauhin's *Pinax*; for these are generally regarded as the most important schemes, except that of Cesalpino, which Ray had not yet studied. In fact both De l'Obel and G. Bauhin are extremely unsound, the former using leaf-structure almost solely, and the latter grouping together Mullein, Primrose, Butterwort and Foxglove; or Ivy, Balsam and Saxifrage:[2] and Ray, probably deliberately, rejected them. Actually the source of his list is the arrangement followed in the forty books of Jean Bauhin's great *Historia Generalis*. He follows these headings almost exactly and often word for word, improving upon them by omitting Bauhin's 'Scented Trees and Shrubs' and 'Malignant and Poisonous Herbs' but omitting, to his loss, the Pod-bearers (mostly Cruciferae) and the Capsule-bearers (mostly Caryophyllaceous). As a 'Method' it is certainly an improvement upon that of his brother, and that Ray chose it is one more proof of his high regard for its author. But it seems clear that as yet Ray had given little attention to taxonomy. His first task was to gain exact knowledge of the plants themselves, and to reveal to others the wealth and interest of their local flora. He had not yet sufficiently studied either species or structure to be ready for the problems of classification.

As it stands the scheme gives more vividly than much description a picture of botanical science at the time when he began his work. He was a scholar and had drawn upon the best available authorities: if this was all that they could provide, he was right in rejecting it. He was too good a scientist by instinct and training to be unaware of its arbitrary and in places fanciful character and at the same time of the great difficulty of creating an adequate substitute for it. At this stage in his career he was wise in preferring an alphabetical order to schemes manifestly defective or hurriedly extemporised.

NOTE. *Ray's garden in Trinity College*

The little plot of ground which Worthington describes and Ray occasionally mentions suggested to him a project which his departure from Cambridge frustrated but of which traces remain in his later work. Writing in February 1660 to Willughby[3] he had mentioned along with the plan for a Phytologia Britannica a 'design to make another catalogue

1 There is little evidence to justify this epithet, though John Johnstone in his *Notitia Regni Vegetabilis* published in 1661 uses a similar arrangement.

2 Cf. *Pinax*, VI, 6 and VIII, 3: in many sections the arrangement seems almost chaotic: e.g. Gentiana, Plantago and Holosteum in V, 5 or Bupleurum, Hypericum, Sedum and Aloe in VII, 5.

3 *Corr.* p. 2.

which I will call Horti Angliae'. This was to be based upon lists obtained from all the noted gardens, of which he had already those of the Oxford garden by Philip Stephens and William Browne, 1658, and of Tradescant's in Lambeth by John Tradescant junior in 1656. Out of his own garden and these he would then make one catalogue, including only such plants as do not grow wild anywhere in England.

With this in view he seems to have prepared a full list of the plants in his own and in other gardens in Cambridge; and according to the article in the *General Dictionary*, published in 1739 and derived from his friend Samuel Dale, this actually existed in manuscript at that date.[1] It did not come into Sloane's hands and is not now traceable.

No doubt the project as planned was incomplete when he left Cambridge. Certainly between 1662 and 1666 he had no opportunity for it. After that he was occupied on the Phytology and many other tasks. Then the idea of a *Hortus* was expanded into that of a *Historia*.

In the *Historia* there are frequent references to his Cambridge garden. Some forty species are definitely mentioned as having been grown in it; and in most cases there is a note as to the source from which he obtained them. 'We grew this for several years at Cambridge from seed sent from London' is a usual form for this note. In three places[2] his friend Edward Morgan, whose garden in Westminster he frequently visited,[3] is mentioned as the sender of plant or seed. Of one plant, an American Daisy apparently undescribed, he writes: 'I received it from a London nurseryman named Walker whose son brought it with other rarities from Virginia';[4] of another, the famous 'Poley Mountain' of the herbalists (evidently *Teucrium polium*), 'we gathered a sprig of it in an Oxford garden, put it in a wallet and carried it for three days in a bag: when we got back to Cambridge we planted it: it rapidly recovered and struck root'.[5] Several plants he received under names different from those used in the *Historia*; a Michaelmas Daisy, 'one like this was sent from London under the name of Spanish Daisy';[6] a Goldenrod, 'we had it from London

---

1 Vol. VIII, p. 692: it is entitled *Catalogus Plantarum non domesticarum quae aluntur Cantabrigiae in Hortis Academicorum et Oppidanorum*: it was possibly modelled upon Dionys Joncquet, *Hortus sive Index Plantarum quas colebat* (Paris, 1659), of which Ray spoke warmly to Worthington (*Diary*, I, p. 334).

2 *H.P.* I, pp. 183, 605, 672.

3 Cf. Gunther, *Early Science in Cambridge*, p. 375, for visit in 1662; *Corr.* p. 5 for 1669(?); and below, p. 150.

4 *H.P.* I, p. 363. Cf. *Garden Book of Sir T. Hanmer*, p. xx, 'June 1655. Walker of St James's has many Virginia plants'.

5 *H.P.* I, p. 524: the identity of Poley Mountain has been variously given; Britten and Holland, *Dictionary of English Plant-names*, p. 385, give *Calamintha acinos* on the strength of Johnson's *Gerard*, p. 676, where London herb-women are said to sell this plant by mistake for P.M. *T. polium* is a variable sub-shrub widely distributed in Southern France and the Mediterranean region.   6 *H.P.* I, p. 269.

under the name of Arnold Villeneuve's Goldenrod';[1] a Lupin, 'sent under
the name of Lupin of Cadiz';[2] and a Wild Leek, 'I grew it at Cambridge
as Roman onion'.[3] There are very few notes of culture: of the 'Tabacco
latifolium' (*Nicotiana tabacum*), 'it sometimes endures a winter as I
proved at Cambridge';[4] and of 'Sanicula montana rotundifolia minor
Hispanica' (presumably the familiar London Pride), 'it likes shade, spreads
easily and winters well'.[5]

Two references[6] to seed received from his Cambridge friend Dent but
collected by 'Mr Covill of Cambridge' in Thrace must refer to a later
garden: for Covel only went to the East in 1670.[7]

There is one reference in the *Historia* which suggests that these records
of his garden were derived from an unpublished catalogue. Under Jasmine
he says: 'some have regarded it as setting no fruit; we formerly followed
them and said the same thing in our Catalogue of Cambridge Plants'.[8]
There is no such remark in the *Cambridge Catalogue* as published.

1 *H.P.* I, p. 279.   2 *H.P.* I, p. 908.   3 *H.P.* II, p. 1121.
4 *H.P.* I, p. 713.   5 *H.P.* II, p. 1047.   6 *H.P.* I, pp. 565, 742.
7 For Covel see below, pp. 214–15.   8 *H.P.* II, p. 1599.

CHAPTER V

THE YEARS OF TRAVEL

Mr Wray hath made a collection of plants, fishes, foules, stones and other rarities which he hath with him.

EDWARD BROWNE from Rome to his father,
Sir T. Browne's Works, I, p. 86.

Before the *Cambridgeshire Catalogue* was out of the press Ray had formed plans for further work. The earliest of his letters to Willughby, dated from Trinity College on 25 February 1659,[1] accompanies a gift of the newly published work and continues with proposals:

> You will remember that we lately, out of 'Gerard', 'Parkinson' and 'Phytologia Britannica', made a collection of rare plants whose places are therein mentioned and ranked them under the several counties. My intention is now to carry on and perfect that design; to which purpose I am now writing to all my friends and acquaintance who are skilful in Herbary to request them this next summer to search diligently his country for plants, and to send me a catalogue of such as they find, together with the places where they grow. In divers counties I have such as are skilful and industrious: for Warwickshire and Nottinghamshire I must beg your assistance.... After that partly by my own search, partly by the mentioned assistance, I shall have got as much information and knowledge of the plants of each county as I can (which will require some years) I do design to put forth a complete P.B.[2] which I hope to bring into as narrow a compass as this book.

He then gives a sketch of its contents and order; outlines a further proposal for a book of English gardens, listing with synonyms all plants grown in them other than natives of Britain; and finally asks Willughby for his comments.

So the project of a complete British Flora at which he had hinted in the Preface to the *Catalogue* was definitely undertaken. His work in collating previous authors had shown him the need and supplied the first steps for a new and correct survey. His pioneer work in Cambridgeshire had fired his enthusiasm, compelled him to master the existing literature, and given him experience in the business of discriminating species. To expand his work was natural, even if the undertaking must have seemed a large task to tackle single-handed; and he evidently had hopes of which we know little[3] from his friends. But to plan it at a time when his future was

1 This is, of course, 1660 new style.

2 Worthington to Hartlib in June 1661 had said: 'I believe he has thoughts to publish a Phytologia Britannica. He has made good preparations for such a work' (*Diary*, I, p. 334).

3 The letters earlier than 1667 include only those to Willughby and P. Courthope, the cousins whom he mentions in the Preface to *C.C.*

beginning to be insecure[1] is proof that he had found an inescapable vocation. During the next ten years, the most crowded and critical in his life,
he carried the plan to its fulfilment, and gathered the material which enabled him to produce the *Catalogue of English Plants* in 1660, the *Synopsis*
in 1690, and the county lists in Camden's *Britannia* in 1695.

He had already before the death of his friend John Nidd made the
journey described for us in his *First Itinerary*. Leaving Cambridge on
9 August 1658 on horse-back and alone he rode to Northampton and 'in
Mr Brooker's garden, saw divers physical plants':[2] thence he went to
Warwick, staying with Sir Henry Newton or Puckering at the Priory. On
18 August he went to Coventry, Ashby-de-la-Zouche, Derby and Ashbourne in the Peak, crossing the Trent at Swarson (Swarkeston) Bridge.
At Ashbourne on the walls of the 'very fair church, built cathedral-wise'
he found *Arabis hirsuta*,[3] a plant of whose identity he had some doubts,
and in the neighbourhood *Erysimum cheiranthoides*,[4] already familiar in
osier holts at Ely. At 'Buxton or Buckstone' next day he inspected Pool's
Hole[5] and had his first experience of botanising in the hills, finding
Alchemilla vulgaris, *Vaccinium vitis-idaea* and *oxycoccus*, *Empetrum
nigrum*, *Lycopodium clavatum* and *selago* 'on the moorish hills hereabouts',
Cochlearia alpina in abundance near Castleton.[6]

On 24 August he directed his course towards North Wales and 'that
night lay at West Chester': he thought poorly of the cathedral—'it hath
little beauty within or without'—but admired the walls and apparently
found *Brassica tenuifolia*[7] upon them. Then by way of Flint, 'Haliwell'
and St Asaph he came to Aberconway and thence over Penmaenmawr to
Bangor and the ferry to Beaumaris. He mentions no plants—the *Itinerary*
at this stage is the briefest record—and as he covered the same ground
four years later the records in the *Catalogue* may apply to either visit: but
as he spent four days in Anglesey he can hardly have failed to use them to
good purpose. On 31 August he went to Carnarvon and next day hiring
a guide started for Snowdon: but 'it rained so hard that I was forced within
a mile of the town to take shelter in a small cottage.... This night I lodged
at Bethkellart.' On 2 September 'I hired another guide: we marched up
on foot about four miles. The top of the mountain was covered with
clouds, so that I lost the prospect of the adjacent country. Divers rare
plants I found on the top and sides of the hill which were then strangers
to me, de quibus consule catalogum.'[8] But to take his advice is to meet

---

1 As his second letter to Courthope, Sept. 1660, reveals (*F.C.* pp. 17–18).
2 *Mem.* p. 121.   3 *C.A.* p. 38.   4 *C.A.* p. 52.
5 His observations on it are recorded in *Misc. Disc.* p. 76: for a description of it
very similar to his in *Mem.* p. 124 cf. Thoresby, *Diary*, pp. 91–2 (July, 1681).
6 *Mem.* p. 126 and *C.A.* p. 75: 'transplanted to a garden it changes its form.'
7 *C.A.* p. 103.   8 *Mem.* p. 129.

the difficulty already mentioned: for two visits to Snowdon took place before the *Catalogue* was published.

Then he struck across to Bala, 'a long and bad way'; and has recorded *Corydalis claviculata* as 'in the hedges as you go from Bala down to Pimblemere':[1] thence to Dolgelly and the ascent of Cader Idris; 'I found no new plants, save the Globe flower [*Trollius europaeus*, Linn.], Cypress moss [*Lycopodium alpinum*, Linn.] and another small club moss with white seeds [*Selaginella selaginoides*]'—*Saxifraga stellaris* recorded from Cader Idris[2] in the *Catalogue* he had no doubt already seen on Snowdon. On 6 September he went on to 'Mahentler' (Machynlleth) to see the silver mills; thence to Llanfair short of Welshpool, 'where I first heard of the death of Oliver L. Protector'; and so to Shrewsbury, Kidderminster, Worcester, Gloucester, Stow-on-the-Wold and St Neots. He got back to Cambridge on 18 September about noon.

In the following year he was busied with the *Catalogue*, and probably confined his excursions to the places in Cambridgeshire of which he has given special lists. But in 1660 there is one mention in a letter to Courthope of activity further afield. 'You do rightly guess', he writes on 26 September,[3] 'that I did accompany Mr Willughby in his travels this summer.' This reference, which seems to stand alone,[4] may explain the curiously conflicting accounts given by the biographers and by the letters of the tour in 1661. Dale,[5] Derham[6] and Pulteney[7] all state that in 1661 Ray with Willughby and others went on the tour to Scotland of which the second printed Itinerary is a record. But the letters, both that to Willughby published by Derham of 14 September[8] and that to Courthope in July,[9] make it clear that in 1661 Willughby was not with him:[10] in the former Ray states: 'On Saturday night last the 7th instant Mr Skippon and myself arrived at Cambridge from a long northern expedition wherein for the most part we followed your footsteps', and in the latter: 'I intend to set out on Monday next [i.e. 26 July], my company is only Mr Skippon and a servant.' It is evident that in 1660 Willughby and Ray did a northern tour similar to that of Ray and Skippon in the following year, and that the biographers, not knowing the letter to Courthope and ignoring that to Willughby owing to Derham's misprint, have confused the two tours.

1 *C.A.* p. 122. 2 *C.A.* p. 85. 3 *F.C.* p. 17.
4 There is a reference in his letter to Willughby of 14 Sept. 1661 (*Corr.* p. 3) to a plant 'which the last year (i.e. 1660) I observed not far from St Neots coming to wait upon you'.
5 *F.C.* p. 4. 6 *Mem.* p. 13. 7 *Sketches of Botany*, I, p. 199.
8 *Phil. Letters*, pp. 358–9; *Corr.* p. 3: by a misprint in Derham this letter is headed 'Mr Ray to Mr Willisel' though the list of errata corrects this to Willughby.
9 *F.C.* pp. 20–1: cf. also *Journal of Botany*, LXXII, p. 220.
10 This is also maintained by Boulger in *D.N.B.* Art. Willughby, and by Gunther, *F.C.* p. 21.

There is in fact abundant proof of a journey by Ray and Willughby, other than those which have been hitherto recorded. In his first Itinerary he speaks of his journey through Wales in 1662 as his 'fourth voyage',[1] though Scott labels the record of it Itinerary III. In the *Catalogue of English Plants*, on p. 103 under the heading Eruca monensis (i.e. *Brassica monensis*, Linn.), Ray writes: 'We found it plentifully, going from the Landing-place at Ramsey to the Town': on p. 112 under Festuca altera (probably *Bromus secalinus*): 'Found among Corn in the Isle of Man', and on p. 113 under Filix florida (*Osmunda regalis*, Linn.): 'It grows very plentifully on the rocks about the Isle of Man.'[2] Similarly, in the *Ornithology*, under the sections on the Razorbill[3] and the 'Puffin of the Isle of Man',[4] there is an account of the cliffs at Spanish Head and of the behaviour of birds on the Calf which is obviously the work of an eyewitness, and a statement that 'Mr Willughby saw and described only a young one [of the Puffin] taken out of the nest'. Finally, in his *Three Physico-Theological Discourses* Ray writes:[5] 'I myself in company with Mr Willughby in the strait between the Isle and Calf of Man took up among the tall *Fuci* growing thick upon the rocks there…Sea-Urchins as big as a man's two fists.' From these references it is evident that the two friends had visited the Isle some time before 1670. Moreover, in the *Itinerary* of his 1662 tour he writes twice of the Erodium 'which I had before observed in the Isle of Man';[6] and this seems to make a visit in 1660 certain. Of the extent of this tour it is impossible to speak with confidence; but Ray's allusion to following Willughby's footsteps and the fact that a large number of typically northern plants were already familiar to him in 1661 make it certain that the tour was extensive. In the letter reporting his discoveries of 1661 he says: 'on this journey I met with but four plants that I had not formerly observed',[7] and of these he gives the names; yet in the Itinerary[8] he mentions finding at Shap *Geranium sylvaticum*, *Carduus heterophyllus*, *Campanula latifolia*, *Primula farinosa*, *Salix pentandra* and *Allosurus crispus*. So, too, on the journey of 1668, when he again visited this area, he lists five or six new species and a number of others 'which I had found before but now have described more accurately',[9] *Impatiens noli-tangere*, *Rubus chamaemorus*, *Rosa villosa*, *Galeopsis dubia*, *Polygonum bistorta*, and *Epipactis purpurata*.[10] All the species named here, and

1 *Mem.* p. 130—in his account of the silver mines which he revisited in 1662.
2 So too *C.A.* p. 133: 'In insulis Monensi et Prestholmensi' (of *Erodium maritimum*), and p. 119: 'Vidimus rupibus submersis innascentem circa insulam Monensem' (of a *Fucus*).   3 P. 324.   4 P. 333.
5 L.c. (3rd ed.), p. 151: so too *Corr.* p. 245.   6 *Mem.* pp. 168, 171.
7 *Corr.* p. 3.   8 *Mem.* p. 163.   9 *Corr.* pp. 26–7.
10 The plant found on this journey at Sheffield is thought to be this species rather than *E. atropurpurea* which he subsequently found at Malham: cf. Lees, *Flora of West Yorks*, p. 434.

many others recorded in the first edition of the *Catalogus Angliae*, must have been first found in 1660: and if in most cases the dating is a matter of inference, in one, *Allium oleraceum* var. *complanatum*, the note: 'This I observed on the scars of the mountains above Settle in Yorkshire: I saw something like, if not identical with, it in fields near Chester *superiore anno*',[1] clearly refers to that year.

The plants mentioned are mostly recorded from Craven; but *P. bistorta*[2] and *E. purpurata*[3] are from 'meadows and pastures about Sheffield', and *I. noli-tangere*[4] is from 'Winandermere in Westmorland not far from Ambleside'. Other references to the Lake district on this visit include those to two fish,[5] that 'called Schelley in Cumberland' (*Coregonus stigmaticus*), which he saw and described from Ullswater at Penrith,[6] and the Char (*Salvelinus willughbeii*), of which he describes two forms from Windermere.[7] The plants include the 'Hooded Water-milfoil' (*Utricularia vulgaris*), 'in a fenny ground near Hawkshead',[8] and the 'Rosewort or Rose-root' (*Sedum roseum*), 'on high mountains... Hardknot etc.'[9] More striking are the references to the 'Hoary Fleabane Mullet' (*Senecio palustris*), 'most in the ditches about Pillin-moss in Lancashire',[10] and to the 'Harestail Rush' (*Eriophorum vaginatum*), 'on mosses and boggy places as Pillin-moss';[11] for this lies west of Garstang and remote from any place visited later. There is at least one note which suggests that he actually covered much of the country traversed in the following year: of 'Pyrola'[12] (the species being probably *P. media* rather than *P. rotundifolia*[13]) he records: 'we found it near Halifax by the way leading to Keighley, and in Northumberland': this is not one of the plants new in 1661, and in 1668 he did not enter Northumbria.

The facts can be interpreted with reasonable probability. On receiving Ray's proposals in February[14] Willughby must have urged an immediate commencement of the task. Although he spent much of the year in Oxford 'to consult the works on natural history in the libraries there',[15] was entered at the Bodleian as a reader,[16] and was certainly living there in September,[17] the summer may well have been free. They probably started

---

1 *C.A.* p. 13: he was never at Settle in the year following a visit to Chester except in 1670–1; and this is too late for *Catalogus Angliae*, ed. 1: the years that best suit this note are 1658–60.
2 *C.A.* p. 44.   3 *C.A.* p. 162.   4 *C.A.* p. 239.
5 Of the tour of 1668 he says: 'In historia piscium nihil promovi': *Corr.* p. 27.
6 *Hist. Pisc.* p. 186.
7 Cf. Catalogue of English Fish appended to *Collection of English Words*.
8 *C.A.* p. 210.
9 *C.A.* p. 264: for Hardknott see note at end of chapter VI.
10 *C.A.* p. 79.   11 *C.A.* p. 180.   12 *C.A.* p. 256.
13 Cf. Lees, *Flora of West Yorks*, p. 322.   14 *Corr.* p. 1.
15 *Mem.* p. 14.   16 Foster, *Alumni Oxon.* IV, p. 1651.
17 Wood, *Fasti*, II, c. 246.

in the later half of June, went north to Halifax and Keighley and perhaps with a detour to Pendle[1] to Settle and the exploration of Ryeloaf ('Hinckle-haugh') and Ingleborough; thence possibly into Northumberland[2] or by Shap to Penrith; so by Ullswater and the Kirkstone to Ambleside and Windermere; then along the ancient road over Wrynose and Hardknott passes, from which they may well have explored some of the Scafell group of mountains to the sea at Ravenglass; thence by boat to Ramsey and a tour of the Isle and Calf of Man, returning from Douglas to the mouth of the Wyre; and so by Pilling Moss to Garstang and the southern road.[3] Such a reconstruction is of course conjectural: but a careful study of the exact wording of the notes in the *Catalogue*—evidence which is too de-tailed to quote at length—lifts it above the level of mere guess-work. They must have finished the journey late in July; for on 1 August,[4] when on his way from Cambridge to wait upon Willughby at Middleton,[5] Ray discovered at Eltisley[6] the plant which afterwards turned out to be *Bupleurum tenuissimum*.

This journey was in fact a turning-point in the lives of both men and an event of importance in the history of science. It initiated a partnership which drew Ray from the study of botany to that of zoology, gave him the means to gather material and the encouragement to use his knowledge, and thus enlarged his life's work from the composition of a *Phytologia Britannica* to that of a *Systema Naturae*.

After his return Ray certainly spent part of September at Black Notley; for on 26 September he wrote to Courthope[7] saying that he had come to Cambridge from Essex on the previous day. But it is tempting to suppose that during August he paid a visit to Thomas Browne,[8] author of *Religio Medici*, at Norwich. In the *Catalogue* under Acorus he says: 'First observed by the famous Mr Tho. Brown M.D. in the river Yare near Norwich and shown to us ten years ago.'[9] Ray is usually accurate in his dating, and this note is borne out by another dealing with *Suaeda maritima*: 'I saw this at Portland, but before that the famous Mr Th. Brown M.D. showed it to me at Norwich collected from the Norfolk coast.'[10] There seems no evidence as to the cause or precise occasion of Ray's visit: but

1 *C.A.* p. 68.

2 The Northumbrian record may refer to 1661; but in that Itinerary he records no Northumbrian plants at all.

3 For an account of the perils of such a journey—'Wrenose and Hardknot which were really mighty dangerous, terrible and tedious'—cf. *Diary of Ralph Thoresby*, I, p. 267 (Sept. 1694).

4 So *Hist. Plant.* I, p. 474.   5 *Corr.* p. 3 and below, p. 120.

6 *C.A.* p. 37.   7 *F.C.* p. 17.

8 He certainly knew Browne at this time, for in *Observations*, p. 237, he speaks of having received from him a picture 'some years before' 1665.

9 *C.A.* p. 7.

10 *C.A.* p. 313. There is another mention of Browne on p. 126.

he certainly became known to Browne (whose eldest son Edward had come up to Trinity in 1657) at about that time.[1]

The upheavals of the Restoration in the University and his own work as tutor and steward kept him anxious and busy during the winter. With the spring he 'made a more narrow search into the country about Cambridge for plants and have discovered in all about twenty-six that are not in our catalogue—some such as I had not seen before, nor mentioned to grow wild in England'.[2] In June he was at Fen Ditton with Worthington, who wrote of it: 'We went to view the herbs in the fields near us. He is perfectly read in the Book of Nature, and all is easy to him.'[3] He planned, however, to go north in July, and though the illness of his pupil Timothy Burrell, Courthope's cousin, compelled a postponement, he left with Mr Philip Skippon,[4] another of his former pupils,[5] and a servant on 26 July.

The journey is described in Ray's *Second Itinerary*, and seems to have been planned rather for the visiting of cathedrals and great churches than for the collection of plants. Peterborough, where the 'choristers made us pay money for coming into the choir with our spurs on';[6] Crowland, where 'there remains only part of the body of the church',[7] and where he found the Golden Dock (*Rumex maritimus*);[8] 'Kirkton' near Boston, which 'gives original and name to that sort of apples which are called at Cambridge corruptly Girton pippins';[9] Lincoln, where 'the choristers had no surplices but only gowns with capes faced with lambskin' and 'we saw the bell, called the great Tom of Lincoln, cast in the year 1610';[10] Kingston-upon-Hull, reached by a five-mile ferry from Barton, and where the church 'hath according to the statute a suffragan bishop';[11] Beverley and its minster; Pontefract and 'great plenty of liquorice'; Leeds, 'a large and rich town which hath a great trade for cloathing'; Knaresborough and the legends of St Robert; 'Herrigate', the 'spaw' and 'the sulphur-well whose water, though it be pellucid enough, yet stinks noisomely like rotten eggs';[12] Ripon, 'the minster much inferior to Beverley', and St Wilfrid's

1 Cf. Edward Browne's letters to his father from Italy in January and April 1665 (*Sir T. Browne's Works*, I, pp. 86, 88).

2 *F.C.* p. 21. A list was given in the letter and this is printed in Gunther, *Early Science in Cambridge*, p. 374: it actually contains twenty-two species, of which two were in fact in *C.C.*

3 To Hartlib (*Misc.* p. 258; *Diary*, I, p. 339): in this same June Isaac Newton came into residence.

4 Lankester, in *Mem.* p. 12, confuses him with his father Sir P. Skippon, author of *A Salve for every Sore*, 1643, and one of Fairfax's chief officers at Naseby—and is followed by Seward, *John Ray*, p. 12.

5 Admitted fellow-commoner at Trinity, 1655, at the same time as Courthope.

6 *Mem.* p. 131.   7 *Mem.* p. 132.

8 So letter to Willughby: *Corr.* p. 3 and *C.A.* p. 188.

9 *Mem.* p. 133.   10 *Mem.* p. 134.

11 *Mem.* p. 140.   12 *Mem.* p. 142.

needle; York, 'the minster a large and stately fabric but in some things inferior to Beverley';[1] and so to Scarborough, where the castle was 'still held with a garrison'. It is a pleasant narrative full of detail, grave and gay, but with only two references to plants, the finding of *Gentiana pneumonanthe*, first in Tattershall Park between Boston and Lincoln and then 'on many heathy grounds both in Lincolnshire and Yorkshire',[2] and 'in a close near the town called Granswick a great store of Carum'[3] (*C. carvi*). They were riding between twenty and thirty miles each day and visiting places of antiquarian and general interest[4] in a country not specially rich in plants or animals; and there can have been little opportunity for close study of the flora.

Along the Yorkshire coast there was more scope. A list of fish, seen or reported, of various kinds of sea-weed with which Ray was beginning to be concerned and of fossils in the lias beds at Whitby, and a record of the finding of the Lesser Tway-blade (*Listera cordata*) on the moor near Freebrough Hill,[5] give interest to the journey: by the bank of the Tees he found one of the few new plants of the journey, *Anthemis tinctoria*, Linn.[6] But from Stockton to 'Duresme', and so to Newcastle, Berwick and Dunbar, there is little to chronicle except an Englishman's first impression of the Scots. 'They cannot endure to hear their country or countrymen spoken against. They have neither good bread, cheese, or drink. They cannot make them, nor will they learn. Their butter is very indifferent, and one would wonder how they could contrive to make it so bad. They use much pottage, made of coal-wort which they call keal, sometimes broth of decorticated barley.'[7] But when on 19 August 'we passed over to the Basse Island' the description of the 'soland geese', the 'scout', the 'cattiwake', the 'scart' and the 'turtle dove'[8] is a sequel to the visit to the Isle of Man; and the plants of the Rock,[9] Wild Beet (*Beta maritima*), Sea Campion (*Silene maritima*), Scurvy-grass (*Cochlearia officinalis*) and especially Tree Mallow (*Lavatera arborea*), must have interested him: for in the *Catalogue of English Plants* there is a note under the last-named saying that he found it on the Bass, arguing that our English species is not the same as that of the two Bauhins, and stating: 'I grew them both for some years in my small garden at Cambridge.'[10]

1 *Mem.* p. 143.   2 *Mem.* p. 133 and *C.A.* p. 130.
3 *Mem.* p. 135: so Worthington to Hartlib, *Misc.* p. 293.
4 Ray's records of this Yorkshire tour are quoted in Gibson's *Camden*—and with more detail than in the *Itinerary*; the Greenlander at Hull, c. 745; the 'vipseys', c. 748; the making of kelp, c. 766; and Worthington records his finding of Roman coins, etc.
5 *Mem.* p. 149; cf. *C.A.* p. 44: it is five miles south of Saltburn.
6 So letter to Willughby: *Corr.* p. 3.   7 *Mem.* p. 153.
8 I.e. Gannet (*Sula bassana*); Guillemot (*Uria aalge*); Kittiwake (*Rissa tridactyla*); Shag (*Phalacrocorax aristotelis*); and Black Guillemot (*Uria grylle*).
9 *Mem.* p. 156.   10 *C.A.* p. 205.

In Leith 'we saw one of those citadels built by the Protector, one of the best fortifications that ever we beheld, passing fair and sumptuous.... This is one of the four forts; the other three are at St Johnston's, Inverness and Ayr'.[1] At Edinburgh 'we went to the Castle—there was then lying in the castle yard an old great iron gun which they call Mount's Meg, and some Meg of Berwick, of a great bore, but the length is not answerable to the bigness'; Heriot's hospital—'it maintained threescore boys who wore blue gowns'; the College, 'for the building of it but mean and of no very great capacity, in both comparable to Caius College in Cambridge'; and 'Argyle and Guthrie, their heads standing on the gates and toll-booth'. 'At the time we were in Scotland divers women were burnt for witches, they reported, to the number of about 120.'

After a day in Edinburgh on 21 August they rode up the Forth to Stirling, and Ray's opinion of the country began to improve; for 'by the way we saw the King's palace at Lithgow, built in the manner of a castle, a very good house, as houses go in Scotland'.[2] Here was a garden and in it 'divers exotick plants, more than one would hope to find in so northerly and cold a country; some such as we had not before seen'. From Stirling they went next day to 'Glascow, which is the second city in Scotland, fair, large and well built, cross-wise, somewhat like unto Oxford'; saw the cathedral—'they call it now the High Kirk and have made in it two preaching places, one in the choir and the other in the body of the church'[3] —and the college, 'a pretty stone building, and not inferior to Wadham and All Souls colleges in Oxon.'. On 23 August they turned south, down the Clyde to Hamilton and on to Douglas: 'A pitiful poor place, scarce an house in it that will keep a man dry in a shower of rain';[4] and on the 24th 'over much hilly ground, the highest place called Anderkin Hill',[5] which 'we judged to be higher than any we had been upon in England or Wales, Snowdon itself not excepted', and where as the *English Catalogue* records[6] Ray found the 'Cypresse-Mosse' (*Lycopodium alpinum*), past 'lead mines at a place called the Lead Hills', to 'Dumfries or (as they spelled it) Drumfrese'. Here they spent the Sunday, and Ray describes the service and was much impressed by the devotion of the people to their church. On 26 August 'we bad farewell to Scotland', arrived at Carlisle and went on through 'Pereth' (Penrith) to Shap. There Ray was able to do some real botanising; and the *Itinerary* closes with a list of some of the plants that he found. Some of these he had already seen in Wales in 1658:

1 *Mem.* p. 156. 2 *Mem.* p. 157.
3 *Mem.* p. 158. 4 *Mem.* p. 160.
5 Now Lowther Hill. The name survives in Enterkinfoot.
6 *C.A.* p. 214: Ray was much attracted by the Club-mosses and described five British species in 1670, two of them being then new: he adds, 'these five mosses are elegant and notable plants and deserve a nobler title'.

some must have been seen on his journey of the previous year: one which he calls Sedi quaedam species flore luteo (*Saxifraga aizoides*) he reports from Shap as a species probably new to science.[1]

On 7 September, having 'travelled about 700 miles in all',[2] he returned with Skippon to Cambridge, found the town distressed by 'an epidemical sickness in the nature of a fever which cuts off many old persons and children', and a week later wrote to Willughby, who had spent the summer at Oxford, the letter describing his travels.[3] In it he mentions the four new plants, *Gentiana pneumonanthe, Saxifraga aizoides, Anthemis tinctoria* and *Rumex maritimus*, and encloses a little branch of a fifth which he had found at Huntingdon, presumably on his homeward road. This is interesting as showing the difficulties of the botanist in those days. He writes of it: 'I found at Huntingdon a plant which the last year I observed not far from St Neots, coming to wait upon you, which puzzles me sore: it is between a Grass and a Caryophyllus, I know not what to call it unless it be Polygono angustissimo folio affinis C.B. but I cannot find that described anywhere'.[4] It is hardly surprising that Babington, who supplied the modern scientific names for the plants mentioned in Lankester's volume and who knew Ray's botanical work very well, could not give an identification here. But in fact the plant is unmistakably placed in the *English Catalogue* under Auricula leporis minima J.B. The Least Hare's-ear or, as it is now styled, *Bupleurum tenuissimum*, Linn.: for in giving localities Ray writes: 'As near Eltisley in the road from Cambridge to St Neots: on a bank by the Northern-roadside a little beyond Huntingdon. At Maldon in Essex in a yard where they build vessels at Full-bridge: at Hastings in Sussex, near the little brook that runs beside the Castle below the bridge, and elsewhere. It flowers and sets seed in August and September'.[5] As the plant is not only difficult to identify, but is normally a seaside species and has never since been reported from Ray's stations, it is not surprising that he hesitated about it: but in view of the fact that he afterwards found it in its normal habitat we may assume that his puzzle was accurately solved.

The two letters to Courthope, both written in October,[6] reveal something of the struggle of his conscience in refusing the living of Kirkby Lonsdale, with all its possibilities of security and botany, and in deciding to decline subscription to the Bartholomew's Act and forfeit his fellowship

1 *C.A.* p. 279: he describes it, and adds 'it is not very different from the small yellow mountain Sengreen of J. Bauhin and Parkinson'.

2 So Worthington to Hartlib, with an account of the journey: *Misc.* p. 293; *Diary*, II, p. 50.

3 Cf. above: it is *Corr.* p. 3.

4 *Corr.* p. 3: it was then in seed, *Hist. Plant.* I, p. 474.   5 *C.A.* p. 37.

6 *F.C.* pp. 22–5. The second is not dated: but the reference to his pupil Tim Burrell implies that the Michaelmas term had just begun. From the letter of 26 Nov. it appears that the young man came up about a month late.

and offices at Trinity. But anxiety did not prevent him from planning further efforts in natural history. 'I do intend', he writes in the second letter, 'the next year "Deo vires et valetudinem largiente" to accompany Mr Willughby into Wales; and if you could so order your affairs as to find time enough to go along with us, there could in my opinion nothing be added to the pleasure and contentment of such a journey. I cannot entertain myself better than to meditate on such a congress.' Poor man, during those months of crisis he needed the refuge of anticipation. Courthope invited him to Danny when he should have given up his work as steward at the end of the Michaelmas term;[1] he spent some time there and went on to Walter Burrell's at Cuckfield.[2] There he discovered a number of plants, *Veronica beccabunga, Cardamine flexuosa, Arabis thaliana, Lathyrus montanus*,[3] *Conopodium denudatum, Luzula pilosa*, and one which he had found and misnamed in the north, apparently *Stachys arvensis*. Leaving Sussex he spent some days in London seeing Morison's work in St James's Park, although Morison himself was away; perhaps meeting Willisel, then employed by him; and inspecting Edward Morgan's garden.[4] But before the end of April he was back for his last term in Cambridge.

There he undertook his final search for fresh additions to the Flora. 'Since I returned hither I have made a ride as far as Kingston Wood in quest of plants. There I discovered, what I never before saw in its pride, growing wild Herba Paris [*Paris quadrifolia*] in many places, but also in Eversden wood in great plenty. I found also there that sort of hairy wood-grass[5] of which I desired T. Burrell in my last to preserve me a pattern, so that I now can ease him of that trouble.'[6] A week later he could add: 'I have been out again in pursuit of plants as far as Gamlingay; there I discovered some that I have elsewhere found in England, others that I never saw before.[7]...I have a collection now of more than forty plants growing wild in Cambridgeshire, more than we have put down in our catalogue.'[8] This work no doubt completed the list contained in the Appendix which he published in the next year, when he wrote on 16 February to Courthope:

1 *F.C.* p. 27.

2 In his *Collection of English Words*, p. 129 (published 1674) Ray speaks of him as 'one of the chief iron-masters in Sussex'; and as 'Esquire, deceased'.

3 Ray never seems to have been sure of the identity of this plant, found near Poynings; cf. *Syn. Brit.* (2nd ed.), p. 187; and Wolley-Dod, *Flora of Sussex*, p. 130.

4 Letter to Courthope from Cambridge 28 April (*Early Science in Cambridge*, p. 375), and cf. Worthington to Hartlib, *Diary*, p. 344, though this probably refers to a similar visit in the previous year. Morgan also grew some of Willisel's plants, cf. *C.A.* p. 206.

5 *Luzula pilosa*, which he had already found at Cuckfield. 6 *F.C.* p. 27.

7 The list is printed by Gunther, 'Letters of Ray to Courthope', in *Journal of Botany*, LXXII, p. 221.

8 *F.C.* p. 29.

'I am intending this spring before I go over to prick a sheet by way of appendix to the Catal. Camb.'[1]

This Appendix[2] of thirteen pages, printed by John Field for the Cambridge bookseller William Morden, actually contains thirty-seven species, two or three being varieties and excluding two mosses and two liverworts; and the list deserves examination as showing both the care of his search and the growth of his knowledge. There are four or five in which he has learned to discriminate their specific distinctions: *Euphorbia peplus* he had already indicated as different from *E. helioscopius*, though in the *Catalogue* he did not list it separately: *Carduus crispus* may also have been included already as a form of *C. lanceolatus*: *Ranunculus acris* and *Cerastium viscosum* he had no doubt confused with their nearest neighbours; and perhaps the same is true of *Ranunculus hirsutus* and *Hypericum hirsutum*. Several are not uncommon, but are easily overlooked, *Arenaria trinervia*, *Galium tricorne*, *Asperula cynanchica*, and the two he claims to be previously undescribed, *Ranunculus parviflorus*[3] and *Carex remota*; or, if conspicuous, are local, *Petroselinum segetum*, *Malva moschata*, *Calamintha nepeta* and *Carex pendula*. Others are confined to the corner of the county at Gamlingay: *Sisymbrium thalianum*, *Trifolium subterraneum*, *Montia fontana*, *Solidago virgaurea*, *Gnaphalium sylvaticum*, *Hieracium murorum* and *Blechnum spicant*. Others are very rare and in some cases now extinct: *Helleborus foetidus*, 'in some hedges of a close near to the moor at Cherry-hinton', where it continued until Babington's time; *Lathyrus nissolia*, 'near Hadnam (Haddenham) in the Isle of Ely', but hardly reported since; *Geum intermedium*, 'this was found in the fields somewhere about the Town and brought into our gardens: I do not remember the place'; *Bupleurum tenuissimum*, which Babington here identified without hesitation; *Hypochaeris maculata*, 'on Gogmagog hills and Newmarket heath'; *Limosella aquatica*, 'in a dried pond'; *Ophrys aranifera*, 'in an old gravel-pit near Shelford by the footway from Trumpington to the church we found hundreds of them'; *Herminium monorchis*, 'in the chalk-pit close at Cherry-hinton pits'; *Ruscus aculeatus*, 'about Anglesey Abbey'; and *Scolopendrium vulgare*, 'on the walls of Hinton church'. The Appendix ends with four pages of emendations, mainly devoted to correcting synonyms given in the *Cambridge Catalogue*. He withdraws *Artemisia campestris* in both its entries: two of the four entries of 'Anagallis aquatica', saying that he had inserted them on the insistence of Mr Nidd; the second *Echium*; the duplicated *Vicia cracca*; and the narrow-leaved *Carduus*

1 In Gunther, *Early Science in Cambridge*, p. 349.
2 A second edition of it was printed in 1685 with a second preface stating that Dent and others have added plants bringing the total of additions to *C.C.* up to sixty: it is not Ray's work, nor is it mentioned in his letters.
3 Now rare, but apparently not uncommon when the country was less shorn.

lanceolatus. He also inserts a long note on *Galium mollugo*; an acknowledgment of error as to *Origanum vulgare* at Cherryhinton; and identifies
his Trinity College *Sagina* with Johnson's Pearlwort. The trivial character
of the alterations is proof, if this were needed, of the quality of his work.

In fact he only stayed in Cambridge for a fortnight or so before beginning the journey of 1662 which was both from the personal and from
the scientific standpoint a most important event in his life. He started with
his mind tolerably convinced as to the impossibility of returning to Cambridge, and in consequence with the future open to him. He had written
on the eve of it to Courthope: 'I hope I may say it without hesitation, I am
deeply sensible of, and most kindly affected with, every courtesy done me,
every civility shewn me; this is the best quality that ever I perceived myself to have':[1] but he was a lonely and reserved man, who had lost the
closest friend of his own age; and it was not clear how far these pupils of
his, young men of a different social status in a day when barriers of class
were high, would accept him. In the same letter to Courthope he had said:
'At Middleton Mr Skippon meets me; I hope his company will not be
unacceptable to Mr Willughby; however, I know not how to reject him.'[2]
The journey might easily prove a failure: it proved in fact a very great
success. Willughby, who had taken his degree in 1655, the year in which
Skippon had come into residence, liked the younger man at once and their
friendship lasted.[3] Ray and Willughby cemented the partnership which
gave Ray his life's work and Willughby his title to fame: before the
journey was over the plans had been made which Derham heard from
Ray's own lips when he visited him at Black Notley on 15 May 1704.
'These two gentlemen,' he writes, 'finding the History of Nature very imperfect, had agreed between themselves, before their travels beyond sea,
to reduce the several tribes of things to a method and to give accurate descriptions of the several species from a strict view of them. And forasmuch
as Mr Willughby's genius lay chiefly to animals, therefore he undertook
the birds, beasts, fishes and insects, as Mr Ray did the vegetables.'[4] No
doubt when Willughby had gone to Oxford in 1660 his own intention to
study natural history had been fixed: no doubt the journey to the north
and the Isle of Man that year with Ray had suggested co-operation: but it
was the tour of 1662 which transformed a common interest into a settled
and defined purpose and led to its fulfilment in the continental travel of the
following years and to the amazing series of books in which the surviving
partner discharged their mutual obligation.

Of the journey the various stages and observations are recorded in the
third and longest of the published *Itineraries*. In this, unlike the others, the

1 *F.C.* p. 29.   2 *F.C.* p. 28: cf. *Early Science in Cambridge*, p. 375.
3 Skippon was one of Willughby's executors.
4 *Mem.* p. 33.

notes on flora and fauna are full and continuous. Willughby was in control for the first and longer period, his energy and enthusiasm were given full scope with the result that visits to famous churches and castles give place to expeditions, in those days quite unprecedented, to little known islands and bits of coast where sea-birds and rare plants were the sole objects of interest. The two friends shared a conviction to which their whole work testifies and which Ray maintained until his death,[1] that, for the naturalist, museum studies and the literature of the subject must be subordinate to first-hand knowledge of the organism in its habitat and wild environment; that classification, important as it was, must take into account the way of life, the function as well as the structure; and that observation of growth and behaviour was of primary value. Few biologists to-day would dis-agree with this belief or refuse to admit that the neglect of it—neglect for which Linnaeus must take a large responsibility[2]—had a damaging effect upon the development of science. The worth of Ray's work and the wis-dom and foresight of his judgments spring largely from his resolve to see the living plant or animal and study it not in a herbarium or a glass case but in the field. The neglect of him is due to the exaggerated emphasis upon problems of taxonomy and the examination of dead specimens which obsessed his successors. We are only beginning to recover from the pre-judice against 'field-naturalists' and to recognise that biology if it is to be true to its name must return to Ray's methods and example.

On Ascension Day, 8 May 1662,[3] Ray left Cambridge for North-ampton. 'Upon the walls I saw growing Saxifraga alba [*Sax. granulata*], Hypericum vulg. [*H. perforatum*], Calamintha vulg. [*C. officinalis*] and Cotyledon vulg. [*C. umbilicus*].' Thence he went by Coventry to Middleton to join the others. On the Monday they started for their tour, going to Sutton Coldfield, where they found *Botrychium lunaria* 'in great plenty'; along the Trent with *Corydalis claviculata* to report; to Stafford, Nantwich, Beeston, where 'we could not find any of those plants which in *Phytol. Brit.* are mentioned to grow wild there, and suppose none such are to be found', Chester, Wrexham, where he found a new Sedge,[4] 'the Molde', Holywell and Denbigh. The Sunday gave them time to botanise; and 'on a bushy hill on the north-west side of the town he found *Hypericum montanum*[5] and *Lithospermum purpureo-caeruleum*',[6] both new to him, 'besides a great many of such as I had elsewhere found', *Tilia parvifolia*,

---

1 Cf. his complaints to Sloane that dried plants however well preserved are of little use without the knowledge of habit, growth and exact locality which 'those that gathered them might easily have given an account of' (*Corr.* pp. 362, 364–5, etc.); and to Petiver that he can no longer visit gardens or do field work (*Corr.* p. 389).

2 Cf. H. T. Pledge, *Science since 1500*, pp. 88–9.

3 A pleasant essay, 'In the footsteps of John Ray', describing part of this tour is in J. Vaughan, *Lighter Studies of a Country Rector*, pp. 21–9.

4 Cf. *C.A.* p. 147.　　　　5 *C.A.* p. 174.　　　　6 *C.A.* pp. 197–8.

Hypericum androsaemum, Potentilla argentea and one which at the time he identified as *Arabis stricta* but afterwards recognised as *Arabis hirsuta*.[1]

On 19 May they went through Bettws to Conway and on over Penmaenmawr, where Ray found *Botrychium lunaria*[2] and an unnamed Sedum, to Bangor. Next day they rode to Carnedd Llewelyn, but 'had not time enough to search the rocks and so found no rare plants there, only Cotyledon hirsuta[3] [*Saxifraga stellaris*] which grows plentifully also upon Snowdon hill'. Proceeding to Beaumaris they crossed next day to Prestholm, now usually called Puffin Island, 'in which we took notice of the ruins of St Sirian's chapel; the tower is yet standing'. There was *Smyrnium olusatrum*[4] 'in great plenty', *Cochlearia officinalis, Crithmum maritimum, Beta maritima* 'and a small sort of Geranium[5] which I had before observed in the Isle of Man' (*Erodium maritimum*), this last a species new to science. Of the birds of the island and of the rest of the tour we shall speak hereafter: they, like the plants, seem to have been the same then as now. Next day they explored the coast of the straits from Llanidan to 'Aber Menai Ferry', finding *Inula crithmoides*,[6] *Asplenium marinum*[7] and *Statice limonium* (he recognised only one species of Sea Lavender, but stated that the plants growing in salt-marshes were larger, those in cliffs and fissures of rock smaller[8]) and a plant which he calls Hyacinthus autumnalis minor, but which he recognised later and in the *English Catalogue* as *Scilla verna*; and on the beaches near the ferry *Diotis maritima*[9] and *Matthiola sinuata*,[10] and also, according to the *Catalogue, Cochlearia danica*.[11] This last he declares to be in his opinion specifically distinct from the large *C. officinalis*; 'for although when transferred to gardens and propagating itself by seed it becomes larger in size, yet it never approximates in any degree to the dimensions of the other species'—a point of some interest as showing the care that Ray took not to multiply species without real cause.

Spending the week-end at Carnarvon and again finding the *Erodium* they went on 26 May to Llanberis and so to Beddgelert. 'By the wayside, near the upper end of Llanberis pool, we saw growing wild Papaver erraticum luteum Cambrobritannicum[12] [*Meconopsis cambrica*], and near the stone tower there a species of Orchis palmata[13] with an odorate flower like to Monorchis',[14] this being of course *Orchis albida*, a plant then undescribed. Next day on Snowdon itself he found *Allosorus crispus*, 'which

---

1 Cf. Babington's note in *Mem.* p. 167 and Ray, *C.A.* p. 38 under Barbarea muralis : 'this I previously confused with Cardamine pumila' (i.e. *A. stricta*): cf. *F.C.* p. 190.
2 *C.A.* p. 199.　　3 *C.A.* p. 85.　　4 *C.A.* p. 167.
5 *C.A.* p. 133.　　6 *C.A.* p. 86.　　7 *C.A.* p. 113.
8 *C.A.* pp. 194–5.　9 *C.A.* p. 136.　10 *C.A.* p. 192.　11 *C.A.* p. 76.
12 *C.A.* p. 27: he notes that it is a perennial, a thing peculiar to this species among poppies.
13 *C.A.* p. 227.　　14 *Mem.* p. 170.

we had before observed in Westmorland', and, according to the *Catalogue, Saxifraga hypnoides*,[1] *Sedum roseum*[2] and four of the Club-mosses.[3] Going on to Aberdaron and Bardsey he found only the plants already noted, but recognised that the Squill had previously been misnamed. On Saturday they rode to Harlech and found *Juncus acutus*,[4] 'of which I suspect *Phyt. Brit.* makes three plants', in the marsh—a locality in which a few plants of it are still to be found, though *J. maritimus* abounds.

The next week's journey, Aberdovey, Plynlimon, Cardigan, 'a poor village called Fishgard where we were put to it for a lodging', St David's and the Head, Haverfordwest, is briefly reported and seems to have been rather unprofitable. The *Catalogue* lists *Antennaria dioica* as 'on the top of Plimllimon-hill':[5] but that is all. At Haverfordwest 'by a little river, which is kept up by banks to drive a mill, grows in great plenty Cyperus etc. q.': this Babington identifies as *Scirpus maritimus*:[6] but in the *Catalogue*, under Cyperus gramineus J.B., this exact locality is given and also 'several places in Warwickshire as in the ditches about Solihull'.[7] Ray's own name for *S. maritimus* is Gramen cyperoides palustre panicula sparsa.[8] The Haverfordwest plant is fully described in *Synopsis Britannica*[9] as Cyperus gramineus and is therefore *Scirpus sylvaticus*, Linn. Here, too, 'in a close near the castle' he noted *Medicago maculata*.

Pembroke yielded nothing and Tenby only a long list of fishes. But on Caldey Island and 'a little island between that and the mainland' (St Margaret's) there was 'great plenty of fowl' and some plants, *Lavatera arborea*[10] 'in great plenty', *Inula crithmoides, Scilla verna*, 'a kind of Tithymalus', probably *Euphorbia paralias*, and others. At Kidwelly 'on the sandy meadows near the town I observed six sorts of rushes', and he mentions *Juncus acutus* and *J. maritimus*; near Aberavon *Senecio palustris*; at Crick, near Caerwent, *Vicia sylvatica*[11] and 'Erythrodanum', that is Madder, *Rubia peregrina*; and at Tintern *Paris quadrifolia* and *Polypodium dryopteris* 'in shady lanes near the Abbey';[12] and so to Gloucester and its cathedral, where they spent the week-end, 14–16 June. On leaving it the party broke up, Willughby going to Dursley, where he found some coins, and then to Malvern, where he fell ill;[13] and Skippon going on with Ray to Alderley and its fossils, and Bristol.[14]

1 *C.A.* p. 279. 2 *C.A.* p. 264. 3 *C.A.* pp. 214–15.
4 *C.A.* pp. 179–80. 5 *C.A.* p. 137. 6 *Mem.* p. 174. 7 *C.A.* p. 90.
8 So *C.A.* p. 147, *C.C.* p. 66, and Babington, *Flora of Cambs*, p. 254.
9 *S.B.* II, p. 271; cf. *H.P.* II, p. 1301: it is the Gramen cyperoides miliaceum of Bauhin's *Pinax* and this is Scirpus sylvaticus of Linnaeus, *Species Plantarum*, I, p. 75.
10 *C.A.* p. 206. 11 *C.A.* pp. 315–16. 12 *C.A.* p. 95. 13 *Corr.* pp. 5, 6.
14 The fact that Skippon and Ray went on alone is not told in the *Itinerary*; and the occasion of Willughby's letter (*Corr.* p. 5) has been obscure. Willughby's departure is told in a letter to Courthope dated 24 July 1662, published in Gunther, *Early Science in Cambridge*, p. 376.

The Avon gorge had long been a botanist's paradise—Turner in 1562, De l'Obel in 1569, Johnson in 1634 and Goodyer in 1638[1] had visited it— and when on 19 June Ray visited St Vincent's rock he 'found several plants more than we had formerly taken notice of'.[2] *Hypericum montanum*, *Hippocrepis comosa* and *Calamintha acinos*, the first three that he mentions, hardly come into this category: but 'Sedum medium' is recognised in the *Catalogue*[3] as a distinct and new species, and is clearly *Sedum rupestre*; and the 'new species of Fern' becomes 'Asplenium sive Ceterach'[4] (*Ceterach officinarum*). In the *Itinerary* he does not mention three of his best finds: *Erodium moschatum*, 'near Bristow on a little Green you pass over going thence to St Vincent's rock',[5] *Scilla autumnalis*, which he afterwards found at the Lizard,[6] and *Trinia glauca*,[7] the most characteristic of them all.

Going on to Keynsham, Bath and Paulton on 20 June Ray reports two species of Spurge: one, which in the *Itinerary* he describes as 'in the cornfields about Camerton'[8] but in the *Catalogue* as 'not far from Kinesham',[9] is *Euphorbia platyphyllus* or *E. stricta*. The other, 'a kind of Tithymalus resembling at first sight the common sun-spurge, but it grows up to a great height, the leaves of a pale green or yellow like Helioscop.; of the figure well near of wood spurge; the stalk sometimes reddish', he apparently failed to identify; for there is no reference to it in the *Catalogue*. It is tempting to suggest that it was *E. pilosa*, which until lately has been regularly found in this exact neighbourhood. Next day they rode to Wookey-hole and Wells and Glastonbury, 'where once again[10] I visited the Torr'. On the ascent and top of Torr-hill he found an *Orobanche*, which he describes with unusual detail and evidently failed to identify: Babington says that it needs examination and suggests *O. barbata*, a form doubtfully distinguishable from *O. hederae*, which is very frequent in Somerset. Similarly, in the marsh he found an unknown 'small umbelliferous plant the leaf somewhat like to one sort of Oenanthe; the flowers very small and consisting of five little white pointed leaves; the umbel, for the most part, had but two spokes of flowers; the seeds striate, a little compressed and but short; the stalk hollow, the root stringy. This is the water parsnip, though it may perhaps be referred to the Oenanthe'. Babington suggests *Apium inundatum*, and this is obviously right; for in the *Catalogue*[11] under Sium minimum there is a full description of the plant, which Ray regards as a species new to science. 'On the top of the hill by the Torr we found a

1 Cf. Gunther, *Early British Botanists*, pp. 75–6.
2 *Mem.* p. 180. 3 *C.A.* p. 280.
4 *C.A.* p. 33. 5 *C.A.* p. 132. 6 *C.A.* p. 172.
7 *C.A.* p. 240. 8 *Mem.* p. 180. 9 *C.A.* p. 299.
10 Here and in the account of Stonehenge (*Mem.* p. 198) is an indication that Ray had previously visited the West country.
11 *C.A.* p. 286.

kind of Vicia[1] with a long white flower'—*Vicia hybrida,* one of his rarest discoveries, for which this was until last century the most famous locality. Finally, from the marshes he reported the radiate form of *Centaurea nigra.*

On the 22nd, between Street and Walton, 'in a close where they digged stone' he found *Trifolium squamosum,*[2] and between Walton and Taunton, 'by a stony lane in the hedge, Lithosperm. minus'. This Babington identifies as *L. officinale*—though Ray has a regular and different name for this plant. Reference to the *Catalogue* shows that L. minus is a synonym for L. flore purpureo, and that its localities are 'near Denbigh-town, also in Somersetshire, not far from Taunton':[3] it is *L. purpureo-caeruleum.* From Wellington to South Molton and Barnstaple they found nothing of importance, but near Bideford, 'all along the hedges in great plenty', *Rubia peregrina,*[4] and between Bideford and Kilkhampton, 'near the rivulets among the moors', *Campanula hederacea,*[5] *Narthecium ossifragum, Scutellaria minor, Hydrocotyle vulgaris* and 'a small sort of Pinguicula which seems to differ specially from the common' (*P. lusitanica,* Linn.): this last is described in the *Catalogue*[6] as a species new to science. On the road to Launceston the abundance of Royal Fern (*Osmunda regalis*) was noted, and a further find of *Campanula hederacea.* But discoveries only came at St Columb 'by the way side in several places' of a plant which he describes in the *Itinerary*[7] as Euphrasia pratensis lutea, but which in the *Catalogue*[8] he named E. lutea latifolia palustris and, claiming it to be not yet described, characterised with care and precision—it is certainly *Bartsia viscosa,* a species confined to the West of England and Ireland; and of another 'which we guess to be Alsine palustris minor serpillifolia' but which in the *Catalogue*[9] is evidently the modern *Illecebrum verticillatum.* 'Camomile',[10] he adds, 'grows in such plenty along the way sides that one may scent it as one rides.' At the quay in Truro—'an indifferent good key'—'grows plentifully a kind of Thlaspi', which is clearly *Lepidium ruderale;*[11] and along the cliffs towards St Ives *Foeniculum officinale* and *Linum bienne.*[12]

On 1 July they rode to the Land's End and on the way found two new and small plants, 'on a moist bank' *Sibthorpia europaea,* then undescribed and not classified by Ray till after 1670,[13] and *Cicendia filiformis* 'on a boggy ground': this he described both in the *Itinerary*[14] and in the *Catalogue*[15] with his usual precision; it also was new to science. He gives the names of some

1 *C.A.* p. 316.   2 *C.A.* pp. 305–6.   3 *C.A.* pp. 197–8.
4 *C.A.* p. 267.   5 *C.A.* p. 52.   6 *C.A.* p. 244.
7 *Mem.* p. 186.   8 *C.A.* pp. 107–8.   9 *C.A.* p. 248.
10 *Anthemis nobilis,* Linn., *C.A.* p. 67.
11 *C.A.* p. 296.   12 *C.A.* p. 196.
13 It appears as Alsine spuria pusilla repens foliis Saxifragae aureae in *C.A.* ed. II, pp. 17–18.
14 *Mem.* p. 189.   15 *C.A.* p. 63.

of the rocks, the Longship, the Armed Knight, 'which they told us fell down about the time the king was beheaded', and the Spanish Lady; heard of one man, Mr Dickan Gwyn, 'who could write the Cornish language'; and recorded a number of fishes and sea-weeds. 'On the beach near Pensance we saw growing many rare plants', *Diotis maritima, Euphorbia peplis*[1] and, according to the *Catalogue*,[2] 'Polygonum marinum', that is *Polygonum raii*; and near Penryn along the hedges *Linaria repens*.[3] At Falmouth, 'a newly erected town', he has nothing to record except a libellous anecdote to explain its older name of Penny-come-quick,[4] and neither Bodmin nor St Neot produced any plants. 'To the south of Saltash' he found *Melittis melissophyllum* 'growing in great plenty': but the week-end there is famous for a description 'of a hurling-play much used in Cornwall', of which 'there are two kinds, the in-hurling and the out-hurling'—a description which should be of interest to the student of the pedigree of football.

On 7 July they entered Devon and 'on the hill which you ascend after you are come over the passage to go to Plymouth' found *Eryngium campestre* in plenty in the spot where it still survives: Ray adds: 'I do not remember to have seen it anywhere else in England.'[5] At Totnes they again found *Melittis*:[6] but Exeter, Sherborne and Salisbury, Amesbury, Winchester and Farnham yielded no records, except antiquarian and ecclesiastical. On 18 July, at Windsor and Eton—'We saw the college which is somewhat inferior to Winchester college'—the *Itinerary* ends.

Ray seems to have gone from London on 24 July when he parted with Skippon to his mother in Essex and to have made an exploration of the Essex coast and river valleys, Maldon, Wormingford, and Colchester.[7] From Notley he wrote the letter to Courthope (no. 11 in the Danny papers[8]) which is obviously misplaced by Dr Gunther, and in which he announced his intention of staying there till Bartholomew Day, the fatal date on which he had to subscribe or forfeit his fellowship, was past. On 28 August he followed this with a longer letter[9] in which he reveals the strength of his feelings and anxieties, cloaking his emotion in the dignity and reticences of Latin. He had had a visit from 'Mr Barnett' (this must be a slip for Barnham) and had promised to go to him at 'Bocton Munchelsey' (Boughton Monchelsea, four miles south of Maidstone), 'though I must confess, were I free I should in many respects prefer Danny'. On 30 August he set out for Cambridge to settle his affairs and dispose of his

1 *C.A.* p. 237. 2 *C.A.* p. 249. 3 *C.A.* p. 195.
4 I.e. Pen y cwm guig: cf. Pope, *Life of Ward*, p. 57, *Sir T. Browne's Works*, I, p. 140: the change was ordered in 1660, cf. *Registers of Falmouth*, I, p. vii.
5 *Mem.* p. 195; *C.A.* p. 105. 6 *C.A.* p. 187.
7 He obtained *Thlaspi arvense*, *C.A.* p. 295 and *Lepidium latifolium*, *C.A.* p. 191, at 'Wormingford near the river Stour' and 'the Hythe at Colchester', where they still abound. 8 *F.C.* p. 25 : its date is 13 Aug. 9 *F.C.* pp. 29–31.

goods. He spent only a few days there; went down to Sir Robert Barnham's in Kent;[1] then to London; and so in the middle of October to a tutorship in the house of Mr Thomas Bacon, Friston Hall, near Saxmundham, where he stayed until the following March. Possibly to this autumn may be assigned some of the plants which bear localities in Kent near Maidstone, Sittingbourne, or Dartford—though the date of Ray's visit is too late in the year for *Neottia nidus-avis*, 'in Cantio prope Maydston',[2] or for the two plants reported from the road between Deptford and Eltham, *Myosurus minimus*[3] and *Trifolium subterraneum*.[4] But at Friston Hall he certainly did some collecting. Saxmundham itself gave him *Thlaspi arvense*[5] and, in the neighbourhood, a species never before described, *Trifolium glomeratum*.[6] His first letter from Suffolk to Courthope describes how on 1 November he 'rode forth to Aldburgh to see those famous Sea Peas, noted by our historians to grow between Orford and Aldburgh upon the shingle or beach of stones by the seaside': and in the *Catalogue* he recounts the narrative quoted by Gesner from a letter of Dr Caius[7] and narrated also by Stow and Camden how in 1555, the year of famine, the inhabitants were fed by the miraculous appearance of these peas. Ray will have none of the miracle, though he will grant that they may have produced a good crop of seed that year: but neither will he hear of Camden's suggestion that the peas are sprung from seeds out of a wreck; for they are good perennials and wholly unlike garden peas. He found also at Orford *Medicago hispida*,[8] 'with a small fruit, somewhat like the nave of a cart-wheel'; at Aldeburgh *Urtica pilulifera*;[9] and possibly on a later exploration of the coast, from Dunwich to Yarmouth, *Sedum anglicum*[10] and *Convolvulus soldanella*,[11] and 'in Lovingland, just over the water from Yarmouth', *Frankenia laevis*.[12]

The work in Suffolk was avowedly a stop-gap. When the offer of it was first made Ray had written: 'I shall balke it…because I have the design of

1 Robert was the eldest son of Sir Francis Barnham, the Parliamentarian who inherited Boughton Monchelsea in 1613, cf. *D.N.B.* Robert opposed Cromwell and was imprisoned in 1649. He represented Maidstone in the Cavalier Parliament and was made a baronet in 1663. His son Robert is presumably the author of the curious letter to Ray (*Corr.* pp. 9–11). If so, he was plainly impressed with Ray's scientific skill: but of the origin of their friendship we know nothing. Derham, *Mem.* p. 22, describes him as a pupil of Ray, but of this there seems to be no other evidence.

2 *C.A.* p. 224.   3 *C.A.* p. 218.

4 *C.A.* p. 306.   5 *C.A.* p. 294.

6 *C.A.* p. 305: this is the identification of Ray's plant in Clarke, *First Records* (2nd ed.), p. 39: the description might apply to *T. suffocatum*, but see below p. 141.

7 Cf. Caius, *De Rar. Animal.* (ed. Venn), p. 63.

8 *C.A.* p. 302.   9 *C.A.* p. 321.

10 *C.A.* p. 280.   11 *C.A.* p. 289.

12 *C.A.* p. 101. Probably these Yarmouth plants were first found on the occasion of the visit to Thomas Browne at Norwich in 1660: cf. *C.A.* p. 79.

travelling hot in my head':[1] and when he got to Friston he declared: 'I have not, nor will I engage myself any longer than till Annunciation next, that I may then be free to wait upon Mr Willughby.'[2] Willughby had already proposed a tour on the Continent in pursuance of their plans for the study of natural history; and they were both trying to persuade Courthope 'to embark in the same bottom' with them: Skippon was to be of the party; and though the 'King of France's designs' might restrict their visit to his country there was, as Willughby argued, 'enough of the world that won't be in his power to disturb'. In fact Ray and Skippon met at Leeds Abbey, Sir William Meredith's house in Kent, where Ray discovered a water-weed in the cistern;[3] and Willughby with Nathaniel Bacon, probably not the son of Ray's host at Friston, but his contemporary and Ray's junior at Trinity,[4] met them at Dover, whence they sailed on 18 April: it is pleasant to suppose that the entry in the *Catalogue* as to the Wild Cabbage (*Brassica oleracea*) 'on Dover cliffs' was due to this occasion.

The record of the tour thus begun was published by Ray in 1673 in the volume entitled *Observations topographical, moral and physiological, made in a journey through part of the Low-countries, Germany, Italy and France: with a Catalogue of Plants not native of England* and reprinted as the first volume of *Mr Ray's Travels* in 1738. In the Preface he naively declares that his own real concern is with the *Catalogue*, 'the number of plants found exceeding my expectation', but that 'considering the paucity of those who delight in studies and enquiries of this nature' he has added a narrative of the tour. This explanation accounts in part at least for the character of the book. It is a mine of information for the historian; for he gives not only full accounts of the government of many cities and the lecture lists of many Universities, but details as to buildings, factories, engineering works, churches and museums: but it is not exciting or particularly attractive. The fact is that he is not acutely interested in the matters that he records: they were put down at Willughby's request and published in order to sell the list of plants. Nor is his style and language racy enough to liven up the story or distinguished enough to give it literary value: he wrote Latin admirably with a delicate sense of rhythm, with dignity of period and felicity of phrase: the English of his letters is appropriate and singularly happy: but in this book he is (we must confess it) often dull and unin-

1 *F.C.* p. 32.   2 *F.C.* p. 33.   3 Cf. *C.A.* p. 12.
4 A Nath. Bacon, probably the son of Nicholas Bacon of Shrubland, Suffolk, was admitted as a pensioner in 1653. The Virginian patriot, also Nathaniel, was, according to *Strange News from Virginia*, a contemporary account of his life, the son of Thomas Bacon of Friston and the young man for whom Ray was responsible there. But he can hardly have been the travel-companion; for he was admitted to Gray's Inn on 22 Nov. 1664: cf. *D.N.B.* There are eight Nathaniel Bacons between 1561 and 1721 in Venn's *Alumni Cantab*.

spired. It must be remembered that he is to be judged rather by the standards of Baedeker than by those of *Eothen*. A still longer account was kept by Skippon and was published in 1732 in *A Collection of Voyages and Travels*.[1]

Calais, Gravelines, Dunkirk, Ostend on horseback; thence by boat to Bruges with a description of canal-locks and a reference to one 'upon Guildford river in Surrey',[2] and then a discussion of buried forests, 'found in places which five hundred years ago were sea'—'which yet is a strange thing considering the novity of the world, the age whereof according to the usual account is not 5600 years'.[3] So to Louvain and full details of the University; Antwerp and a list of plants in a priest's garden; Middelburgh, Breda, Dordrecht ('streets so clean that a man may walk them in slippers without wetting his foot'),[4] Rotterdam, Delft and the Museum of Jan Van der Meer[5] ('Dens Hippopotami—though it be a question whether there be any such animal'),[6] Leyden, its University and the bird grove at Sevenhuys, Haarlem, Amsterdam, Utrecht and another University lecture list; and a description not too flattering of the Dutch. 'Puddings and brawns are dishes proper to England'—'those great round cheeses coloured red on the outside'—'their strong beer (thick beer they call it and well they may)'—'their houses are kept clean with extraordinary niceness...nay some are so curious as to take down the very tiles of their pent-houses and cleanse them'—'in most of the cities there are a great number of chiming little bells which seldom rest: but for rings of great bells all Europe cannot show so many as England alone, so that it might well be called the Ringing Island'.

Hertogenbosch, Eindhoven, Maastricht, Liége ('near this city are gotten coals in great plenty but they lie very deep under ground; some of us went down into a pit 150 yards deep: they use for fuel round balls made of clay with a certain proportion of coal beaten small, tempered together and dried in the sun which they call Hotshots'),[7] 'the Spaw', Limbourg, Aachen with the 'iron crown and the sword of Charlemagne, as also the gospels said to be written by the Evangelists' own hands', Cologne ('this night we lodged at a pitiful poor walled town called Brisaca where we first began to have feather-beds laid upon us instead of blankets'),[8] Coblenz, and so up the Rhine 'in a boat drawn by men' and so slowly that 'we observed many plants in the cornfields, meadows, rocks, woods and sands by the river-side' of which a list is given.[9] And so to Mainz and Frankfort.

1 Vol. VI, pp. 359–736, with the title *A Journey through part of the Low Countries, Germany, Italy and France by Philip Skippon Esq: afterwards knighted, in company with the celebrated Mr Ray*, etc. This is useful for the later stages of their tour.

2 *Obs.* p. 3.   3 *Obs.* p. 8.   4 *Obs.* p. 23.

5 The apothecary, not his famous fellow-townsman, the painter.

6 *Obs.* p. 27: he learnt better, cf. *Syn. Quad.* pp. 123–5.

7 *Obs.* p. 58.   8 *Obs.* p. 74.   9 *Obs.* p. 77.

The journey so far had taken exactly three months, and on 17 July they left Frankfort on horseback for Worms, Speyer, Mannheim and Heidelberg. Here the great tun, two unicorn's horns which are 'the horns of a fish of the cetaceous kind, not of a quadruped as is vulgarly but erroneously thought',[1] and the University receive special notice. Thence by coach to Strasburg, Basle and its University, Baden, Zürich, Constance, Munich and Augsburg, where they spent a week and found many rare plants. So to Nürnberg and Altdorf, where finding some fossils Ray introduces a long and important digression in which he states his conviction that these 'were originally the shells or bones of living fishes and other animals bred in the sea'[2]—in those days an opinion by no means general; and thence, having met Moritz Hoffmann the botanist, to Ratisbon (Regensburg). On 11 September they hired a boat for Vienna, where they saw 'the Silurus or sheat-fish, the greatest of all fresh-water fish that we have seen', in the market, and the Emperor, 'of a mean stature and dark complexion, his under lip thick and hanging down a little, much like his effigies on his coin'.[3] And then by coach to Venice. At Schadwien 'our coachman hired ten oxen to draw his coach to the top of the hills' and Ray botanised; they proceeded by the upper Mur through Bruck and Knittelfeld to the Drave at Villach and so by Pontebba to the plain of Friuli; on 6 October they came down to Treviso and went on by boat.[4]

Nearly fifty pages are devoted to an account of the government of Venice, of the ludicrum or imp in a jar and of the making of soap: but they spent the winter at Padua, where Ray worked hard attending lectures in anatomy by Pietro Marchetti.[5] In February they went on to Vicenza, Verona, Mantua and by water to Ferrara and Bologna, where they visited the museum of Aldrovandi and the garden of Giacomo Zanoni, but did not find the great Malpighi at the University.[6] So, early in March, they came by way of Modena, Parma, Piacenza, Crema, Brescia and Bergamo to Milan, Turin and Genoa, where the coast yielded many plants; by boat to Spezia and so to Lucca, Pisa and Leghorn, where they stayed some days before sailing to Naples. There they visited Pozzuoli and Baiae,

1 *Obs.* p. 83: in this connection he quotes 'The History of the Antilles written in French by R.F. of Tertre', i.e. Jean Baptiste du Tertre, *Histoire générale des A.*, Paris, 1667–71.   2 *Obs.* p. 120. See below, pp. 422–6.

3 *Obs.* p. 141, cf. Edward Browne's description, *Journey* (London, 1685), pp. 140, 154.

4 The tour through Germany is relatively lacking in detail, especially in lists of plants and accounts of fishes and birds. Ray explains this in a letter to Robinson, *Corr.* pp. 165–6: 'I did describe most of the animals we met with in our travels; but all my notes of high and low Germany were unfortunately lost.'

5 See below, pp. 314, 373. For Ray's notes cf. *Phil. Trans.* XXV, no. 307, pp. 2283–303.

6 Malpighi's name occurs in the list of lecturers with a note 'abs. cum reser.', *Obs.* p. 232. He was professor at Messina 1662–6.

climbed Vesuvius, 'which seemed not to threaten any future eruption',[1] and were greatly impressed by the virtuosi of the academy: 'a man could scarce hope to find such a knot of ingenious persons and of that latitude and freedom of judgment in so remote a part of Europe and in the communion of such a church'.[2] One of them, a doctor from Cosenza, T. Corneli, wrote about the Manna Tree (*Fraxinus ornus*) in December to Ray at Rome.[3]

At this point the party divided, Willughby and Bacon staying in Naples and Ray going on with Skippon, who was his sole companion for the whole of the rest of the tour. Derham in his *Life of Ray* made a mistake in stating that it was at Montpellier more than twelve months later that Willughby left:[4] and this mistake has been copied by all subsequent writers. Skippon's narrative states the truth explicitly.[5] But in fact Derham himself printed in the *Philosophical Letters* clear evidence of the correct story. For on 5 June 1664 Skippon wrote to Willughby 'since we left you and Mr Bacon at Naples we have seen Messina and Malta':[6] and this is corroborated not only by Edward Browne's letter to his father describing his meeting with Ray and Skippon in Rome and their journey to Switzerland,[7] but by the fact that Willughby was back in England before the end of 1664 and reported the results of his journey to the Royal Society on 4 January 1665.[8] Willughby stayed for a time at Naples and then travelled to Rome,[9] where he may have stayed some months: for his account of his Spanish journey only begins at Bagnols on 31 August. This record, printed at the end of Ray's *Observations*, describes his tour[10] up to his arrival at Fuenterabia on 13 November; and a letter to Ray shows that he went straight back from there by Bayonne, Bordeaux, Poitiers, Blois and Orléans.[11] No doubt anxiety about his father's health had much to do with his desire to get back: but in view of the credit given to him for the

1 *Obs.* p. 275.   2 *Obs.* p. 272.

3 *Corr.* p. 6: the year must be 1665, not 1663—the error being doubtless a misprint in Derham, *Phil. Letters*, p. 9—for Corneli, cf. *Obs.* pp. 272, 410, and a letter of Sir John Finch to Prince Leopold from Naples in Nov. 1663, where Corneli is described as 'a mathematician and physician...a follower of Descartes and a great favourer of things new': cf. A. Malloch, *Finch and Baines*, p. 41.

4 *Mem.* p. 16.   5 *Journey*, p. 609.

6 *Phil. Letters*, p. 361. Skippon wrote to report that they had seen Pietro Castelli's manuscript book on Insects which Willughby had apparently wished to purchase: cf. *Journey*, p. 614 and *Corr.* p. 24.

7 Letter of 2 April 1665 (*Works of Sir T. Browne*).

8 Birch, *History of the R.S.* II, p. 3.

9 In *Hist. Plant.* I, p. 526 Ray records a plant found by Willughby on this journey.

10 The only reference to any Spanish observation is in *Hist. Insect.* p. 62 to a locust at Cardona—this having been visited 7–9 Sept.

11 *Corr.* pp. 7–9.

Histories of Birds and Fishes it is important to recognise that for the most fruitful part of the continental journeys Ray and Skippon were alone.[1]

On 2 May they arrived by sea at Messina: 'it is a proverb, A Messana Assai polvere, pulce et putane', but there and in Sicily they collected a great number of new plants; for, trying to get by boat to Malta, they had to wait several days owing to bad weather and in miserable conditions at Syracuse, Cape Passero and elsewhere along the coast. On 13 May, when the wind abated, they crossed to the island, taking several 'sea-tortoises' on the way, one of which had 'two great bunches of those they call Bernacle-shells growing to his back.... The opinion of a bird breeding in them, and of which Michael Meyerus hath written a whole book, is false and frivolous'.[2] Of Malta he gives a long account derived mainly from Giovanni Francesco Abela's *Malta illustrata*, 'written in Italian and published in Malta 1646',[3] but he adds a long list of fossils,[4] a mention of the Maltese dogs, now extinct, and cats, and of a few plants with a discussion of St Paul's viper: 'whether there be venomous beasts or no I am sure there are venomous insects here, the very biting or stinging of the gnats being more virulent than in other places: I do not remember that in England a gnat did ever cause a swelling in the skin of my face, but there it left a mark behind it that was not out for a month after'.[5]

From Malta they returned to Catania, whence on 20 May they ascended Etna,[6] up to the snow-line, and then to Messina, where they visited Reggio and saw the catching of sword-fish; took boat and coasted along Italy to Salerno, whence they revisited Naples; after a fortnight they embarked for Leghorn, landing each night and collecting many plants; and went on to Florence, where Ray had an attack of fever:[7] they spent the summer there while Willughby was making his tour in Spain. On 1 September they started south to Rome, where they stayed till 24 January 1665. 'Rome is a place not only well worth the seeing, but very convenient to sojourn in...the present Romans seemed to me in their houses and

1 Bacon left Willughby in North Italy and intended to rejoin Ray in Rome; but caught smallpox and only met them in Venice in March 1665 (*Journey*, p. 694). He left them for England at Geneva on 20 April.

2 *Obs.* p. 291: the book is Count Michael Maier's *Tractatus de Volucri Arborea*, Frankfurt, 1619; cf. J. Read, *Prelude to Chemistry*, pp. 232–3.

3 So Ray: this book was actually published in 1647 by Bonacota as *Della Descrittione di Malta*.

4 Skippon presented specimens of these fossils to the Royal Society: cf. Grew, *Musaeum*, pp. 256–7.

5 *Obs.* p. 313: cf. 'gnats called moschetti', Skippon, *Journey*, p. 623.

6 He refers to the great eruption of 1669 'since our being there'.

7 *Hist. Plant.* I, p. 646: he was treated with cucumber pulp by an English doctor there named Kirton, certainly John Kirton, M.D. of Padua, physician to Robert Dudley, Duke of Northumberland, whom he assisted in his chemical operations in Italy: Kirton was still at Florence 1673, cf. Foster, *Alumni Oxon.* II, p. 859 and Wood, *Fasti*, I, p. 467.

furniture, in their diet, in their manners and customs, and in their very
pronunciation, so liquid, plain and distinct, more to symbolise and agree
with us English than any other people of Italy.'¹ Besides accounts of the
buildings and archaeology he gives a long description of the wines sold
there and of the provision markets: 'Plenty there is of wild fowl of all
sorts...of small birds the greatest plenty I have anywhere seen...one
would think that in a short time they should destroy all the birds of these
kinds in the country: they spare not the least and most innocent birds
e.g. robin-redbreasts, finches, titmice, wagtails, wrens.' Fish are also
listed both from fresh water and from the sea; as also wild boars, venison
and the porcupine. His Histories of Birds, Fishes and Animals show how
rich a hunting-ground he had discovered. An interesting side-light is
thrown upon his activities in a letter of Edward Browne, whose party was
also in Rome at this time and who wrote on 16 January to his father:
'Mr Ray hath made a collection of plants, fishes, fowls, stones and other
rarities which he hath with him, and Mr Skippon besides a great many
which he hath sent home, though they had the ill fortune to lose one
venture with a servant of theirs who is now slave in Tunis.'² During this
visit Ray observed and checked the course of a comet from 20 till 29
December—observations which demonstrate an interest in astronomy and
were published after his death by the Royal Society.³

From Rome they travelled up the Via Flaminia through Narni, Spoleto,
Foligno, Ancona, Rimini, Ravenna, Faenza and Bologna to Venice; and
at this point in the record chapters⁴ on the 'Abilities, Manners and
Customs' and upon the 'Meats, Fruits, Sallets etc.' of the Italians, are interpolated. Some of these are of interest—'in Lombardy, Tartufale a kind
of subterraneous mushroom,...the way to get them is to turn swine into a
field where they grow who find them by the smell, and set one to follow
the swine and gather them up'⁵—'snails, especially the great whitish or
ash-coloured shell-snails which we have since found plentifully upon the
downs near Dorking in Surrey' (*Helix pomatia*, still found there)—
'several herbs which are not yet or have been but lately used in England,
Selleri which is nothing else but sweet smallage, fennel; artichoke they eat
raw; in Sicily at the highest village upon Mount Etna stalks of a tall prickly
thistle'—'Buffles are common beasts in Italy and they make use of them
to draw their wains'.

1 *Obs.* p. 368.
2 *Sir T. Browne's Works* (ed. Wilkin), I, p. 86: Skippon had apparently not sent his
plants; for there is mention of a dried specimen gathered in Germany and preserved by
him in *Hist. Plant.* I, p. 813.
3 *Phil. Trans.* XXV, pp. 2350–2: Edward Browne refers to the same comet: cf. *Sir
T. Browne's Works*, I, p. 84: so does Skippon, *Journey*, p. 669: it was noted at Cambridge
by Samuel Newton, *Diary*, p. 7 and in London by Pepys, *Diary*, 17 and 24 Dec.
4 *Obs.* pp. 392–411.   5 *Obs.* p. 403.

From Venice, by Treviso, Trent, and Bolzano, they entered the Lower Engadine by Tauffers and the Münsterthal, noting 'the peculiar language which they call Romansch'[1] and the chamois horns in the houses and passing Buffalora down to Zernetz and Zuoz and a 'little place called Ponte': 'our faces were so hacked and burned (if I may take leave to use that word) by the cold that we looked like so many gypsies':[2] Skippon's account makes it clear that the journey, especially after leaving Ponte, was very rigorous.[3] Crossing the Albula to Bergün they went on to Chur, Wallenstadt and down the lake to Wesen and Glarus and over the snow to Brunnen, 'beside the Lake of Lucerne'; then to Lucerne and by Zug, where he saw a 30 lb. pike and many other fish,[4] to Zürich. On 7 April they left and rode through Aarau to Berne ('here they keep five or six bears in a pit: these I observed to climb the fir-trees growing in the pit and delighting to sit on the tops of them like birds on a perch'),[5] to Fribourg, Lausanne and Geneva, where they arrived on 20 April, spent the whole of the summer and did much very successful botanising on Mont Salève and the Jura.[6]

In late July they went on into France, staying a few days at Lyons and at Grenoble, visiting the Grande Chartreuse and then proceeding by Orange and Avignon to Montpellier. This was then a great centre of intellectual life and botanical studies; and they met a number of Englishmen, William Croone, the famous doctor, Sir Thomas Crew,[7] whose pictures of birds Ray saw and afterwards mentions, Peter Vivian, a fellow of Trinity who must have known Ray at Cambridge, 'Dr Moulins a Scotchman', arrested shortly afterwards in Paris, who may be the James Molins, M.A. of Aberdeen, who was incorporated at Trinity in 1667, 'Mr Jessop', who must surely be the Francis Jessop of Broomhall with whom Ray stayed in 1668, Samuel Howlett, elected a fellow of St John's College, Cambridge, in 1664, and Martin Lister, also a fellow of St John's, who became one of his closest friends.[8] According to Skippon, Chiquenau (probably Chicoyneau) was professor of botany:[9] but Pierre Magnol, a young man but 'a better herbarist', made a deeper impression upon Ray.[10] He also met the great Nicolaus Stensen (Steno), the Danish scientist and geological

---

1 *Obs.* p. 412.   2 *Obs.* p. 417.

3 The winter of 1664–5 was exceptionally severe: Ray records that it damaged many thousand Olive-trees in Provence: cf. *Hist. Plant.* II, p. 1542. Skippon, *Journey*, p. 696, says that Ray lost his sight for some days and that his fingers were so benumbed as to be useless.

4 *Obs.* p. 430.   5 *Obs.* p. 431.
6 The list fills pp. 436–45.   7 Second Baron Crew of Stene, 1679–97.
8 For the list of English visitors cf. *Journey*, p. 714.

9 Presumably father of François (1672–1752), prefect of the gardens at Montpellier. Edward Browne refers to him as at Montpellier early in May, 1665, *Sir T. Browne's Works*, I, p. 103.   10 *Corr.* p. 134.

pioneer, who was then under thirty but already famous as an anatomist; and was present when he dissected an ox's head and demonstrated the existence of the 'ductus salivaris'.[1] In September they visited Frontignan and Martigues, and in December explored Provence, visiting Arles and the Camargue, spending a few days in Marseilles, where Ray paid special attention to the fish-market, and returning to Montpellier for Christmas. It is possible that in January they visited Bordeaux in company with Croone.[2]

On 1 February 1666 the French king ordered all Englishmen to leave France within three months. In consequence Ray, with Skippon, Vivian and others, left Montpellier on 26 February for Lyons, where they found Lister and Jessop. Ray went on with Skippon and Lister on 6 March to Paris. There they stayed some three weeks, meeting Dionys Joncquet, professor in the Jardin royal, whose *Hortus sive Index Plantarum* Ray already knew[3] and who had just published the first part of his *Hortus Regius*; Guy Crescent Fagon, 'a very ingenious person and skilful herbarist who had the greatest hand in the editing of the Catalogue of the Physic Garden then put forth and was employed in the laboratory and apothecary shop';[4] and Léon Marchand, who 'has the best hortus siccus that we ever saw',[5] and whom Ray described as 'the best herbarist I met with in France'.[6] On 1 April Ray and two others left by 'Chasse-marée', a fish cart from Calais; Skippon and Lister, not liking that mode of travel, stayed a few days longer, and reached England on 8 April.

The meeting with Lister, though we know of it only from casual allusions by Skippon and in letters,[7] was an event of importance; for the consequent friendship between the two men had a large influence upon them both. Lister had entered St John's in 1655, and as his uncle Sir Matthew Lister had been a physician to Charles I he was given a fellowship there by royal mandate at the Restoration. There is no evidence that he had met Ray until they began to botanise together at Montpellier. Though they returned to England independently, Ray visited him in Cambridge soon afterwards[8] and thenceforth a regular correspondence, now published in full by Gunther,[9] was maintained between them. Lister was at this time

1 This, his great discovery, was made in Amsterdam and published in 1662 in his *Observationes Anatomicae*.

2 Worthington, *Diary*, II, p. 203, says that Ray was in Bordeaux with Croone then: Croone proposed Ray as curator of the Royal Society's Museum in 1669, cf. Birch, *History of the R.S.* II, p. 355.

3 Worthington, *Diary*, I, p. 334, says 'he commends one Joncquet for a copious collection of plants': this in June 1661. Edward Browne attended Joncquet's lectures in June 1664, *Journal*, ed. G. L. Keynes, p. 24.

4 *F.C.* p. 138: he afterwards became superintendent.   5 *Journey*, p. 730.

6 *Corr.* p. 134: Ray received seeds from him (*Hist. Plant.* I, p. 318) and exchanged letters (*F.C.* p. 85).

7 Cf. *Corr.* pp. 31, 62, 107, and *C.A.* p. 122.

8 *F.C.* p. 111.   9 In *F.C.* pp. 110–37.

living in his uncle's house at Burwell near Louth in Lincolnshire, and studying spiders. He published a paper on them in 1668, and shortly after, probably in 1669, made his first scientifically useful discovery in the matter of the threads and flight of gossamer-spiders—he and Edward Hulse writing to Ray about it almost simultaneously[1] and with embarrassing results.[2] He had married in 1669 and bought Carlton Hall near Skipton in Yorkshire, whence he reported several rare plants, including *Polemonium coeruleum* found at Malham Cove.[3] In 1670 he took a house 'without Michaelgate Bar' in the city of York, where he set up in practice as a doctor. In 1671 he was elected to the Royal Society;[4] in 1682 translated into English and methodised Goedart's work on Insects; and in 1685 published his own chief book, the *Historia Conchyliorum*, a very important study of shells, consisting of over a thousand drawings made by his daughter Susannah and his wife Anna.

That he owed a great debt to Ray's friendship is evident from the letters. Ray, who was always inclined to overrate his friends, encouraged him in his efforts to write Latin,[5] to study science, to develop his knowledge of the Exanguia: only slowly did it dawn upon him that Lister did not quite reciprocate his affection; and until the end he thought of him as a great savant. Lister at first genuinely responded, took a real interest in their common studies, and when he moved to London in 1684 was ready to help in practical matters like the publication of the *History of Fishes*. But he gives the impression of having always an eye to his own interests, of caring too much for the credit that would accrue to himself and of never being quite so learned as he imagined. Gradually the friendship faded and in the last phase, when Ray was old and ill and isolated at Dewlands and Lister was the fashionable London doctor, Ray can only surmise that he had never been forgiven for giving Hulse as the first observer of gossamers.[6] Nevertheless, he suppressed a paper of his own 'On Respiration' which put out views opposed to Lister's.[7]

Scientifically Lister's best work was done in the study of Exanguia,[8] and his first book, *Historiae Animalium Angliae Tres Tractatus*, published in 1678, reveals his merits and defects. The first Tract deals with Spiders; and in the Preface he declares that ten years previously he had described most of the insects of the island and paid special attention to Spiders, going to Lincolnshire and devoting to them 'not merely hours or a few days but many

1 Cf. *Corr.* pp. 57–9; the date of Lister's letter (*Corr.* p. 31) and of Hulse's is lost, but they were earlier than 1670.
2 Cf. *Corr.* pp. 77–85. Ray ascribed priority to Hulse: Lister resented this: Ray published a full statement in *Phil. Trans.* v, no. 65, pp. 2103–4.
3 *Corr.* p. 57.   4 On 3 Nov. (Birch, *History of the R.S.* II, p. 435).
5 *F.C.* p. 115: in spite of Ray's compliments his Latin was often faulty.
6 *F.C.* p. 300, letter of Dec. 1694 to Robinson.   7 *F.C.* pp. 299–302.
8 Cf. *Corr.* pp. 11–13 (the date should be 1668) and 24.

months' and revising his descriptions in the next winter. It is typical of the man that after this one season's study he should write: 'I do not want anyone to think that I have described absolutely all the species: but I make bold to say that no one can find casually in this country any new species not described by me.' The second Tract deals with land and fresh-water Molluscs, the subject to which his *History* and two later books on their anatomy, published in 1695 and 1696, were devoted. The third deals with sea-shells and 'stones figured like them', and here he laid it down that all 'formed stones', fossil shells and palaeontological remains are 'lapides sui generis', a position from which he obstinately refused to move although it was rejected by the best minds of the time.[1]

He was in fact an observer rather than a thinker, a man of energy and mental alertness who in his younger days did good work as a pioneer. This and his influential connections gave him a great reputation and might have led him to a brilliant career. But he was too well satisfied with his own achievements, too lacking in humour or appreciation of others, to overcome the natural conservatism and pugnacity of his temperament. After 1680 he did nothing of importance: in his own profession he came into open conflict with Thomas Sydenham and the progressive medical school: in the Royal Society his influence coincided with and contributed to a period of decline: and in his last years he became a mockery to his critics and an obstacle to the development of science.[2] If the invitation which he gave to Ray after Willughby's death in 1672 to make his home with him in York[3] had been accepted, his record might have been vastly different.

In any case his support during the most difficult period of Ray's life was invaluable. It carried Ray over the period between his early friendships with Willughby, Wilkins and Skippon, and his later ones with Dale, Robinson and Sloane. It was by no means the least of the many benefits of his wander-years.

Note. *Ray's Herbarium*

In the Botanical Department of the British Museum are twenty books of different sizes containing some thirty sheets apiece and numbered by the letters of the alphabet. They contain a collection of dried plants sewn on to the paper and neatly labelled. They formed Ray's Hortus siccus.

1 He had done some field-work in geology, and in March 1684 had laid before the Royal Society a proposal for maps giving the character of the soil: cf. Weld, *History of the R.S.* I, pp. 290–1.

2 It is unfortunate that the notice of Lister in Munk, *Royal College of Physicians* is defective and that the article in the *D.N.B.* copied from it exaggerates its defects: thus both state that he became F.R.S. on the recommendation of Edward Lhwyd (who at that time was a child of ten years old) and the *D.N.B.* states that he adopted Ray's views of fossils. The best memoir is in *Yorks. Archaeological Review*, II, pp. 297–320.   3 Cf. Derham, *Life of Ray*, p. 29.

Most of the specimens are relics of his continental tour. Switzerland and especially the Jura, Italy, and Sicily, with definite localities corresponding to those mentioned in *Observations*, are well represented: but there are some from Belgium, Holland and Germany.

British plants are not numerous and few of them have localities attached: but among these are several that fix the identity of species recorded in the *Catalogue* or *Itineraries*. These are *Trifolium glomeratum*, from Messina and with a note that it is also found in Suffolk;[1] *Draba incana*, 'in the hill country of Craven';[2] *Illecebrum verticillatum*, which he calls Polygonum verticillatum J.B., 'in marshes near the extreme corner of Cornwall';[3] *Lathyrus hirsutus*, 'in Hockley parish';[4] *Galium boreale*, which he calls Rubia cruciata sive laevis quadrifolia, 'by a little brook near Orton';[5] *Saxifraga aizoides*, 'upon the skirts of Ingleborough Hill plentifully';[6] *Cornus suecica*, 'upon the north end of the highest of Cheviot Hills plentifully';[7] *Potentilla fruticosa*, 'on the banks of the Tees not far from Greta Bridge'.[8]

A notice of this collection appeared in the *Journal of Botany*, I, p. 32 (1863), where it was stated to have been bequeathed by Dale to the Apothecaries' Company in 1739 and stored in the Orange House of the Chelsea Physic Garden with the collections of Dale and of Rand, Petiver's successor. Thence it was transferred to the British Museum in 1862.

An account of it was contributed to the *Journal*, VIII, pp. 82–4, by Henry Trimen; and from this the particulars here given are derived. In the *Journal*, XXXI, pp. 107–9, James Britten gave a résumé of this article and added the detail derived from Peter Kalm, the Swedish botanist, who saw it in 1748 and recorded that the mutilation of some of the sheets was due to Sherard, who had borrowed it from Dale. A further brief notice of it is contained in Seward's *John Ray*, p. 25.

In addition there is in the possession of Lord Middleton some part at least of the joint collection made by Ray and Willughby. This is preserved in a copy of Ray's *Historia Plantarum* bound up in five volumes and interleaved with sheets of coarse paper on to which are sewn dried plants. Many of these are cut out of earlier sheets and are named either in Willughby's or Ray's hand from G. Bauhin's *Pinax*. In all cases Ray has added a reference to the appropriate book, section and page of the *Pinax*. In a few cases notes are added, e.g. *Scilla autumnalis* 'at the Lizard point' (in F. W.'s hand) or *Scilla verna* (?) 'found near Ligorn on sandy hillocks; the flower was past when we were there in April' (J.R.). There is no evidence to show when or by whom these specimens were remounted.

1 *Obs.* p. 280; *Cat. Stirp. in Ext.* p. 108; *C.A.* p. 305.
2 *Corr.* p. 26; *C.A.* p. 296.   3 *Mem.* p. 186; *C.A.* p. 248.
4 *C.A.* p. 190.   5 *Corr.* p. 26; *C.A.* p. 268.
6 *Mem.* p. 163; *C.A.* p. 279.   7 *C.A.* II, p. 67.   8 *C.A.* II, p. 228.

CHAPTER VI

THE *ENGLISH CATALOGUE*

He hath spared neither pains nor cost, travelling himself through all the considerable parts of this Kingdom, and so viewing and gathering himself almost all the plants here described.

Philosophical Transactions, reviewing *Catalogus Angliae* (v, no. 63, p. 2058).

The return from his three years on the Continent may be said to mark the close of Ray's apprenticeship to his work in science. The tour had given him a great range of material, larger perhaps than that of any previous botanist except De l'Ecluse, and including a knowledge of animals, birds, reptiles, and fishes such as no other Englishman had ever acquired. In addition it had given him status in the world of learning and a measure of confidence in his own capacity. The shy student who in 1660 had hesitated to send a copy of his *Cambridge Catalogue* to Hartlib was now the friend of Hoffmann and Corneli and Marchetti, of Steno and Magnol and Marchand; and could thenceforth exchange opinions with the leaders of contemporary research on level terms. In some sense the rest of his life was the examination and exposition of the data thus obtained. Though he did in fact undertake a large amount of further field-work and never gave up the desire for it, this, except in insects and to a less extent in cryptogams, was only supplementary. His main activity was devoted to the great series of books in which his results thus far obtained were published.

Unfortunately the evidence for the events of the years after his return is by no means either adequate or easy to use. There are no itineraries nor any sort of continuous record. Letters, though thanks to the correspondence with Lister not too few, are casual and allusive. Movement is almost continuous; for he had no fixed home. The whole story is difficult to fit together; and for the first year the only real record is the long letter to Lister dated 18 June 1667, contained partly in *Correspondence*,[1] partly in *Further Correspondence*,[2] and quoted at some length by Derham.[3]

The rest of 1666 Ray seems to have spent in Essex or with his friends in Sussex; but he certainly paid a visit to Cambridge, apparently staying with Lister at St John's and meeting his colleague Samuel Howlett, with whom he had come back from Paris. He describes himself as having bestowed his spare hours 'in reading over such books of Natural Philosophy as came out since my being abroad, viz. Mr Hooke's *Micrographia*, Mr Boyle's *Usefulness of Nature Philosophy*, *Origin of Forms*, *Hydrostatical Paradoxes*, Sydenham *De febribus*, the *Philosophical Transactions*, the business

1 *Corr.* pp. 13–14.　　2 *F.C.* pp. 111–13.　　3 *Mem.* pp. 17–18.

about Greatrakes, turning over Kircher's *Mundus subterraneus*, etc.'[1] The
visits to Sussex may well be responsible for some of the records in the
Catalogue: *Hieracium murorum*,[2] whose beautifully marked leaves he notes
at Cuckfield, where he probably stayed with the Burrells; *Epilobium
angustifolium*,[3] then very much rarer than now, 'in a wood near Troiers
in Mayfield parish'; *Asperula cynanchica*,[4] 'on Sussex downs'; *Peucedanum officinale*,[5] 'in the marsh-ditches near Shoreham'; *Bupleurum
tenuissimum*,[6] 'near the little brook that runs beside the castle, below the
bridge, at Hastings'.

Most of the winter he spent with Willughby at Middleton,[7] arranging
the extensive collections which they had made on the Continent, and
spending a short period in compiling 'the tables of plants, quadrupeds,
birds, fishes, etc. for the use of the *Universal Character*' which his friend
John Wilkins, soon to be Bishop of Chester, was about to publish.
Of these works we shall treat later. With them was the 'gathering up into
a Catalogue all such plants as I had at any time found growing wild in
England, not in order to the present publishing of them, but for my own
use'. He adds: 'Possibly one day they may see light, at present the world
is glutted with Dr Merret's bungling Pinax. I resolve never to put out
anything which is not as perfect as possible for me to make it.' His request
in the next sentence that Lister 'would take a little pains this summer about
grasses, that so we might compare notes; for I would fain clear and complete their history' shows that he knew where his researches needed
supplementing. Probably the work thus described kept him busy all
through the spring of 1667: for there is no hint of any journeyings except
a brief visit to Cambridge in the letter of 18 June, in which the record of
the previous months is given.

On 25 June he and Willughby started on their second journey to the
West—plants and fishes being their principal objective. Unfortunately the
Itinerary kept by Ray, though known to Scott when he published the
others, did not seem to him of sufficient independent interest, 'nothing
more than some short hints by way of diary'.[8] He has preserved in footnotes one or two extracts: and in the *Life* we are told that they 'travelled
through the counties of Worcester, Hereford, Gloucester, Somerset and
Devon into Cornwall as far as the Land's End, where they arrived August
the 17th; and then returned through Hants to London on September the

1 *F.C.* p. 112.   2 *C.A.* p. 255.   3 *C.A.* p. 204.
4 *C.A.* p. 268.   5 *C.A.* p. 240.   6 *C.A.* p. 37.
7 A panelled room overlooking the courtyard on the north side is still by tradition
assigned to Ray. During his stay there he assisted in the collection and transcription
of 'Memoirs and Observations taken out of old muniments'. A small 4to volume in
MS. with this title still preserved by Lord Middleton contains abstracts from Richard II
to Edward IV in his hand. There is also a copy of a charter of King John: cf. *Hist.
MSS. Commission, MSS. of Lord Middleton*, pp. 269–71 and 37.   8 *Mem.* p. 120.

13th following'.[1] To the outward journey must belong the discovery of *Mentha rotundifolia*, 'growing by the river's side at Lydbrook near Ross in Herefordshire plentifully';[2] they seem to have spent some time at Penzance, and when at Land's End he also visited the Lizard and observed certain plants mentioned in a fragment from the *Itinerary*,[3] *Herniaria glabra*, 'in great plenty', *Asparagus officinalis*, 'on the cliffs',[4] and near them *Scilla autumnalis*:[5] he adds, 'on Goonhilly Downs near the Lizard Point is a kind of heath which I have not elsewhere seen in England' (*Erica vagans*), which becomes in the *Catalogue* 'Erica foliis Corios multi-flora, by the way-side going from Helston to the Lizard-point plentifully'. On the way back it is probable that he paid a visit to Portland; for of a Sea-weed, which he calls 'Long Sea Lace', he writes: 'About Weymouth and Portland-island I have found them growing to stones',[6] though the other reference, under *Suaeda maritima*, 'I observed it many years ago growing in abundance on the isthmus that connects the isle of Portland with the Dorset coast',[7] seems strange when applied to a visit in 1667. Then they struck inland to Blandford, staying at Bryanston,[8] went through Wimborne to Christchurch, visited the Isle of Wight and returned to London from Portsmouth. For the *Catalogue* includes, under *Ulmus montanus*, 'it grows in the hedges by the high-way between Christchurch and Lymington in the New Forest';[9] under *Lavatera arborea*, 'I have observed it in many places by the sea as at Hurstcastle near the Isle of Wight';[10] under *Mercurialis annua*, 'on the sea-beach near Ryde in the Isle of Wight plentifully';[11] and under 'Lappa minor', that is *Xanthium strumarium*, Linn., 'it occurs rarely among us. I once found it in the road from Portsmouth to London, some three miles from Portsmouth'.[12]

After this tour Ray went to Notley, and was taken seriously ill. Lister's letter of 21 September, expressing grave anxiety about his health,[13] and his reply indicate a fear that his lungs were permanently affected. He had planned to go to London on 7 October to stay with his friend George Horsnell, the doctor, 'at his house next door to the Rose tavern in Cursitors Alley',[14] but postponed his visit until the time of his election and admission to the Royal Society on 7 November.[15]

1 *Mem.* p. 21.   2 *C.A.* p. 208.   3 *Mem.* p. 190 note.
4 *C.A.* p. 31.   5 *C.A.* p. 172.   6 *C.A.* p. 120.   7 *C.A.* p. 313.
8 The evidence is in *Hist. Insect.* p. 75.   9 *C.A.* p. 320.
10 *C.A.* p. 206.   11 *C.A.* p. 208. He notes carefully that it is dioecious.
12 *C.A.* p. 39.   13 See above, p. 62.
14 *F.C.* p. 116: Horsnell is recorded as the discoverer of *Polygonatum multiflorum*, *C.A.* p. 248, and of a locality for *Dianthus deltoides*, p. 60: he seems to have been a cousin of Philip Skippon; for in a letter to Ray dated London, 22 Jan. 1662 and printed in *Phil. Trans.* XXIV, no. 302, p. 2077, Skippon describes how he and 'my cozen Horsnel' visited a patient.
15 He was excused from payment of his subscription on Wilkins's proposal on 20 June 1668: cf. Birch, *History of the R.S.* II, pp. 203, 207, 300.

His election, despite the aristocratic character of the Society, can hardly have been unexpected. Willughby had been an original member: with Wilkins, the chief mover and first secretary, who had proposed him for election on 31 October, he had been acquainted ever since 1660, and his work during the previous winter for the *Real Character* had strengthened the connection: Skippon had just been elected. He had in fact received through Willughby a request from Wilkins[1] during his time abroad that one or other of them should proceed to Teneriffe, ascend the Peak, and make certain observations suggested by the Society—all expenses being paid and the necessary instruments being sent out.[2] His election would strengthen the Society at a time when it was just beginning to 'pass through so many censures'[3] and to feel the need of more members capable of providing programmes for its meetings. Henry Oldenburg, Wilkins's colleague who did most of the business and all the correspondence, had been arrested in June and had spent two months in the Tower on suspicion of espionage:[4] Robert Hooke, jealous of Oldenburg, resentful of patronage and exploitation but too vain to refuse occasions for it, was unable to keep the virtuosi permanently amused: the Fire of London had occasioned a move of the meetings to Arundel House with consequent dependence upon the Howards and suspicion of Popish leanings: and though Henry Stubbe had not yet published his attack upon Sprat's *History*, criticism was in the air.[5] A few more members capable of first-hand observational work would give the Society a fresh lease of life.

Of the character and achievements of the Society as well as of its status and personnel we have exact contemporary evidence. For in this same year Thomas Sprat, afterwards Bishop of Rochester and Dean of Westminster, had published his *History*. This, as he explained in the Advertisement to the Reader, because of the cavils of detractors is written 'not altogether in the way of a plain History but sometimes of an Apology'. It is in fact a defence of the new or experimental philosophy, which he attributes to Lord Bacon and to the group whose first meetings were in 1638, as against the Aristotelian and Scholastic tradition, and of the purpose and performance of the Society as against the charges of hindering education, unsettling the established order and undermining the Christian religion. His book makes it clear that the opposition came from the champions of the old learning and the old

1 *Corr.* p. 9.

2 Questions relating to observations at Teneriffe, propounded by Brouncker and Boyle, had been registered by the Society in Jan. 1661 (Weld, *History of the R.S.* I, pp. 97–100).

3 From Evelyn's dedication to Charles II of the second edition of his *Sylva* in 1669.

4 Cf. Weld, *History of the R.S.* I, pp. 201–3.

5 Cf. e.g. Lower, *De Corde*, London, 1669, dedication to Millington, pp. 7, 8 ; P. du Moulin to Boyle (*Boyle's Works*, ed. 1772, VI, p. 579), etc.

régime,[1] who having crushed the critics of the Restoration by the Cavalier Parliament and the Act of Uniformity, were suspicious of a Society which stood for novelty and experiment and had arisen under the Commonwealth. That the criticism was neither widespread nor intelligent is probable from the vagueness and inconsistency of its accusations.

The list of members given here in full contains nearly two hundred names, most of them, as is carefully explained, 'gentlemen'.[2] 'In this number perhaps there may some be found whose employments will not give them leave to promote these studies with their own hands...it being their part to contribute jointly towards the charges and to pass judgment on what others shall try.'[3] But if some of them, noblemen and bishops, were rather patrons than performers, there was a goodly number of virtuosi or talented amateurs and sufficient physicians, engineers and other trained students. Sprat prints selected samples of their papers and a special eulogy on Dr Christopher Wren.[4]

In fact the Society was vital and varied in its activities. It had survived the Plague and the Fire, had reassembled in good strength and had already built up a world-wide reputation. *Philosophical Transactions*, started on 6 March 1665, was edited by Oldenburg as the sole source of his income and appeared on the first Monday of every month. These and Birch's *History*, which is virtually a résumé of the minutes of the weekly meetings, give us very full knowledge of its discussions and experiments. They range from the invention of an easy coach with four springs[5] or of a megaphone[6] to the record of a deer shot at Farnham with a broad-headed arrow by Sir Paul Neile[7] and to the presentation of a gun for experiments to improve artillery from the king.[8] But most of them were concerned either with physics or with physiology, with air-compression and blood-transfusion, with apparatus or vivisections. Skippon, who seems to have spent the winter after Ray's election in London, reported to him the events of their meetings in a series of letters, several of which found favour with Derham.[9] Ray himself, writing to Lister from London on 23 November about a Snail with a reversed spiral shell, says: 'I have never found one such as you describe....It deserves to be preserved, described and

---

1 Statements like those of J. W. Draper, *History of the Conflict between Religion and Science*, p. 307 or A. D. White, *Warfare of Science with Theology*, I, p. 41, that attack on the Royal Society was religious or theological, are unsubstantiated by the evidence.

2 Sprat, *History*, p. 67: this list is carried on by that in T. Thomson, *History of the R.S.* pp. xxi–lxix.

3 P. 433.

4 Pp. 311–17.

5 Birch, *History of the R.S.* II, p. 30.

6 L.c. II, p. 440.

7 L.c. II, p. 187.

8 L.c. II, p. 24.

9 *Corr.* pp. 22–3.

painted. If my stay in this city had not been uncertain I would have asked it of you to show to the Royal Society of which I have lately become a member'.[1]

He went to the Burrells at Cuckfield in Sussex almost immediately after this letter, and while there received from Skippon a message urging him to translate Wilkins's *Essay on a Real Character* into Latin: 'he is confident no man can translate his book better than yourself'.[2] So he devoted the winter to it—and wasted his pains. During it his friend Robert Barnham died;[3] and in the spring of 1668, when he left Cuckfield, his first visit was to the parents and widow at Boughton Monchelsea. On the way he wrote to Lister from London on 19 April[4] an interesting letter about books on insects and the Castelli pictures which he had seen in Sicily.[5] He spent three weeks in Kent, finding *Lathraea squamaria* 'on a ditch-bank at Bredgate [Bredgar] near Sittingbourne'[6] and perhaps some other of his Kentish plants.

There followed, as he confessed to Lister on 26 July, three months of perpetual motion.

I went back to London: four days later I left London and went to Essex on business [probably in connection with the purchase of Dewlands[7]]: then after a short week I start off for Haslingfield [the seat of Sir Thomas Wendy, who had married Willughby's sister Letitia] to meet Mr Francis Willughby there and welcome his newly married wife [Emma, the second daughter and co-heiress of Sir Henry[8] Barnard]....I did not stay long—though while there I ran over to Cambridge to greet old friends: then I return the third time to London on necessary business [actually he attended a meeting of the Royal Society on 11 June].[9] From London to Essex, from Essex to Haslingfield. Then I accompanied Mr Willughby when he went home to Middleton. Even then I did not stop, but a few days later took the road again and made for the north. My next halt was at Sheffield in the county of York. There my old friend Mr Jessop received me most kindly, and I am now staying with him. You ask why? You know that I am a botanist and meditating a British flora. I planned to give this summer to North Yorkshire and Westmorland. I have done it, spending a fortnight on my travels....I have got back to Sheffield and propose to stay here till the equinox.[10]

1 *Corr.* p. 22: he sent to Oldenburg a letter from Lister dealing with this snail-shell, printed in *Phil. Trans.* IV, no. 50, pp. 1011–16.

2 *Corr.* p. 22.   3 *F.C.* p. 116.   4 *Corr.* p. 24.

5 Pietro Castelli, of Rome and first superintendent of the gardens at Messina, had left pictures of insects to his nephew which Skippon thought of buying: cf. *Journey*, pp. 613–14, and *Phil. Letters*, p. 361.   6 *C.A.* p. 92.

7 The title-deeds for the land were dated 1669: so Mrs Pattisson, whose husband owned it when she wrote, in *Englishwoman's Magazine*, II, p. 274.

8 So Epitaph of F. W. in Middleton church and Le Neve, *Knights*, p. 316; not Thomas, as *D.N.B.*

9 Birch, *History of the R.S.* II, p. 295. At this meeting he promised to help Daniel Coxe in his intended history of vegetables.

10 *F.C.* p. 117 and *Corr.* p. 25 : the letter is in Latin.

Of Francis Jessop we know of no contact with Ray, although he speaks of him as an old friend, before their meeting at Montpellier and Lyons.[1] He had been born in 1638 in Broomhall, the house built in Henry VIII's time, now a district in the city. His mother was a daughter of Sir Francis South, and writing to Lister Ray speaks of a Mr South, a cousin of Jessop, who was a fellow-townsman of Lister's family.[2] He seems to have been made known to Willughby by Ray and perhaps stayed at Middleton in 1670.[3] He certainly sent there a number of birds which he procured and stuffed;[4] and the two men became friends. He lived all his life in Sheffield, received the freedom of the Cutlers' Company, and collected a considerable library.[5] His chief interest seems to have been in mathematics, for Ray speaks of him as making good progress during his stay,[6] and after his departure Jessop propounded to him a problem in Conics.[7] In 1687 he sent to Aston a tract of 30 pages entitled *Propositiones Hydrostaticae ad illustrandum Aristarchi systema*, which was submitted to the Royal Society and printed.[8] But he was also interested in chemistry, and in 1670 was writing to Ray about experiments in the distilling of pismires—that is, with formic acid—and with vinegar, in which he was working with Samuel and John Fisher.[9] These Ray reported to the Society by letter in January and February 1671 and at a meeting attended by him and Willughby on 18 November.[10] He certainly kept in touch with Jessop until he settled at Black Notley, and probably until Jessop's death in 1691.

After returning from his fortnight's tour, which according to Dale was ended by an attack of measles,[11] he wrote a long account of his finds to Lister on 10 September.[12] The new species were *Draba incana*, 'in damp places on the flanks of Ingleborough and Hincklehaugh';[13] *Saxifraga oppositifolia*, 'on the north side of the summit of Ingleborough';[14] *Thlaspi alpestre*, 'around Settle and Ingleborough and elsewhere in Craven';[15] *Bartsia alpina*, 'near Orton by a stream running across the road to Crosby',[16] a species then undescribed; 'a new species of Sedum or Cotyledon', presumably *Saxifraga stellaris* (not in fact new to him), 'on the rocks of Wrenose';[17] and, as we learn from the *Catalogue*, *Galium boreale*

1 *Journey*, pp. 714, 724.   2 *F.C.* p. 119: cf. Hunter, *Life of O. Heywood*, p. 293.
3 *F.C.* p. 48.   4 Cf. *Corr.* p. 33 and *Phil. Trans.* x, no. 117, p. 393.
5 For details cf. J. Hunter, *Hallamshire*, pp. 209–15.
6 *Corr.* p. 25.   7 *Corr.* pp. 33–4.
8 Birch, *History of the R.S.* IV, p. 556.
9 *Corr.* pp. 67–72 and pp. 91–3: the Fishers were both physicians in Sheffield, sons of James Fisher, vicar there, but ejected in 1662: for their chemistry cf. *Obs.* p. 202.
10 *F.C.* pp. 50–4, 63; *Phil. Trans.* v, no. 68, pp. 2063–6.
11 *F.C.* p. 5.   12 *Corr.* pp. 26–7.   13 *C.A.* pp. 296–7.
14 *C.A.* II, p. 269, not in *C.A.* 1st ed.
15 *C.A.* p. 49.   16 *C.A.* p. 86.
17 *C.A.* p. 85: he certainly got as far as this and searched the Furness fells: cf. *Corr.* p. 45.

near Orton and at Windermere.[1] In addition he had been shown *Meum athamanticum*, 'on the road between Sedbergh and Orton plentifully';[2] and by 'Mr Witham', probably Gilbert Witham of Garforth,[3] 'in Haselwood woods' near Tadcaster *Actaea spicata*[4] and *Pyrola minor* or possibly *P. secunda*.[5] He had also done more accurate descriptions of some found previously and sends a list of these and others 'common in these parts but nowhere else so far as I know in England', including *Rubus chamaemorus*, of which he gives a discussion.[6]

After this very profitable tour and some discoveries in the neighbourhood of Sheffield, *Campanula hederacea*,[7] *Polygonum bistorta*[8] and *Epipactis purpurata*,[9] he spent Michaelmas and all October at Notley and then joined Willughby at Middleton. He was at work on his *Collection of English Proverbs* and on the *Catalogue*; Willughby was in poor health and tied by the birth of his first-born; collecting had to be given up. Instead we find him conducting experiments on the rise of sap, discussing spiders and ants with Lister, and when taken to Chester in April 1669[10] by Wilkins buying a porpoise from a fisherman and dissecting it.

There are two other visits which certainly belong to the period between the return from France and the completion of the *Catalogue*, which took place in May, and which may probably be assigned to 1669.

The first is a visit to Charles Howard, brother of the Duke of Norfolk and an original Fellow of the Royal Society, who owned the Manor at Dorking, or as Ray always calls it Darking, in Surrey.[11] In the *Observations*,[12] writing of edible snails and 'the great whitish or ash-coloured snail-shells' (*Helix pomatia*), he says that he has since found them plentifully on the downs near Darking, 'whither as we were informed by Charles Howard Esq. they were brought from beyond seas'.[13] In the *Historia*

1 *C.A.* pp. 268–9, where it is listed as Rubia erecta quadrifolia J.B.: in *Corr.* p. 26 he calls it Galium cruciatum J.B. and it is identified (wrongly) by Lankester with *G. cruciatum*, which Ray always calls Cruciata and knew well.

2 *C.A.* p. 209.

3 B.A. Trinity Coll. 1634–5: he had already helped Merret in the *Pinax*: cf. Preface, where he is 'Ds Witham, Eboracensis'.

4 *C.A.* p. 71.

5 *C.A.* p. 256: the note makes clear that the plant is not *P. rotundifolia* or *media*; cf. also *H.P.* II, p. 1233: Lees, *Flora of West Yorks*, p. 322, says that it is probably *P. minor*.

6 *C.A.* pp. 68–9.   7 *C.A.* p. 52.   8 *C.A.* p. 44.

9 *C.A.* p. 162.   10 Cf. *Hist. Pisc.* p. 32.

11 Aubrey, *Natural History and Antiquities of Surrey*, IV, pp. 148–9, 164–6, has an account of this in 1673 under 'Darking or Dorkyng'. The long 'Hope' or valley planted and terraced and called Dibdin (Deepdene), the laboratory and gardens of this 'Christian philosopher' are ecstatically described: 'I can never expect any enjoyment beyond it but the Kingdom of Heaven.' Cf. also *The House of Howard*, II, p. 584.

12 P. 404.

13 The species is still abundant along the line of the North Downs.

Plantarum, dealing with St John's-worts, he records having seen in the gardens of Charles Howard a species which he had found in Sicily: 'it was not yet in flower when we were there in the middle of May'.[1] He also notes in the *Catalogue* that he found *Cardamine amara* 'in boggy and watery places near Darking in Surrey'.[2]

The second is told in an undated letter to Willughby, which Derham prints in his Appendix.[3] In it Ray is discussing the changes and variations in plants grown from seeds, and confessing that he is 'now almost convinced that plants can degenerate and change their species[4] within the limits of their genus or tribe'. He had been visiting Oxford and been impressed by the work of Jacob Bobart junior and William Browne of Magdalen College, 'one of the best botanists of his time',[5] at the Botanical Garden, of which Bobart had just succeeded his father as custodian. He goes on to express his conviction that the small forms of white and yellow Waterlily do not in fact exist, as had been stated, on the Thames; and then describes his own movements. 'On Monday I proceeded to Caversham, a village about a mile from Reading, and not far from there on a chalk hill rising above the Thames found Orchis anthropophoros [*O. militaris*][6] from the instructions of Mr Browne: it was not common, and was the same as I had found at Geneva.' This find is recorded in the *Catalogue*[7] together with *Orchis ustulata* and *Herminium monorchis*[8]—though in writing he declares that it was 'the sole fruit of his journey'. He then spent a day in inspecting Mr Edward Morgan's garden at Westminster, which he had previously visited in 1662 and from which, as we have seen, he got plants for his own garden in Trinity.[9] Morgan had been one of the companions of Thomas Johnson on his tour of North Wales in 1639. He had been in London since before the composition of How's *Phytologia*, and according to Merret's *Pinax* had found the polyanthus 'in great Woolver Wood in Warwickshire' and transplanted it. Evelyn visited his garden on 10 June 1658. His 'hortus siccus' is still preserved in the Bodleian and his plants were removed in 1676 to the newly founded Apothecaries' Garden at Chelsea.[10]

1 II, pp. 1018–19.   2 *C.A.* p. 220.
3 *Phil.* Letters, pp. 369–70; *Corr.* pp. 4, 5.
4 The evidence that led him to this doubt is given in the Preface to *Catalogus Stirpium in Exteris* published with *Observations* in 1673: he refers to his visit to Oxford as two years previously and describes a Cowslip giving rise to Primrose and Oxlip: see below, p. 174. In *C.A.* p. 330 he assigns this visit to 'the previous year'.
5 Chief author of *Catalogus Horti Botanici Oxoniensis*, 1658; a fellow in 1657 till his death in 1678: he is quoted *Syn. Brit.* II, p. 277.
6 The note in *Correspondence* gives *Aceras anthropophora*: but Browne's record in Merret, *Pinax*, and his specimens are clearly *O. militaris*: and Caversham was the locality for this species not for *Aceras anthropophora* (cf. Druce, *Flora of Oxfordshire*, p. 291).
7 *C.A.* p. 225.   8 *C.A.* p. 227.   9 Cf. above, p. 109.
10 Cf. Gunther, *Early British Botanists*, pp. 351–4.

He moved back into Wales and was still alive in 1685 at the garden at Bodysgallen in Carnarvonshire.[1]

Ray then goes on to tell how on Wednesday he went on foot into Kent to Rough hill, now Rowhill, south of Dartford to find plants which Willisel said that he had found there. He failed to see them or to hear of them from the country folk, but discovered instead *Ajuga chamaepitys*,[2] *Bupleurum rotundifolium* and a new species to him, which Jacob Bobart had shown him in his own collection at Oxford, *Geranium columbinum*.[3] The letter is interesting not only for its botany and for Ray's hesitations about the rigid fixity of species, but as introducing Thomas Willisel, whose career G. S. Boulger has immortalised in the *Dictionary of National Biography* and who was closely associated with Ray during the next few years. He was a man of the humblest origin who had collected simples professionally and was possessed of real genius for the discovery and discrimination of plants. Ray, who had met him first in London[4] when he visited Morison's garden in St James's Park in April 1662, had described him to Worthington as 'a soldier (sometime belonging to Lambert) who having taken a great affection to the botanical studies hath arrived to a great knowledge in plants; and is sent by Dr Morison into several parts beyond sea to make a collection for that garden'.[5] In acknowledging his help in the *Catalogue* he describes him as 'a person employed by the Royal Society in the search of Natural Rarities, both animals, plants and minerals; the fittest man for such a purpose that I know in England, both for his skill and industry'.[6] The *Catalogue* in fact contains at least twenty references to him. He died in 1675 in Jamaica and years later in 1692 Ray wrote of him: 'Tom Willisel's loss I cannot remember without some trouble.'[7] 'Had God granted him life and health he would have made great discoveries and highly improved natural history. Very few species would have escaped his notice: he was indefatigable and could endure hardship and live as well upon oatcake and whig[8] as another man upon flesh and wine, and ramble over hills and mountains and woods and plains.'[9]

1 Cf. Gunther, *Life of Lhwyd*, p. 74: Lhwyd describes him as 'a man equally commendable for his good life and indefatigable industry'.
2 *C.A.* p. 69.
3 *C.A.* p. 131, where it is fully described and said to have been seen in abundance at Swanley near Dartford.
4 *C.A.* p. 186.
5 So Worthington to Hartlib (*Misc.* p. 261): John Aubrey's racy description of him (*Natural History of Wilts*, ed. Britton, p. 48) is quoted in *D.N.B.*
6 *C.A.* p. 340: he was appointed as from 25 March 1668 at a salary of thirty pounds a year on 20 May and given a certificate of his employment on 10 June (Birch, *History of the R.S.* II, pp. 371, 378). He brought a collection of plants, birds and fishes from Scotland to the Society on 21 Oct. 1669 (l.c. p. 398: and Weld, *History of the R.S.* I, p. 224).
7 *F.C.* p. 233. 8 I.e. buttermilk. 9 *F.C.* p. 235.

The year 1669, even if these visits belong to it, is one of the least eventful in his career. Only one expedition is clearly recorded, when on 14 October with Willughby 'he rode to see the famous fir trees some two miles and a half distant from Newport in a village called Wareton [Wharton] in Shropshire'. Derham has preserved his note on these trees: 'There are of them thirty-five in number very tall and strait without any boughs till the top. The greatest and which seems to have been the mother of the rest we found to be fourteen feet and a half round the body and they say fifty-six yards high which to me seemed not incredible';[1] and Lankester has printed a long note recording that all the trees had disappeared by 1776. The interest of the visit arises out of the question of the species and origin of these firs. Ray includes them in his *Catalogus Angliae* as 'Abies Ger. Park. alba sive femina J.B.', records them as 'said to be common in the Scottish Alps'; but adds that 'these Scottish trees are more likely to be the species generally called Picea'.[2] Plot in his *History of Staffordshire*[3] quotes Ray as identifying the Wharton trees with Picea; and Ray in his *Historia Plantarum* points out Plot's error. He then adds: 'These trees were very well known to me. I saw and studied them more than once and collected their seeds for our garden. In my Catalogue I referred them to the Abies.'[4] Further, he identifies Abies and the Wharton trees with the Silver Fir (*Abies pectinata*), whose leaves are green above and brilliantly white beneath, whose cones are upright and whose habitat is Bavaria and the Alps, and distinguishes them from the Picea of Parkinson, the Norway Spruce (*Picea excelsa*), whose leaves are green and cones hang down, a native of Scandinavia.[5] In his *Synopsis Britannicarum*[6] he includes both species as British, though he admits doubt as to the Wharton specimens being truly indigenous.

Other letters of this year make it clear that the *Catalogue* is being written; on 10 December 1669 he sent it to Lister for comment;[7] and on getting it back wrote that he had 'procured some considerable experiments and observations, medical, from Dr Needham and some other ingenious physicians of my acquaintance, so that I have not yet sent it to be printed'.[8] It was not until 29 June 1670 that he could say:

My *Catalogue* is in the press but not yet finished. About a fortnight agone I received 7 sheets, containing 112 pages, which I guess will make about a third part of the body of the *Catalogue* so that the whole will be 335 pages without

1 *Mem.* p. 25.   2 *C.A.* p. 1.   3 Pp. 208–9.
4 *H.P.* II, p. 1395.   5 *L.c.* p. 1396.   6 *S.B.* II, p. 287.
7 *Corr.* p. 47. He replied on 22 Dec.: 'I read it greedily and am extremely pleased that you have added the particular uses [i.e. medical value] to the titles. I have no additions to make you an offer of, only I read it not without pen, ink and paper by me, on which I now and then scrawled something and have taken the boldness to send it you' (*Corr.* p. 50).
8 *Corr.* p. 53.

Preface, Appendix and Index; yet I hope it will be not so gross as to be altogether unportable for the pocket. The letter and paper I like well, and the correcting is tolerable, much better than I expected at London.... Some things that came to my hands and memory after I had parted with the copy I have been forced to add by way of Appendix.[1]

In the same letter[2] he sends a brief report of the plants found that season, *Agrostis spica-venti*, 'in sandy grounds',[3] *Silene nutans*, 'in flower and seed all about Nottingham Castle',[4] and *Dianthus deltoides*. On 17 July he sends a long list of rarities brought by Thomas Willisel and tells Lister not to think of buying the *Catalogue*. On 22 August he sends him a copy of it, 'not very big in bulk, but still smaller in worth'. 'I don't know whether it will please others; it certainly displeases me: for I never satisfy myself in what I write.'[5] It had been published by Martyn in London and was dedicated to Francis Willughby, 'his most honoured friend and patron (Maecenas)'. Ten years' work was finished and the plants of England had been surveyed: but Ray, who in this same letter announces the dropping of the 'W' in the spelling of his name and his unwillingness to undertake the lucrative offer of a tutorship abroad, was only at the beginning of his scientific career.

To complete the record of his travels only one more remains. After a severe attack of jaundice in the spring of 1671[6] he was sufficiently recovered to 'begin a simpling voyage into the north, taking Thomas Willisel along with me'[7] in the first week of July. Willisel had brought him certain plants which he included in the Appendix of the *Catalogue*,[8] including two which Skippon[9] had handed on to him, *Artemisia campestris*[10] and *Veronica triphyllos*,[11] found by Willisel near Barton Mills and at other places in the Breck district. But he always wished to see his plants for himself and went back to Yorkshire, Westmorland and Northumberland for this purpose. The only extract from the *Itinerary*[12] covers 22–24 July and tells of his finding *Asperugo procumbens*[13] on Holy Island, and *Sisymbrium irio*[14] on the walls of Berwick. Of this plant, of which Berwick is now the only British locality, he already had experience in London: 'in

1 *F.C.* p. 126.   2 *Corr.* p. 60.
3 *C.A.* p. 137: 'T. Willisel brought it collected round London'; *C.A.* II, p. 136, where is added 'in 1670 I observed it in the sandy grounds about Nottingham and in 1671 in the corn about Barton Mills in Suffolk'.
4 *C.A.* p. 202: 'on the walls of Nottingham Castle found by T. Willisel': in *Corr.* p. 56 he records his own sight of it in April 1670.
5 *Corr.* pp. 61–2.
6 *Mem.* p. 26; *Corr.* pp. 84, 86.   7 *Corr.* p. 87.
8 *C.A.* pp. 325, 339–40.
9 *Corr.* p. 85; this letter should have been inserted earlier by Lankester.
10 *C.A.* p. 2.   11 *C.A.* p. 340.
12 *Mem.* p. 151 note.   13 *C.A.* II, p. 19.
14 *C.A.* II, p. 100.

various places on mounds of earth between the City and Kensington: after the City was burnt, it came up in 1667 and 1668 in the greatest plenty within the walls on the rubbish-heaps around St Paul's Cathedral'.[1] 'About two miles from Berwick, in a boggy ground, not far from the road leading to Edinburgh', he found *Tofieldia palustris*,[2] a plant new to science, and next day at 'Scrammerston Mill about a mile and a half distant, a watermill between the saltpans and Berwick, upon the sandy beach by the seaside' *Mertensia maritima*,[3] of which previously he had only seen the dried specimen supplied in 1670 by Willisel.

From the second edition of the *Catalogue* it appears that several of his other ambitions were fulfilled, and the references give a clue to his route. This seems to have been by Halifax, Heptonstall and Burnley to Settle; to Shap and the east side of Ullswater; to Penrith, Alston ('Osten') and Hexham; to Newcastle and thence by Wooler and the Cheviot to Holy Island and Berwick; and back by Bamborough and the Tees to Greta Bridge. On the way north he found at last *Polemonium caeruleum*[4] which Lister had found and reported to him in 1670,[5] which by an error of printing had been omitted in the *Catalogue*,[6] and which 'grows about Malham Cove, a place so remarkable that it is esteemed one of the wonders of Craven, in a wood on the left hand of the water as you go to the Cove from Malham plentifully [it is still abundant there but on the opposite slope] and also at Cordill or the Whern [Gordale Scar], a remarkable cove where comes out a great stream of water, near the said Malham'. Of other plants he records *Epipactis atropurpurea*,[7] 'on the sides of the mountains near Malham 4 miles from Settle in great plenty'; *Polygonum viviparum*,[8] 'shown me this year (1671) by T.W. in a mountainous pasture about a mile and a half from a village called Wharfe, not far from the footway leading thence to Settle plentifully'; *Arenaria verna*,[9] 'on the mountains about Settle plentifully'; *Polygonatum anceps*,[10] 'this year (1671) shown me by T.W. growing on the ledges of the scars or cliffs near Wharfe and Settle plentifully'—it is still there though less common; *Sedum villosum*,[11] 'on the moist springs about Ingleborough Hill as you go from the hill towards Horton in Ribblesdale, and Hartside Hill near Gamblesby on the way to Osten, Cumberland'; *Alchemilla alpina*,[12] 'this year (1671) on a mountain in Westmorland beside a great pool or lake called Hulls-water about two miles from Pereth [Penrith], just over against Wethermellock [Watermillock] on the west side'; *Lobelia dortmanna*,[13] 'in a pool or lake

1 *C.A.* p. 104.   2 *C.A.* ii, p. 30.   3 *C.A.* ii, p. 93.
4 *C.A.* ii, p. 299.   5 *Corr.* p. 57.   6 *Corr.* p. 64.
7 *C.A.* ii, pp. 157–8: he unites this with the *Epipactis* previously found near Sheffield (*C.A.* p. 162 and above, p. 114): but Lees, *Flora of West Yorks*, p. 434, argues that the Sheffield plant was *E. purpurata*.
8 *C.A.* ii, p. 42.   9 *C.A.* ii, p. 35.   10 *C.A.* ii, p. 238.
11 *C.A.* ii, p. 270.   12 *C.A.* ii, p. 11.   13 *C.A.* ii, p. 132.

called Hulls-water that divides Westmorland from Cumberland, 3 miles from Pereth plentifully'; and finally *Cornus suecica*[1] and a Willow-herb, probably *Epilobium alsinefolium*,[2] 'among the rocks on the West' and 'in the rivulets on the sides of the Cheviot Hills'. He finished his tour with the record of a plant new to science, *Potentilla fruticosa*,[3] 'on the south bank of the river Tees, near a village called Thorpe, and below Eggleston Abbey'—still almost its only habitat in Britain: this had been found and was now shown him by his friend Ralph Johnson, Vicar of Brignall, who had also discovered *Gagea lutea*[4] 'in the skirts of the woods' about Greta Bridge. It was a notable climax to what proved to be his last 'simpling voyage'.

Much of the winter he spent at Middleton, preparing two more books, the *Observations* and his *Collection of English Words*, for the press and helping Willughby in his various interests. During it he spent a month in London and attended three meetings of the Royal Society, on 9 and 23 November and 7 December. After this he paid a long visit to Bishop Wilkins at Chester, revising his Latin version of the *Real Character* and staying at least from 18 December[5] till February, when he was 'on the coast of Lancashire about Liverpool' and recorded an observation of great flocks of Knots.[6] He returned to Warwickshire by 2 March 1672.[7]

When, late in May, he was proposing to start on a botanical expedition to the West country 'which will despatch the search of England', his plans were smashed by Willughby's illness and death, on 3 July.[8] His duty to his friend put a stop to his active work in the field; and when Wilkins also died on 19 November his travelling except for a few short visits to London[9] came to an end.

His journeys gave him the data for his life's work. The continental tour, with its crowded years of observation, collection and study, made him familiar with the flora and fauna and to a less degree with the geology of Western Europe. It also introduced him to the centres of learning and many of the leading experts. The fruit of these researches was harvested at intervals during the next thirty years in the series of volumes which helped to lay the foundations of zoology. The tours in Britain had a more immediate sequel in the appearance of the two editions of the *Catalogus Plantarum Angliae et Insularum Adjacentium*, published by John Martyn, from which much of the record of his activities has been drawn.

1 *C.A.* II, p. 67.　　　2 *C.A.* II, p. 194.　　　3 *C.A.* II, p. 228.
4 *C.A.* II, p. 219: Ray had previously noted this discovery: *Obs.* p. 250 note. Johnson had sent him 'bulbuli' of it: *H.P.* II, p. 1154.
5 *Corr.* p. 93.　　　6 *Ornithology*, p. 302.
7 *F.C.* p. 130.　　　8 *F.C.* p. 131.
9 He was with Wilkins in Tillotson's house on 18 Nov. and visited London again in 1673.

The volume was of the same shape and style as its predecessor, but with some fifty more pages, arranged in the same alphabetical order with synonyms, localities and notes. To mark its origin plants found in Cambridgeshire have a capital 'C' prefixed to them. There is a short list of authorities and a considerable appendix consisting of various observations and prescriptions supplied by Edward Hulse, Thomas Willisel and George Bowles.[1] Hulse, already mentioned in connection with the threads of gossamer spiders, was a friend of Cambridge days. He had come up to Emmanuel College in 1653 and been elected a fellow in 1658, but was ejected at the same time as Ray. In 1668 he had gone to Leyden and taken his doctorate there. He became a medical man of some distinction. George Bowles of Chislehurst had been a companion of Thomas Johnson, contributing records to his *Mercurius*, and was both a well-qualified apothecary and a skilled botanist. There is no evidence that he was personally known to Ray, but he held an honoured place in scientific circles.

In the Preface Ray states carefully the purpose and character of the book. He had been asked, on a visit to Cambridge, for a new edition of the *Cambridge Catalogue*, but had already undertaken a Catalogue of the whole country. With this object he had explored it from 'Promontorium Belerium' (Land's End) to Carlisle and Berwick in order to see the flora in its habitat and to check previous records. As a result he had personally seen practically all the plants now listed except a few from L'Obel and Johnson marked with an asterisk. This study has been necessary because of the brevity and insufficiency of the descriptions of the older botanists and the consequent and reckless multiplying of species. The commoner plants have especially suffered: 'the histories of the Grasses, for example, or the Hawkweeds, Oraches, Docks are seriously confused and obscured'. To disentangle the confusion requires special aptitudes, intelligence and memory, knowledge of Greek and Latin, a keen eye for observation and for discrimination, and patience to explore characteristics. Gerard and Parkinson were not capable of doing more than increasing their readers' perplexity; and Johnson was killed before his work was done.

Ray has adopted an alphabetical arrangement, partly to save space and time, partly because he hopes to produce a general Method shortly. Synonyms he has given from Gerard, Parkinson and the Bauhins. The medical notes he has derived from reliable doctors and botanists with very many observations of his own or of his friends. Because drugs affect different people differently, doctors should use as large a variety as possible.

---

1 He supplied the records of *Orchis hircina*, 'in the lane that leads from Crayford to Dartford'; and *Senecio saracenicus*, 'in Shropshire in the way as one goeth from Dudson in the parish of Cherbury to Guarthlow' (Dudston and Gwarthlow near Montgomery), *C.A.* pp. 341–7: cf. *Mercurius*, pt II, pp. 13, 27–8.

He is deeply indebted to Schroeder,[1] even where the debt is not acknowledged, for his judicious epitome of previous information on the subject.

He apologises for seeming to 'thrust his sickle into another's harvest' by
stating that though not himself a doctor he has been a student of herbalism
from childhood, has long been familiar with botanical literature, and has
enough experience of medical practice to recognise the potentialities of each
plant. 'Yet in these matters I am less a doctor than a natural philosopher.'

The importance of such work is obvious, though it is still too often
treated with contempt as a waste of time. Creation is not purposeless:
how then can we refuse to investigate or contemplate it? Plants are of
value for food or medicine: yet the older scholars failed both in observation
and in description; and the moderns have too often followed them. It is
not enough to study plants in general: each single species requires investigation; for often between plants closely akin there is remarkable
difference of character and use. We must learn first to define and recognise
each species and then to discover its peculiar properties. 'It is surely not
amiss to spend a whole lifetime if we can discover a medicine or two to
alleviate or cure the deadly diseases that afflict us.'

He then deals with two further subjects; first, the problem of what
constitutes a species, a matter to which he returned in the Preface to his
Continental Catalogue in 1673 and in his paper to the Royal Society in
1674; and secondly, his rejection of more than thirty species listed as
British by other writers, on the ground that they are either incapable of
enduring our climate, or garden escapes, or not to be found in the places
from which they are recorded. He concludes with an assurance that he
does not claim to have discovered all the native plants, since some like
Purple Spurge, Field Eringo and Nottingham Catchfly are confined to a
particular locality, and that he cannot expect to have escaped mistakes.

Enough has been said in the narrative of his journeys to give an idea
of the thoroughness of his exploration;[2] and we have indicated that his
identifications and arrangement showed a real improvement upon the
Cambridge Catalogue. The attention to pharmacology is a marked feature
of this book. An index of diseases with lists of plants appropriate to their
treatment fills ten pages at the end: and the notes in the *Catalogue* show that
he drew not only upon Schroeder but from a large number of other authorities. Thus he cites Thomas Sydenham's *De Febribus*,[3] which he had studied

1 Johann Schroeder's *Pharmacopeia Medico-Chymica* was published at Lyons in
1649 and frequently reprinted, an English version, *The Compleat Chymical Dispensatory*,
translated by William Rowland, being published in 1669. Book IV in it contains the
Phytologia, an index of plants with their uses.

2 We may add the point to which Gunawardena draws special attention, the valuable
note on the parasitism of the Dodder based upon experiments of Parkinson and by
himself: cf. *C.A.* II, p. 85.

3 *C.A.* p. 3: Sydenham is of course the most famous of all contemporary doctors.

on his return from the Continent,[1] two treatises of Robert Boyle,[2] Walter Charleton's *De Lithiasi*,[3] Thomas Willis's *De Scorbuto*,[4] an unpublished manuscript of 'Dr Bates',[5] and a special prescription for swellings of the breast from Percival Willughby,[6] which he says is also unpublished. He draws largely upon his own friends, Hulse,[7] Walter Needham[8] and Martin Lister.[9] Of botanists he quotes J. Bauhin,[10] Bock,[11] Camerarius,[12] Cesalpino,[13] d'Aléchamps,[14] de l'Ecluse,[15] de l'Obel,[16] Dodoens,[17] Fuchs,[18] and Mattioli[19] for various prescriptions. Many of the traditional uses he rejects as ill-authenticated or superstitious;[20] and he protests against the frauds of herb-women.[21] Among his own notes[22] is a long report of the case of 'a noble lady personally known' to him whose eye was treated with belladonna, the fresh leaf of 'Solanum lethale' (*Atropa belladonna*); he records the dilatation of the pupil and had evidently kept full particulars. He includes a number of interesting records not bearing strictly upon medical uses. Thus he notes that Hazel charcoal is used by painters,[23] Shave-grass (*Equisetum hiemale*) for scouring pots and pans,[24] powdered Spindle-tree (*Euonymus europaeus*) for destroying lice,[25] and lichen for a purple dye;[26] that Ptissana, which the French call 'tisane', is made from barley, liquorice and grapes;[27] and that, according to Lister, inhaling the smoke of lettuce produces the same sensation of giddiness as a first attempt to inhale tobacco.[28] He quotes d'Aléchamps for the use of fern roots instead of bread in Neustria,[29] and Olaus Worm's *Museum* for an experiment to disprove the belief that cups of ivy-wood had the property of separating water from wine.[30] There is also a long note on the

1 Cf. *F.C.* p. 112.   2 *C.A.* pp. 160, 233, 238, 247.
3 Author of the *Onomasticon*; see below, p. 310, quoted *C.A.* p. 91.
4 Published Oxford 1667; *C.A.* p. 189.
5 Probably George Bate, physician to Charles I, Cromwell, and Charles II, though he died in 1669.
6 Sixth son of Sir Percival Willughby of Wollaton and uncle of Ray's friend: he practised in Derby and lived until 1685.

7 E.g. pp. 94, 205, 253, 254.   8 E.g. pp. 2, 10, 25, 30, 41, etc.
9 E.g. pp. 4, 47, 53, 122, 185, etc.   10 E.g. pp. 38, 42, 51, 68, 70, etc.
11 E.g. pp. 14, 21, 25, 40, 56, etc.   12 E.g. pp. 56, 60, 80, 88, 295.
13 Apparently only on p. 4.   14 P. 280.
15 Pp. 69, 102.   16 E.g. pp. 28, 47, 93, 104, 164, etc.
17 E.g. pp. 9, 46, 51, 68, 70, etc.   18 P. 46.
19 E.g. pp. 30, 36, 64, 72, 82, etc.   20 E.g. pp. 81, 265.
21 *C.A.* p. 223; they sold the dangerous roots of the Hemlock Dropwort (*Oenanthe crocata*) as roots of Peony. This fact is derived from Johnson, *Gerard's Herbal*, p. 1060. Ray says that some disputed the poisonous character of these roots; so he leaves the matter for future investigation, which he carried out and reported to the Royal Society in *Phil. Trans.* XIX, p. 634: cf. letter to Sloane, *Corr.* pp. 313–15.

22 *C.A.* pp. 287–8.   23 *C.A.* p. 84.
24 P. 100.   25 P. 106.
26 P. 194.   27 P. 171.
28 P. 185.   29 P. 115.   30 P. 159.

'manna' on Ash trees, in which he quotes Altomar and his own Neapolitan friend Tomasino Corneli.[1]

The labour which these notes must have involved will be clear from this summary of their sources. If his efforts seem disproportionate or unnecessary, we must remember the fantastic and horrible drugs that were still in general use—the best doctor in Braintree[2] regularly used essence of woodlice and peacock's dung[3]—and the extent to which magic, astrology and alchemy still entered into medical practice; and remember too that England was ague-ridden and that quinine (Jesuits' bark) was just beginning to be employed for its cure.[4] Whatever the positive value of his conclusions, a matter which only an elaborate research into pharmacology could decide, it is certain that he did important service in promoting the scientific investigation of the medical use of plants, and in combating its age-long association with superstition. He rejected outright the popular doctrine of signatures and the belief that the phases of the moon or the influence of the planets affected the gathering of herbs. He pleaded for exact observation and experiment, for the clinical work which Sydenham was commencing to adopt. He did his best to quote only authorities whom he believed to be reliable and to supplement their findings by first-hand evidence. Some day, when the history of medicine is fully studied, he will receive proper appreciation as one of the great pioneers of an intelligent knowledge of the pharmacopoeia.

The second edition of the *Catalogue* was published in 1677. It only differs from the first in its additional records of species, most of which have been already quoted, and of medical uses; in the absorption of the Appendix into the Catalogue; in the insertion of two plates; and in the provision of a four-page list of the new plants and of nine more added by Dent, the Cambridge doctor. At the end there is a list of errata, a compliment to the printer and a statement that the Method promised in the Preface had been prepared and sent to Wilkins, but that since his death no one had been found to finish and publish the fuller version of the Essay which he had projected.

As we shall see, Ray produced a fuller Method a few years later. He had by his journeys and Catalogues discharged the first of his undertakings, the 'design to put forth a complete Phytologia Britannica'.[5]

---

1 P. 118: cf. Corneli's letter in *Corr.* p. 6, the date of which must be a year too early.

2 Benjamin Allen, whose Commonplace books have been recovered and contain this evidence: cf. articles by Miller Christy in *Essex Naturalist*, XVI, pp. 145–75 and XVII, pp. 1–14.

3 Both and many other such are in Schroeder—woodlice for colic and jaundice, peacock's dung for megrims and epilepsy: *Pharmac.* V, pp. 317, 303.

4 Cf. the references to its use and dangers in Ray's letters, especially *F.C.* pp. 22–3 to Courthope, who had asked his advice about it.

5 *Corr.* p. 2.

NOTE. *The topography of Ray's journeys*

The student who desires to reconstruct the conditions of travel in England in Ray's time or to trace the naturalist's footsteps has two invaluable contemporary helps, John Ogilby's survey of the chief roads undertaken at the command of Charles II and published with elaborate plans in folio in 1675,[1] and the excellent maps of the counties prepared by Robert Morden for Gibson's sumptuous edition of Camden's *Britannia* in 1695. Ogilby is of little use for the Welsh and Pennine explorations: but he shows the routes by which communications normally proceeded and in his detailed descriptions gives a good account of the countryside. Supplemented by the allusions in Pepys and Evelyn and other adventurers on the roads the resulting picture reveals how remarkable were these to us commonplace voyages. Considering the excessive badness of the tracks, the disturbed state of the country, and the perils from highwaymen and footpads, it is remarkable that Ray carried through his excursions into the wildest parts of Britain without hindrance or other hardship than that of poor accommodation.

To identify the places visited is not usually difficult. The names are frequently spelled phonetically (Ray's spelling is often variable) and it sometimes requires research and local knowledge to find the modern equivalents: 'Bocton' in Kent, for example, is only fixed by a single reference to its full name in Camden's plant-lists. But in general, as the existence of ancient churches shows, the village sites of England have hardly changed since the seventeenth century. Contemporary maps are widely different from ours for the industrial areas and the great towns: for the rest of the country they give a faithful and still accurate picture. The distribution of the rural population was determined by its need of a market-town within the reach of a farm-cart; and the horse travelled almost as far then as now.

There are two points at which precise identification is not too easy.

One is the 'high hill' near Settle which Ray calls Hincklehaugh, or as he once spells it Hinckell-hoe.[2] Modern maps seem to have no knowledge of such a name; nor is it known in the district. The obvious suggestion might well be made that it is only a variant form of Ingleborough; but Ray in one place mentions the two together as distinct localities,[3] and in another describes Hincklehaugh as 'overhanging the town of Settle'[4] and

1 This was the first 'road-book' and appeared as the *Travellers Guide* in many condensed and revised editions.
2 *Corr.* p. 44.
3 *C.A.* p. 297: the plant is *Draba incana*.
4 To Lister, *Corr.* p. 28: here he does not mention its name, but this is fixed by the record in *Hist. Insect.* p. 273.

as 'about three miles east' of it.[1] This suggests that it is the gritstone eminence on one of the lines of the Craven fault now called Ryeloaf and 1794 feet high; and enquiries establish that this hill was formerly called Inglehow.[2]

The other problem is more interesting, and concerns his knowledge of the mountains in the English Lake district.

It is a curious fact that all the references to the localities of mountain plants mentioned in the *English Catalogue*, apart from those to Snowdon and Cader Idris, are to 'Hardknot and Wrenose'. It is the more curious when we discover on referring to the county maps in Gibson's edition of Camden's *Britannia*, the edition to which Ray contributed lists of the rarer plants, that in the Lancashire map 'Wrenose Hill' and in the Cumberland map 'Upon Wrenose' and 'Hard Knot Hill' are the only mountains named in this area. They are drawn as high and isolated peaks, Hard Knot on the site of Scafell and Wrenose at the junction of the three counties, with a building on its summit; but Scafell itself and the mass of mountains round it are completely ignored. In the text of Camden the Esk is said to rise 'at the foot of Hardknott, a steep ragged mountain on the top whereof were lately dug up huge stones and the foundation of a castle not without great admiration, considering the mountain is so steep that one can hardly get up it'.[3] It is clear that neither Camden nor Robert Morden, who drew the maps in 1695, nor presumably Ray,[4] knew the names or geography of any of the mountains, Scafell, Bow Fell and Crinkle Crags, which lie behind the foot-hill still called Hardknott and the slopes of Wrynose[5] Fell.

The explanation is obvious to anyone who knows the district or consults a modern map. The ancient Roman road from the harbour of Ravenglass (Clanoventa)[6] and Muncaster ran up the Esk and at Hardknott camp struck over the Hardknott Pass to the valley of the Duddon, followed it under Wrynose Breast to the Wrynose Pass and so to the Brathay and

1 *C.A.* p. 68, describing *Rubus chamaemorus*.

2 I am indebted for this identification to Mr Frederic Riley of Settle: cf. his book, *The Settle District*, p. 139.

3 Gibson, *Camden*, c. 820: the Roman camp upon Hardknott is pictured as the frontispiece of Collingwood's *Roman Britain*.

4 Commenting on the saying
     'Pendle, Penigent and Ingleborough
     'Are the three highest hills all England thorow'
he says: 'These are indeed the highest hills in England, not comprehending Wales. But in Wales I think Snowdon, Caderidris and Plimllymon are higher': *Collection of Proverbs*, p. 255.

5 Ray's spelling is more correct. 'Wrynose is a vulgar error: the natives call the place Ráynus...a clipped form of the raven's pass Ravenhause': H. H. Symonds, *Walking in the Lake District*, p. 39.

6 Perhaps created a naval base by Agricola: Collingwood and Myres, *Roman Britain*, p. 114.

Little Langdale, going on by way of Colwith, Skelwith and Clappergate to Ambleside (Galava), and over High Street to Ullswater. Ravenglass, 'a haven for ships' as Camden calls it, was the natural port in the seventeenth century for travellers crossing from Ramsey in the Isle of Man. It seems almost certain that in 1660 Ray and Willughby, who certainly landed at Ramsey, sailed from Ravenglass, having ridden up the line of the Roman road. Ray, judging by the plants recorded, must have explored some of the mountains north of the route and named the locality by the only places that he knew, the two passes 'Hardknot and Wrenose'. It is also probable that on their return they landed on the Lancashire coast near the modern Fleetwood at the mouth of the Wyre, perhaps at Hackensall, and explored Pilling Moss, which lay between it and the main road at Garstang.

There is no evidence that on his later journey Ray got further than Ambleside. He came from Sedbergh to Orton, then to Penrith and so down Ullswater, where he mentions 'Wethermellock' and explored 'an isle called Householm', and from Ambleside did an expedition to Wrynose and the Furness Fells.

CHAPTER VII

THE YEARS OF VARIED OUTPUT

We owe much more than is intimated to the indefatigable industry of Mr John Ray, a person of polite and incomparable learning and of a most exquisite judgment especially in the History of Nature.

Philosophical Transactions, reviewing *Historia Piscium* (xv, no. 178, p. 1301).

The death of Francis Willughby, on 3 July 1672, in his thirty-seventh year, was a blow to Ray more severe even than the loss of his fellowship at Cambridge. On the earlier occasion he had foreseen his fate and chosen it under the constraint of conscience but with open eyes. It had led to the formation of a plan and a partnership, the worth of which was tested and approved in the next decade. Now almost without warning the partnership was broken and the plan imperilled. Ray's feelings are revealed in the prayer that he offered in the family after its bereavement[1] and in the noble Preface to the *Ornithology*: they are not less plainly shown in his refusal to abandon their joint purpose, in his acceptance of its fulfilment as a debt of honour to his dead friend, in the energy with which he took up the immediate task of caring for the children and perpetuating the work entrusted to his charge.

The loss of Willughby was inevitably irreparable. His portraits, the picture probably by Gerard Soest, engraved rather badly by Lizars,[2] and the bust by Roubiliac in Trinity College, bear out the testimony of his partner. He was a man of noble quality, high-minded, sensitive, alert, and energetic, one who could not spare himself, but in Derham's quaint phrase 'prosecuted his design with as great application as if he had had to get his bread thereby'.[3] Little as we know of him or of his career, it is evident that though he belonged by birth and training to the great world of politics and fashion and felt an interest in problems of education and government outside the range of Ray's concerns, he had found in the company of students who formed the Royal Society an environment in which he preferred to spend his life, and in the comradeship of Ray an inspiration to exact scientific research. It is difficult for us to imagine the fascination of the new worlds of exploration, knowledge and ideas opening up before the 'virtuosi' of that time, or the eagerness with which young and ardent minds turned from the stifling repressions of Puritanism and the still more stifling licence of the Restoration to the pursuit of natural philosophy. Willughby had the gifts, the temperament and opportunities for such

1 *Mem.* pp. 57–8.   2 In *Jardine's Naturalist's Library*, XVI.   3 *Mem.* p. 35.

adventures. Grounded in the traditional curriculum of Classics and Mathe-matics, the pupil of Duport[1] and the friend of Barrow,[2] he found new interests with the learned and versatile Wilkins and in the group gathered by him. His first interest seems to have been in archaeology and the studies developed by Camden;[3] but when in 1660 he went to Oxford for serious study of the libraries and collections he was evidently committed to scientific pursuits; and though Ray in the *Cambridge Catalogue* does not hesitate to question his interpretations he can describe his abilities and skill in the highest terms.[4] His chief interest was evidently in insects, but he did a good deal both of field-work and of dissection in the study of birds, mammals and fishes. Ray left the insects largely to him,[5] but shared with him the other subjects,[6] dissecting a bird and a mammal at Padua and a porpoise at Chester[7] and obtaining fishes through Dent[8] and birds from him and Jessop.[9]

In 1668 he married Emma, the second daughter and co-heiress of Sir Henry Barnard; and children followed rapidly,[10] Francis, who died in his twentieth year after being made a baronet in his eleventh, Cassandra, who married the Duke of Chandos, and Thomas, who inherited the estates and was made Lord Middleton by Queen Anne. In the spring of 1669 he was taken seriously ill while on a visit with Ray to Wilkins at Chester: Ray describes it as a tertian fever complicated by malignant symptoms[11] and says the paroxysms were so severe that it was a fortnight before he could be put into a coach and taken home. His health continued poor and he spent the summer quietly at Middleton: but at the end of 1670 Lister wrote to Ray: 'I am very glad Mr Willughby is near well again.... Methinks he is very valetudinary and you have often alarmed me with his illnesses.'[12] In 1671 he received by bequest from Sir William Willughby, who claimed kinship with him,[13] an additional estate in Nottinghamshire and lands in Yorkshire; the bequest was disputed by the son of Sir Wolstan Dixie of Market Bosworth, who had married Sir William's sister; a quarrel was forced upon Willughby and gave him a vast amount of business.[14] But at the close of the year he was in London[15] with Ray at the Royal Society and

1 Professor of Greek, Cambridge, 1639–54: tutor of Trinity College and of F.W.: cf. J. F. Denham, 'Memoir of F.W.' (*Jardine's Naturalist's Library*, XVI, p. 58).
2 Cf. letters of Barrow to F.W.; Derham, *Phil. Letters*, pp. 360, 362–5.
3 Cf. an early letter to Courthope in Gunther, *Early Science in Cambridge*, p. 343.
4 *C.C.* Preface, p. 7.   5 Cf. *Corr.* p. 30.   6 *Corr.* p. 39.
7 *F.C.* 58–62, 86–7.   8 *Corr.* p. 15.   9 *Corr.* pp. 16–17, 33.
10 Cf. letter of F.W. to Courthope: *F.C.* p. 36: 'God has blest us with a boy and girl worth both the Indies and I hope there is a plentiful ovarium left.'
11 *F.C.* pp. 122–3.   12 *Corr.* p. 73.
13 Ray questions the accuracy of the claim: *F.C.* p. 128.
14 *F.C.* p. 129.
15 Meetings of 9 Nov. (*F.C.* p. 63), of 23 Nov. and 7 Dec. with Ray (Birch, *History of the R.S.* II, pp. 495, 499).

'meditating a voyage into the New World, that he might, as far as in him lay, perfect his history of animals'.[1] In December Ray went to Wilkins at Chester, and was there until the end of February, when he joined Willughby at Middleton.[2] On 8 May Willughby was at the Royal Society;[3] and Ray was planning a final tour. Then, on 17 June 1672, he wrote to Lister that Willughby was dangerously ill and that his own 'simpling voyage' had been postponed.[4] 'He fell into a pleurisy, which terminated in that kind of fever called Catarrhalis, within less than a month after he took to his bed':[5] he died on 3 July, aged 37.

Ray, who in 1670 had refused an invitation to act as tutor and companion to three young men of noble birth on a continental tour at the salary of £100 a year with all expenses paid,[6] now had also to refuse to join his friend Lister at York;[7] for by Willughby's will he was left as one of five executors, the others being his brother-in-law Sir Thomas Wendy, Mr Barnard,[8] Mr Skippon and Mr Jessop, and in addition given the care of the education of his two sons[9] and an annuity of £60 a year.[10] The obligation to carry on and publish Willughby's work was an additional, perhaps a stronger, tie. 'I am', as he wrote from Middleton to Courthope,[11] 'like now to set up my staff here, at least so long as my old lady[12] lives.'

In this resolution he was confirmed by the death of his other great friend, Bishop Wilkins of Chester, four months after that of Willughby. After his removal from Trinity in 1660 Wilkins had made his peace with Charles II and been made Rector of Cranford in Middlesex and two years later of St Lawrence Jewry. He made his home in London; on the incorporation of the Royal Society became its secretary; and in the Great Fire of London suffered the loss of his vicarage and library and of the manuscript of his *Essay towards a Real Character*, a work to which Ray was soon afterwards asked to contribute. In 1668 he was made Bishop of Chester; and there, as we have seen, Ray was a regular visitor, spending several months at a time in the Palace.

Wilkins, whose temper was generous and views liberal, specially befriended the victims of the Act of Uniformity and had joined Sir Matthew Hale in his effort to secure their comprehension within the Church. It is evident that he did his best to persuade Ray to reconsider his attitude and

1 *Orn.* Preface.
2 The evidence is *Corr.* p. 93, letter of 18 Dec. from Chester; *Orn.* pp. 302 and 366, birds seen there in February, and *F.C.* p. 130, letter of 2 March mentioning return to Middleton.
3 Birch, l.c. III, p. 49   4 *F.C.* p. 131.
5 *Orn.* Preface.   6 *Corr.* p. 65.   7 *Mem.* p. 29.
8 Probably his father-in-law, who was not knighted till 1677.
9 *Mem.* p. 36.   10 *Mem.* p. 28.   11 *F.C.* p. 38.
12 Presumably Willughby's mother, the Lady Cassandra, daughter of the Earl of Londonderry.

accept preferment; for one of Ray's early letters to Lister states his inability to comply.[1] But the friendship of the two men was based primarily upon common interest in science and cemented by Ray's work in translating the *Real Character* into Latin and by his high regard and affection for the bishop. Wilkins, travelling to and fro between London and Chester along Watling Street, passed within a few miles of Middleton and apparently made a habit of breaking his journey there;[2] and Ray was constantly moving from the one place to the other. When in the autumn of 1672 Wilkins was taken ill in the London house of John Tillotson, husband of his step-daughter Elizabeth French, and afterwards Archbishop of Canterbury, Ray hastened to him, saw him on 18 November and heard of his death next day. It cut him off from what had been almost a second home, and made permanent residence at Middleton and devotion to Willughby's interests an obvious course.

So Middleton became for the next three and a half years his fixed abode. It was a very attractive estate with a lake in the grounds and the 'vivarium'[3] or small menagerie in which Willughby had housed his animals, with the collections made on their travels and arranged in the winter after their return and with the notes, memoranda and library gathered for their work. The parish contained only a handful of people, a church going back to Norman times and at this period served from Tamworth (where Ray's friend Samuel Langley,[4] formerly a fellow of Christ's College (1644–9), was incumbent) by George Antrobus, Master of the Grammar School there and a man of some distinction.[5] Though the neighbourhood was, as Ray afterwards noted, conspicuously lacking in plants of any rarity, the materials and library were a great resource to him.

His position in the household had during Willughby's life been easy. It was, as the whole literature of the period shows, the custom for every large landowner and most of the smaller squires to have a resident chaplain.[6] Many of the dispossessed clergy, both the royalists in 1644 and the nonconformists in 1662, had found shelter in this way. Willughby, in inviting Ray to his home, was conforming to custom. How far Ray acted as chaplain is unknown: he certainly did so with the Bacons at Friston,[7] and probably took prayers daily at Middleton. But the genuine friendship

1 31 Oct. 1668: *Corr.* p. 30.
2 Cf. letters of 6 Dec. 1668 (*Corr.* p. 36) and of 28 April 1670 (*Corr.* p. 55).
3 Ray alludes to it as the locality of a fungus: *Syn. Brit.* II, pp. 16–17.
4 Cf. *Hist. Pisc.* p. 221, and below, pp. 355–6.
5 Derham, *Mem.* p. 37, calls him minister of the parish: he may have had regular charge of Middleton. He and Langley are both mentioned in a paper submitted to the Royal Society in 1684: Birch, *History of the R.S.* IV, pp. 279–80.
6 Macaulay, *History of England*, I, pp. 160–2, deals at length with the custom and its effects.
7 Cf. Ray's letter to Courthope: Gunther, *Early Science in Cambridge*, p. 346.

between the men saved the position from the indignities too often attaching to it. Ray was eager to help in his friend's business and hobbies, but there was no question of social disparagement. Even after Willughby's death his mother's affection obviously saved him from unpleasantness. But Mrs Willughby, who seems to have been devoid of finer feelings, evidently regarded him as the typical 'levite' of the period and did her best to put him in his place. He foresaw that his position would be precarious as soon as Lady Cassandra died.

Nevertheless he was prepared to fulfil his trust and had already found Middleton a good place for work. Apart from certain researches into the family archives[1] and the preparation of the Tables and Latin version for the *Real Character*[2] he had produced in 1670 in addition to the *Catalogus Angliae*,[3] the first edition of his *Collection of English Proverbs*. The title-page gives a summary of its character—'digested into a convenient method for the finding any one upon occasion [they are in fact arranged in groups according to their general subject, and within the group alphabetically according to the most significant word] with short Annotations [sometimes needed to explain a peculiar word or custom] whereunto are added Local Proverbs with their explications, old proverbial Rhythmes, less known or exotick proverbial sentences, and Scottish Proverbs'. It is a remarkable collection, drawn largely from the native wit of the countryside and the wisdom of the common people, sometimes coarse in its language and in one case, for which Ray made a dignified apology in the second edition,[4] indecent, but generally apposite and often strikingly shrewd. He divided it into three main headings; by subject, proverbs referring to three great human experiences; by form, proverbs that are complete sentences, phrases or similes; by locality, proverbs in dialect or containing local allusions.

The Preface explains the sources from which it was drawn; local proverbs out of Thomas Fuller's *Worthies of England*, these being largely derived from John Heywood; catalogues compiled out of various printed collections, foreign and English; English saws derived from his own knowledge; and Scotch proverbs collected by David Fergusson, minister of Dunfermline, these having been published in 1641, though Fergusson died in 1598. Then he gives a list of the books that he has used; the *Children's Dictionary*; William Camden's *Remaines*—the second edition published in 1614 and containing nearly four hundred proverbs; John Clarke's (or as Ray calls him Clerk's) collection, the *Paroemiologia*; an

---

1 Cf. *F.C.* p. 40, describing the 'memoirs' found at Middleton, many of the documents in which are transcribed in Ray's handwriting.

2 *Corr.* p. 55.   3 *F.C.* p. 126.

4 *Collection*, 2nd ed. pp. 57–8: there is also an admirable defence of 'slovenly and dirty words' in the Preface.

'Alphabetical collection by N. R. Gent.', that is *Proverbs English, French, Dutch, Italian and Spanish: all Englished and alphabetically digested*;[1] George Herbert's *Jacula Prudentum*, printed seven years after his death in Witt's *Recreations*, and separately in 1651; a collection by Robert Codrington, published in 1664 as an appendix to his edition of Hawkins's *Youth's Behaviour*, and separately in 1672; and *Paroemiographia*, published by James Howell as an appendix to his *Lexicon Tetraglotton* in 1659.

Out of this mass of material Ray made what Mrs Heseltine describes as 'a manageable book, one of the best and most useful compilations...it is annotated as no former collection had been; and the notes learned, leisurely and genial, are still invaluable for the study of dialect and folklore'.[2] He states that he began the work ten years before its publication. He must have gathered the items slowly, tracing them to their original sources and storing them in his capacious memory. Possibly his experience of Cambridge warned him that the lore of the village compared favourably with the platitudes of the schools. In any case he must have worked it over, gradually shaping it into order, until William Morden, the Cambridge bookseller who had produced the Appendix to his *Cambridge Catalogue* in 1663, urged him to issue it and 'promised to get it well printed'[3] by John Hayes, the printer to the University. It appeared with the inscription 'by J. R., M.A. and Fellow of the Royal Society'; and was so successful that a second edition was required in 1677[4] and was produced with Ray's name in full and with large additions in 1678.

The Preface to this second edition contains the names of many friends who had helped him with new material. Some of these are familiar; Francis Jessop of Broomhall; George Antrobus, 'Master of the free school at Tamworth', who had married him; Francis Brokesby of Rowley, his colleague at Trinity; Sir Philip Skippon, now knighted; and Andrew Paschall of Chedsey in Somerset, who discussed with him subjects that ranged from tidal senaries to honey-dew and manna. But others only touch his life in this connection; Robert Sheringham of Caius College, fellow from 1626 to 1651 and again from 1660 to 1678, and author of *De Anglorum Gentis Origine* and other works; Michael Biddulph of Polesworth in Warwickshire, who died before the edition was published and whose son, also Michael, was afterwards Member of Parliament for Tamworth;[5] 'Mr Newton of Leicester', doubtless John Newton, Vicar of St Margaret's and afterwards of St Martin's, Leicester and a fellow of Clare Hall; and 'my worthy friend Mr Richard Kidder, Rector of Rayn in Essex' (Rayne near Braintree), who had been a fellow of Emmanuel

1 Published London, 1659.
2 In Introduction to *Oxford Dictionary of English Proverbs*, p. xxiii.
3 *Corr.* p. 55.　　　　　　　　4 *F.C.* p. 136.
5 Cf. Venn, *Alumni Cantab.* I, p. 149.

College and was one of the few who like Ray refused to accept the pro-visions of the Act of Uniformity—though he afterwards became Bishop of Bath and Wells. Kidder supplied the collection of Hebrew proverbs printed at the end of the book. Along with the material derived from these sources there is, as Mrs Heseltine fully recognises and as a perusal of the *Oxford Dictionary* proves, a great number of proverbs of all kinds for which Ray is himself the earliest authority; and most of these have the authentic ring of ripe and rural wisdom. The book was one of the most popular of Ray's works, reaching a fifth edition in 1813.

It is probable that by this time he had also written the two other small books which he published at the expense of Thomas Burrell, cousin to Peter Courthope,[1] his *Collection of English Words* and his *Dictionariolum trilingue*: for the former consisted of material gathered in his expeditions, some of it written as early as 1667; and the latter was largely a by-product of his and Willughby's researches into the Latin names of birds and beasts.

The *Collection* was in fact published in 1673,[2] a tiny volume, containing the first serious attempt to gather and preserve the folk-speech and to dis-tinguish the local dialects of England. Ray was, as we have before ob-served, very apologetic about this strange and original hobby of his, and inclined to defend it to his friends solely on the ground of its usefulness to travellers in outlandish parts. In this first edition the lists of words are not very long; and in order to increase and commend the book he added to them his first catalogues of English birds and English fishes, and also accounts of various mining and industrial activities. But a beginning had been made, and when in 1691 a second edition appeared, the help and interest of a few friends and his own fuller knowledge enabled him to dis-pense with the catalogues and produce a real dictionary of dialect. When Skeat published his edition of Ray in 1874 he said: 'It may very safely be said that on the whole Ray's is the most important book ever published on the subject of English dialects, with the sole exception of such publications as belong to the present century.'[3]

With the catalogues of birds and fishes we shall be concerned hereafter.[4] They and the Tables which he helped Willughby to prepare for Bishop Wilkins represent his first drafts of material for the *Ornithology* and the *Historia Piscium*. The other supplement stands apart from the rest of Ray's work and has never attracted attention. It describes in some detail the

1 Cf. *F.C.* p. 37: 'I was bold to dedicate a small trifle to you [P. C.] which I printed chiefly to gratify your cousin, T. B., and at his instance'—referring to the *Collection of English Words*.

2 It must be the unnamed book by Ray which Hooke bought on 26 Nov. for one shilling: *Diary*, p. 72: it was 'printed by H. Bruges for Tho. Burrell at the Golden Ball under St Dunstan's Church in Fleet Street'. The date on the title-page is 1674.

3 Introd. p. v.   4 Cf. below, pp. 322-4 and 343-4.

processes for mining, smelting and working silver, tin, iron, vitriol, red lead, alum and salt; and also contains notes on wire-work and from Walter Burrell on agriculture. These he had observed and recorded on his travels. Silver mills he had studied at Machynlleth in September 1658[1] and again in June 1662;[2] tin in Cornwall on his second and more leisurely visit; iron during his visits to Sussex, where the Burrells were noted iron-masters; vitriol or copperas (ferrous sulphate) at 'Bricklesey' (Brightling-sea) in Essex from stones collected on the coast of Sheppey;[3] minium or red lead, very briefly described, without locality; alum at Whitby on his visit in 1661;[4] salt at 'Namptwych' in Cheshire and Droitwich in Worcestershire, and by a different process from sea-sand in Lancashire; and wire-work at Tintern in Monmouthshire, visited in 1662.[5] The re-cords are of interest not only on account of the curious technical terms employed which Skeat, who edited them with an introduction and notes in 1874, has carefully collected, but as explaining clearly and simply the methods in use before the age of steam and elaborate machinery.

The *Collection* itself is in the first edition divided into North-country words (pp. 1–56) and South and East country (pp. 57–80). Ray, in the prefatory address 'to the Reader', claims to have derived the greatest part of them from his friends and to have drawn most of the information on etymology, 'for want of sufficient skill in the Saxon, Dutch and Danish languages', from Skinner's *Etymologicon Linguae Anglicanae*[6] or from Somner's *Saxon Dictionary*.[7] He is very modest, indeed almost apologetic, about the results.

The considerations which induced me to make them public were: first, be-cause I knew not of anything that hath been already done in this kind. 2. because I conceive they may be of some use to them who shall have occasion to travel the Northern Counties, in helping them to understand the common language there. 3. because they may afford some diversion to the curious and give them occasion of making many considerable remarks:

and

I am sensible that this Collection is far from perfect, not containing one moiety of the local words used in all the several counties of England. But it is as full as I can at present easily make it, and may give occasion to the curious in each County to supply what are wanting and so make the work complete.

1 *Mem.* p. 130.   2 *Mem.* p. 172.
3 Sloane alludes to these 'copperas-stones or Pyrites' as found on Sheppey after north-easterly storms: letter to Ray, *Corr.* p. 186.
4 *Mem.* p. 148.   5 *Mem.* p. 178.
6 Stephen Skinner, a doctor, left treatises on etymology edited after his death by Thomas Henshaw and published with this title in 1671. Skeat is very critical of their value, but does not blame Ray for accepting them.
7 William Somner published his *Dictionarium Saxonico-Latino-Anglicum* in 1659 on receiving the Spelman lectureship in Anglo-Saxon at Cambridge.

In these respects at least his hopes were not wholly disappointed. The little book sold well: a measure of interest was aroused: further lists were sent to him for the second edition, which he actually produced in 1691: and a contribution of the highest value was made to the study of the English tongue and to etymology in general. Skeat,[1] who never praised easily and is almost savagely critical of the pioneers of philology, commends 'this work of our honoured countryman John Ray' and names him 'the remote originator of the English Dialect Society'.

One small point which Skeat ignores may be mentioned, Ray's curious belief that Hebrew roots had influenced the speech of England. One or two instances of this occur in the *Collection*. It is similar to the theory developed in the *Cambridge Catalogue*[2] that Greek and Latin words were derived from a Hebrew source. Both are of course quite illusory—due to the fact that he knew no Anglo-Saxon or Sanskrit and shared the common belief in the age and perhaps original universality of the language of Israel.

The *Dictionariolum*, called by Derham and in later editions *Nomenclator Classicus*, appeared in 1675 with the imprint 'Typis Andr. Clark, impensis Tho. Burrell', though Smith, the bookseller, in a list of Ray's works in 1692 gives the date as 1672.[3] Less permanently valuable, it was much more successful, going into four editions in Ray's lifetime, in 1689, 1696 and 1703, and many others later.[4] Indeed in days when Latin was still the common speech of educated men this little vocabulary in three columns, English, Latin and Greek—'serviceable', as Derham says, 'not only to schoolboys but to the amendment of our Dictionaries and Lexicons'[5]— must have been indispensable. Ray's account of his intention in the Preface is that he found the names of birds, insects and plants very confused; 'what I chiefly minded was to correct such manifest mistakes as the little insight I have in the history of plants and animals enabled me to discern in former nomenclatures'. But in fact he produced a very complete catalogue of stars, metals, plants, beasts, birds, fishes, insects, parts of the body, diseases, meat, drink, clothes, buildings, God, spirits, faculties, kindred, household utensils, school, church, husbandry, warfare, shipping, arts, measures and numbers—and all in 91 pages.

1 His edition was published for the English Dialect Society, Series B. Reprinted Glossaries XV–XVII.

2 *C.C.* pp. 59–61 (Part II)—a note on the derivation of hyssop, see above, p. 101. The theory of a Hebrew origin for language was strongly held by Joseph Mead, the famous scholar who influenced Henry More and the Cambridge Platonists: he had a book of 'Hebrew radices with the Greek, Latin and English words derived from them': Worthington, Life of Mead in *Works*, I, p. iv: cf. L. T. More, *Isaac Newton*, p. 25.

3 Derham also gives 1672 as the date and says that it was for the use of Willughby's sons. Hooke bought a copy for 8*d.* at Martyn's on 8 March 1675: *Diary*, p. 151.

4 An eighth edition appeared in 1736.   5 *Mem.* p. 28.

Along with these was the book that he was completing in the spring of 1672[1] before Willughby's death, the *Observations Topographical, Moral and Physiological*, printed by Andrew Clark and published by John Martyn, from which we have drawn the story of their journeys on the Continent. This was produced, partly at least, as Ray admitted in the Preface, in order to sell the *Catalogue of Foreign Plants* bound up with it, which he had evidently completed not long after their return. Dedicating the book to Philip Skippon,[2] he writes: 'Being huddled up in some haste upon a deliberate perusal of them [the *Observations*] I find the phrase and language in many places less ornate and in some scarce congruous. But my main aim having been to render all things perspicuous and intelligible (which I hope I have in some measure effected) I was less attentive to grammatical and euphonical niceties'[3]—a criticism which we saw was not unjust. Of the *Catalogue* he adds that he has had it by him for many years, hoping to travel abroad again; but that the deaths of friends and his own age have now made this impossible.

Whatever the defects of the *Observations*, the *Catalogue* is, as continental botanists were quick to recognise, a very remarkable achievement. It is the more remarkable when we remember that until his tour began Ray had been intensively studying the plants of his own country and had no experience except through books and gardens of the Mediterranean or Alpine floras. No doubt in trying to establish the identity and synonyms of British species and to sift the medleys of Gerard and Parkinson he had surveyed the literature and records of north-western Europe. But his studies can hardly have qualified him to note and name the vegetation of Italy, Sicily and Malta or of the Jura and Mont Salève. Probably, just as he spent the winter of 1664–5 studying fish in Rome, he devoted that of 1665–6 to the classification of his notes and herbarium at Montpellier; and in the spring certainly spent some time in the gardens in Paris. These two were the chief centres in France for botanical and indeed scientific research. He would have books and collections and competent colleagues available. His mastery of the subject and of Latin would give him international status and a welcome; and he evidently had happy memories of his visit. In his earliest letter to Tancred Robinson,[4] who was then working in France, he writes:

Monsieur Marchand [Léon Marchand, the doctor of Paris] and Dr Magnol of Montpellier were the most skilful herbarists I met with in France.... M. Fagon

1 Cf. *F.C.* p. 131: he was preparing it in May; in Oct. Oldenburg wrote to Lister that they would soon 'have finished by the press Mr Willughby's and Mr Ray's Voyages, of which I have already seen divers sheets': *F.C.* p. 64. Hooke bought it at Martyn's for 6s. on 26 Feb. 1673: *Diary*, p. 26.

2 He had meant it for Wilkins, but on his death offered it to his own companion.

3 *Obs.* Dedication, p. 3 : John Locke, *Catalogue of Books of Travels*, disputes this.

4 *Corr.* p. 134 and *F.C.* p. 138: letter of 27 July 1683.

[Guy Crescent Fagon, afterwards superintendent of the Jardin des Plantes], a very ingenious person who had the greatest hand in the editing of the Catalogue of the Physic Garden then put forth, was employed in the Laboratory and Apothecary shop belonging to the Physic Garden. M. Marchand I hear is dead; and Dr Magnol has since published a Catalogue of the plants growing wild about Montpellier which I want and know not how to procure here.[1]

That both *Catalogue* and *Observations* were based upon and largely consisted of actual diaries, itineraries and daily notes, such as he kept on his journeys in Britain, is clear enough. To enlarge them by the insertion of documents like the syllabuses of University lectures or by discussions like those upon fossils or local government would not have been a very laborious task; and at Middleton, where the extensive collections of their travels were housed, the material for amplifying the record was available. This is clearly the method pursued in the *Catalogue*. On his journey he had compiled on the spot the lists of plants found at each place as printed in *Observations*; for he adds notes of species which he had seen at home, as, for example, of *Trifolium glomeratum* at Messina: 'I found this lately in England'[2]—actually at Saxmundham in 1662.[3] In preparing the full record he rearranged these local lists alphabetically, added synonyms from the chief authors, and marked with an obelus some[4] of those that he had recorded in Britain. This work cannot have been done earlier than 1668, for *Actaea spicata*,[5] *Galium boreale*[6] and *Thlaspi alpestre*,[7] first discovered during his northern tour of that year, are marked; and perhaps not before 1670, for *Silene nutans*,[8] listed at Geneva without note, has in the *Catalogue* an obelus and the note: 'I observed it lately wild in England'—actually Willisel[9] found it on the walls of Nottingham Castle in 1669 and Ray saw it there in April 1670. Nor can the work be later than this year; for plants like *Alchemilla alpina*, first recorded on his tour in 1671, have no obelus. The clearest indication of date in the text is a reference in the Preface to his visit to Bobart at Oxford as having been paid two years ago; and this visit was probably in May 1669.

1 In *Hist. Plant.* I, Explic. Nom. Ray gives Magnol's *Botanicum Monspeliense* as published at Lyons in 1676: in it Ray's name and writings are not mentioned.
2 *Obs.* p. 280. 3 *C.A.* p. 305 and above, p. 130.
4 Almost all the plants found early in his work are unmarked: e.g. in *Obs.* pp. 264–5 in the list at Leghorn he notes six plants as 'growing wild in England but more rarely', yet none of these has its obelus in the *Catalogue*. Those marked are plants found in 1668 with the exception of *Erica vagans* (*C.E.* p. 43: 'I found it not so long ago in Cornwall', actually in 1667; cf. *C.A.* p. 172) and *Orchis militaris* (*C.E.* p. 79) found at Caversham in 1666 (*C.A.* p. 225 and *Corr.* p. 5).
5 *C.E.* p. 30; *C.A.* p. 71. 6 *C.E.* p. 93; *C.A.* pp. 268–9.
7 *C.E.* p. 21; *C.A.* p. 49. 8 *Obs.* p. 437; *C.E.* p. 70; *C.A.* p. 202.
9 Four more of Willisel's plants, *Veronica triphyllos* (*C.E.* p. 7; *C.A.* p. 340); *Polygonatum anceps* (*C.E.* pp. 86–7; *C.A.* p. 248); *Ornithogalum pyrenaicum* (*C.E.* p. 15; *C.A.* p. 228); *Panicum crus-galli* (*C.E.* p. 50; *C.A.* p. 148), are marked.

Prefixed to the *Catalogue* is a Preface which is in part a first draft of the paper on 'The Specific Differences of Plants' which he presented to the Royal Society in 1674. This part will be considered later, but there are a few other points in it which deserve mention. He begins with a passage which shows how deeply he had appreciated the novelty and riches of continental vegetation.

Whether my readers will enjoy these bare lists of names, I do not know: to me to gaze at the plants themselves freely growing on the lavish bosom of mother earth was an unbelievable delight; I can say with Clusius that I was as pleased to find for the first time a new plant as if I had received a fortune; to discover very many daily that were unknown to me and strangers to our Britain was an ample reward for travel.

He then notes the much greater number of species to be found in the south, and the presence of northern species at higher elevations among them; and comments specially upon the inexhaustible botanical treasures of the Alps. Turning to more general questions he raises the problem of spontaneous generation and admits the possibility of it in the case of Fungi and Algae, where there are no seeds, and of the Mistletoe, which sometimes springs from the lower side of a branch—an admission which he soon withdrew and which proves that the work on this Preface is early.[1] After dealing with his main theme, the criteria of specific distinctness, he has a very interesting paragraph raising the question whether species themselves are immutable. He declares that 'a true transformation of species cannot be denied unless we set aside the evidence of first-hand and reliable witnesses'; and quotes Olaus Worm's *Museum* for the occurrence of seeds of barley and rye in a single ear, and of Johnson's edition of Gerard, which refers to Goodyer for an ear of white wheat containing three or four grains of oats. Whatever be thought of these citations, the fact which he brings forward himself is full of interest. Jacob Bobart the younger had assured him at Oxford that from seed of the Cowslip he had raised both Primroses and Oxlips. These plants certainly hybridise so easily that many botanists have treated them as a single species. But Ray must surely have been misinformed. Cowslip crossed with Primrose might produce the hybrid which superficially resembles the true Oxlip, but could not produce pure Primrose. Possibly Bobart had sown seed of the hybrid Oxlip and obtained the natural result, Primroses, Cowslips and hybrids. In his paper to the Royal Society and in subsequent treatment of the subject Ray does not press the matter. But, as he had admitted to Willughby,[2] for a time his confidence in the permanence of specific distinctions was shaken.

1 In all his later writings he strenuously opposed the idea of spontaneous generation: cf. for fungi, *F.C.* p. 287, for algae and mistletoe, *F.C.* pp. 282–3.

2 Cf. *Corr.* pp. 4, 5, an undated letter describing this visit: and *C.A.* pp. 328–30.

The *Catalogue* itself is as he had described it little more than a bare list of names. In his tour, especially during the long stay at Geneva 'in the proper season for simpling', he had mainly followed Jean Bauhin both for names and for localities. When he found species unrecorded by him he gave them names and brief but sufficient descriptions of his own. Considering how thoroughly the area had already been worked by native and visiting botanists his success in finding novelties is very remarkable. But when he conflated the local records the descriptions were omitted and the only indication of new species is the absence of initials after the title. In consequence the precise identity of what he is recording can only be fixed by reference to *Observations*. For students familiar with the existing literature and visiting the places that he had explored, his work would be useful. For others it is little more than a testimony to his diligence, though its accuracy can be substantiated if it is checked against modern lists. For himself (and this is perhaps its greatest value) it was an ideal preparation for the tasks to which he set himself later. He had seen a larger number of plants in their wild state than any contemporary botanist and made himself a master of his subject: he had added a considerable number of new species to the European flora and collated all the authorities in identifying them: he had gained an international reputation as a scientist and established friendly contacts with all the leaders of learning: he was qualified for what he always regarded as his life's work. But for some years after the appearance of the *Catalogue* he had little time or opportunity to make use of his experience for his special subject.

When his friend's death threw upon him responsibilities that put an end to his field-work, the opportunity for writing and research might seem to have been enlarged. His duties as resident executor involved much business for a few months. His resolve to publish books based on Willughby's material compelled him to turn aside for nearly ten years from botany. But the children were still babies and needed nurses and governesses rather than a tutor. He had little to distract him; and apart from occasional visits to London or to his mother he seems to have been sedentary and free for study. The *Ornithology* and *History of Fishes* show how good a use he made of his time. He might well have settled down to an academic and literary life, dwelling in the homes of wealthy patrons, accepting their discourtesies and dependent upon their forbearance.

Fortunately for his future happiness and work, his character and temperament were not such as could be fully satisfied by cold and impersonal activities. Scholarship, the devotion to study in the library or the field, did not supplant his strong human sympathies, his desire for friendship and affection. It is not for nothing that he claimed gratitude to his friends as his one good quality.[1] There is, as we have seen, in all his re-

1 To Courthope: *F.C.* p. 29.

lationships a warmth of appreciation and a sensitiveness to the worth of others strong enough to make the life of a solitary student impossible for him.[1] He had already complained whimsically to Lister in 1669 that all his friends were getting married—'what is to become of me?'[2]—and when the shock of Willughby's death revealed to him his loneliness, he 'cast his eyes on' a lady of twenty years, Margaret Oakeley, a member of the household at Middleton, apparently governess to the children, and daughter of John Oakeley of Launton near Bicester in Oxfordshire,[3] a 'gentleman', as Derham is careful to add, 'of a younger branch of a family of that name in Shropshire'.[4]

It is one of our most serious losses that we know almost nothing of the lady or of Ray's feelings. Derham, in editing the Lister letters, not only thought that of 30 May 1673 in which Ray told his friend of his hopes of marriage not worthy of publication but only left a sentence of epitome.[5] The preservation of half a dozen brief notes in Ray's handwriting on the back of Lister's reply reveals to us the struggle in his mind. They are cryptic little sentences, jotted down as he tried to straighten out with himself the tangled web of his duty; and we are too ignorant of the circumstances at Middleton, too uninformed by Ray's habitual reticence about his inner life, to interpret them with confidence. But they disclose an outline of his dilemma. His mother, living at the house which he had built and bought for her,[6] the Dewlands, was wholly dependent upon him. His own health was precarious—he was a bad life, and the constant journeys on horseback from Warwickshire to Essex would soon be too much for him. Was it fair to ask a girl of twenty to share it? Moreover, as his fourth point puts it—'You brought up in a different way and not likely to love my prayers'—there was evidently a disparity, whether social or theological. Then if they had children, they 'will never delight in my company for that I shall be old before they come to years of discretion'—a fear which the eleven years' delay before the birth of his eldest emphasised, but which his last book proves to have been to some degree groundless. Finally, there was the position at Middleton. Lady Cassandra was old and frail; Willughby's widow disapproved of Ray and all his works; if he asked Miss Oakeley in marriage and she refused, his loneliness would be complete. However, he took the venture. She agreed. They were

1 This constantly appears in his letters, even in those edited by Derham where such passages were strictly eliminated: cf. *Corr.* p. 13.

2 *F.C.* p. 123.

3 Perhaps a sister or aunt of the William Oakeley, 'clerk of Oxfordshire', whose sons William and Richard were sizars at Jesus College, Cambridge, in 1719 and 1727.

4 *Mem.* p. 29.   5 *F.C.* p. 131.

6 His father had died in 1655 and it is probable that he built the house soon after: but apparently the land was not purchased until 1669: so *Englishwoman's Magazine*, II, p. 274, where this is said to be proved by the title-deeds.

married in Middleton church by the Rev. George Antrobus on 5 June 1673.[1]

The marriage was obviously happy. During the first six years, when they had no fixed home, the position must have been very difficult. There is perhaps a hint in a single sentence of the *Synopsis Quadrupedum*[2] that the middle-aged bachelor with his austere outlook and frugal habits did not adjust himself easily to the intimacies of marriage. But the quality of his work even in the early years and, after they had settled at Dewlands, his manifest contentment and refusal of all opportunities for a change, prove that the partnership was stable and satisfying. His letters and the many references in his unfinished *History of Insects* bear out the allusions by Allen and Dale and give us a picture of a family life simple, united, undisturbed alike by the upheavals of contemporary affairs or the friction of ill-mated personalities. That his wife ever shared or understood his scientific interests is improbable. The few letters from her to Sloane[3] after Ray's death give an impression of practicality and common sense rather than of intellectual power; and she evidently had little say in the disposal of his collections and uncompleted work. But with a small house, very limited means, four young children, and a husband plagued with sores and a weak digestion, it is no small achievement to have enabled him to carry through the vast programme of writing and research that he accomplished in the last two decades of the century. In those years he wrote many thousands of pages for publication and many hundreds of letters to correspondents and friends; he produced constantly new editions of his previous books, the massive volumes of the *History of Plants*, the series of *Synopses* on beasts, reptiles, birds and fishes, the philosophical and religious treatises and their successive revisions; he undertook the collection and breeding of some three hundred species of local lepidoptera—and every entomologist knows the clutter of food-plants and apparatus involved; and he kept his mind free, adventurous and receptive. In addition, during the last ten years his ulcers needed daily dressing, so that the whole morning was often occupied on it, and he and the family were often seriously ill. In spite of it guests were welcomed; the Rays played their part in the small society of the neighbourhood; the children were brought up; the household arrangements went smoothly. Margaret Ray may have been a commonplace person and her marriage no theme for romance. But she must have been a woman of fine character and great if mainly domestic ability. Life can never have been easy or idle for her; and in her husband's achievement is her testimonial and her reward. He needed a helpmeet: she gave him his need to the full.

At the time of his marriage it was evident that life in the household after Willughby's death was not easy. The young widow had no interest in her

---

1 *Mem.* p. 29.   2 *S.Q.* p. 43.   3 Printed in *Corr.* pp. 476–80.

husband's scientific pursuits or scientific friends. She may well have found the influence of a learned, reserved and serious-minded scholar tiresome and over-exacting; and her mother-in-law, for whom Ray had a warm affection, was perhaps not too easy in her advocacy of him. Bishop Wilkins's death had broken the link between Middleton and the learned world; and when Lady Cassandra also died, on 25 July 1675, a change was speedily made. The children were taken out of Ray's tuition; and he and his wife had to leave Middleton. The little paper of Latin instructions for the training of his pupils which Derham found and preserved[1] shows in what spirit he had understood his task, and Thomas, the younger and only surviving brother, stated after Ray's death that 'I must always have a great respect to his memory though on some accounts I had not many obligations to him'.[2] Mrs Willughby soon after her mother-in-law's death married the notorious but immensely wealthy Josiah Child,[3] removed to Wanstead, and disappeared from any contact with her former husband's friends.

By this time work on the Latin version of the *Ornithology* was finished; and progress had been made with the Fishes. The Rays went to Coleshill in the winter of 1675–6 and in April settled at Sutton Coldfield,[4] only four miles from Middleton, evidently unwilling to make a complete break. Early in 1677 the children were transferred to Wanstead, and his reason for staying in the neighbourhood being removed Ray decided to go to his own country. Late in that year his 'honoured friend' Edward Bullock lent him his house, Faulkbourne Hall, near to Witham and only a few miles from Black Notley. His position there is not quite clear[5] and seems to have been uncomfortable. The constant moves and maintenance of his wife and mother had evidently reduced him to real poverty: he was too sensitive and too independent not to dislike living on charity: in any case the house was too big.[6] But he stayed there until Elizabeth Ray died on 15 March 1679 'in her house on Dewlands in the hall chamber about three of the

1 *Mem.* pp. 35–7: note.

2 Letter of 14 Sept. 1705; Boulger in *Proc. of Essex Field Club*. T.W. was only five when Ray left.

3 He rose from an apprentice to be victualler to the fleet and Chairman of the East India Company: Baronet 1678. He bought Wanstead Abbey and lived there. Derham in abstract of Ray's letter to Lister of Jan. 1677 says: 'he sordidly covetous'. He and his brother John were responsible for some of the worst pages in the record of British dealings with India. Cf. Macaulay, *History*, IV, pp. 134–7, etc.

4 Ray always spells it Cofield.

5 Cf. *F.C.* p. 136: Gunther, *F.C.* p. 157, apparently following Boulger, *Proc. of Essex Field Club*, IV, p. clx, says that he was tutor to Bullock's son, and suggests that Bullock knew him through Aubrey by way of the Wylds. It seems more probable that Bullock knew Ray as a neighbour, asked him to look after his house in his own absence, and introduced him to Aubrey—though they must have been acquainted through the Royal Society.

6 The tower and western rooms of the present house are of the late fifteenth century: the rest is later than Ray's time.

clock in the afternoon, aged as I suppose seventy-eight'.[1] In June 1679 he and his wife moved to Black Notley, 'where I intend, God willing, to settle for the short pittance of time I have yet to live in this world'.

This was in fact what he had already resolved. During his time at Sutton Coldfield he had refused a travelling tutorship, offered through John Aubrey, to the son of Sir Francis Rolle.[2] Soon after, as we learn from Hooke's *Diary*, he was approached about the Secretaryship of the Royal Society, vacant owing to the death of Oldenburg on 5 September 1677. During the following weeks there was much discussion, and though on 28 September Ray came up to London, was given a definite invitation and declined it,[3] Aubrey and others did not stop pressing him to change his mind.[4] He must have seemed an obvious selection: but he was not prepared to accept. Debarred by his ordination from secular posts and by his conscience from ecclesiastical preferment, he had nevertheless obtained in the homestead in his native village all that his ambition desired. 'Divinity', he wrote to Aubrey, 'is my profession':[5] if men would not let him teach it in their schools or preach it in their pulpits except at a price which his conscience could not pay, he could at least pursue his calling in the study of the works of the Lord, 'sought out by all them that have pleasure therein', and devote himself to the production of the books which Plukenet declared to be 'the best medium to reach heaven, better than the divinity of the schools'.

NOTE. *Ray's house called Dewlands*

Particulars as to the structure and arrangement of the house were given me by Mrs Turner, who lived in it with her husband and family in 1900 at the time of the fire and is still (1941) living in a new house on the site.

It was in fact simply a larger edition of the smithy cottage, built like it of lath and plaster set in oak frames, and with a fourth room at the back turning the plain oblong into the L-shape so familiar in Essex houses of this period. The cottage consisted of three rooms, a hall or kitchen in the middle with a store-room or laundry to the right and a parlour to the left, built round the great chimney which served hall and parlour and rose directly in front of the entrance door: behind the chimney was a built-in staircase approached from the hall and leading to a landing with two doors into the chambers over the hall and parlour; the third chamber, beyond that over the hall, was originally approached by a second stair from the

1 *Mem.* p. 37.
2 *F.C.* p. 158: he was of East Tytherley, Hants, knighted in 1665 (Le Neve, *Knights*, p. 189).
3 *Diary of Robert Hooke*, p. 316.
4 Cf. his letter of 3 March to Aubrey, *F.C.* p. 159: this Aubrey showed to Hooke, *Diary*, p. 348.   5 *F.C.* p. 163.

laundry. Dewlands similarly had its hall in the centre with kitchen to the right and parlour to the left. But here the main door was into a lobby between kitchen and hall, and the great fireplace stood between these rooms: behind it in the kitchen was a stair to the bedroom above and to a passage giving access to the chamber over the hall. On the left of the hall and approached by a door from it was a passage running from the front of the house to a door leading into the room at the back described as the brewhouse or scullery and fitted with coppers: a door on the far side of this passage led into the parlour, and beyond it beside the passage was a staircase giving access to the chambers over the parlour and brewhouse, and to two cupboards or closets which may well have been used as additional bedrooms: other cupboards traditionally housed Ray's books and collections. The parlour had its own fireplace and chimney; so had the large room over the brewhouse, and this was said to have been Ray's study and library.[1]

Remains of the old barn, shown in the engraving of Dewlands, still survive, though the building has fallen into ruin.

How much land belonged to Ray is uncertain except for the garden and orchard.[2] Morant in 1768 speaks of 'Balls, a farm of about 24 *l* a year, formerly belonging to the great Mr Ray'[3] and in 1706 Mrs Ray told Sloane that 'Mr Ray did not leave £40 per year among us all out of which taxes, repairs and quit-rents make a great hole.'[4] The property is a little farm on the left of the road from Black Notley to Braintree and is described in a surrender of 1807 as 'all that messuage or tenement with the barns, stables and out buildings and seven closes of land thereto belonging called Balls'. It was copyhold, and is now called Notley Lodge Farm. But according to the Court Rolls Balls belonged to the Rev. John Palmer and Susannah his wife from 1687 till 1725. So Morant is in error.

But at the Court on 18 March 1705 it is recorded that 'John Ray, Clerk, who held...by the annual rent of 20 s/ certain lands and tenements had died' and that Margaret, Catharine and Jane Ray his daughters pay the relief. Presumably this property is the 'Bird's lands' which he left to them in his will. At the same Court Margaret Ray, widow, was ordered to mend and scour her ditch between the lands of William Rawlins and Thomas Simpson.[5]

---

1 J. Vaughan, in an essay 'Essex and Early Botanists' in his *Wild-flowers of Selborne*, pp. 128–9, says that the room at the back over the scullery was Ray's study, as the warmest in the house; and his sister, Miss E. Vaughan, tells me that when he visited Dewlands in 1899 this was the clear tradition. This is corroborated by the note in *Corr.* p. 489.

2 See below, pp. 202–3.   3 *History of Essex*, II, p. 124.
4 *Corr.* p. 478.   5 I owe this information to Mr Alfred Hills.

CHAPTER VIII

THE STRUCTURE AND CLASSIFICATION
OF PLANTS

Likewise their Order and Kindred: for the adjusting whereof our Learned
Countryman Mr Ray and Dr Morison, have both taken very laudable pains.
NEHEMIAH GREW in 'An Idea of a Philosophical History of Plants,' Jan. 1673
(*Anatomy of Plants*, p. 1).

The move from Middleton, though it released Ray from other ties and on
his mother's death enabled him to devote twenty-five years to study and
writing, had one serious effect. It cut him off from the collections and
notes gathered during his continental tour, from books and other aids to
his work. In the same letter to Aubrey he had written: 'Mr Willughby's
library remains at his house at Middleton for the use of his son and heir',
and when he was publishing the *History of Fishes*[1] he complained that it
was now impossible for him to get at the records or material.[2] Later on
Sloane was very good in sending him literature: the Braintree carrier was
constantly taking parcels to and fro into which the kindly Irishman would
put a present of sugar for the family. But the want of access to other books[3]
tended to confine Ray to botany, where he had a library of his own which
he had refused even in the years of his homelessness to part with.

Thus as soon as he settled at Dewlands he returned to his proper subject, and took up the matter of classification. This he had hardly yet
touched. The *Cambridge Catalogue* contained only the briefest outline of
a scheme; and his next years were devoted to field-work and direct study
of living plants. But in 1666, shortly after his return from the Continent,
he received an invitation that he felt bound to accept. On 20 October of
that year Bishop Wilkins wrote to Willughby a letter in which, after explaining that the Great Fire of London had destroyed all the impression
of his *Essay* yet printed, except his own proof-copy, and the manuscript
of the remainder, he asked help in improving it.

I must desire your best assistance [he wrote] for the regular enumeration and
defining of all the families of plants and animals. I thought to have found great

---

1 He seems to have finished the manuscript before he came to Dewlands—certainly
some years before it was published. Child at one time seemed willing to help in its
publication, but refused to let his wife spend money on it: *F.C.* p. 136. See below,
p. 349.

2 *Corr.* p. 165.

3 He must in fact have bought regularly; for when he died some 1500 volumes were
in his library.

benefit in this kind by Dr Merret's late book,[1] but it hath not answered my expectation; nor do I know any person in this nation who is so well able to assist in such matters as yourself, especially if we could procure Mr Ray's company to join in it....I would earnestly desire we might have Mr Ray's help if you can contrive it. If I could fully satisfy myself in the methodical enumeration of such things, I would put out the next edition in folio with handsome cuts of all such things as are fit to be represented in figure.[2]

Wilkins, who had known Willughby before leaving Oxford and must have seen something of Ray during his brief tenure of the Mastership of Trinity,[3] is here referring to the Tables of plants and animals that have been already mentioned. They were designed for his last and most enterprising work, the *Essay towards a Real Character and Philosophical Language*, actually published by the Royal Society in 1668. This was an ingenious but premature attempt to produce a universal and simplified speech; and the mere planning of such a project is proof of Wilkins's imaginative acuteness and practical audacity. He realised that Latin was going out of use and that the growth of international trade and contacts created the need for a language easily constructed and generally known. Classifications of flora and fauna were required in order to provide root-syllables for each group in accordance with the general method of his vocabulary. The attempt, like its successors, was unsuccessful and the book almost stillborn. But Willughby, with whom Ray spent most of that winter, undertook the joint task.[4]

The work had to be done hurriedly—Ray afterwards declared that he had to complete the plants in the space of three weeks—for Wilkins's book was already being printed. Moreover, the Bishop imposed severe conditions upon his contributor.

I was constrained [wrote Ray][5] in arranging the Tables not to follow the lead of nature, but to accommodate the plants to the author's prescribed system. This demanded that I should divide herbs into three squadrons or kinds as nearly equal as possible; then that I should split up each squadron into nine 'differences' as he called them, that is subordinate kinds, in such wise that the plants ordered under each 'difference' should not exceed a fixed number; finally that I should join pairs of plants together or arrange them in couples.

What possible hope was there [he adds] that a method of that sort would be satisfactory, and not manifestly imperfect and ridiculous? I frankly and openly admit that it was; for I care for truth more than for my own reputation.

1 'Merret's bungling Pinax', as Ray called it: *F.C.* p. 112, for which see above, pp. 77, 143.
2 Derham, *Phil. Letters*, p. 366: in the Epistle to the Reader prefixed to the Essay Wilkins wrote: 'As for those most difficult tables of plants I have received assistance from Mr John Ray who besides his other general knowledge hath with great success applied himself to the cultivating of that part of learning.'
3 He became Master on 17 August 1659 (Cooper, *Annals*, III, p. 474).
4 Cf. letter to Lister, June 1667: *F.C.* p. 112.
5 Letter to Lister (in Latin), 7 May 1669: *Corr.* pp. 41-2.

Wilkins no doubt had in mind the practical requirements of his system, a master syllable for each squadron, one of nine variant syllables to follow it, and a suffix to denote the species: his interest was philological not scientific. Moreover, like many such amateurs[1] he was fascinated by the harmonies of nature and a mystic sense of the significance of numbers; and thought that his trinitarian arrangement, three squadrons and three times three 'differences', corresponded to the symmetry inherent in the nature of things. That Ray managed both to comply with his demands and to produce a classification superior to anything till then published is in itself no mean achievement. If it was done in three weeks, at least he did not waste much of his time on it.

The Tables,[2] which are given in full (though the threefold grouping is obscured) in *Makers of British Botany*,[3] need not be reproduced here. The arrangement is exceedingly ingenious: it fastens upon many points valid for taxonomy: it establishes a number of genuinely natural groups. But that is all that can be said. It is artificial and made to order; and, as the book to which it was attached attracted so little attention that the Latin version prepared by Ray with immense labour has never yet been printed, it disappeared swiftly into oblivion.

Unfortunately it did not do so without one evil consequence. Robert Morison of Aberdeen, who had been wounded with Montrose at Brig o' Dee in September 1644 and then for sixteen years had been working at botany in exile in France, had been recalled by Charles II in 1660, made King's Physician and Professor of Botany and put in charge of the Physic Garden in St James's Park.[4] He was a man of the Court and had suffered and been rewarded for the royal cause; and his natural arrogance[5] was heightened by his experiences. In 1669, the year after the publication of Wilkins's book, he published his *Praeludia Botanica*. It began with a magniloquent but ungrammatical eulogy of his own attainments and intentions and Latin poems in his own praise. Its contents consisted of three parts; a list of the plants cultivated at Blois; a list of the 'hallucinations' of Gaspard Bauhin in his *Pinax* and criticisms of Jean Bauhin—an 'opus invidiosum' as Haller called it;[6] and in the third part a dialogue between

1 A striking example of a similar abstract and numerical classification can be found in the Quinary System of W. S. Macleay the entomologist and its application to ornithology by Swainson and Vigors: in it both fives and nines appear and genera are multiplied arbitrarily for the sake of symmetry.

2 *Real Character*, pp. 67–120, the first three pages being introductory.

3 Edited by F. W. Oliver (1913), pp. 29–31.

4 Cf. Pepys, *Diary*, 19 April 1664, for his visit to this garden; and Ray's reference to his own visit: Gunther, *Early Science in Cambridge*, p. 375.

5 As his biographer says 'a plain, downright, honest man...fuit vir qui ficum ficum vocavit'—a type still familiar: cf. *Hist. Plant. Oxon.* III, and Blair, *Botanick Essays*, p. 90.   6 Cf. Vines and Druce, *The Morisonian Herbarium*, p. xxxiv.

'Botanographus Regius' and ·'a Fellow of the Royal Society' on the classification of plants. In this he dismissed as wholly useless the efforts of all his predecessors and attacked with special acrimony the method recently.put out by an unnamed Fellow of the Society. 'There is displayed a method in tables arranging classes of plants from the resemblance of their leaves.' '*Ego tantum confusum chaos: illic de plantis legi, nec quicquam didici, ut monstrabo tibi et lapsus et confusionem, alias. Quia degrediendum nunc, non est*',[1] which despite its impossible punctuation presumably means: 'I saw there only a chaotic muddle of plants: I learnt nothing: I will show you the faults and confusion some other time. We must not waste time on them now.' Nothing even in that age of plain speaking could be more brutal. The two men had apparently never met: but Ray had called upon him in his London house in the spring of 1662, and though he was away had been shown his garden and plants[2] and had previously given to Worthington an appreciative account derived from Willughby of his work in St James's Park.[3]

Ray was cut to the quick. He hated controversy, never in any published work disparaged anyone, indeed was not only scrupulous in acknowledging merit and obligation, but often far too generous in his estimate of other and of younger men. The attack was as unfair as it was unexpected; for the form of the leaf had only been used in one of the three squadrons and there, in the main, only when a genuine relationship could be established on other grounds. No doubt his consciousness that the work was artificial and only undertaken under pressure increased his sensitiveness. In any case and for the only time in his history he poured out his feelings to Lister in the letter already quoted.[4] 'Dr Morison in a little book lately published, called *Praeludia Botanica*, has taken exception to the Tables and, without naming him, to their author, whether deservedly or unworthily it is for others to decide.' Then after explaining the conditions under which the Tables had been composed, he adds:

Nevertheless I despise that particular writer with good cause. Although he is so ill-equipped that he cannot even write decent Latin, he flatters himself in such bad taste and is so impenetrably conceited that he scorns men a thousand times more learned than himself and thinks himself unfairly treated because he has not been promoted long ago to a professorial chair.[5] But as long as he sneers so fatuously at the Royal Society, he makes himself ridiculous to all sane and decent-minded people.

1 Morison, *Praeludia Botanica*, p. 476. No wonder that Ray complained that Morison was 'no grammarian and full of errors': *F.C.* p. 286.
2 Cf. letter to Courthope in Gunther, *Early Science in Cambridge*, p. 375: cf. above, pp. 121, 151.
3 Letters of Worthington to Hartlib, July and Aug. 1661: *Misc.* pp. 261, 267.
4 *Corr.* pp. 41–2.
5 Morison was elected to the Chair of Botany in Oxford in Dec. 1669.

But in public he said nothing. In the Preface of his *Methodus*, after expressing his debt to Cesalpino and Jung, there is a dignified acknowledgement of obligation to 'Robert Morison, Doctor and Professor of Botany at Oxford, lest I should convict myself of depriving anyone of his meed of praise or of decking myself in borrowed plumes'.[1] When preparing his *Historia Plantarum* he tells Robinson that he had hesitated to continue it because 'he had some expectation of Dr Morison's work';[2] and later that he has decided to continue it because no Englishman since Turner has attempted it—'Dr Morison is a Scotchman, so I make not him an English herbarist nor pass any sentence on his performances: *judicio stetque cadatque tuo*'.[3] In the *Historia* itself there is a tribute in the Preface[4] to him and to the Bauhins, whose work Ray greatly and rightly admired, and in the List of Authorities a note of Morison's corrections of Gaspard and after mention of Morison's books a brief comment expressing regret at his criticism of others and disappointment at his own work. It is only in the last of his botanical writings that Ray alludes to his own grievance. Speaking of the Tables, he says:

Grievously offended by this Dr Robert Morison of Aberdeen, afraid that he might lose some of the reputation—a great and not undeserved reputation among botanists—won by the publication of specimens of the Method which he used to boast that he had derived not from books but from nature, and resenting the fact that I had put my sickle into his harvest, castigated in unworthy fashion the Tables and their unnamed author.[5]

Yet Morison[6] would not let the matter rest. Patrick Blair of Dundee, his champion, in his *Botanick Essays*[7] records that in his public lectures he used to condemn Ray as one who 'studied plants more in his closet than in gardens and fields' and adds 'this was the ground of contention betwixt them...and it was the tartness of this severe though true reflection which created such a resentment in Mr Ray against Dr Morison'.[8] 'It was a great failing in Mr Ray to have such a resentment since 'twas Mr Ray who first gave the ground of offence.'[9] We may regret that the two men were estranged: but seeing that Morison had published nothing when Ray's Tables appeared and made a gross and gratuitous attack upon him, the

1 *Meth. Plant.* Preface, p. 7. Ray is surely glancing not only at Morison's insistence on his originality, but at his suspected and now proven plagiarisms.

2 *Corr.* p. 146.   3 *Corr.* p. 156.

4 *H.P.* I, Preface, p. 3.

5 *Meth. Emend.* Preface, pp. 1, 2.

6 This account of him is chiefly derived from Prof. Vines's work, the essay in *Makers of British Botany* and the introduction to the *Morisonian Herbarium*.

7 Published in 1720: few more misleading and unfair criticisms can ever have been printed than Blair's attack upon Ray: l.c. pp. 76–101.

8 *Botanick Essays*, p. 100.

9 L.c. p. 91: Blair is wholly ignorant of the facts.

quarrel was not of Ray's making; and if Morison accused him of indifference to field-work, he was guilty of a foolish falsehood. Ray had seen the vast majority of British and a great number of continental species in their wild surroundings, constantly insisted that closet knowledge of dried specimens was insufficient, and used his little garden to the full. Morison had done some field-work in France but in Britain never moved outside the Physic Garden and Whitehall.

Poor Morison, his arrogance brought him little profit. The Royal Society ignored him; Oxford got tired of his 'harangues' and made fun of his Scottish accent;[1] the king left his salary unpaid until he was compelled to beg money from the archbishop;[2] the new method which he claimed to have discovered from nature was found to have been copied from others: his exposition of it in the one volume of the *History* that he lived to publish was confused and inconsistent; his work was carried on by Jacob Bobart the younger, and when it appeared had incorporated much from Ray. His career was cut short in 1683 by an accident. While crossing the Strand from St Martin's Lane to Northumberland House he was knocked down by a coach, fractured his skull and died in his house in Green Street, Leicester Fields. Tournefort,[3] who from his connection with the French Court was eager to commend him, declared that he would have wished to praise him if he had not so extravagantly praised himself; and that it ill became him to depreciate and ignore Gesner, Cesalpino and Colonna, seeing that he had copied whole pages out of the works of the two last. Linnaeus,[4] who gave him credit for reviving the interest in classification, yet concluded that 'all that is good in Morison is taken from Cesalpino, from whose guidance he wanders in pursuit of affinities rather than of characters'. The discovery of Morison's copy of Cesalpino heavily annotated in his own hand gives the lie to his explicit repudiation of indebtedness and to his claim to have drawn his method solely from nature.

From this rather unsavoury topic, made the more unpleasant by the efforts of Blair and others to use it against Ray, we can turn to the preparation of his first serious essay in classification, the *Methodus Plantarum Nova*.

Immediately after the compilation of the Tables Ray had begun a series of investigations which gave him a deeper insight into the principles which determine taxonomy. The work for Wilkins had proved that he realised the need to take the whole structure of the plant into account rather than to fasten, as Cesalpino and Morison did, upon a single character. But to

1 Cf. passages from Wood, *Life and Times*, III, pp. 17, 49, quoted by Vines, *Morisonian Herbarium*, p. xxvii.
2 Cf. letter in Vines, l.c. pp. xxvi–vii.
3 *D.N.B.* Art. on Morison, quoting Tournefort.
4 Cf. *Correspondence of Linnaeus*, II, p. 281, in Vines, l.c. p. xxxviii.

do so involved a more searching enquiry into the physiology and growth of plants than he had yet undertaken. To this he now devoted himself, with results that appeared from time to time in the records of the Royal Society. It is to these papers that Ray first owed his reputation as more than an enthusiastic collector and observer: in them he made a great contribution to botanical science.

Before describing them mention should be made of a colleague in the Society whose work, though it never came to fruition, may have encouraged Ray in his researches. Daniel Coxe[1] had gone up to Jesus College, Cambridge, in 1659, had been a scholar there in 1661 and was elected F.R.S. in 1664. Later in life he wrote on chemistry and medicine and became a distinguished colonial adventurer. But at first he showed a strong interest in botany. On 19 April 1665 he submitted to the Society a paper of 'Enquiries concerning Vegetables' printed by Birch[2] and showing considerable acumen. His opening sentence: 'It is impossible to compose an exact history of vegetation till we understand the nature of the ground as a matrix', and the expansion of it, in comment upon the relation between soil and plant, raises an issue familiar to us but unrecognised except by popular systems of classification at his time. He follows it with a list of seventy-eight questions dealing mainly with crops and fruit-trees, but also with the distribution of seeds, the variation of flower-colour, the seed-leaves, and with other questions of growth and environment. Apparently he intended to follow up this paper with a history of vegetables, and on 11 June 1668 Ray was asked by the Society to assist in the production of this work and promised to contribute what he could. Whether the two men ever collaborated is unknown: there are no references to it in Ray's writings: but Coxe's paper may well have suggested to him certain lines of study. Certainly it was at this time that his own researches began.

The earliest of his published results is the famous record of 'Experiments concerning the Motion of the Sap in Trees', communicated by Oldenburg on 10 June 1669 and printed immediately afterwards in *Philosophical Transactions*, IV, no. 48, pp. 963–5.[3] Ray and Willughby had bled a number of trees, Birch, Sycamore, Walnut and Willow, at Middleton in the early spring and summer of 1669 under different climatic conditions and with carefully noted results. They established the presumption of movement by simple but ingeniously planned experiments which not only led them on to more elaborate investigations, but aroused much interest and started other workers, Martin Lister always ready for

1 Cf. Venn, *Alumni Cantab.* I, p. 408.
2 *History of the R.S.* II, pp. 32–40.
3 Reprinted in *F.C.* pp. 45–7: further letters from Ray and Willughby are in *Phil. Trans.* V, no. 57, pp. 1166–7; no. 58, pp. 1191–2; no. 68, pp. 2069–70; VI, no. 70, pp. 2125–6.

new hobbies, John Beale, author of *Aphorisms concerning Cider*, and Ezerel (or Israel) Tonge, then rector of a church in London and afterwards notorious as the ally of Titus Oates, the first of whose three papers on the subject appeared in the *Transactions* in the spring of 1670.[1] It is also at least possible that this awakening of enthusiasm encouraged Nehemiah Grew, who had been working at the structure of plants since 1664, to submit his first paper 'The Anatomy of Vegetables Begun', to the Royal Society in May 1671 and to publish it before the end of the year.[2]

This paper of Ray's, though its experiments were conducted with elementary equipment and yielded no very important results, was in fact the first systematic attempt to study the physiology of the living plant, and thus opened up a new field of research and gave a new direction to botanical science. Its reception reveals both the reason for its popularity and the influence which for many years distorted the treatment of the subject. On hearing it the Society requested its authors to experiment 'whether there be any circulation of the juice in vegetables as there is of blood in animals'.[3] Harvey's great discovery supplied an analogy which excited attention, even if it gave a bias to study and led to interpretations which were not in fact correct.[4] It was not until the work of Stephen Hales,[5] whose *Vegetable Staticks* was published in 1727, that theories of circulation were discredited: and Hales drew his love of botany from Ray's work, explored Cambridgeshire with his *Catalogue* in hand,[6] and took his researches as his starting-point.

These first enquiries did not contribute directly to classification. The subject which Ray next laid before the Society led to one of his most important and enduring achievements in it. This was his paper, 'A discourse on the Seeds of Plants', presented in accordance with the regulation engaging its members, if able and willing, 'to entertain the Society once a year with a discourse grounded upon experiment'.[7] He sent this and another contribution on 'The Specific Differences of Plants' to Oldenburg on 30 November 1674, remarking that though the latter had been in part already published in the prefaces to the two Catalogues, the former was 'but inchoate and imperfect', and adding: 'I hope (God willing) next spring to prosecute and perfect my design of distinguishing plants by the

---

1 It seems to have been actually written in 1661: cf. the account in Worthington, *Diary*, II, p. 69.

2 Cf. Oldenburg's letter in *Corr.* p. 98.

3 Birch, *History of the R.S.* II, p. 382.

4 Thus e.g. Blair, *Botanick Essays* (1720), devotes his last paper to proving 'the circulation of the sap in plants to be the same with that of the blood in animals'.

5 Born in 1677; fellow of Corpus Christi College, Cambridge, 1703.

6 Cf. the account of Hales's simpling in Stukeley's *Diary*, quoted by Gunther, *Early Science in Cambridge*, p. 383.

7 Cf. Birch, *History of the R.S.* III, p. 162 and Ray's reply, *F.C.* p. 67.

content of the seed. I should not have presented them so soon to the Society but that I have nothing better to send, and am unwilling to be wanting in the carrying on so good a design as they have now set on foot'.[1] The papers were read and registered on 17 December;[2] and on the 21st Oldenburg wrote to give him the thanks of the Society.[3]

Despite Ray's apologies his study of seeds had already disclosed to him a decisive distinction in the character of the seed-leaf. 'The greatest number of plants spring out of the earth with two leaves, for the most part of a different figure from the succeeding leaves...the seed-leaves are nothing else but the two lobes of the seed'—though as yet he hardly emphasises its importance, this is the recognition of the dicotyledonous plant. 'Of seeds that spring out of the earth with leaves like the succeeding...nor have their pulp divided into lobes'—these are the monocotyledons. The paper contains a large number of observations as to the form and growth of the seeds in these two classes; and it is clear that he has not yet fully grasped their significance: but the basic facts have been observed and recorded, and a great step taken towards a scientific classification. This paper, published a year before the first part of Malpighi's *Anatome Plantarum*,[4] gives Ray a plain priority.

The second, on the specific differences of plants, is as he confessed a condensed version of opinions already expressed in his Catalogue, principally in the *Catalogus Stirpium in Exteris Regionibus*, published with his *Observations* in the previous year. But the importance of the subject not only for botanical and biological studies in general, but for his work in classification, is so great that the paper deserves careful attention. His opening sentence explains this:

Having observed that most herbarists, mistaking many accidents for notes of specific distinction, which indeed are not, have unnecessarily multiplied beings, contrary to that well-known philosophic precept; I think it may not be unuseful, in order to the determining of the number of species more certainly and agreeably to nature, to enumerate such accidents and then give my reasons why I judge them not sufficient to infer a specific difference.

Hitherto, in fact, no attempt had been made by any student to define what constituted a species. Mouffet in dealing with insects had classed larvae and imagines as belonging almost to separate orders; Rondelet had separated fishes on grounds of sex or age or local variation; Gerard and Parkinson had multiplied the number of species almost arbitrarily and even the Bauhins had included many mere varieties and not a few repetitions. Criteria were sorely needed and these Ray set himself to supply.

He first makes a list of the accidents; differences in size, scent and taste; in the colour and shape of the root, as in carrots; in the number of angles

1 *F.C.* p. 68. 2 Printed in Birch, l.c. III, pp. 162–73 and *F.C.* pp. 70–83.
3 *Corr.* p. 114. 4 Published in two parts in 1675 and 1679.

in the stalk, as in purple loosestrife; in the variegation or curling of the leaf; in the colour, doubleness, gemination ('such as we call hose in hose'), fistulousness, as in the double-daisy, and proliferousness, as in the childing daisy, of the flower; in the size and taste of the fruit and the colour of the seed. All these he regards as insufficient to prove distinctness. Of size, though he grants that there are limits, yet between plants of the same species there is 'a very great latitude, of ten sometimes to one, wholly to be imputed to the soil, the season, the climate or some other external circumstances': this he illustrates by reference to the transplanting of mountain species and the differences in sheep or horses. Variegation of leaves 'is only a symptom of a morbid constitution'. Diversity of colour is due either to climate ('many animals on high mountains as in northern countries are not rarely found white, as for example bears, foxes, hares, ravens,[1] blackbirds') or to nourishment, as illustrated by the great variability of domesticated animals and birds. Other varieties cannot be distinct species because they 'spring frequently from the seed of the same individual plant' and often degenerate 'if they stand long in one place without culture': they can also be artificially produced. It is one of the outstanding qualities of Ray's achievement that he refused to base specific differences on other than structural qualities. He laid down the lines for all his work in this paper.

A third paper, which is at once an abridgement and a more orderly treatment of the same theme, is that 'on the number of plants'. Ray, as his covering letter to Oldenburg indicates,[2] was dissatisfied with the form of the papers on seeds and species; and this may well be a more leisurely version of the latter, intended for the Royal Society but when finished too similar to its predecessor to be worth publication. It seems to have been found by Derham among Ray's papers after his death, and was first printed at the end of the *Philosophical Letters*.[3] It picks up and restates, often in the same words, the main arguments of the earlier paper, and as Gunawardena has argued[4] must certainly have been written between its date and that of the *Methodus*. Its only importance is in its final section, that dealing with the possible disappearance of species. Previously Ray had confined himself to a sentence stating that 'as is generally acknowledged by philosophers and might be proved also by divine authority, God having finished his work of creation' the number of species is 'in nature fixed and determinate'. Now he drops all appeal to authority and argues instead that '1. though it is absolutely and physically possible yet it is

1 The inclusion of the white Raven is probably due to the specimen bred in Cumberland and seen on exhibit by Evelyn, *Diary*, 4 Oct. 1658 and later in St James's Park, *Diary*, 22 Feb. 1665.   2 *F.C.* pp. 67–8.
3 P. 344: Lankester has printed it at the end of the *Memorials*, pp. 207–14.
4 L.c. pp. 139–40.

highly improbable that any species should be lost. 2. though some species should be destroyed yet it is impossible morally that any man should be sure thereof.' His view is based upon the fact that he does not know of any species confined to a single locality, and upon the difficulty that considering the size of the world it would be impossible to prove its extinction. Thus though he still clings to the idea of a creation once for all, he no longer states it on a priori and dogmatic grounds.

The publication of these papers proves that even during the years when he was devoting himself to discharging his debt to Willughby he had never given up his primary interest and was steadily preparing himself for a return to it. As early as 1674, in the dedication of his *Collection of English Words*, he had declared his intention of producing 'a Method and History of Plants'; and when he got to Faulkbourne and had made good progress with the *Historia Piscium*, he wrote a long letter to John Aubrey in which he outlined his plans for future work. Aubrey, who was then living with Hooke at Gresham College and devoting himself to the Royal Society, had as we have seen suggested a tutorship to Ray[1] and then asked him to become Secretary of the Society. Deeply touched by an offer which he nevertheless felt obliged to decline, Ray explained his refusal by a survey of the position of botanical studies.[2]

There has not to my knowledge been published any general History of Plants since Bauhinus'. Yet is not that the most full and comprehensive book of that subject extant; Parkinson's *Theatre* and *Garden* together containing many more species: and no wonder, seeing Bauhinus his History, though published since, was written before Parkinson's.[3] Caspar Bauhinus (younger brother to John) his *Theatrum Botanicum* of which the *Pinax* is extant would have been the most perfect and complete work of the kind...but since his death there has been only the first tome printed, his son and executors finding, I suppose, the charge of proceeding with the edition of the next too immense for them....I suppose that his History together with what Fabius Columna,[4] Prosper Alpinus,[5] Cornutus,[6] our Parkinson, Geor. Marggravius, Dr Morison and Paulo Boccone have written concerning plants, doth complete the History of all yet extant, except some few stragglers whose descriptions are scattered here and there in narrations of voyages and travels, catalogues of gardens and musea, philosophic transactions etc. I here make no mention of Johnson's *Dendrologia*[7] because I have not read and examined it and because in his other works I find him a mere plagiary and

1 Cf. *F.C.* p. 158, the first letter in the series.   2 *F.C.* pp. 159–61.

3 Jean Bauhin's *Historia Universalis* was published in 1651, Parkinson's *Theatrum* in 1640.

4 Fabio Colonna wrote *Phytobasanos*, Naples 1592, *Ecphrasis*, Rome 1616.

5 Sixth superintendent of the Garden at Padua; wrote *De Plantis Aegypti* 1592 and *De Balsamo Dialogus* 1591, a revised edition of both appearing in 1640, and *De Plantis Exoticis*, published posthumously by his son in 1627.

6 Jacques Cornut, a doctor of Paris, published *Canadensium Plantarum Historia* in 1635.

7 John Johnstone, *Dendrographia* 1662.

compiler of other men's labours. Our Parkinson I put into the number because there be many plants in his work not elsewhere described.

After this survey he makes a request and a promise. 'Concerning the improvement of the *Real Character* I have nothing to add, having not lately spent many thoughts in the consideration of it' (this refers to the Latin translation of Wilkins's Essay which Ray had made in 1667–8 but which owing to Wilkins's death had remained, and still remains, unpublished: for it Ray had prepared a revised and improved classification of plants).

One favour in reference to the Latin copy of that book I have to beg of you: that is the use of the Tables of plants for some short time. The reason of this request is because I intend shortly to publish a General Method of Plants which I hold myself obliged by promise to do and have been lately solicited by friends speedily to make good my word. Those Tables[1] I drew up for the Bishop with all the exactness I could, with subserviency to his design, but have not myself any copy of them. They would much ease and assist me in the design I am upon. ...I suppose there are among the Bishop's papers those very Tables I sent him written with my own hand, so that if they can be found the fair copy of the work need not be mutilated by taking them out there and sending them. If you please to send them by Jo. Fox, one of the Braintree carriers who inns at the Pewter-pot in Leadenhall Street and comes out of town on Friday morning weekly, they will be sure to come to me: or if it be more easy, do but send them to Mr Martyn's the bookseller and he will take care to convey them hither.

Judging by the two next letters Ray's directions were useless: Aubrey failed to find them.

Nevertheless once he had settled into his own house at Black Notley the work on the *Methodus* went forward rapidly. It was published by Henry Faithorne and John Kersey with the date 1682 on the title-page, a slim octavo volume of 166 pages not including the preface and short introduction or the very full index. The dedication was to Charles Hatton,[2] the younger son of Christopher Baron Hatton, and the friend also of Dr Morison.[3] Visiting Bologna in 1684 Tancred Robinson presented a copy of it to Malpighi and reported that he 'expressed a great respect for you and is not a little proud of the character you gave him in it'.[4]

The Preface is so admirable a statement of the limitations of any classification and of the particular conditions under which Ray carried

1 An account of these Tables and of the reasons which had prevented their publication is given on the last page of *Catalogus Angliae*, 2nd ed. 1677.

2 Ray's first mention of 'his honoured friend' is in 1678: *F.C.* p. 162: Hatton's interest in plants has left no memorial except Morton's record of his successful sowing of Mistletoe (*Nat. Hist. of Northants.* p. 378).

3 Hatton sent him the MS. of Boccone's *Icones Plantarum Siciliae*, published in 1674.

4 *Corr.* p. 142.

through his work that its contents are best summarised by translating his own words.

The number and variety of plants inevitably produce a sense of confusion in the mind of the student: but nothing is more helpful to clear understanding, prompt recognition and sound memory than a well-ordered arrangement into classes, primary and subordinate. A Method seemed to me useful to botanists, especially beginners; I promised long ago to produce and publish one, and have now done so at the request of some friends. But I would not have my readers expect something perfect or complete; something which would divide all plants so exactly as to include every species without leaving any in positions anomalous or peculiar; something which would so define each genus by its own characteristics that no species be left, so to speak, homeless or be found common to many genera. Nature does not permit anything of the sort. Nature, as the saying goes, makes no jumps and passes from extreme to extreme only through a mean. She always produces species intermediate between higher and lower types, species of doubtful classification linking one type with another and having something in common with both—as for example the so-called zoophytes between plants and animals.

In any case I dare not promise even so perfect a Method as nature permits—that is not the task of one man or of one age—but only such as I can accomplish in my present circumstances; and these are not too favourable. I have not myself seen or described all the species of plants now known. I live in the country far from London and Oxford and have no Botanical Gardens near enough to visit. I have neither time nor means for discovering, procuring and cultivating plants. Moreover botanical descriptions often omit or slur over the essential points that decide classification, flowers and seeds, calyces and seed-vessels. So I have sometimes had to follow conjectures and set down rather what I surmised than what I knew.

He then explains that there are three main standards of classification—first from locality; then from use as food or drug or ornament; then 'from the likeness and agreement of the principal parts, root, flower and its cup, seed and its vessel'. The first and second are useless: they tear apart what are obviously akin, and unite what obviously differ. The third is far the best and most congruous with nature. It is notable that in this section he agrees with Morison in his Dialogue,[1] though he, unlike Morison, does not name the authors of the methods that he rejects—d'Aléchamps in his *Historia Plantarum* and Besler in the *Hortus Eystettensis*.

A warm tribute is then paid to Andrea Cesalpino, 'the first so far as I know to classify plants by the number of seeds and seed-vessels developed from each flower and from the position of the *corculum seminale* or point at which germination starts'. He prints Cesalpino's Synopsis at the end of his book and refers the reader to it for a statement of the reasons which prevent him from accepting it. But he admits a great obligation, even though for himself he is convinced that corolla and calyx must also

1 *Praeludia Botanica.*

be taken into account, and that the position of leaves on the stalk cannot be altogether neglected.

After a similar tribute to Jung and the *Isagoge Phytoscopica* and to Colonna and the annotations to *Res Medica Novae Hispaniae*, and a mention of Morison and the *Praeludia* and *Historia Universalis*, he claims that his work is not a compilation but is drawn from original observations and yields a new Method. In calling it a General Method he must not be supposed to include all genera of plants; for not half of these are yet known or described.

Finally, he draws attention to his use of the term petal, which following Colonna he applies to the leaf of a flower, and to the distinction between trees or shrubs and herbs, which is that the former produce buds, that is new shoots, 'foetus novellos', annually in summer, buds covered with scales during the winter. Then he asks his readers to point out any blunders or mistakes that they detect with a view to their correction: 'provided this is not done merely for contentiousness or vilification, it will be a kindness very welcome to me'.

The prefatory matter is concluded with a list of the characteristics of the chief kinds (orders as we should call them) of herbs; and with a division of herbs into three classes—'with no or imperfect flowers', 'with perfect flowers and bare seeds', 'with perfect flowers and seed-vessels'—and these into subordinate divisions 'as an aid to memory'. The insertion of these tables is not explained: and in this book the classes are not stressed, though Ray constantly returns to them later. Evidently as printed here the threefold division is at its earliest stage; for he has not yet adopted the primary distinction into dicotyledons and monocotyledons and so groups the Grasses with the *Apetala* in Class I and the Bulbs with the six-petalled in Class III.

The book itself opens with five introductory essays, the first three of which are an expanded and revised version of his Royal Society papers. The first discusses seeds, their size and number, the length of their viability[1] and means of germination, the analogy between the development of vegetable seeds and of the animal ovum, and finally reproduction by cuttings struck in the earth,[2] a method which he recommends to fruit-growers. The second deals with the seed-leaves, which he describes as peculiar—in size as smaller, in shape as usually undivided, in surface as hairless, in position as opposite, and in structure as pulpy and lacking fibres. These essays are the prelude and largely repeat his earlier paper. The third

[1] Here he quotes Morison, though without discussing the case of *Sisymbrium irio* after the Great Fire in 1666 with which Morison deals in *Praeludia Botanica*. Ray takes instead the rapid appearance of *Brassica sinapis* on dykes in the fens.

[2] He quotes Lauremberg, *Horticult.* ch. XXI, i, and describes a particular method of setting such slips.

expounds the structure of the seed and its embryo, Ray remarking that in all the seeds that he has dissected the embryo is always discoverable, though in some it is fully formed and plainly visible, in others less perfect so that its parts are hard to distinguish. He then lays down a division between the seeds which consist of two lobes corresponding to two seed-leaves, though some seeds of this kind do not actually send up the lobes above ground in the form of leaves, and the seeds of which the seed-plant does not consist of two seed-leaves with radicle and bud. He claims that 'from this difference in the seeds can be derived a general distinction of plants, a distinction in my judgment the first and by far the best of all—that is into those which have a seed-plant with two leaves or dilobate, and those whose seed-plant is analogous to the adult'. Here is the division into Dicotyledons and Monocotyledons which all subsequent botanists have adopted. He then goes on to divide the dilobate seeds into those in which the seed-plant or embryo fills up the whole seed (exalbuminous) and those in which besides the seed-plant there is contained pulp or marrow (albuminous). These in their different forms he then describes at length, introducing a plate from Malpighi of the seeds of Sycamore and Radish to illustrate two of his types. He follows this with a quotation from Malpighi's *Anatome Plantarum*, Part II, saying that it has only come into his hands after his own account has been written—it was in fact published in 1679. In this the word cotyledon is regularly used for what Ray had called lobe: it relates experiments proving that, in Ray's words, the cotyledons supply nourishment to the plant as the placenta does to the foetus in mammals. He then goes on to the monocotyledons and describes the germination of Grasses, of the Fir and of Liliaceous plants, Iris, Arum, Asparagus 'and as I think Peony and Cyclamen'. The essay is literally an epoch-making piece of work: for its details he had the support of the greatest physiologist of the time: but the credit for discerning the significance of these details is Ray's alone. If he had had like Morison or Malpighi a post in a Physic Garden or a University, he would have been able to expand his experiments and discover that Peony and Cyclamen do not belong to the monocotyledons and to fulfil his desire to investigate the whole range of seed-structure. What he had reported to the Royal Society he has now recognised as a principle of profound importance, restated far more systematically and found supported by Malpighi.[1]

The two final sections deal with matter explanatory of the tables of classification. Section iv is concerned with flowers and derived in part from Jung, whose definition Ray quotes and modifies. Flowers consist of

[1] It is interesting that the earliest publishers' review, *Weekly Memorials for the Ingenious*, issued by Faithorne and Kersey and bound up as a volume in 1683, in a notice of the *Methodus Plantarum*, 26 June 1682, draws attention to the seed-leaves and calls them cotyledons.

petals, stamens and style, of which the style is not shed: a flower possessing all these is called perfect: one lacking petals and consisting either of stamens alone or stamens and style is imperfect. Imperfect or stamineous flowers often have perianths that look like petals, such as Buckwheat and Persicaria. Knapweed, Scabious and others are mistakenly called stamineous: they are composites, filled not with stamens but with tubular flowers. Perfect flowers are divided into simple and composite, the latter consisting of a number of florets either as in Scabious with a perianth or without it. If simple they have either one petal or many, the former often so divided as to appear polypetalous but falling singly as in Borage and Bugloss. Each division can be either symmetrical or asymmetrical, instances of the four types being Bindweed and Tobacco, Dead-nettle and Horehound, Stock and Pink, Bean and Pea.

Section v deals with the division into Trees, Shrubs, Sub-shrubs and Herbs, a division which he describes as 'popular and accidental rather than accurate and philosophical'. He points out that trees and shrubs are in fact inseparable, since the same species is sometimes one sometimes the other, and that some plants obviously herbs are not less obviously akin to shrubs. He does not here mention the point about buds made in the Preface, but nevertheless thinks it best to maintain the three classes, dropping sub-shrubs, in deference to common usage, to the approval of most botanists and to the fact that this classification is obvious and not easily replaced.

In view of the full abstract of Ray's system at this stage of its development printed in *Makers of British Botany*[1] it is unnecessary to give a synopsis of it here. In his three sections he treats of Trees in eight tables, Shrubs in six and Herbs in forty-seven, the sequence being broken by brief general statements upon some main divisions, Compositae, Umbelliferae, Monopetalae (*Campanulaceae* and *Scrophulariaceae*, of which he gives alternative groupings in tables 26–7 and 28), Tetrapetalae et Siliquosae (*Cruciferae*, three tables), Papilionaceae and Bulbosae. Appended to many of the tables are notes either to explain the position of species like Sorbus torminalis and aucuparia[2] which might come either into the Pomiferae or the Bacciferae, or to comment upon those of special interest, like the note on the Cocos palm or the following: 'To this table belongs the Cacao tree which produces many nuts enclosed in a common envelope: from this comes that notable paste or confection called Succolata or Chocolate from which is prepared a very wholesome drink in daily use among the Indian and Spanish inhabitants of America.'[3]

1 Pp. 32–4, edited by F. W. Oliver, the essay on Morison and Ray being by S. H. Vines: it gives a careful record of Ray's work in classification but otherwise is rather an apology for Morison than an account of Ray as a botanist.

2 Pp. 31 and 35.

3 Chocolate as a beverage is advertised in London as early as 1657: cf. *Public Advertiser* for 16 June. Pepys drank it, *Diary*, 24 Nov. 1664.

In the first two sections, Trees and Shrubs, he is content to accept a slightly modified form of the traditional classification, basing their arrangement upon the type of fruit. This results in strange anomalies, the Fig among the Apples and Oranges; the Date-palm along with the Olive and the Cherry: the Myrtle, the Yew and the Elder in the same table; Walnut, Almond, Oak and Chestnut together: and the Birch and Alder among the Conifers. Shrubs suffer almost equally ill. He puts all the papilionaceous together, but by adopting thorny or thornless as criteria makes havoc of tables ii and iii. The last table consists of Shrubs which are obviously similar to particular herbs and which he proposes to transfer to their kindred.

Herbs he arranges as follows:[1]

(I) Imperfect flowers. (A) ? No seeds. 1. Algae (including Corals), Fungi, Lichens; 2. Mosses and Horsetails.

(B) Tiny seeds. 3. Ferns (he notes that what botanists reckon the flowers of the Osmunda are merely unexpanded leaves).

(II) Dicotyledons. (A) Imperfect flowers, large seeds. 4. Sexu distinctae, dioecious plants, *Urticaceae* and Mercurialis; 5. Sexu carentes, *Chenopodiaceae* and others, e.g. Alchemilla, *Scleranthus*, and *Lappa*;[2] 6. *Polygonaceae*, including 'Rhabarbarum' (Rhubarb) and the Docks with a note that 'Potamogiton angustifolium' (*Polygonum amphibium*) is a Persicaria.[3] (B) Perfect flowers. (*a*) Compound. 7–12. Compositae (six tables mainly corresponding to natural groups); 13. *Dipsacaceae* (including Eryngium and *Armeria*); (*b*) Simple, with naked seeds. 14. *Valerianaceae* (with a mixture of others, Thalictrum, Fumaria, Limonium, Circaea, etc); 15–16. Umbelliferae (with note that *Bupleurum*, Astrantia and Sanicula belong here); 17. Stellatae (*Rubiaceae*); 18. Asperifoliae (*Boraginaceae*); 19–20. Verticillatae (*Labiatae*) (two tables, shrubby and herbaceous); 21–2. Semine nudo Polyspermae (*Ranunculaceae* with *Malvaceae* and *Rosaceae*); (*c*) Simple, with seed-vessel. 23. Pomiferae (*Cucurbitaceae*); 24. Bacciferae (mixed table of berry-bearing plants); 25. Multisiliquae (*Crassulaceae* and folliculate *Ranunculaceae*);[4] 26. Flore monopetalo (*Campanulaceae*, Gentiana[5] and others); 27–8. *Scrophulariaceae* and others; 29–31. Flore tetrapetalo siliquosae (*Cruciferae*) (three tables with some overlapping); 32. *Papaveraceae*, *Euphorbiaceae* and others very mixed; 33–6. Papilionaceae (four tables beginning with climbers and ending with clovers); 37. Flore pentapetalo, *Caryophyllaceae*, *Cistaceae*, *Hypericaceae*, *Lythraceae*; 38. *Violaceae*, *Resedaceae*, *Geraniaceae*, and

---

1 The Latin names used by Ray are not italicised.
2 But not, as Vines states (*Makers of British Botany*, p. 32), *Artemisia*.
3 Cf. above, p. 87 and *C.C.* p. 124.
4 Here, in spite of his contention about its seed-leaves, he places Paeonia.
5 This and Nicotiana appear also in 28.

others; 39. A mixed table of plants with five petals united at the base, e.g. *Primula, Erythraea*, Verbascum.[1]

(III) Monocotyledons. (A) Imperfect flowers. 40–1. *Gramineae*; 42. *Juncaceae, Cyperaceae, Typhaceae*. (B) Perfect flowers. 43–5. Bulbosae (*Liliaceae, Iridaceae* and *Amaryllidaceae*) (three tables, strong-smelling; scented; scentless); 46. Bulbosis affines (*Orobanchaceae, Orchidaceae* and others including Cyclamen and Pyrola); 47. Anomalae et sui generis, aquatic and terrestrial—such plants as cannot be placed in any previous category.

To comment in detail is unnecessary, but a few notes will not be out of place. At the start it is clear that marine studies have hardly yet begun. Ray collected a number of Algae and recognised that they stand apart from all 'true' plants: but though he mentioned the animal affinities of zoophytes he counts corals, 'seaweeds with hard and stony structure', as plants and regards the sponges as a sort of fungoid algae. Here, as also among the Cryptogams, on which he did much work later, he has few data available. Among flowering plants he is on surer ground. A miscellaneous table like 14 shows how the category 'naked solitary seeds one to each flower' misleads him: but he admits insufficient knowledge of Glaux and that Agrimonia, unlike Circaea, has a seed-vessel rather than a seed and so does not properly belong here. Under Umbelliferae he adds a note: 'so called from the umbella [*sic*] with which women shelter their faces from the sun'; and in it next to Cicuta he lists 'Millefolium aquaticum', the plant which appears in the *Cambridge Catalogue* and which there as also in *Historia Plantarum*[2] seems to be *Myriophyllum verticillatum*.[3] With the Bedstraws, Borages and Labiates he is very successful; and in the next two, tables 21 and 22, divided by the structure of the calyx, if he includes the Mallows, at least he groups Helleborus and Adonis with Ranunculus, *Geum* with *Potentilla*, Clematis with Anemone—and Ulmaria. The stress upon seeds leads to terrible confusion in table 24, though even there Asparagus, Polygonatum and Convallaria come together and Paris is not far off. The next table is interesting: similarity of seed-vessel brings Sedum and Sempervivum together in a first group, and Aconitum, Delphinium, Aquilegia, Paeonia, Caltha, *Trollius* and *Eranthis* in a third (this being a distinct achievement): but in the middle are *Butomus*, Abutilon and Helleborus niger. In table 27 he groups *Impatiens* and Pinguicula with Linaria and adds a note that *L. elatine* and *L. cymbalaria*[4] are to be included. The three tables dealing with Crucifers are very successful, and in

---

1 Listed also in 28.   2 II, p. 1322.

3 If this reference stood alone, we should identify M. aquaticum with *Oenanthe fluviatilis*.

4 This species was first introduced by William Coys in his garden at Stubbers in Essex: cf. Gunther, *Early British Botanists*, p. 313.

one of them he uses the seed-leaf as a character. Table 32 by its heading includes a number of anomalous species: but the division of the Poppies which brings Chelidonium maius into relation with *Glaucium* is good; *Oenothera* is grouped with *Epilobium*, and the species of Veronica, put under Alsine in the *Cambridge Catalogue,* are now restored to their proper kindred. There are some interlopers among the *Caryophyllaceae*—Rock-roses, St John's worts and Yellow Loose-strife—but Alsine, now curtailed, is homogeneous, and he places *Sagina* and *Stellaria* with it: he groups *Lythrum* here in a six-petalled class and in a note says that *L. hyssopifolia* is stated by Dr Morison to have five petals but by J. Bauhin to have six—'I have not observed it closely enough to decide between them'.[1] The structure of the seed-vessel leads him to put Parnassia next to *Saxifraga* in table 38, but to follow it with Viola, Reseda, and Geranium, a not uninteresting sequence. The final table of the Pentapetaloidae is admittedly mixed; and the notes appended to it make clear Ray's dissatisfaction and suggest an alternative classification: he admits that he has not yet studied the seed-vessels of this group and regards his present arrangement as tentative. The same is true of the last table, 47, when having arranged the Monocotyledons satisfactorily he puts together a number of peculiar species, many of them like Cuscuta and Adoxa[2] hard to place, and some exotic plants of whose structure he has no adequate knowledge. On the last page of the book is a note of plants omitted, among which is the Opuntia or Ficus indica, the Prickly Pear, which by this time had become naturalised in the Campagna and which Ray in reply to Robinson[3] had declared to be of American origin.

In the majority of cases, as will be seen, he is successful in making his tables coincide with natural families. He has at least eliminated almost entirely the purely superficial characteristics which had done duty previously, and unlike any of his predecessors has ceased to use mere habits like climbing or mere qualities like hairiness or flower-colour. Where he goes wrong is almost always in cases where a more minute attention to structure is needed, and usually in plants which require categories to themselves. Even to-day it is not obvious to the beginner why *Ranunculus* should be classed with *Aquilegia* and separated from *Potentilla*; or why *Veronica* should belong to the same family as *Antirrhinum*. His work is of course based almost entirely upon European, indeed even British, plants; for the large collections or careful accounts of Indian and American species then being gathered were not yet available to him. When allowance is made for the

1 P. 141.
2 In *Syn. Brit.* p. 98 he groups it with the Bacciferae and near to *Cornus suecica.*
3 Cf. Robinson's report and Ray's reply: *Corr.* pp. 143 and 147.

lack of microscopical study[1] and of a wide range of specimens it will be admitted that he has made a promising start.

His own sketch of Cesalpino's system and his comments upon it make clear his grounds for rejecting the principle of classification by seeds alone. A comparison with Morison's scheme,[2] which was in fact modified by Bobart in accordance with Ray's work, is very much in his favour. His own earlier Tables published in the *Real Character* prove how greatly his ideas have advanced. No one will pretend that his 'Method' is satisfactory: but if it be judged in relation to his predecessors and to the extent of contemporary knowledge, it will be clear that Linnaeus's strictures to Haller, 'in the knowledge of generic principles less than nothing',[3] are neither generous nor even just. It could easily be argued that Ray in fact laid down lines of classification more in accord with genuinely scientific and evolutionary principles than those of his illustrious successor.

NOTE. *Ray's relations with Nehemiah Grew*

It is a curious fact that we should know almost nothing of the relations existing between the two men who contributed most to botany in England at this time. Both were educated at Cambridge; for Grew matriculated at Pembroke in 1659: both were members of the Royal Society, Grew being elected in 1671: both were investigating the same problems, and have left correct and colourless references to one another in their writings. They must surely have met; they should surely have been friends, for they were men of high character, similar interests and generous temper. Yet neither of them at any point shows any personal feeling or indeed any special concern for the other.

In Morison's case we know the quarrel and its effects: but as between Ray and Grew there is no clear sign of any breach or of any envy or antagonism. In the *Historia Plantarum*[4] there is one criticism of an opinion expressed by Grew in his *Musaeum*: but the references to his work on the anatomy of plants are frequent and eulogistic, 'noster Grevius' and Malpighi being constantly mentioned side by side. In Grew's works the *Musaeum* freely acknowledges indebtedness to the *Ornithology*, and in the *Idea* tribute is paid to Ray for his work in classification. But Grew's own obligations are to Glisson the great doctor, to Highmore and Sharrock for their books on the culture and propagation of plants, to Hooke for his *Micrographia*, and to his brother-in-law Dr Henry Sampson.

---

1 Gunawardena strongly criticises Ray for his failure to use a microscope; for he assumes from *Obs.* p. 122, where he speaks of studying fossils with a microscope, that he like Hooke used a glass of high power. In fact Ray calls any lens a microscope, cf. *F.C.* p. 175; and never possessed more than a small magnifying glass.

2 Also set out in *Makers of British Botany*, pp. 29–31.

3 Smith, *Correspondence of Linnaeus*, II, p. 280.   4 *H.P.* II, p. 1674.

When Mrs Arber drew my attention to the method or 'key' proposed and described by Grew for the identification of plants, its resemblance to the tables in Ray's *Methodus* and to the rules for classification in his *Methodus Emendata* suggested that Ray had been influenced by it: but although this is possible it is by no means proven or, I think, probable; for many of the points, e.g. number of petals or flower-colour, which Grew selects are expressly rejected by Ray; and by 1676, the date when this paper of Grew's was read, Ray's own type of table had been already developed.

Similarly, it might be expected that when Grew wrote his *Cosmologia Sacra* in 1702 he would show some consciousness of the existence of Ray's *Wisdom of God* which covers the same ground. Derham's *Physico-Theology* is manifestly dependent upon Ray; and Ray acknowledges his own connection with Henry More's *Antidote*. But Grew's book, though its arguments are often almost identical, contains no clear evidence of connection with Ray's work, and may well be wholly independent. Certainly its general philosophical position is different from that of Ray.

It would be pleasant to establish proof of acquaintance between the two greatest of English botanists: it is difficult to believe that there was no close contact between them: but the evidence indicates that they lived and worked on lines that were strictly parallel.

CHAPTER IX

THE *HISTORY OF PLANTS*

Our countryman, the excellent Mr Ray, is the only describer that conveys some
precise idea in every term or word.

GILBERT WHITE to DAINES BARRINGTON, Letter X, 1 August 1771.

The publication of the *Methodus* marks the close of the first half of Ray's
career as a scientist. By it he had fulfilled a threefold obligation laid upon
him by the embarrassing request of Bishop Wilkins, by the failure of the
Tables, and by the requests of his friends; and so had completed the first
phase of his botanical studies. Hitherto his life had been unsettled: he had
turned his enforced homelessness to good account, collected a mass of
material, discharged his debt to Willughby, and done good service to
British botany. But opportunities for large-scale work in his own field
had been scanty. He was not master of his time or circumstances, and
could not settle down to uninterrupted study or plan a long piece of research. Only when he had finally moved to Dewlands and renounced all
prospect of further travel or of promotion could he begin the larger tasks
for which his experience fitted him. Lists of synonyms and localities and
a scheme of classification were, as he now realised, the proper prelude to
a larger undertaking, a History of Plants which should not merely catalogue but describe, which should not be confined to Britain or Western
Europe but include all known species, which should not be alphabetical
or arbitrary in its arrangement but should illustrate, expand and modify
his Method. The book was actually begun before the end of 1682.[1]

In his own house and his own neighbourhood steady and ordered work
became possible. The house[2] itself, though built for his mother, was large
enough for his needs. It was a long low two-storeyed L-shaped building,
standing back from the Braintree-Witham road, with a good strip of
garden in front, and a barn, a cottage and outhouses on the south side.
There was not much room for growing plants—'a cold soil and an ill-
situated place'[3] exposed to north and east winds. But there was an
orchard[4] in which, as he reported,[5] *Euphorbia platyphyllus* 'comes up

1 Cf. *F.C.* p. 289.

2 It was burnt down in 1900: a photograph, taken shortly before, is reproduced in
Essex Naturalist, XVII, p. 1; and there is an engraving, in *Corr.* p. 459 and *F.C.* p. 8,
presented in 1848 to the Ray Society: cf. note on pp. 179–80 above.

3 *Corr.* p. 179.

4 An ancient pear tree traditionally planted by Ray stood there till recently: cf.
Vaughan, *Wild-flowers of Selborne*, p. 84.

5 *Syn. Brit.* ed. II, p. 184: G. S. Gibson, *Flora of Essex*, p. 274, says that a specimen
gathered by Dale in Ray's orchard is in Buddle's Herbarium, *Herb. Sloane*, vol. 123, p. 30.

spontaneously', and a poultry-yard in which he observed the behaviour of chicks when a hawk came into sight.[1]

Here he seems to have spent almost the whole of the next twenty-five years in an intense activity of study and writing, broken only by occasional visits to London and at first by a few excursions into the neighbourhood, or by the coming of occasional guests. For the first ten years his health seems to have been good; and though for the years 1679–82 letters are almost entirely lacking,[2] all the evidence points to a decade of happiness and reflects his satisfaction at having gained a home, an income tiny but sufficient, and opportunity for uninterrupted devotion to his scientific interests. But, as usual, neither books nor letters give us any clear picture of his life or household. No doubt the tale would have been a chronicle of small doings; the to and fro of the postboy and carrier; the settlement of domestic ways and means; the petty business of the village; the ordered routine of the house, the regular hours for meals and prayers and corre-spondence, and the ceaseless energy of planning, composing and writing the long series of books which were then begun. We can infer it from hints and from his output: but none of his visitors seems to have put an impression of it on record.

During this first decade at Dewlands his children were born. We can gather the dates of their arrival from his letters to Robinson: but the man who after eleven years of married life could announce his fatherhood to his best friend in an apology for the slow progress of his work: 'I have also lately been a little disturbed and interrupted by the indisposition of my wife who was yesterday delivered of two children at a birth, both females',[3] is not a promising subject for intimate biography. Margaret and Mary, the twins, born on 12 August 1684; Catharine on 3 April 1687; Jane on 10 February 1689—we know their names and that Skippon was Catharine's godfather[4] and apparently Robinson Jane's,[5] but of their baby-hood nothing else except that Jane suffered from fits while teething.[6] But we must not infer from this reticence that he had become an inhuman recluse. As soon as the children were old enough there is abundant evi-dence of the depth of his affection and comradeship. We get glimpses from the letters of the family eating Sloane's health[7] in the sweetmeats that he constantly supplied, or playing with Lhwyd's fossils[8] and Aubrey's magnifying glass:[9] and, as we shall see, the unfinished *History of Insects* lifts the veil and shows us the old naturalist encouraging his girls to bring

1 *Wisdom of God*, ed. III, pp. 178–9.
2 For these years there are no certainly dated letters: the correspondence with Lister ceased in 1677 save for one in 1689; that with Robinson and Sloane does not begin till 1683; there are no letters to Aubrey between 1678 and 1687.
3 *F.C.* p. 141.   4 *F.C.* p. 291.   5 *F.C.* p. 292.
6 *F.C.* p. 293.   7 *Corr.* p. 337.
8 *F.C.* p. 261.   9 *F.C.* p. 175.

him in Burnished Brass moths from his garden[1] and Geometer caterpillars from the big oak near the house.[2] And when Mary died in 1698 his reticence gave way and he reveals to his friends the poignancy of his grief.

The neighbourhood was, as he confessed to Aubrey, 'barren of wits, here being but few of the gentry or clergy who mind anything that is ingenious'.[3] His friend Edward Bullock of Faulkbourne Hall was seldom at home and died before 1692; and his son, also Edward, treated the family at Dewlands with marked coolness.[4] The owner of Black Notley Hall was an absentee: the rector, Joseph Plume, was elderly: James Coker, the retired grocer of Braintree who lived at Plumtrees, was the only man of means whose name has come down to us as a friend of the Rays. Society was certainly not very lively.

Braintree was however within a short walk and was a place of more than local importance. The anonymous historian of Essex in 1770 calls it 'a great thoroughfare from London into Suffolk and Norfolk; the Norwich, Bury and Sudbury stage coaches pass daily through it', and adds: 'the principal manufacture is long baize which employs many hands: here is a market every Wednesday: two fairs are held here annually'.[5] In Ray's time it was a centre of the cloth industry and communications with London were regular and reliable. Here he made several friends. Robert Middleton, who had succeeded John Cardell, the successor of Robert Carr, as Vicar in 1678, stayed until he went down into Sussex in 1690, came back on visits in 1694[6] and 1697,[7] and was evidently a good neighbour. He had been a sizar at Christ's College, graduating in 1653, so that his acquaintance with Ray may have begun in Cambridge. Certainly it became intimate; for when he accepted the living of Cuckfield Ray wrote to his old pupil Timothy Burrell introducing him as 'One of the best ministers and the best men that I know'.[8] Moreover, the town was fortunate in its doctors; and one of them, Samuel Dale, speedily became infected by Ray's enthusiasms, eagerly assisted in his botanical work, and took the place in his later years which Willughby had held earlier.

Dr Boulger and the Essex County Society have done good work in recovering for us a knowledge of Dale, a man who played an honourable

1 *Hist. Ins.* p. 182.
2 Blown down in the great gale of Nov. 1703: *Corr.* p. 438.
3 *F.C.* p. 181.
4 *F.C.* p. 182: he married Mary, daughter of Sir Josiah Child, which may partially explain his treatment of the Rays.
5 *New History of Essex*, I, pp. 413–14.
6 Gunther, *Early Science in Cambridge*, p. 354: Gunther has reproduced in facsimile two letters from Ray to Burrell, preserved in the Library of Trinity College.
7 When he stood sponsor to Allen's son: Miller Christy, *Essex Naturalist*, XVII, p. 149.
8 Gunther, l.c. p. 351.

part in the life of the community and was proud to act as the friend and helper of the illustrious naturalist.

Born in 1658 or 1659, and probably in Whitechapel, he may have been connected with Braintree through his father's silk-trade. He was apprenticed to an apothecary there in 1674 for eight years; and seems to have begun to practise independently soon after Ray settled in Black Notley. Under Ray's guidance, as he declares, he was initiated into scientific work, travelled over much of East Anglia in search of plants, and raised many difficult species from seed in his garden. In 1685 he showed Ray the berries of *Adoxa moschatellina*;[1] in 1691 he brought to him the first seed-bearing specimens of *Zostera marina*;[2] and after a visit to Cambridgeshire induced Ray to make two breaches of his own principle by duplicating on his evidence *Linum perenne*[3] and *Geranium sanguineum*.[4] In 1693 he published the first edition of his *Pharmacologia seu Manuductio ad Materiam medicam*, an annotated list of mineral, vegetable and animal drugs, which was revised and reissued in 1705 and 1737: the Preface contains a double tribute to the influence and help of his friend and neighbour 'of whom I am justly proud'; and the book itself contains ample evidence of Ray's co-operation. When Ray turned to the study of insects Dale became his chief assistant,[5] and in 1705 was asked by Sloane to complete the unfinished History. His reply is characteristic:

I heartily thank you for your good opinion of my ability to perfect Mr Ray's Historia Insectorum; I must confess my inclination is good to serve both the widow and the public, but believe this undertaking to be above my sphere. Were it only to finish the English part I do not doubt but with your kind assistance to do it (being better acquainted with Mr Ray's insects than any other man), but the exotic part I cannot fathom, nor am I master of so good a language as anything joined to Mr Ray's would deserve.[6]

All Ray's collections had in fact been sent to him shortly before the great naturalist's death: he prepared the catalogue for the sale of his library: he was indefatigable on behalf of the family: but to act as the literary heir and executor of his friend was unfortunately a task which he was too modest to undertake.

1 *H.P.* I, p. 685.   2 Cf. *F.C.* p. 220 and *Syn. Brit.* 2nd ed. p. 7.

3 In *Syn. Brit.* 1st ed. p. 157 he rejected the procumbent form as only described by Morison, 'and by him neither fully nor accurately'. In *Syn. Brit.* II, pp. 220–1 he separates the erect from the procumbent, saying that the latter had been observed by Dale and that he had seen both species grown from seed in Dale's garden: he added: 'I would not affirm with confidence that they differ'.

4 *Syn. Brit.* II, p. 219: he adds a species with larger and paler flowers and more deeply divided leaves 'found by Mr Dale on the banks of the Devil's Ditch towards Reche'.

5 *Corr.* p. 448 and *Hist. Ins.* passim.

6 Boulger, *Transactions of Essex Field Club*, IV, p. clxii. Dale was not a great scholar: Ray had asked Robinson to translate the preface of the *Pharmacologia* into Latin for him, since his (Ray's) Latin would be recognised: cf. *F.C.* p. 298.

Before Ray's death Dale had become deeply interested in the study of fossils, encouraged by his discovery of abundant organic remains in the cliff at Harwich in 1696.[1] Ray put him into touch with Lhwyd, who was then preparing his *Lithophylacium*; and apparently an interchange of letters and specimens took place between them.[2] These studies Dale published in 1730 in his much amplified and very interesting edition of Silas Taylor's *History of Harwich*,[3] which contains a judicious and accurate list of flora and fauna as well as an account of the fossil-beds.

He was, as his portrait[4] and record prove, a man of robust physique and character and considerable intellectual gifts. Though his practice was widespread and he spent much time in travel, this did not prevent him from taking an active part in local government and affairs. He was for many years a trusted member of the Council or Vestry of Twenty-four, and in 1706 founded an Independent Chapel in Braintree.[5] He lived in a house at Bocking End,[6] was twice married and had a large family. Visiting London not infrequently[7] he kept Ray in touch with friends and interests there. The short but valuable Life, the first to be published, that appeared in the *Compleat History of Europe for the year 1706*, though unsigned, is by him. It is the document printed by Gunther,[8] and now in the Bodleian.

Of the other Braintree doctor, Joshua Draper, we know nothing until in 1689[9] he was joined by the young Cambridge graduate Benjamin Allen. Dickens has immortalised his name; but the man himself has only become familiar to students of the period by the work of Dr Boulger and Mr Miller Christy's articles on his Commonplace books. He had been a pupil of Thomas Gale at St Paul's School before going to Queens' College in 1681, and, in spite of the fact that he never learnt to write intelligible English, was evidently intelligent, alert and self-confident. Early in 1690 we find him treating Jenny, Ray's baby daughter, for what he termed an epilepsy, and Ray himself for a sharp attack of pneumonia. Ray reported to Robinson[10] that he used Riverius's[11] method for Jenny; and his own notes of both cases have been preserved.[12] Shortly afterwards he married

1 So, apparently, *F.C.* p. 301: Ray mentioned it to Lhwyd in March 1698; *F.C.* pp. 269, 277.

2 Cf. *F.C.* p. 282, and Gunther, *Life of Lhwyd*, pp. 463, 466.

3 Quoted in manuscript in Gibson's *Camden*, c. 359: the appendix containing Dale's lists etc. is pp. 257–456.

4 Reproduced in Gunther, *F.C.* p. 108.

5 Enlarged and largely rebuilt in 1818.

6 A Tudor House, remodelled in the eighteenth century, beyond the White Hart Inn, now occupied by Holmes and Hills, Solicitors.

7 *Corr.* pp. 268 (1693), 410 (1702).

8 *F.C.* pp. 4–7.   9 *F.C.* p. 293.   10 L.c.

11 Lazare Rivière, author of *Praxis medica*, 1644, *Observationes medicae*, 1646, and *Institutionum medicinae lib. v*, 1655, all translated by Culpeper in 1658.

12 Cf. Miller Christy, *Essex Naturalist*, XVII, p. 9.

Katharine Draper and developed a large practice, living in the large red-brick house now the Constitutional Club. Ray, who was apt to take his friends at their own valuation, formed a high opinion of his abilities, and this was enhanced by his discovery of the male Glow-worm in 1692[1] and of the Death-watch Beetle (*Arnobium tessellatum*).[2] He examined the 'wells of Epsham',[3] also in 1692, in order to discover the composition of Epsom salts, and in 1699 published 'at his own charge'[4] *The Natural History of Chalybeate and Purging Waters*, a hastily written book for the badness of which he naively but quite justly apologises in the Preface. He certainly collaborated with Ray in zoology[5] and seems to have been a man versatile but superficial, eager for knowledge but credulous and lacking in judgment—probably a bright boy who had never learnt the difference between first-rate and second-rate attainments and was not improved by success in a small provincial town. That he was intimate with the Rays is proved by the fact that in 1697 Mrs Ray was godmother and Ray proxy godfather[6] for his eldest boy, Thomas: but Mary Ray's death in 1698 shook their confidence in him and caused a measure of estrangement. He seems to have been jealous of Dale but to have had a genuine admiration for Ray, strong enough to lead him to choose a grave next to his in Black Notley churchyard.

The neighbourhood, barren though it might be of wits, thus gave him some enlargement of his family circle. But it was from the larger world that he gained shortly after settling at Dewlands the two friends with whom the history of his last years is most intimately connected, Tancred Robinson and Hans Sloane.

Robinson was the second son of a well-to-do Turkey merchant of Yorkshire and was apparently at school in York, where he met Martin Lister, who had set up in practice there in 1670. William, his elder brother, went up to St John's College, probably on Lister's advice, in 1671, and Tancred followed in 1673. He took his M.B. in 1679 and then proceeded to Paris, where he attended the botanical lectures of Tournefort and the anatomy school presided over by Duverney. In 1681 he appears to have sent his first letter to Ray,[7] though only an epitome by Derham of Ray's answer has been preserved. In 1683 Ray sent him a list of queries to Paris

1 *Meth. Ins.* p. 9; *Corr.* p. 252; *F.C.* p. 229.
2 Cf. *Hist. Ins.* p. 167; *Corr.* p. 400.
3 *F.C.* p. 175: Ray reminds Aubrey that he had met Allen at Dewlands in 1692.
4 So Ray to Robinson, *F.C.* p. 303.
5 Cf. *Syn. Quad.* p. 93 for their dissecting together.
6 A fact of some interest in view of Calamy's statement that Ray 'accounted it an error to have sponsors' (*Continuation of Account of Ministers Ejected*, I, p. 122)—a statement which Calamy recognises as inconsistent with Ray's dying confession. In view of this incident Calamy probably exaggerates.
7 *F.C.* p. 286: he seems to have sent some books to Ray.

and in July received a long reply.[1] Soon after he moved to Montpellier
for further study, and thence to Bologna, where he met Malpighi, to Rome
and to Naples. Early in 1684 he went to Venice, thence in April to Geneva,
and so in the summer to Leyden, where he found the *Hortus Malabaricus*
then being produced by the co-operation of various botanists. By 1 August
he was in London, living in chambers in the Temple, where he invited
Ray to visit him at any time. He was immediately elected to the Royal
Society and in February 1685 stimulated it to undertake the publication
of the *Historia Piscium* which Ray had completed some years before. He
was so energetic in his efforts and correspondence that in the same year
when William Musgrave ceased to be Secretary he was elected as colleague
to Francis Aston. Unfortunately a few weeks later Aston's quarrel and
violent resignation[2] led the Society to dispense with paid secretaries and
carry on with Honorary Secretaries (Sir John Hoskyns and Thomas Gale)
and a Clerk (Edmund Halley)—an arrangement which shortly preceded
the suspension of the *Philosophical Transactions* in December 1687, and the
temporary decline of interest. But Robinson kept in touch with scientists
and was the chief link between them and Ray. He visited Black Notley
in May 1686;[3] and Ray stayed with him in London in September
1687.[4]

It is one of the most serious losses to the student of Ray that some three
hundred of his letters to Robinson which were sent to Derham and
epitomised by him have been lost. For Ray, who in the Preface to the
second edition of the *Synopsis Britannicarum*[5] called him 'Amicorum
Alpha', wrote more fully and on a wider range of subjects to him than to
any other of his correspondents. The epitomes give us useful dates and
allusions, but are inevitably lifeless and often irritating; for they constantly
mention subjects of vital importance but leave us ignorant as to what was
said about them. Plainly Robinson, who in his letter from Montpellier had
described himself as 'one that does really admire you above the rest of
mankind',[6] never lost his belief in Ray's unique qualifications and continually urged him to produce a complete Systema Naturae. Of his own
capacity it is not easy to judge: for he seems to have published very little
and is known to us chiefly through his letters. He seems to have had the
qualities rather of an editor and adviser than of an original thinker or
research worker, and was hardly himself in the first rank as a scientist. But
he combined a wide range of interests with a sound and discriminating

1 *Corr.* pp. 132–3: as an indication of the postal arrangements of the time we may
note that Robinson's letter is dated Paris, 12 July and Ray's answer Black Notley,
27 July.

2 The story of the dispute and the scene in which it ended is in Birch, *History of
the R.S.* IV, pp. 449–54, and Weld, *History of the R.S.* I, p. 303: cf. below pp. 353–4.

3 *F.C.* p. 289.   4 *F.C.* p. 291.
5 Final paragraph.   6 *Corr.* p. 138.

judgment, and by the energy of his character, the extent of his correspondence, and the generosity of his temper he exerted a large influence upon the scientific movement. In his own profession he made a considerable reputation, became physician to George I and practised until 1745–6. Ray regularly discussed with him the plans and details of all his work, submitted to him the manuscripts of his books, benefited greatly by his comments and suggestions, and gratefully accepted his help in dealing with publishers and preparing volumes for the press.[1]

Sloane, who had been born in Co. Down in 1660, had gone to Paris for medical studies and had met Robinson there: through him he had been made known to Ray.[2] The story in the *Dictionary of National Biography* that he had met Ray before going abroad is apparently based upon the mistaken dating of a letter from Ray to him printed by Lankester as of 8 June 1681.[3] This is manifestly a copyist's error, almost certainly for 1691; for in it Ray mentions 'the little tract which I ordered Mr Smith[4] to present you with', a reference to the *Wisdom of God*; he expresses regret at not having seen Sloane's 'rare collection of plants', those which he brought back from Jamaica in 1689; he urges him to publish his observations, a suggestion acted upon in the Jamaica Catalogue of 1696; he invites him to Black Notley in terms which imply that he has already been there; he signs himself 'Your affectionate friend and humble servant', a form which he certainly could not have used in 1681; and he addresses the letter to him 'at the Duchess of Albemarle's'—the Duke with whom he visited Jamaica in 1687–9 had died on the island; Sloane settled there on his return from the island;[5] the only other letter sent by Ray with this signature and address is that of 25 May 1692.[6]

It was actually in 1684 after Sloane's return from Paris and Montpellier that the two men began to correspond;[7] and their first letters show how eagerly Sloane offered help to the production of the *Historia Plantarum*. He was an enthusiastic but not always judicious naturalist, a born collector, who like Robinson had met William Courten alias Charleton[8] in Paris, a liberal and energetic friend of science. His services to the Royal Society, which he resuscitated when he became its Secretary in 1693;[9] to the Chelsea

1 In *Syn. Brit.* Preface Ray calls him the 'midwife' whose care has provided a healthy delivery for his book.

2 *Corr.* p. 338.   3 *Corr.* p. 130.

4 Smith published nothing for Ray till 1690.

5 Hooke visited him there in Aug. 1689: cf. Gunther, *Early Science in Oxford*, x, p. 140.

6 *Corr.* pp. 249–51.   7 The first letter is 11 Nov.: *Corr.* pp. 157–8.

8 Courten sent Ray the Macreuse: *Corr.* pp. 135, 147; see below, pp. 333–4: he set up a museum in the Temple and Sloane afterwards bought his collections.

9 *Phil. Trans.* were revived in 1691 but only three numbers, 192–4, were published: no. 195 appeared in 1692: regular publication only began in 1693. For Sloane's success cf. Weld, *History of the R.S.* I, pp. 356–8.

Garden, in which he became interested as soon as he settled in London; and to the nation, by the sale to it of the huge collections which formed the first British Museum, have secured for him a lasting memorial. It says much for his goodness of heart that in spite of his great commitments and medical practice he was never too busy to visit, correspond with and prescribe for the humble household at Dewlands. Unfortunately the long series of letters from Ray to him printed by Lankester and preserved in the British Museum are largely concerned with details of health and treatment and in other respects are less informative than those to Robinson. But of his affection and generosity they give abundant proof; and it is fitting that from his death-bed Ray addressed him as the 'Best of Friends'.[1]

The encouragement of these two young, brilliant and ardent disciples, coming at a time when Ray might easily have accepted the belief that his life's work was done and have settled down to domesticity and the village, gave him twenty years of ceaseless and seldom equalled activity. His self-depreciation might, and did, protest that he was unfit for the tasks they laid upon him; his consciousness of age, ill-health, loss of memory and failing powers might drive him to expostulate; his isolation from museums and libraries, from London and the learned, might convince him, give him grounds for convincing them, that their demands were unreasonable. But they refused to accept his excuses or to abate their faith in his capabilities; and between them saw to it that his life was filled to its limit. They were both men of singular energy, in touch with the world of science and culture, and in their own field eminent: but they both realised and were proud to demonstrate that at Dewlands was their master, the man who had a great gift to make to the world and whom they could enable to finish his work. That they should have stood beside him without jealousy, indeed with an unfailing generosity to him and to one another, so that in all the mass of correspondence there is never a hint of strain or disappointment, testifies to the quality of their relationship and explains its creativity. Ray constantly expresses his wonder that these friends of his, so busy and so distinguished, should be content to give him their trust and affection: but when he was almost broken by overwork and pain the fact of their friendship remained an unfailing support and inspiration. It is sometimes hard as one reads the letters of those heroic years not to feel resentment against those who could drive the old and invalid scientist so hard. But it is abundantly clear that his astonishing output could never have been produced without the spur of their confidence and the help of their contributions.

Sloane, as we shall see, began at once to offer suggestions for the *History of Plants*, drawing upon his knowledge of Tournefort and the resources of the London gardens. His loans of books and specimens, his

1 *Corr.* p. 459.

sweetmeats and venison, his prescriptions and visits were unfailing and increasingly valued. Robinson's interest was less exclusively botanical. From the first he supplied projects and problems, keeping the General History of Nature steadily in view and assisting it by his readiness to discuss and suggest. In 1685, when he became a Secretary of the Royal Society, he not only revived Oldenburg's practice of corresponding with the more eminent Fellows, but constantly referred to Ray the questions with which the Society was concerned; and these were a varied and curious collection.

Thus it was to Black Notley that the first sample of Maple Sugar known in this country was sent with a request that Ray would test it and investigate native trees to see whether similar products could be obtained by bleeding them. 'It was sent from Canada', said Robinson, 'where the savages prepare it out of the juice of the wounded maple, eight pints whereof affords a pound of sugar.... The Indians of Canada have practised this time out of mind; the French begin now to refine it and to make great advantage.'[1] Ray, declaring himself too busy to try experiments in tapping British trees and sceptical of any success, got 'a friend and neighbour, apothecary', probably Dale,[2] to test the sample and reported favourably upon it.[3] Shortly after reports about the obtaining of gum-lacquer from the Hornbeam, the action of vegetable poisons upon the blood,[4] and the growth of coffee[5] were submitted to him. Sloane was not slow to follow Robinson's example: the discovery of bituminous earth 'not far from Moorfields near the new square in Hockesden' (Hoxton as we call it) had excited him, and he wrote twice and with eagerness: Ray, however, declared his belief that the phenomenon was due to the burning of a paintshop which had stood there.[6] A less sensational report soon arrived of the adulteration of Jesuits' bark (quinine) with black cherry bark dipped in aloes.[7] Some of the enquiries suggest that the Society was occasionally hoaxed. As early as 1671 Ray had been asked to investigate the circumstantial story of a maggot which by proper feeding grew as big as a man's thigh.[8] But truth was stranger than fiction in the pleasant business of the Surinam snail (*Bulimus haemastomus*). Lhwyd wrote to Ray[9] that William Charleton (Courten the collector) had exhibited this in 1690: 'it

1 Letter of 16 March 1685: *Corr.* p. 162: cf. *H.P.* II, p. 1701.
2 It might also be 'Jacobus noster Garretus pharmacopœus'—James Garret—of *Hist. Ins.* p. 64, if this is the Black Notley Garret and not the famous London apothecary of an earlier day.
3 *Corr.* pp. 163, 165; cf. Gunther, *Early Science in Oxford*, XII, p. 87.
4 *Corr.* pp. 174–5.   5 *Corr.* p. 193.
6 *Corr.* pp. 177–80: cf. letter of Halley to Wallis: Gunther, *Early Science in Oxford*, XII, p. 114.
7 *Corr.* p. 190.   8 *F.C.* p. 63.
9 *Corr.* p. 212; and Gunther, *Life of Lhwyd*, p. 99.

was not above the bigness of a pullet's egg, yet laid an egg as big as a sparrow, and the young one hatched of it was twice as big as the egg'. Ray replied very sceptically;[1] and demanded confirmation of this portent from Robinson. He answered: 'Mr Charleton has it with eggs and young ones which are the wonder of our philosophers here; but I being naturally too jealous do almost suspect (though I durst never declare my suspicion) that the eggs and young ones have been severally and very artificially added to the snail.'[2] This did not prevent Lister from including a description and plate of the shell with egg and young in his *Exercitatio Anatomica*,[3] published in 1694: and he was probably right.

Our first knowledge of the preparation of the *Historia Plantarum* comes from the letters of the two new friends. In the spring of 1683 Robinson wrote to Lister that his 'friend Mr Ray is about writing a General Herbal which must needs be very accurate'.[4] In April 1684 he wrote to Ray: 'I am overjoyed that so vast a memory, so exact a judgment, and so universal a knowledge will be employed in compiling a general history of plants, an undertaking fit only for your extraordinary talents. I am in great hopes (because I wish it very impatiently) that you will bestow on the world a general history of nature (if God Almighty bless you with health and a long life).'[5] Though Ray, overwhelmed by this outburst of hero-worship, replied: 'As for my intended history, I am now sensible I have undertaken a task beyond my strength, and yet it is *res integra*. I have not yet proceeded so far but I can, without inconvenience, give it over'; he was not untouched by such confidence, and added: 'Yours and some other friends' opinions and expectations from me do inspire me with such force and courage as not to despair of my abilities, but to contemn all difficulties and contend even to excel and outdo myself.' He then explains that he had had hopes of Dr Morison's work, but since these have not been fulfilled and the work itself has been stopped, he is 'inclinable to listen to the solicitations of friends'—'that whereas those of other nations are busy and active in this kind it might appear that the English are not altogether idle or asleep but do at least endeavour to contribute something'.[6]

Similarly, in the autumn of 1684 Sloane wrote to express his delight at 'so useful a work as the History of Plants done by you will be'. 'I am sure', he added, 'we want it extremely and that it will be very much esteemed by the botanists beyond sea, particularly M. Tournefort, the king's professor at Paris, with whom I correspond, who told me that he desired it extremely and that he had a very great respect and honour for you.'[7] Sloane at once offered to do all in his power to help either by pro-

---

1 *F.C.* p. 207.   2 *Corr.* p. 214.
3 *Exer. Anat.* p. 133 and Tab. vi: cf. A. H. Cooke, *Molluscs*, p. 124.
4 *F.C.* p. 136.   5 *Corr.* p. 141.   6 *Corr.* p. 146.   7 *Corr.* p. 156.

curing books or by sending specimens of plants: and he was as good as his word.[1] As a start he enclosed a list of rarities in the gardens of Fulham and Chelsea,[2] and gave the interesting news of

Mr Watts [the gardener at Chelsea] having a new contrivance (at least in this country); he makes under the floor of his greenhouse a great fireplace with grate, ash-hole etc. and conveys the warmth through the whole house by tunnels, so that he hopes, by the help of weather-glasses within, to bring or keep the air at what degree of warmth he pleases, letting in upon occasion the outward air by the windows. He thinks by this means to make an artificial spring, summer or winter.[3]

Letters urging various inclusions and additions follow regularly, and Watts's artifice is reported to have saved many rarities during the next and severe winter.[4] Ray in reply to this enthusiasm[5] expostulated that his

reasons for attempting this work were 1. To satisfy the importunity of some friends who solicited me to undertake it. 2. To give some light to young students in the reading and comparing other herbarists.... 3. To alleviate the charge of such as are not able to purchase many books: to which end I endeavour an enumeration of all the species already described and published. 4. To facilitate the learning of plants without a guide or demonstrator, by so methodising them and giving such certain and obvious characteristic notes of the genera that it shall not be difficult for any man to find out infallibly any plant especially being assisted by the figure of it. And lastly because no man of our nation hath lately attempted such a work; and those that formerly did, excepting Dr Turner, were not sufficiently qualified.[6]

But the task grew and grew.

In the letter just quoted Ray had expressed a hope, which was not in fact fulfilled, that his great work would be illustrated with plates. Pulteney in the first volume of his *Sketches of Botany*[7] devotes a chapter to the story of the woodcuts used in the early herbals, of their collection by Plantin, their transference from book to book, and of the use made of them by Gerard and Johnson. Faulty identification and in some cases faulty drawing limited their value, but Ray was no doubt right in saying that most people 'looked upon a history of plants without figures as a book of geography without maps'.[8] Morison had felt this so strongly that he had reduced himself almost to penury in order to provide proper illustrations, copper-plates not woodcuts, for his work. Such expenditure was plainly outside Ray's means. Mrs Willughby, who had paid for the plates in the

1 Many such are mentioned in the *Historia Plantarum*: particularly interesting is that which enabled Ray to correct his identification of *Astragalus danicus*, *H.P.* I, p. 939.

2 He sent seeds in 1684 which Ray grew and described next year: cf. *H.P.* I, pp. 165, 228, 330, 362. John Watts was in charge there 1680–92.

3 *Corr.* p. 158.   4 *Corr.* p. 161.

5 He was well aware of Sloane's 'volatile temper': *F.C.* p. 293.

6 *Corr.* pp. 160–1.   7 *Sketches*, I, pp. 155–63.   8 *Corr.* p. 155.

Ornithology, had since her second marriage refused to help with those for the *Historia Piscium*.[1] The Royal Society felt the strain upon its resources to be too severe for any further outlay; and there was no generous benefactor to do for the Plants what Pepys and his colleagues were doing for the Fishes. Very reluctantly Ray yielded to the force of circumstances, and after much thought about the pictures ('I am so teased about cuts for my History of Plants, all my friends condemning wooden and telling me I had better print it without any[2]') decided to give up the project, consoling himself that the book would then come out much sooner and be smaller and less expensive. But until the last days of his life he never ceased to hope that by some means plates could be provided; and at the end of the first volume an appeal is printed asking for subscriptions and proposing to issue the plates family by family as funds permitted. Unhappily money was not forthcoming.

Apart from this question the letters to Robinson during his brief Secretaryship of the Royal Society (the only ones of the whole correspondence that have been preserved entire[3]) are full of references to the work. Robinson himself was to contribute a section on the chemical analysis of plants; and this is discussed at some length both as to its general treatment of the subject[4] and later as to its style and latinity.[5] Then the matter of inviting subscriptions arises. Ray had sent a preliminary account of the book to Faithorne, who was already publishing the *Historia Piscium*: 'I thought he intended no more than to show it to particular friends and acquaintances.'[6] Faithorne and Mott the printer issued it as a public appeal for subscribers: Ray expostulated that he disliked the practice, that to invite subscriptions for a work not yet finished was doubtfully honest, and that he could not be sure of finishing the books in the specified time. Finally among various references to materials for the work is one of special interest to Dr John Covel, from whom Ray hoped to receive help as to the plants of 'Thrace, Greece and Asia the Less'.[7]

Covel had gone up to Christ's College in 1654 and gained a fellowship there in 1661. He was evidently not known to Ray at that time, but devoting himself to medicine and the study of plants he became well acquainted with his friend Peter Dent, the Cambridge pharmacist. In 1670 he was appointed chaplain to the Levant Company and went out to

---

1 *F.C.* p. 136.

2 *F.C.* p. 146, letter of 12 May: on 21 May Aston (Sec. of the R.S.) wrote to W. Musgrave: 'Mr Ray's *History of Plants* being designed to be printed with old figures, we have prevailed that it may be printed without figures....I believe it will be an incomparable book': Gunther, *Early Science in Oxford*, XII, pp. 92–3.

3 Selections from them were printed by Derham and in *Corr.* pp. 134–77; the remainder are in *F.C.* pp. 138–54.

4 *Corr.* pp. 166–7.   5 *F.C.* pp. 152–3.
6 *F.C.* pp. 147–9.   7 *F.C.* p. 145.

Constantinople, where he interested himself in the Orthodox church, and in· the antiquities and plants of the Near East, producing the records printed in 1893 by the Hakluyt Society. He returned in 1677 and Ray appears to have obtained from him through Dent seeds of certain Thracian plants.[1] In 1681 he was appointed to succeed Ken as chaplain to the Princess of Orange at the Hague. In April 1685 Ray wrote that he hoped to get Covel to contribute to the *History*, since he had 'formerly had a transcript of a letter of his from Constantinople containing some descriptions and more names of plants with references to his own draughts'. He adds: 'I was informed by Mr Dent that he was to come over with the Princess of Orange and that he (Mr Dent) intended to wait upon him at London, and that he would make me a visit on his return home and give me an account of what he had effected.'[2] In fact just at this time a letter of Covel's to England fell into the hands of William in which the chaplain spoke very indiscreetly of William's treatment of his wife: Covel was dismissed from his post summarily; and neither Ray nor Dent managed to get into touch with him. In 1688 he succeeded Cudworth as Master of Christ's and grew many of the rarer species of British plants in his garden there.[3]

The letters dealing with these matters were written in April and May 1685, and on 25 May Robinson submitted draft proposals for the printing of a General History of Plants to the Council of the Royal Society, and obtained their assent to the project. Ray was still engaged on the final revision and printing of the *Historia Piscium* and was now committed to preparing a still larger work for the press. Suggestions from Robinson on both subjects were poured upon him; and in his replies details about plates and corrections of the fishes are sandwiched in with minute discussions of the format and contents of the new volumes. 'Truly I am straitened for time'[4] he declared on 14 September; but that did not prevent him from dealing with Bondt's 'heathenish figures...of the Congri',[5] with 'a parcel of dried plants from Dr Sloane', with 'Wepfer's philosophy concerning poisonous plants',[6] and, as we have already mentioned, with Robinson's own section on chemical analysis.

1 Cf. *H.P.* I, pp. 565, 742.   2 *F.C.* p. 145.

3 This fact is derived from MS. notes dated 1702 in a copy of Ray's *Synopsis Britannicarum*, ed. II, in my possession. For Covel's interest in geology, cf. Gunther, *Life of Lhwyd*, p. 469: 'he has engaged the lime-kiln men of Cherryhinton to save Fossils'. He also showed Lhwyd 'the only collection of gold coins of the later Greek emperors I ever saw...also a paper of red pearl of the size of hazel kernels taken out of Pinnae in the Isle of Lemnos'. For von Uffenbach's account of Covel, whom he visited in 1710, cf. J. E. B. Mayor, *Cambridge under Queen Anne*, pp. 147–52, etc.

4 *F.C.* p. 151.   5 *F.C.* p. 153.

6 *Corr.* p. 176—the same letter: the allusion is to Johann Jakob Wepfer, *Cicutae aquaticae Historia*, published at Basle in 1679.

This letter is unfortunately the last that has been preserved entire, and for the next five years we have only Derham's epitomes of the correspondence.[1] On 15 September the Royal Society issued an instruction signed by S. Pepys, P.R.S., to Henry Faithorne, their printer, to put the *Historia Plantarum* in hand. A few days later Ray received and refused an offer of preferment. The fishes were practically finished, and copy for the plants was beginning to be despatched to London. Each week for the next six months the carrier seems to have taken up a bundle of manuscript for the printers, Dale helping with the checking of it at Black Notley and Robinson revising it in London.

The strain threatened to be intolerable. It did in fact produce the first outbreak of a trouble from which Ray was thereafter seldom free. The second of the October epitomes contains the ominous word 'diarrhoea'; the fourth adds: 'His medicine in his diarrhoea. Physic a conjectural science'—a remark the truth of which he had reason to emphasise during the next twenty years; the fifth, of 26 October, notes that he was recovering. The work does not seem to have been interrupted: but his health was beginning to pay the price. The weekly letters to Robinson continue, but the epitomes are laconic and monotonous. 'Hieracium. Three years in completing his first volume of Plants'—so in December: 'Copy for the press'; 'Copy sent of Verticillates'; 'Additions of plants'; 'A plant or two'; 'A Legume'; 'Astragalus etc.'—so the progress of the printing is noted. By May we read 'Index to 1 Vol.', and in June the book was published and he presented copies of it to his friends.[2] By August he was busy sending off copy for Vol. II.

This first volume, a very large folio, containing 22 unnumbered and 984 numbered pages, printed in a small type, is one of those works which by their sheer mass and magnitude create a sense of awe. In these days of small books and co-operative effort it seems hardly credible that such a tome can be the product of a single author and three years of writing. As one studies the endless series of descriptions, the multitude of references, the concentrated attention bestowed upon each separate species, and the innumerable points of detail discussed in the appended notes, the impression is deepened. When one reads his confession in the Preface—

I know that there are other species, new and undescribed, in the gardens of Universities and of the great: these must some day be published: I have dealt chiefly with those already recorded: even here I am conscious of omissions through lack of enquiry, negligence, forgetfulness or haste: my readers will perhaps notice more such: what else can be expected from one mere man who had not even a secretary but must needs plough the whole field with his own hand

1 *F.C.* pp. 288–93.
2 Faithorne presented a bound copy on the best paper to the Royal Society on 7 July: *F.C.* p. 97.

—one can only be amazed at his fortitude and perseverance. If he had lived in a cloister or a library, the work would still have been heroic: he did it in a cottage with few books, dependent upon a rather unreliable carrier and on the good offices of friends in London: he was nearly sixty, already in indifferent health, and the four baby girls were born during its production. There have been giants in the earth: and on the evidence of these books Ray would have a claim to stand among them.

Yet in them and in all his references to them there is not the slightest sign of self-congratulation: he does not imagine, as he states and proves in the dedication to Charles Hatton,[1] that he has done anything remarkable. Rather he is glad that before he is too old he has been able to return from the 'alien habitations and by-ways' into which he has for so long been drawn to his own road of research, his own proper business. Genuinely and very generously he expresses his obligations to others. He rehearses the names of those who since Jean Bauhin and John Parkinson have added to botanical knowledge; mentions the two Bauhins, Colonna and De l'Ecluse as those whose descriptions he has most freely followed; gives a list of his personal helpers, William Bowle, M.D., deceased, Peter Dent, physician of Cambridge, Ralph Johnson, Martin Lister, M.D., Walter Needham, M.D., James Newton, Joshua Palmer, M.D., Philip Skippon, knight, Barnham Soame, Thomas Willisel, deceased: and then picks out four for special gratitude, Edward Hulse, M.D., Tancred Robinson, M.D., Hans Sloane, M.D., and Samuel Dale, 'doctor and physician, my neighbour and friend at Braintree, who checked the synonyms, corrected errors and supplied omissions'. The Preface contains little else except a couple of paragraphs explaining his method and a third stating his reasons for rejecting the external aspect of seeds and seed-vessels as the sole criterion in classification.

Of the friends here mentioned William Bowle is the only one unknown to fame unless he is identical with Dr George Bowles, the recorder of the Lizard Orchid, who had contributed largely to the notes on the uses and properties of plants in the *Catalogus Angliae*.[2] If not, it is tempting to suppose that he was the doctor and friend of Ray's boyhood and responsible for some of his interest in plants: but in fact his contribution to the subject or to this book is obscure. Barnham Soame is almost equally unknown.[3] He lived at Little Thurlow in Suffolk, and had a son of the same name who graduated at St John's College, Cambridge, in 1681–2 and became a fellow there in 1693 after taking his M.D.:[4] presumably he is the Mr Soames whom according to Skippon they met in Rome in 1664 and in

1 He declares here, as in his letters to Robinson (*F.C.* pp. 287, 302), that Hatton, whose wishes he regards as commands, had instigated the work.
2 Cf. above, p. 156.　　3 He contributes references to obscure medical writers.
4 Venn, *Alumni Cantab.* IV, p. 119: he incorporated at Oxford in 1695.

Venice in 1665 and with whom they went to Tivoli:[1] possibly he was connected with his friends, the Barnhams of Boughton Monchelsea. Joshua
Palmer, another new name, came to St Catharine's in 1675 from Wanlip
in Leicestershire and went out to study at Leyden in 1682[2] under the great
Hermann, who had a strong connection with English doctors and botanists.
On his return he seems to have settled in his native county, and though a
member of the Royal College of Physicians to have been absorbed in local
practice.[3] James Newton, another who makes his first appearance here as
one of Ray's helpers, is a man of more distinction. He was still young,
though the date given for his birth in the *Dictionary of National Biography*,
1670(?), must be nearly a decade too late; for he states in his posthumous
book, *A Compleat Herbal*,[4] that he had begun the search for plants in 1680,
and was certainly known to Ray as a competent and travelled botanist in
1685.[5] From the records of his finds he obviously toured many parts of
the country and did good work: and at some time between 1685 and 1690
visited Ray and went in his company to the estuary of the Blackwater at
Tollesbury.[6] Unlike the other three already mentioned he appears constantly as a helper in Ray's later botanical works. He kept a private
lunatic asylum near Islington, and died at an advanced age in 1750: but his
writings on plants, unpublished at the time of his death, were according
to the Preface of the Herbal 'the work of his younger days'.[7]

While acknowledging these obligations the Preface also explains Ray's
reasons for writing, his method of classification, and the scope and
character of the work. He had begun to collect material for it at the suggestion of Francis Willughby; but on his death, remembering that medicine was not his profession and that the task would take more time and
labour than he could devote to it, he dropped the project. During
Morison's life he was content to leave the matter to him. On his death
Hatton and others urged him to undertake it and he consented. A new
History was plainly needed. Since the time of J. Bauhin and Parkinson
numbers of plants had been discovered and described by the explorers of
America and the East as well as by European botanists. As proof of this
he gives a list of the chief workers, which is an indication of the progress
of science since the time of his *Cambridge Catalogue*, and to which we refer
later. Having undertaken the work he hastened to discharge it before
old age made it impossible.

1 *Journey*, pp. 650, 506, 674.
2 He sent particulars of plants in the garden there: cf. *H.P.* I, p. 206; and MS. notes
from Hermann, l.c. pp. 335, 376.
3 Munk, *College of Physicians*, I, p. 429.
4 Published 1752.   5 *Corr.* p. 139 and *H.P.* I, p. 817.
6 *Syn. Brit.* II, p. 196.
7 In *Syn. Brit.* II, p. 9 Ray refers the student of sea-weeds and mosses to 'Mr
Newton's History shortly to be published'.

'Method is the mother of memory: memory the mother of the muses.' To method therefore he devoted close attention, trying not only to formulate a correct definition and arrangement of the 'genera' (that is, the large groups or orders), but to marshal correctly the 'species subalternae' (or genera) and the 'species infimae' (or species). He then briefly recapitulates the reasons for preferring his own Method as already published, though he alters it by uniting Shrubs with Trees and beginning not with Trees but with Algae.[1]

Finally, he describes what he has done. 'First I gathered all the plants scattered over many volumes into tribes and groups, separate members into a single body....I thought it unnecessary to examine every early publication, for the Bauhin brothers had done this thoroughly. But I worked over all the discoveries of subsequent authors as well as my own British and continental records.' He concludes by insisting that the work cannot be taken as complete; for vast areas still remain to be explored and no one man can cover even the existing material without omissions. For accurate descriptions he has relied most upon the Bauhins, Colonna and De l'Ecluse. Very many plants are inadequately described: these he has tried to record more exactly; but many must remain obscure. Some of these he has included, but with special care to avoid duplication. Synonyms he has drawn mainly from the Bauhins, Gerard and Parkinson.

The bibliography which follows the Preface gives additional proof of the extent to which Ray had increased his botanical researches since the publication of the *Cambridge Catalogue*, and of the growth of science. A few of the names in the Cambridge list—Aetius, Pietro Gassendi, the *Opus Pandectarium*, and some of the Garden Catalogues—are omitted. The additions are numerous, either foreign botanists whose work was irrelevant to his British lists or of authors subsequent to 1660. The chief of them are mentioned also in the Preface. There is Jakob de Bondt of Leyden, who had travelled through India and been a doctor at Batavia from 1625 till his death in 1631 and whose *Historia Naturalis Indiae orientalis* was published at Amsterdam in six volumes. This Willem Piso of Leyden and Amsterdam edited and combined with his own work on the West Indies to form the *De Indiae utriusque Re Naturali* (Amsterdam, 1658), of which volumes IV, VI and XII deal with plants. Georg Marcgraf, a German, went with Piso as physician to Prince Moritz of Nassau into South America and died there at the age of thirty-four in 1644; his *Historia Rerum Naturalium Brasiliensium* in eight volumes, the first three devoted to plants, was published at Amsterdam in 1648. Francisco Hernandez'[2] *Plantas y Animales de la Nueva España* was published in

1 There are a few other changes, but these are relatively unimportant.
2 These four, Bondt, Piso, Marcgraf and Hernandez, had been already used by Ray in his *Ornithology*.

Mexico in 1615, and was edited and translated by Antonio Reccho at Rome in 1651. Jakob Breyne of Dantzig produced his *Exoticarum Plantarum Centuria prima* in 1678, and was reported by Robinson[1] to be hard at work on the second century in 1684. Paolo Boccone, whose *Icones et Descriptiones rariorum Plantarum* appeared in 1674, had been in correspondence with Ray at that time.[2] Giacomo Zanoni, keeper of the Physic Garden at Bologna, published his *Historia Botanica* in Italian in 1675. Robinson, who reported his death to Ray from Geneva in 1684,[3] had a copy of his book sent to Faithorne in London together with Pierre Magnol's *Catalogus Monspeliensis*, published in 1676, for Ray's use. Christian Mentzel,[4] physician to the Elector of Brandenburg, published an *Index Nominum Plantarum* at Berlin in 1682. Abraham Munting, Professor of Botany at Gröningen, published a history of plants in Dutch in 1672 which Ray lists but did not see until 1697, when Smith the publisher sent it to him in the new and larger edition of 1696: 'it will be of little use being written in a language I understand not and because the Latin names are his own without synonyms.'[5] He also wrote *De vera antiquorum herba Britannica* (Amsterdam, 1681), which Ray apparently saw in January 1685.[6] The *Hortus Malabaricus*, sumptuously produced with life-size engravings and names in four languages by Hendrik Rheede van Draakenstein and others, especially Arnold Syen of Leyden and Jan Commelin, had already run to five volumes (1678–84) and was expected to be completed in thirteen:[7] Ray commented upon it that 'they might have thrust it into a quarter of the compass and rendered it more useful: the like may be said of Breynius' book'.[8] In addition to these Ray also mentions in the Preface Prospero Alpino, whose *De Plantis Aegypti* (Venice, 1592) and *Dialogus de Balsamo* (Venice, 1591) had been revised and enlarged with observations by Johannes Vesling at Padua[9] in 1640; Jacques Philippe Cornut, author of the *Historia Plantarum Canadensium* (Paris, 1635); and John Parkinson.

In the Preface he makes special reference to Robert Morison, 'an Aberdonian Scot, a man of great reputation for botanical learning and long skilled in the study and culture of plants', whose History had been interrupted by his debts and untimely death. The first volume of this, published at Oxford in 1680, with a fine series of engraved plates each bearing the

1 *Corr.* p. 143: he visited Ray at Black Notley in April 1703, and was elected F.R.S. in May.
2 Ray wrote to him but did not sign the letter: he replied to Oldenburg, *Corr.* p. 108: cf. *F.C.* pp. 66 (to Oldenburg), 133 (to Lister)—all in Sept. 1674.
3 *Corr.* p. 141.   4 He is known also in connection with fossils: *F.C.* p. 213.
5 To Sloane, *Corr.* p. 316.
6 *F.C.* p. 287: he quotes from it, *H.P.* I, pp. 172, 175–7, etc.
7 So Robinson, *Corr.* p. 143: twelve volumes were actually published.
8 *Corr.* p. 146.   9 Ray quotes his epitaph there, *Obs.* pp. 214–15.

name and arms of its donor from Prince Rupert and James Duke of Monmouth through a long list of noblemen, gentry, doctors, and apothecaries, is included in Ray's bibliography with the significant note:

While he confined himself within limits proper to his ability, to drawing up catalogues of gardens, defining generic characters and correcting the 'hallucinations' of other botanists, his work was meritorious; when his self-satisfaction and contempt of others led him to overstep these limits and attempt a Universal History of Plants, he injured his reputation and failed to fulfil expectations.

In the bibliography he mentions John Evelyn's *Sylva*, John Johnstone's *Dendrographia*[1] (Frankfort, 1662), Robert Sibbald's *Prodromus Historiae Naturalis Scotiae* (Edinburgh, 1684), and Sir George Wheeler's *Journey into Greece* (London, 1682);[2] and the work of Marcello Malpighi and Nehemiah Grew on the Anatomy of Plants and Joachim Jung's *Isagoge Phytoscopica* (Hamburg, 1679), which he had seen in manuscript at Cambridge. How much he owed to the three last named appears plainly in his very generous and constant references to them in the introductory essays of his book.

After a full but concise glossary of technical terms used in botanical descriptions and an index of the chief genera in volume 1 follow these essays, occupying pp. 1–58 and constituting by common consent a remarkable survey of the physiology and structure, reproduction, characteristics and classification of plants.

He has displayed the rare talent requisite to bring scattered observations into one point of view, here may be found the principal discoveries made by Caesalpinus, Columna, Grew, Malpighi and Jungius in addition to those made by Ray himself; and in this way resulted the most complete treatise which had yet appeared on vegetation in general; and it must be remarked that, although this work may not have been very frequently quoted, yet it is through it that the doctrines of these authors were made common and became as it were popular in the science; and on this account we believe that the best monument that could be erected to the memory of Ray would be the republication of this part of his work separately. These writings formed an epoch in the history of botany.[3]

Two qualities stand out in these essays. The first is Ray's genius for surveying the whole field of current botanical knowledge and selecting from it the contributions of permanent value. Just as in the identification and description of plants he had fastened unerringly upon the works of

1 Ray lists it as *Dendrologia*.

2 From this Ray extracted a list of plants for his *Collection of Curious Travels*, pp. 364–7: Wheeler or Wheler travelled with J. Spon of Lyons, who published a separate account, *Voyage de Grèce*, Lyons, 1678.

3 Cuvier and Thouars in *Biographie Universelle*, Paris, 1843, XXXV, pp. 252–6, quoted in *Mem.* pp. 103–4.

the two Bauhins, Colonna and De l'Ecluse—an admirable though for an
Englishman of the time a remarkable selection—so in physiology he chose
the definitions of Jung, most philosophic of all botanists, and the researches of Malpighi and Grew and to a less extent of Robert Sharrock,[1]
supplemented them by his own observations and experiments, and presented the whole as a consistent and very thorough account of vegetable
life. The second is his unfailing generosity in looking for the best, not the
faults, in others and acknowledging the worth of their work. No doubt
by admitting obligation even when the debt was trifling, he obscured the
originality and scope of his own studies. If like almost every one of his
contemporaries and too many of his successors, not excluding Linnaeus,
he had been willing to appropriate and build upon the results obtained by
other scholars, his own reputation would have stood higher: no one would
then have thought of putting him on a level with Morison. But plagiarism
of any sort was an abomination to him: indeed he preferred to give more
than their strict due not only to his friends like Willughby or Sloane but
to workers of the past, strangers whom he had never known and men
who disparaged and criticised his own labours. He never advertises
his own discoveries or parades his own research: he never mentions
those whose errors he rejects except in cases of notorious arrogance and
pretension.

In addition, he has here as elsewhere a brilliant power of clear and
succinct style. He quotes and summarises the findings of others fully and
fairly: his own he condenses drastically except in sections like that on
cotyledons, where fullness is indispensable. The mass of information in
these thirty chapters is masterfully handled; the evidence is skilfully set
out and its bearing on the sequence of the argument clearly shown; the
result is an exposition of the best theoretical and experimental knowledge
of the time.

The material is so closely packed that to do justice to it would involve
a translation of most of the chapters. An idea of their scope can be
suggested by quoting their headings and giving a few notes on points of
special interest.

1. What is a plant? He quotes Jung's definition and comments on the
physiology of the Sensitive Plant (*Mimosa pudica*), of the closing of
leguminous leaves and of many flowers, discussing the effects of pressure,
temperature, light and moisture.[2]

2. The parts of plants in general.

1 Fellow of New College; afterwards Archdeacon of Winchester; died 1684; author
of *History of Improvement and Propagation of Vegetables*, Oxford, 1660: 'very knowing
in vegetables and all pertaining thereunto': Wood, *Athenae*, IV, p. 147.

2 Cf. Sachs, *History of Botany*, pp. 536–8: he gives Ray priority and high praise for
this discussion.

3. The roots of plants. Jung's definition with exposition from Grew's *The Comparative Anatomy of Roots*.

4. The stalks of plants, from Malpighi and Grew.

5. The movement of the sap. A digest and reinterpretation of his own experiments in the light of the physiological studies of Malpighi and Grew.[1]

6. The yearly growth of the trunk. A condensed version of his note in the *Cambridge Catalogue* with reference to Malpighi.

7. The differences of stalks, from Jung with some additions. Definitions and examples.

8. The buds, from Malpighi and Grew.

9. The leaves of plants, from Jung and others. Definitions; physiology; a discussion of the influence of light on the colour of leaves—this being original.

10. The flowers and their parts. Important discussion of stamens: he states Grew's claim that they are male sex-organs, supports this by reference to dioecious plants, willows, hops, etc., and to observations on the fertilisation of date-palms; says that it requires further confirmation; and admits it only as probable.

11. The differences of flowers, from Jung with additions. He gives a table of classification.

12. The fruits and seeds. He quotes Cesalpino, Malpighi, Lauremberg, and Grew.

13. The seeds of plants, general observations. Number, size, viability and germination: he has proof of germination after five years but thinks longer periods probable: he denies spontaneous generation.

14. The seed-leaves. His own description as in the *Methodus*.

15. The seed-plant and contents of the seed. His paper to the Royal Society as amplified in the *Methodus*.[2]

16. The secondary parts of plants, tendrils, thorns, etc. Partly from Malpighi.

17. The nourishment of plants. He discusses nutrition through the roots; the passage of air and water through the porous fibres; and the method of the ascent of sap. He refers to Grew and Malpighi.

18. The sowing and propagation of plants. Mainly practical, raising of seeds, cuttings, layering, etc. Further discussion of spontaneous generation.

19. Grafting. Description of different modes.

20. The specific differences of plants. He defines a species as a plant that comes true from seed; and revises his paper to the Royal Society and Preface to the *Catalogus Angliae*.[3]

---

1 Cf. Sachs, l.c. p. 471: he does not realise that Ray's experiments were earlier than the work of Lister or Grew and overestimates Malpighi's physiological work.

2 Cf. above, pp. 194–5.   3 Cf. above, pp. 189–90.

21. The transmutation of species in plants. He quotes much evidence, including a case concerning cabbages tried at Westminster; rejects much of it; admits that some of it constrains him to admit the possibility of transmutation; but suggests that it only occurs between species closely related and possibly not specifically distinct.

22. The stature and size of plants. A long list of trees of record size.

23. The age and duration of plants. Annual, biennial, permanent and perennial.

24. The qualities and uses of plants in food and medicine. He condenses his reasons for rejecting the doctrine of signatures; suggests that caterpillars of the same species feeding on different plants indicate similarity of quality;[1] and describes differences of savour, quoting from Grew.

25. The locality of plants. Parasitic, aquatic, alpine, etc.

26. The classification of plants in general. He rejects all systems not based upon the structure of the principal parts.

27. The classification of plants into trees and shrubs, sub-shrubs and herbs, and the subordinate orders. He accepts two main classes; divides herbs into dicotyledons and monocotyledons; and follows his *Methodus* with slight modifications.

28. The collection, drying and preservation of plants. He rejects all astrological notions as superstitious, and gives practical directions.

29. The chemical analysis of plants. Contributed by Tancred Robinson.[2]

30. The diseases of plants and their cures. Rust, mildew, galls, the last from Malpighi.

After this first book there follow 924 pages of descriptions, beginning with the Corals, Sea-weeds, Fungi, and Mosses (59 pages in Book II) and the Ferns (34 pages in Book III) and ending with the vast tribe of the Leguminosae (100 pages in Book XVIII). Of the first sections the contents are few and casual: hardly any serious work had been done and large tracts of vegetable life were almost unexplored. But with Book IV the flowering plants begin, and at once the character of the work changes. Ray's long and intensive study of the British flora had given him an exact knowledge of the identity of a large number of plants and a clear insight into the characteristics that determine specific differences. He had also a surprisingly wide personal acquaintance with the flora of Western Europe: he had himself seen a majority of the plants described by other writers and had discovered a considerable number of new species. More-

1 An original and profitable suggestion: he excepts omnivorous caterpillars and illustrates from larva of *Hipocrita jacobaeae* feeding on Ragwort and Groundsel.

2 Ray acknowledges and comments upon this in letter of 29 April 1685 (*Corr.* p. 166): Robinson agreed to certain excisions 19 May (*Corr.* p. 170).

over, in his efforts to name his finds he had gained a thorough knowledge of the literature, and a shrewd insight into the merits and weaknesses of his authorities. He was thus able to describe a sufficient number of typical species clearly, to arrange cognate forms round them in appropriate groups, and to insert other records with a reasonable degree of confidence. In easily recognisable families, where he has a large amount of homogeneous material to deal with, as for example in the Umbelliferae or Leguminosae, the arrangement is highly successful and the treatment thorough and competent. In a difficult genus like *Trifolium*, where he had had exceptional opportunities for study, did not have to rely on the vague descriptions of others, and could exercise his own admirable powers of discrimination, the result is masterly. Other families, in which the characteristics are less obvious, or the number of known species small, or his own experience defective, are less satisfactorily handled; the classification becomes incoherent; the descriptions lose their precision; and only species of marked individuality are easily definable.

In general his own work stands out as scientifically on an altogether different level from that of his sources. When he has himself seen and identified a species it is usually easy to recognise: when he has described it, it is almost always fixed beyond doubt. His descriptions are masterpieces of brevity and completeness. He first gives the chief characteristics of the plant as compared with its neighbours; then describes root, stalk, leaves, flowers and seeds; then records its time of flowering, its habit, annual or perennial, and its localities, both general—meadows, hills, marshes, etc.—and particular—often in minute detail; finally, he adds notes of quality and medicinal uses. The complete accounts are so good as to make it regrettable that they are almost confined to species otherwise nondescript. For the vast majority of plants he prefers to quote the records of his predecessors, not indeed with any strict regard for priority, but selecting the best and with a scrupulous acknowledgement of its origin. He often condenses and sometimes emends or amplifies; but even so the work is seldom up to his own standard, and is often slipshod and defective. Many of these records are so vague as to make the identity of the plant uncertain. In all the groups there are a number of entries so indeterminate as to be valueless; and he constantly apologises for their insertion 'lest the history should be incomplete', 'lest I should omit a real species', or 'in order that further investigation may be made'. We may admire his respect for the labours of others and his hatred of plagiarism; but it is disappointing to find inferior definitions of plants with which he was thoroughly familiar inserted in place of his own far more adequate accounts.

Outside the flora of Western Europe botanical knowledge was in its infancy. Considering the extent to which the world had been opened up for trade, it is remarkable that the records of Asiatic and American species should be so few. Hernandez, Marcgraf, Cornut furnish a number of descriptions, generally difficult to interpret; John Tradescant the elder is credited with several introductions; John Banister[1] and a few others are occasionally mentioned; Ray himself had grown a Virginian Persicaria,[2] three American Golden-rods,[3] a Virginian Sneezewort,[4] the Tobacco-plant[5] and two or three others, and mentions the Virginian Aster,[6] the 'Nasturtium indicum' from Peru,[7] and two Evening Primroses[8] and the Potato[9] as thoroughly established; he refers occasionally to certain gardens and catalogues; otherwise the New World is hardly known. Bondt and sometimes the *Hortus Malabaricus* for India; Covel[10] and Wheeler for the Near East; Alpino for Egypt and Breyne for the Cape of Good Hope; these supply a few not very reliable records. Of the wealth and characteristics of tropical vegetation there is as yet hardly a hint. Indeed, apart from the Gourds, Melons and Cucumbers, which form a large and separate group, all 'exotics' are arranged in close if often strained attachment with families whose typical representatives are British.

If Ray had been able to visit the larger gardens himself, this defect (which he fully recognised) might have been lessened. His records of his own plants at Cambridge, a reference to an Astrantia in Charles Hatton's garden in London,[11] and the use which he had made of the royal garden at Montpellier, show how much he lost by not having access to the plants grown by Bobart at Oxford or by Watts at Chelsea. With the development of greenhouses and stoves and under the influence of wealthy patrons the herb gardens of Parkinson and his generation were developing into great collections, carefully tended and constantly increased.[12] Sloane might send him lists of these[13] and the dried sprigs of which he makes occasional mention; Morison's records of species raised from seed at Oxford, though brief and usually without notes of origin or locality, gave him a number of entries; the Catalogues of foreign gardens, and especially the 1665 edition of the Paris list,[14] were of some value: but, as he regularly insisted, descriptions and dried specimens could not make up for ignorance of the living plant, or enable a confident identification of its relationship. 'Dried

---

1 As a very recent explorer: *H.P.* I, p. 674.   2 P. 183.
3 Pp. 278–9.   4 P. 363.   5 P. 713.
6 One of Tradescant's introductions, p. 269.
7 P. 487: he notes that it has nothing in common with the genus *Nasturtium* except its savour!
8 P. 862: he classifies them with the Willow-herbs.
9 P. 675.   10 Cf. *F.C.* pp. 145–6, 149.   11 P. 475.
12 Thus in Dec. 1684 Sloane had written: 'There is a vast number of East and West India seeds come over this year': *Corr.* p. 159.
13 E.g. letter of 11 Nov. 1684: *Corr.* p. 157.
14 He had visited the garden at that time and the list stimulated his memory.

specimens cannot represent all the principal parts, flower, seed-vessel or fruit, and seed: and I have seen none of late discovery growing in gardens, not having ability to take journeys to visit them'—those words written much later are true even of this first volume of his great work.

Some of his notes are interesting. More than thirty species are mentioned as having been grown in his Cambridge garden;[1] and it is probable that he had by him the unpublished manuscript of his *Catalogue*. References to Black Notley are few, though he records that in 1685 he grew from seed said to be that of the 'Virginian Macock or Pumpion' a plant 'as like the common pumpkin as one egg is like another',[2] and that in the past season he grew a gourd from seed sent by Sloane.[3] Of Tomatoes, 'Poma amoris', he writes 'in our garden they germinate late but sometimes do well, flowering in summer and bearing fruit in autumn, though they are killed at once by the first frosts';[4] and of Strawberries that they are now served not only with sugar and wine but 'cum lactis cremore'—though he prefers the older fashion.[5] He identifies 'the Irish Chambroch' with the Purple Clover (*Trifolium pratense*);[6] and after citing Marcgraf's description of 'Ipecacoanna' (*Psychotria ipecacuanha*) adds 'per vomitum purgat'.[7] There is a discussion of the various uses of Tobacco with mention of Johann Neander's *Tabacologia* (Leyden, 1622) and Raphael Thorius's hymn (London 1627);[8] a description of the preparation of Woad,[9] based upon Wedel but supplemented by an account supplied to him by a friend; and a shorter record of the use and effects of opium.[10] The medical notes are largely derived from Hulse, from Robinson's notebooks, from Palmer's records of Hermann's lectures and from Soame's reading of medical writers.

The preparation of the second volume involved as great a strain as its predecessor. Some of the work had been already done; for the final decision as to the division of the volumes was only taken in January 1686, and in the notice of Dent's discovery of *Malaxis paludosa* at Gamlingay the date is given 'as in this year 1684'.[11] In June 1685 he wrote that he had not yet finished the Grasses nor 'meddled with the Trees and Shrubs'.[12] All through 1686 the epitomes tell of weekly letters to Robinson full of reports and queries. The plants were all sent off by December, and filled up to page 1348. The 'Dendrology' followed during the next eight months; and must have proved an even more difficult business. In the plants the bulk of the records dealt with European species, most of them familiar to him. The trees were nearly all exotic; their value for fruit and

1 See above, pp. 108–10.   2 P. 641.
3 P. 643: this apparently died: *Corr.* p. 179.
4 P. 675.   5 P. 610.
6 P. 944: Lhwyd in 1699 wrote 'their Shamrug is the common Clover': *Phil. Trans.* XXVII, no. 335, p. 506.
7 P. 669.   8 P. 714.   9 P. 842.
10 P. 854.   11 *H.P.* II, p. 1233.   12 *F.C.* p. 149.

timber and for planting in the parks of the rich had drawn attention to them in countries where the general flora was still almost unknown. Travellers' reports, collections of nuts and dried fruits, samples of foreign woods, medicines like quinine, foodstuffs like sago, beverages like chocolate and coffee provided material that had to be examined and correlated. The *Hortus Malabaricus* was a mine of information; but for the Americas there was as yet no such survey. Even among the plants it is obvious that Ray moves confidently among orders which he knew well and at first hand, but hesitates over those like Ferns or Gentians which he had little opportunity of studying in the field. For trees he is much more dependent upon book knowledge; and the handicap is manifest. Moreover, the adoption of a system of classification by fruit, though he improved upon his *Methodus* by segregating the Palms and breaking up the other orders into small groups, made arrangement inevitably artificial. It is testimony to the soundness of his insight that he nevertheless managed to produce a treatise which covers the known species, avoids fable and duplication, and in a large majority of cases enables the various forms to be identified.

Robinson supplied him in April 1687 with a catalogue of the cultivated varieties of Apples, Pears and other home-grown fruit; and was diligent in suggesting sources of information and sending books from London. Sloane provided lists from Chelsea and kept him in touch with Watts; but in September he went to Jamaica with the Duke of Albemarle and was away for two years. Neither they nor Dale nor indeed any British botanist had expert knowledge or enough capacity to help in the enormous labour of collating the available material. He had to carry through the task single-handed and at high speed; and at a time when the political situation was sufficiently critical and distracting to make any continuous effort difficult. The Dendrology, which occupies pp. 1349–1848 of the volume, is, as he recognised in the Preface, imperfect: it remains an achievement not only remarkable in its completeness and full of interest in its detail but, considering the circumstances of its production, literally amazing.

That the cost was not small is evident from a single item in the epitome. On 8 April we read: 'his herpes and pain cured by Hollyhock leaves boiled in may-butter'—a first reference to the sores on his legs which made the later years of his life a purgatory and which the curious treatments recommended by his medical advisers merely inflamed.

In July we read: 'Shall finish in a month'; and in early September, when at last the script was complete, he visited Robinson in London and spent a week inspecting the collection of foreign seeds and nuts in William Courten's museum in the Middle Temple and of exotic trees and shrubs in Henry Compton's gardens at Fulham Palace.

Courten, grandson of Sir William the great silk-merchant and coloniser of the Barbadoes, was the collector to whom we have already referred in the matter of the Surinam snail. He had been abroad since 1670 and had met Robinson in Paris, whence at his request he had sent Ray specimens of the Macreuse or Scoter duck in 1684.[1] He had then come to London, calling himself Charleton, and had set up what Ray describes as 'a repository of rare and select objects of natural history and art so curiously and elegantly arranged and preserved that you could hardly find the like in all Europe'.[2] This had already supplied material for the *Historia Piscium*; and was afterwards acquired by his friend Sloane and formed an important part of the original British Museum. Ray in the epitome of the letter of 19 September, thanking Robinson for the visit, speaks of him as Charleton, but in the *History* regularly calls him Courtine. Three notes of his specimens occur in the earlier part of the book, two being bean-pods[3] and the third a dried Tamarind:[4] a list of other vegetable curiosities is printed as Chapter XII of Book XXXII.

Compton, whose remarkable career is one of the strangest in the history of the Church of England—he was the sixth son of Spencer Compton, second Earl of Northampton,[5] was a soldier in 1660, ordained in 1662, and made Bishop of Oxford in 1674 and then of London in 1675—had been suspended by the High Court of Ecclesiastical Commission in 1686 for refusing to inhibit John Sharp, afterwards Archbishop of York. Deprived of the exercise of his functions, but not of the temporalities of his see, he had devoted himself to the planting of his gardens at Fulham, and with the aid of an expert gardener, George London, had successfully established many rarities imported from America and collected the finest arboretum in England. Ray evidently spent some time there and paid a glowing tribute to the 'Maecenas of botany'. Some of these treasures, a Cactus which Ray calls 'Echinomelocactus',[6] an American Holly,[7] a Tulip-tree sent by Banister from Virginia,[8] a Genista from Africa[9] and several others,[10] had already been reported to him by Robinson: but the list compiled by him after his visit, and including the Tulip-tree, a Virginian Plane and fifteen others, is printed separately as Chapter XI of Book XXXII.[11]

After this holiday Ray attacked the index and found it tedious; it actually fills forty-five columns: on 25 November he thanks Robinson for his work on the errata: a few days later the volume was ready for publication. On 6 December Robinson wrote to Sloane in Jamaica telling him of the lamentable decline of the Royal Society and announcing that a copy of Ray's book

1 *Corr.* pp. 147–8 and below, p. 334. 2 *H.P.* II, p. 1800.
3 *H.P.* II, pp. 1741, 1777. 4 P. 1749.
5 Killed at Hopton Heath March 1643 as general of Charles I.
6 *H.P.* II, p. 1467. 7 P. 1623. 8 P. 1690.
9 P. 1732. 10 Pp. 1664, 1752. 11 Pp. 1798–9.

is waiting to be sent to him by the first shipping: on 8 April 1688 he told him that it had been despatched by Captain Brooks; that he would find in it a large Appendix; and that 'Mr Ray designs to publish a second Appendix as soon as sufficient materials come to him'.[1] In fact this second volume contains 8 prefatory pages, 985–1944 of descriptions and 35 of indices. For each volume Ray seems to have received £30 in cash and twenty free copies.[2]

The book, although as he insists only the second half of a single treatise bound up separately for convenience, yet begins with a Preface; and this is remarkable for its testimony to the effect upon him of the recent discoveries of tropical vegetation.

Its first paragraphs are devoted to the explanation of certain features of the work that had evoked comment. He has deliberately abstained from controversy about names and synonyms and from attempts to identify the plants vaguely described by early writers: rather he has selected and used the names commonly employed. These names he has not altered even when he has transferred a plant from one group to another: only in the case of Indian and native names he has sometimes added descriptive titles. Occasionally through carelessness he has omitted to state the source of a description, especially in the case of some taken from J. Bauhin. In quoting authorities he has been scrupulous in copying their exact words lest he should misrepresent their meaning. In the matter of foreign fruits which De l'Ecluse, J. Bauhin, J. de Laet and Olaus Worm have so sedulously collected he has not the experience to assign them to the proper trees or even to classify them with confidence: he has listed them and asks those who have access to collections to reduce them to order. Errata due either to the printers or to himself are inevitable; 'in so large a work an elderly man tired out with long-drawn labour gets drowsy or writes hurriedly what ought to be struck out'. Apart from such slips there are mistakes in arrangement in plants that he has never seen or only seen years ago when he never dreamed of producing a History. The length of the Appendix is partly due to oversights, but much more to the very many new plants discovered since the first volume was published.

In this connection he pays tribute to Paul Hermann of Leyden for his *Catalogue* of the physic garden there. Many of the plants in it are in fact grown in the gardens in London; but the expectation of Hermann's work, added to his own difficulties about visiting these gardens, prevented him from trying to anticipate it by independent descriptions of them. That Hermann, who had spent some years in India, was preparing a *Musaeum Zeylanicum* and had gathered a mass of material for it[3] is an event of the

---

1 *F.C.* p. 154.   2 So letter to Sloane: *Corr.* p. 365.

3 Hermann died in 1695 before this project was accomplished: his notes and collections were used by J. Burmann in his *Thesaurus Zeylanicus*, Amsterdam, 1737, and by Linnaeus, *Flora Zeylanica*.

highest importance for natural history. He then praises Hendrik van Rheede van Draakenstein for his vast efforts and expenditure over the *Hortus Malabaricus* and John Banister, who after studying the plants of the West Indies is now in Virginia working at the natural history of that country. These works have revealed the differences between the European and the tropical flora, differences so amazing that 'if a man were carried there in his sleep he would not believe his eyes when he woke up'.

Who would believe that in the single province of Malabar, and that not so very large, there are found more than three hundred wild trees and shrubs, and probably many more?...If any European travelling through woods saw the bark of trees shining at night so brightly as to light up the path and enable him to read letters, would he not be astounded? Yet in America we have witnesses to this, John Earl of Carbery lately Governor of Jamaica[1] and now President of the Royal Society and Thomas Trapham M.D.[2] now practising in the island.... This too is remarkable and peculiar to India that there are trees that shed their leaves when they flower and again burst into leaf when they set fruit....Who would believe that there is a botanical Europe in the middle of tropical India?— so that notable traveller Bernier assures us. In the land of the Mogul (Mogorum Imperator) he crossed high mountains; on their southern slopes Palms, Pepper, Sugar-canes, every thing Indian and exotic; after six or seven hours travel, beyond the summits, a wholly new scene, a swift passage into Europe, Oaks, Elms, Pines, Hyssop, Rosemary....And these strangers you can see not only in books but transplanted into our own gardens.

He then notes the remarkable progress in botany in Europe during the past twenty years through the work of Paolo Boccone, Jakob Breyne, Pierre Magnol and Joseph Tournefort. 'The botanists of our day must press forward vigilantly and vigorously, setting no limit to their researches but those which nature herself has fixed: they must take the whole world for their field.'

Finally, he offers special thanks to four fresh helpers, William Courten, of whom we have already spoken, Leonard Plukenet, Samuel Doody and James Petiver.

The Preface ends with a warm acknowledgement of Robinson's unfailing assistance, and with a list of books omitted from the bibliography or published since its appearance. Among these are Christopher Merret's *Pinax*, Robert Plot's *Histories of Oxfordshire* and *Staffordshire*, James Sutherland's *Hortus Edinburgensis*, and Grew's *Musaeum Regalis Societatis*: and in addition to Hermann, Jan Commelin's *Catalogus Hollandiae*

1 He went in company with Henry Morgan the buccaneer in 1674 and returned with an unsavoury reputation in 1678. During his absence Dryden dedicated to him the licentious drama *Limberham*.

2 Son of Cromwell's surgeon-in-chief: demy of Magdalen College 1654–61; at Leyden 1663; died in the earthquake in Jamaica 1692. Ray drew Sloane's attention to this statement in a letter sent just before his departure July 1687: *Corr.* p. 195.

and work in the *Hortus Malabaricus*, Moritz Hoffmann's Altdorf Catalogues, and the books of Giovanni Battista Trionfetti of Rome and Paul Ammann of Leipzig.

There is appended to the Preface a description of the 'Merry Tree' (*Prunus avium*) and a list of plants extracted from the long letter of Thomas Lawson of Great Strickland, sent to Ray on 9 April 1688.[1]

Of the helpers now mentioned for the first time Plukenet was a fellow-student, probably at Westminster School, of Courten and Robert Uvedale, the great horticulturist. At this time he was living at Westminster, where he had a small garden, and at Horn Hill near Chalfont St Peter's, Buckinghamshire, where he had a farm;[2] but had not become known as a botanist. In 1689 he attracted the attention of Queen Mary, who put him in charge of the gardens at Hampton Court with the title of Regius Professor. His name is mentioned by Ray to Sloane as early as 1685,[3] as well as in the misdated letter in which he is said to be drawing plants *in piccolo* 'using a small scale and thrusting many species into a plate'.[4] In June 1690 he sent Ray a long criticism of his *Synopsis*;[5] and in July 1692 replied rather irritably to Ray's comments upon his *Phytographia*, of which he had presented him with a copy.[6] Ray wrote warmly of him in his *Fasciculus* and *Synopsis*, but later admitted that he was 'a man reserved, jealous of his reputation, and none of the best natured, not to give him a worse character being my friend'[7] and 'as ill-natured and liable to mistakes, however confident and self-conceited he may be'[8]—opinions very generally shared. Plukenet produced many books illustrating his own collections. Although embittered and very critical of Sloane[9] and Petiver, he seems to have had a real regard for Ray; and the famous saying 'the best medium to reach heaven, better than the Divinity of the Schools' is his comment upon the *Historia Plantarum*.

Doody was the son of an apothecary and worked in his father's shop in the Strand. He came into touch with Robinson soon after the latter's return from abroad; and our first knowledge of him is of an excursion which the two took in August 1686 'five or six hours down the river',

1 *Corr.* pp. 197–210.

2 *Syn. Brit.* p. 22.   3 *Corr.* p. 139.

4 *Corr.* p. 130: see above, p. 209.   5 *Corr.* pp. 214–24.

6 *Corr.* pp. 232–7: Ray commented very favourably upon it in *Phil. Trans.* XVII, no. 193, p. 528.

7 *Corr.* p. 307.   8 *Corr.* p. 371.

9 Trouble seems to have arisen over the plants of Jamaica: Sloane prepared a book on these after his visit; Plukenet sent Ray a list of Jamaican plants: of this Ray wrote to Sloane in 1696 (*Corr.* pp. 299–300): 'I fear there are many mistakes in his work.... He is a man of punctilio, a little conceited and opinionated, and such men are incapable of advice, especially reprehension.' Nevertheless in his *Mantissa* Plukenet referred to Sloane's *Catalogue* and to his work with the most savage contempt: cf. *Mantissa*, pp. 18, 22, 33, 39, etc.

when they 'found many rare plants upon the chalk hills and marshes near Gravesend'.[1] A Commonplace book by him dealing with botany and dated 1687 is in the Sloane collection.[2] He was something of an expert on cryptogams and supplied Ray with information and lists for his later works. In 1692 he was appointed Watts's successor at Chelsea,[3] and in 1694 was proposing the publication of a Bryologia:[4] but he seems to have been dilatory in matters of business[5] and liable to bouts of intemperance.

Petiver, also an apothecary, had been educated at Rugby and apprenticed to the chemist of St Bartholomew's Hospital in 1683. He was established in premises in Aldersgate in 1692 and lived there all his life. He does not seem to have been well known to Ray until the close of the century, and his name occurs in the letters first in 1696.[6] Although he compiled the list of Middlesex plants for Gibson's edition of Camden's *Britannia* and by that time had amassed a herbarium of more than 5000 specimens, he was not a great botanist: indeed, it was his interest in entomology rather than in plants that brought him and Ray into close association: during the last years, when the *History of Insects* was in preparation, the two men became intimate. Petiver was rather a collector than a scientist—an enthusiastic collector of whom Ray wrote: 'he hath the greatest correspondence both in East and West Indies of any man in Europe'.[7] His books were mainly a series of plates and descriptions dealing with his specimens, the *Museum* published between 1695 and 1703 and the *Gazophylacium* between 1702 and 1709:[8] but he also produced two volumes of copper-plates illustrating Ray's *Historia Plantarum* in 1713 and 1715. His collections were bought by Sloane and so passed into the British Museum. Both he and Doody became Fellows of the Royal Society in 1695 when Sloane was trying to revive its activities.

Thomas Lawson, whose contribution arrived almost too late for publication, was an older and more learned man. Born near Settle and educated at Giggleswick he entered Christ's College in 1650 at the age of nineteen and may perhaps have known Ray at Cambridge. He had been appointed minister of the parish of Rampside, Lancashire, but after a visit from George Fox in 1652 resigned his living and became a Quaker. For some years he lived in hardship, but in 1658 married, settled at Great Strickland in Westmorland and set up a school there. It is tempting to suppose that Ray may have visited him there on some of his journeys and been shown the local plants. But the first mention of him in the letters is in connection with Plukenet and Newton in 1685.[9] In addition to the long

1 *Corr.* p. 187.   2 Sloane MS. 3361.
3 *F.C.* p. 175.   4 *F.C.* p. 247.
5 Cf. Ray's complaint with regard to the second edition of the *Synopsis*: *F.C.* p. 261.
6 *Corr.* p. 303.   7 *F.C.* p. 279.
8 For details of their sales cf. below, p. 417 n. 2.   9 *Corr.* p. 139.

list already mentioned he sent *Cystopteris fragilis* to Ray some time before June 1689[1] and a packet of fossils[2] later in the same year. He died in 1691. Ray praises him as 'an industrious diligent and skilful botanist' in the Preface of his *Synopsis Britannicarum*.

The volume itself begins with the two last tribes of Dicotyledons—the large assemblage of Caryophyllaceous and other five-petalled plants in Book XIX and the Primulaceae and other 'pentapetaloids' in Book XX. The 'Clove-July-Flower' is the first species. There is a note of the introduction of *Lychnis chalcedonica*[3] into horticulture from Constantinople and of the double Campions (*L. alba, dioica*, 'called Red Bachelors Buttons'[4] and *flos-cuculi*[5]) grown in gardens. Of 'Gossipium aut Xylum, called Cotoneum by recent writers from its resemblance to the Cotton Apple', which he rightly places between Geranium and Althaea, he records that *G. herbaceum* is grown freely in Malta and Sicily and also as Robinson testifies in the Campagna, and that its cotton is used for undergarments, padding and the staunching of wounds.[6] He has a longer note on the uses of Flax (*Linum usitatissimum*), quoting Pliny and J. Bauhin and ending with the statement that 'printers' ink, atramentum typographicum, is made of lamp-black and linseed oil'.[7]

Book XXI[8] begins the Monocotyledons, 'herbs whose seed-plant is not two-leaved or rather not possessed of two cotyledons'. He starts with the Crown Imperial (*Fritillaria imperialis*) and other Fritillaries, one of which, *F. lutea*, he had grown at Cambridge;[9] the Lilies, Garlics, and Daffodils. Of a Daffodil, apparently then a new species, he writes that he received the bulb from his 'learned and esteemed friend', Robert Uvedale,[10] the famous gardener and schoolmaster of Enfield who had entered Trinity College in 1659 and been elected a fellow in 1664. In the case of these, as of the Tulips and Irises, wherein cultivation had already produced much variation, he faces more plainly than in any of his earlier writings the question of specific distinctions. He refuses to list and describe the 'indefinite and annually increasing number' of horticulturists' forms on the ground that 'as we have often stated, the number of true species in nature is fixed and limited and, as we may reasonably believe, constant and unchangeable from the first creation to the present day'.[11] So he states the conviction which he shared with all his contemporaries, with Linnaeus and almost all biologists until the nineteenth century: it is the answer to his own doubt about the transmutation of species and to any suggestion that he was a forerunner of an evolutionary doctrine.

---

1 *F.C.* p. 187.　　　2 *F.C.* p. 203.　　　3 P. 992.　　　4 P. 994.
5 P. 1000.　　　6 P. 1064.　　　7 Pp. 1072–3.
8 By an error of numeration two books are numbered XXI, that of the Bulbosae and that of the Bulbosis affines.
9 P. 1107.　　　10 P. 1130.　　　11 P. 1188.

Of irises he knew *Iris susiana*[1] from Constantinople and *I. tuberosa*,[2] which he ascribes to Arabia and the East, and grew *I. pallida*[3] at Dewlands. The 'most common purple Flower-de-luce'[4]—even then called *Iris germanica*—seems to have been known as a wild plant of hillside thickets, though he gives no localities. He proposes the name Hemerocallis for the Day-lily,[5] and records the growing of Cannas in hot-houses, 'hypocausta';[6] and speaks of *Cyclamen europaeum*, which he still groups among Monocotyledons, as grown in English gardens.[7] The Grasses and Sedges, apart from the British species, are not of great interest; but he has notes on the use of the 'Spanish Reed' (*Arundo donax*) for fishing-rods, of the 'Writers' Reed' for pens by those Easterns who do not use quills, and of Cane for seating chairs.[8] Ginger[9] (*Zingiber officinale*), which he calls Zinziber and classes next to the Bamboos; Vanilla[10] (probably *Vanilla planifolia*), whose affinities with the Orchids Hernandez' description did not enable him to recognise but which was already used for flavouring chocolate; and the Pine-apple[11] (*Ananas sativa*), which Evelyn first tasted at the table of Charles II,[12] also receive comment. The Book finishes with a series of plants to be inserted earlier, and some like the Piperaceae whose flowers no one had then described.[13]

Then begins the Dendrologia. The Palms he claims to have rearranged; and there is a mass of material from the *Hortus Malabaricus*, from Piso and Marcgraf, and a note on the Sago Palm (probably *Metroxylon sagus*) from Robert Knox, who having been held prisoner in Ceylon from 1659 to 1679 had published the *Historical Relation of the Island* in 1681.[14] Of the Date Palm (*Phoenix dactylifera*) he recounts the evidence of Theophrastus as to the dust of the male flowers increasing the fertility of the female and adds: 'our opinion, as stated in the first Book, is that the apices supported by the stamens take the place of male seed in plants and serve for fertilising the females'[15]—a stronger statement than he had previously made. Of the 'Coco or Cokernut' (*Cocos nucifera*) he says 'it is the most useful of all trees; from it are derived wine and milk and oil and sugar and vinegar and solid food and thread and needles and girdles and clothes and cups and spoons and other vessels and mats and baskets and sunshades and paper'.[16] He finishes the Palms with the Musaceae, the Plantain-tree and the 'Bonana-tree'.

In the next Book, which begins with the Walnut, Hazel, Beech, Chestnut and Oak, he has a number of references to John Evelyn's *Sylva* of

1 P. 1181: according to Linnaeus, *Spec. Plant.* I, p. 55, this came from Constantinople to Belgium in 1573.

2 P. 1190.	3 P. 1182.	4 P. 1180.
5 P. 1191.	6 P. 1202.	7 P. 1206.
8 Pp. 1276–7.	9 Pp. 1314–15.	10 P. 1330.
11 P. 1333.	12 *Diary* for 19 Aug. 1668.	13 Pp. 1341–4.
14 P. 1367.	15 P. 1353.	16 P. 1359.

which the first edition had been published in 1664 and a second much enlarged in 1669. There is no evidence that the two men ever met; and Evelyn, though an original member of the Royal Society, was rather a virtuoso than a scientist: but in the second edition of the *Sylva*, when dealing with the rising of sap in trees, he seems plainly to refer to the experiments of Willughby and Ray,[1] and is himself cited by Ray in regard to Hunting Spiders[2] and again as to the claim of the Elm (*Ulmus campestris*) to be indigenous in Britain.[3]

Book xxv deals first with Conifers, and the discussion of the Silver Fir (*Abies pectinata*) and Spruce (*Picea excelsa*) has been already quoted.[4] On Evelyn's evidence he treats the Scotch Fir (*Pinus sylvestris*) as a British native together with the Yew (*Taxus baccata*) and two Junipers (*J. communis* and *sibirica*): he notes that the Arbor vitae of Canada (*Thuya occidentalis*) is now common in English gardens and that the Cypress (*Cupressus sempervirens*) was almost exterminated by the bitter frost of 1683–4: but he knows nothing of the introduction of the Larch (*Larix europaea*) or of the Cedar of Lebanon (*Cedrus libani*). He inserts the Alder (*Alnus glutinosa*) and Birch (*Betula alba*) among the Conifers:[5] but devotes the latter part of the book to Poplars and Willows, Elms, Hornbeam and Mulberry, noting that though he has tried many other food-plants for the Silk-worm (*Bombyx mori*), a diet of Mulberry leaves is alone productive of silk.[6]

He deals next with fruit-trees and first with the Figs, taking from the *Hortus Malabaricus* descriptions of various kinds of 'Ficus indica' or Banyan (though he never uses the word[7]) and from Vesling of the 'Sycomore of Egypt' (*Ficus sycomorus*). Under the Apple he gives Robinson's list of the various English Apples which 'if our palate is any judge take the prize for excellence from all Europe'. This includes twenty-one sorts of dessert apples, among them the Golden Pippin, Summer Pearmain, Codling and Costard Apple 'so famous that apple-sellers are still called Costardmongers'; six sorts of local apples, 'the Harvey apple, named from its inventor Dr Gabriel Harvey,[8] being the delight of Cambridge'; thirty-one of keeping apples; and twenty of cider apples. Of Pears he lists sixty-one named kinds, divided into wall-fruit, orchard-fruit, fruit for cooking and fruit for perry. Quinces were then specially used for 'marmelata', which

---

1 *Sylva*, ed. II, pp. 73–4.   2 *Corr.* p. 29.

3 *F.C.* p. 166: Evelyn (*Sylva*, ch. iv) regarded it as introduced; Ray (*H.P.* II, p. 1427) disputes this.

4 P. 1395: see above, p. 152.   5 Pp. 1409–10.   6 P. 1430.

7 Ray gives the Indian names from the *Hortus Malabaricus*.

8 Of Christ's, Pembroke and Trinity Hall (elected Master there 1585); the friend of Spenser: Harvey apples were among those planted in 1674 in Sir Thomas Sclater's orchard at Linton and listed in the 'Catley book': cf. Gunther, *Early Science in Cambridge*, p. 450.

he derives from the Spanish Marmellos or Quince. Guava trees (*Psidium guaiava*, of the Myrtaceae) had already been brought from Barbados and seen by him in English shrubberies; as had also the Pomegranate (*Punica granatum*) from Provence. Of the Cactaceae, interpolated here, he has a long account of the 'Prickly Pear' (*Opuntia vulgaris*), which had already run wild in Italy[1] and been seen by him in the Belvidere gardens: to this he appends a list of species from Hernandez ending in the 'Nopalno-cheztli' (*Nopalea coccinellifera*) and the Cochineal insect (*Dactylopius coccus*) feeding on it, which on the evidence of Robinson he assigns to the 'Lady-cows' or Coccinellid beetles:[2] he then sets out a number of species from Piso, including the 'Prickly Torch-plant' grown by Watts at Chelsea and a Hedgehog Cactus belonging to Bishop Compton. There follow the Roses, a medley of wild species, mostly British, and garden forms, the Red and the White, the Damask, York and Lancaster, Provence, Cinnamon, Musk and Monthly; the Lotus and a number of Indian berry-bearing trees; the Currants, Bilberries, Honeysuckles and more Indians; the Myrtles and Ivy; and, finally, the Jamaican All-spice and Cloves (*Pimenta officinalis* and *Eugenia aromatica*).

In Book XXVII he turns from berries to drupes, or rather to 'trees in which the flower adheres to the base of the fruit'; and begins with the Apricot, Peach, Nectarine, Almonds, Nutmeg (*Myristica fragrans*) and Plums. To the 'Zizyphus or Jujube tree' (*Z. sativa*), which he had seen wild in Calabria and of which he had bought and disliked the fruit in Venice, he appends a note on 'lacca',[3] lacquer, hesitating whether it is a resin or perhaps a product of insects. Cherries, Olive (*Olea europaea*), Mango (*Mangifera indica*),[4] a number of other Indian trees, and finally the Cinnamons, with a long discussion whether the Cinnamon of the ancients is or is not the same as the Cassia or as the Cinnamon of his own time.[5]

In Book XXVIII are the smaller drupes; the Sassafras (*S. officinale*) and others; the Arbutus (*A. unedo*), which he records as wild in western Ireland; the Terebinth and Mastich (*Pistacia terebinthus* and *P. lentiscus*); and the Mistletoe (*Viscum album*), seen on Maple and Ash in the park at Middleton, on Hazel near Braintree,[6] on Lime, Elm, Willow, Rowan, and Buckthorn, on Apples, Pears, Hawthorn and rarely Oak,[7] which he declares is not spontaneously generated as some of his contemporaries

1 Cf. Robinson's letter of April 1684: *Corr.* p. 143.
2 It is a Hemipteron of the family Coccidae.
3 The genuine Chinese lacquer is the sap of *Rhus vernicifera*, not of a Zizyphus.
4 He notes 'its wood is used with Sandal-wood for cremation'.
5 *Cinnamomum cassia* of the Lauraceae is called Cassia; but botanically Cassia is a leguminous genus.
6 Cf. *Corr.* p. 60.
7 P. 1583; cf. the list supplied to him by Willisel: cf. Derham, *Phil. Letters*, p. 351. He adds White Beam, but omits Maple and Pear.

affirmed, but sprung from seeds evacuated by birds. Then come the 'Phillyrea or Mock-privet' (*P. latifolia*) seen in Etruria; the 'Water Elder' (*Viburnum opulus*) and its double variety only found in gardens; the Mezereon (*Daphne mezereum*), which he had seen in Germany and at Geneva, but did not accept as wild in England; the European and 'Virginian or Bucks'-horn Sumachs' (*Rhus coriaria* and *R. typhina*); the Jasmines, Privets, Barberries and Elders; and then a long section on the Vine. He first describes different species, the Currant Vine of Corinth, the 'Parsley Vine' called Canadian, the Vine from which are dried 'uvae insolatae, the Raisins of the Sun' imported from Spain, and the Wild Vine of Virginia. Then he gives a list of eleven Grape-vines specially prized in England; and a description of the various methods of growth. Then follows a list of the most famous wines, French, Spanish and Portuguese, and German; and finally paragraphs on 'aqua vitae' (brandy), Vinegar and Tartar. The 'Thee or Tea plant' (*Camellia thea*) follows the Vine. He refers to the authors who have dealt with it, J. Breyne in his first Century, Pechlin in his Tractate of 1684, Dufour in his book on Coffee, Tea and Chocolate translated in 1685, and to an English Treatise of 1682; but prefers to all these the *Observations* of Wilhelm ten Rhyne, physician to the Emperor of Japan, from whom he quotes an account of its culture and preparation. Its locality is China and Japan; and though describing its use as an infusion in hot water as commonly known in Europe, he does not regard it as a safe beverage or as having the beneficial results ascribed to it in the East.[1] The Spindle-tree (*Euonymus europaeus*), Holly, Buckthorn (*Rhamnus catharticus*), Capers (*Capparis spinosa*), which he had seen at Toulon,[2] and Crowberry (*Empetrum nigrum*), and finally the Blackberry and Raspberry, which does not seem to have been then cultivated, complete the Book.

The next contains among many other exotics the Citron, Lemon and Orange (*Citrus medica* and *limonum*, and *C. aurantium* and *sinensis*), each with a number of named varieties, including the Lisbon or Chinese Orange; the 'Mammee or Toddy-tree' (*Mammea americana*), whose fruit and growth he describes recognisably though he erroneously says that toddy-wine is produced from it when tapped;[3] the Calabash tree (*Crescentia cujete*), and the 'Caco tree' (*Theobroma cacao*), on which he quotes William Hughes, the gardener and author of books on vines and gardening, who had been a filibuster and in 1672 published *The American Physitian or a Treatise of the Roots, Plants, Trees growing in America*, and to which he appends a long account of Chocolate both as a confection and as a beverage, with a list of plants added to it for flavouring.

1 Tea-drinking was of recent introduction; cf. Pepys's account of his first taste of it on 25 Sept. 1660.
2 *Obs.* p. 464, where their pickling is described.
3 It is the product of a Palm, *Caryota urens*.

Book XXX is a rather miscellaneous collection of foreign plants, including Camphor from Japan (*Cinnamomum camphora*) and from Sumatra, which he thinks the same as that from Borneo (*Dryobalanops aromatica*); Storax (*Styrax officinalis*), which he saw around Rome; Bladder-nut (*Staphylea pinnata*), which he found in the hedges at Pontefract but does not regard as native to Britain; Pistachio nut (*Pistacia vera*), described by Rauwolf, which he distinguishes from Terebinth; Manihot or Yucca (*Manihot utilissima*), of which he gives a long account from Piso and from which 'the American natives from Florida to the Straits of Magellan derive the bread called Cassava'[1] and also 'Mandihoca' and 'Tipioca'; the Bay (*Laurus nobilis*) and other Laurels, and the 'Bon, Ban, Buna or Bunnu' from which Coffee is prepared.

Of this tree (*Coffea arabica*) ignorance was still almost complete. In 1687 Robinson had written to say that 'M. Bernier who passed the Red Sea into Arabia doth affirm that the Arabs assured him that the Coffee fruit was sown every year under trees up which it did climb' and that therefore 'Alpinus never saw the true coffee plant'. Nevertheless, Ray rightly quotes Alpino that 'it is like a Euonymus but with thicker, harder, greener and evergreen leaves'; and Dufour, the merchant of Lyons, author of a book on coffee,[2] who described it as like a small cherry-tree with slender branches and small leaves. The flower, according to Ray, was still unknown. He gives a long account of the preparation and qualities of the berry; and declares that in all Cairo, though there were a thousand coffee-houses, there were only two men who could roast it properly. He comments upon the rapid spread of coffee-drinking in Europe, upon the revenue derived from it in Britain, and upon the fact that London[3] has now as many coffee-houses as Cairo; and expresses wonder that one small country can supply so huge a demand.

After dealing with the Box (*Buxus sempervirens*), the Limes, 'Venice Sumach' (*Rhus cotinus*), Ashes, Maples, Tamarisks, Plane, then grown at St Albans, Christ's Thorn (*Paliurus australis*), common in hedges near Verona and Montpellier, 'Widow-wail' (*Cneorum tricoccon*), seen between Montpellier and Frontignan, and fifteen Heathers, he devotes Book XXXI to the 'Siliquosae' and begins with those with papilionaceous flowers. The Judas-tree (*Cercis siliquastrum*) he had seen on the banks of the Tiber, the Carob (*Ceratonia siliqua*) in Sicily, 'Pseudo-acacia americana Robini' (*Robinia pseudo-acacia*) in English gardens, one of his four Coluteas (*C. arborescens*) on the slopes and crater of Vesuvius, the 'Bean-trefoil-tree not stinking'[4] (*Cytisus laburnum*) on the Jura and in Savoy.

1 P. 1683.
2 Philippe Sylvestre Dufour, *De l'usage du caphé, du thé et du chocolate*, Lyons, 1671, translated into English by John Chamberlayn, London, 1685.
3 The first had only been opened in 1652.   4 P. 1721.

The section closes with a number of Brooms, including *Spartium junceum*, 'very common in Italy, Sicily and Provence'.[1]

The rest of the Book contains a number of exotics; the Brazil Wood (*Haematoxylon campechianum*), whose 'red dye is known to all the world';[2] Tamarind (*Tamarindus indica*), which he declares to be native to both East and West Indies; Balsam (probably *Commiphora opobalsamum*), of whose perfumes he gives particulars; 'Pipe tree or Lilac' (*Syringa vulgaris*), of which he describes both blue and white forms, admitting that he knows them only in gardens but had seen the white near Mont Salève far from any house; Oleander or Rose-bay (*Nerium oleander*), which he had seen near the foot of Etna;[3] and Mangrove (*Rhizophora mangle*), whose 'fruit is food for land-crabs rather than for men'.[4]

The final Book is entitled 'Anomalous and little known trees and parts of trees', and contains an immense number of notes which give a clear idea of the difficulties with which Ray had to contend. It begins with plants mentioned by Breyne, Piso and Marcgraf, and the *Hortus Malabaricus* and other sources, but so incompletely known or described that he has not been able to place them in their appropriate families. The most interesting is the 'Arbor febrifuga Quinquina or Jesuits' Powder',[5] of which he says that Charles Goodall, the pupil and successor of Walter Needham, had shown him the drawing of a branch sent from Italy to the Royal Society. From this and from other fragmentary accounts he thinks it like a plum or pomegranate.[6] Of its use he gives an account from Sydenham and Robert Tabor and from Jacques Spon, the French physician and traveller. After it come lists of the trees grown by Bishop Compton at Fulham and the fruits in William Courten's museum; and then a long series of records of the parts collected or used; timber like Sandalwood, which he calls 'Saunders', and Ebony; fruits like Nux vomica[7]—twenty-one pages of them ending in a list from Grew's *Catalogue of the Museum of the Royal Society*; and finally the resins used in medicine, Frankincense[8] and Myrrh,[9] Ammoniac[10] and Asa foetida,[11] Benjamin[12] and Liquidambar, this last from the American species, *L. styraciflua*, which he had already listed as 'Styrax Virginiana' among those grown by Compton.[13]

1 P. 1726.
2 P. 1737: he here treats 'Brazil Wood' and 'Logwood' as distinct, though both are used for dyeing: later in his Preface to Sloane's Jamaican Catalogue he accepts Logwood as 'Lignum campechianum': *Corr.* p. 467.
3 Cf. *Cat. Ext.* p. 85.   4 P. 1772.   5 P. 1796.
6 The various species of *Cinchona* are in fact Rubiaceous.
7 The seed of *Strychnos nux-vomica*.
8 From species of *Boswellia*.   9 From *Commiphora myrrha*.
10 From an Umbelliferous plant, *Dorema ammoniacum*.
11 From *Ferula foetida*, also Umbelliferous.
12 From *Styrax benzoin*.   13 Cf. p. 1799.

The Appendix, which fills pp. 1849 to 1944, gives perhaps more clearly than anything else an impression of the scrupulous diligence with which Ray pursued his task. It contains at least 388 separate additions, one or two like the account of the Dodders[1] omitted by oversight, many more corrections of classification or description based upon further knowledge, many more records of species reported since the completion of the script, but most short notes supplied by friends or drawn from books not previously searched, and including new synonyms, points of structure and growth, localities, and details of habit and use. At the end is a list of Emendanda; and catalogues of plants mentioned by Zanoni, Breyne and Hernandez, which he had not been able to classify and include in the body of the work. Among these is a list of Virginian plants observed by John Banister and evidently sent by him to Bishop Compton in 1680; and this is supplemented by the additional names of those whose seeds Banister had sent to Ray himself in 1687.[2]

Of the number of species actually treated in the two volumes widely different statements have been given. In fact it is not easy to speak with certainty. He numbers serially within the group or genus each form that he regards as specifically distinct, but includes many others without numbering them which he yet might accept as good species. He also prints a number of lists without descriptions, some of them of garden forms like the Anemones or Tulips, but others of wild plants catalogued by one of his authorities. If we include only such as he certainly knew and treated as distinct, the total is just over 6100, a number showing small increase upon the 6000 of Bauhin's *Pinax* unless we remember how drastically Ray pruned the lists of his predecessors by the conditions which he laid down for specific distinctness. Of nearly all these he has got an adequate description, notes of locality often full and exact, details of habit and of use, and generally a discussion of its relationship to kindred forms. In some of them the treatment develops into a concise essay, like those on Tobacco and Coffee, the Apple and the Vine.

NOTE. *Ray's wine-list. H.P.* II, p. 1617 *translated*

It is not my intention to recount all the kinds of wine mentioned by authors old and new: it is enough to indicate briefly some of those used and valued among us to-day. These are—

From France: 1. 'Parisinum', light and short-lived, when well matured of pleasant flavour and fit for the connoisseur, *Vin de Paris, Champagne*:[3]

1 To be inserted at the end of Book xx.
2 So epitome of letter to Robinson, June 1687: *F.C.* p. 291. Banister was killed by a fall from rocks before his return to England. In 1698 Ray refers to his 'unhappy and untimely death': *Corr.* p. 357.
3 Should surely come under next heading.

2. 'Campanum', the finest of all French wines:

3. 'Burdegalense' or 'Gravianum', *Common Claret Wine or Graves Wine*:

4. 'Burgundiacum', *Burgundy Wine*; among Burgundies Beaune holds the first place according to the proverb 'Vinum Belnense super omnia recense':

5. 'Album vulgare', *Common White Wine*:

6. 'Frontinianum' (of Frontignan), full-bodied and musk-scented:[1]

7. 'Aurelianum' (of Orleans), valued in France as generous and wholesome:

8. *Hermitage.*

From Spain the best known are 1. 'Canarinum', *Canary Sack*, appraised above all, generous and mellow, nature's own friend:

2. 'Malacense', *Malaga Sack*, richer and heavier than the last-named, but followed by headache if taken freely:

3. 'Xeranum', *Sherry Sack*, harsher and less pleasant to the palate:

4. 'Alonense', *Alicant Wine*, red and heavy, pleasant to the palate but bad for the head and digestion; not very unlike this is the kind that our countrymen call *Tent Wine*—I do not know the reason for the name.

From Portugal is imported 'Portuense', *Porto-port Wine*, but less often.

The wine of the island of Madeira is mostly exported to America.

From Italy few are imported: 1. From Etruria red Florentine wine, very digestible and wholesome with meals:

2. Verdea so-called, also from Etruria, white, sweet, mellow, more palatable than digestible:

In Italy itself the most famous[2] are—at Venice, 'Malvaticum', Malmsey, like Canary, not Italian but from Crete; Red 'Vicentinum', called *dolce & picante*;[3] in Etruria 'Faliscinum' or of Monte Fiascone; at Rome Alban, once famous, and Monte Pulciano; Syracusan from Sicily; at Naples the red wine called Lachrymae Christi, and a Greek wine grown on the slopes of mount Vesuvius.

From Crete is brought Red 'Moschatellinum', *Red Muscadine.*

From Germany only the Rhenish wines or those grown near the Rhine are imported to us, as *Neccar Wine*, 'Rinchovianum' (Rheingau), 'Hochamore' (Hochheimer), 'Mosellanum', and 'Baccherac'.[4]

Nowadays the dry wines like the red French and Florentine which taste rather harsh and astringent are thought more wholesome and digestive than the white ones which are mellow and sweet.

[Names which Ray prints as English are in italics.]

1 Cf. *Obs.* p. 460: 'the so famous muscat wine'.
2 He gave a rather longer list in *Obs.* p. 363.
3 Cf. *Obs.* p. 217, where it is described as 'very rich and gustful'.
4 Cf. *Obs.* p. 76 for wines of 'Baccharach' and 'Rhincow'.

CHAPTER X

THE FLORA OF BRITAIN

Of all the systematical and practical floras of any country the second edition of
Ray's Synopsis is the most perfect that ever came under our observation.
SIR JAMES E. SMITH, Rees' Cyclopaedia, vol. XXIX.

The *Historia Plantarum*, for all its massive learning, did not achieve the
success that it deserved. Ray himself, writing in 1689 to Lhwyd, said 'as
for cuts for my History of Plants there are none to be expected; the book
sells not so well as to encourage the undertakers to be at any further
charge about it. The times indeed of late have not been very propitious
to the booksellers' trade';[1] and Sir James Smith, writing more than a
century later, declared that though 'so ample a transcript of the practical
knowledge of such a botanist cannot but be a treasure, yet it is now much
neglected, few persons being learned enough to use it with facility for
want of figures and a popular nomenclature'.[2] It was in fact handicapped,
as Ray had foreseen, by the lack of plates; and its bulk made it necessarily
a book for the few. It is a monument to its author's greatness, and prepared the way for Linnaeus. But it could hardly do more.

For indeed it was singularly unfortunate in the time of its appearance.
Ray lived and worked unmoved by the upheavals of the period: but the
men to whom such a book could appeal belonged to the great world; and
in Britain that world was in confusion. Indeed it is surprising that in the
year of the Revolution Ray's volume should have been printed and published without interruption: that it should have failed to attract public
attention was only to be expected.

Moreover, the Royal Society had fallen upon evil days. Shortly after
the order to print had been given Robinson had resigned the secretaryship.
Pepys ceased to be President at the Annual Meeting in 1686, John
Vaughan, third Earl of Carbery, whom Pepys had previously styled 'one
of the lewdest fellows of the age',[3] being his successor till 1689, when
Thomas Herbert, Earl of Pembroke, held office for a year, during which
he never attended a single meeting. In 1687, on 6 December, Robinson
wrote to Sloane: 'the Royal Society declines apace, not one correspondent
in being. The revenue is settled upon Mr Hooke....Mr Ray's second
volume is now finished and I have one for you which I will order

---

1 *F.C.* p. 191.   2 *Mem.* p. 77.

3 *Diary*, 16 Nov. 1667: for a more cautious verdict cf. Weld, *History of the R.S.* I,
p. 317: for his infamous treatment of Lhwyd, to whom he pledged his support and then
repudiated his signature, cf. Gunther, *Life of Lhwyd*, pp. 298, 535.

Mr Faithorne to send you by the first shipping'[1]—Sloane being then in Jamaica. There was no money in the Society's funds, no readiness to take office, no opportunity for scientific work. Even the publication of *Philosophical Transactions* was suspended. It was not until national life had settled down and Sloane took over the secretaryship in 1693 that the Society began to recover.

Ray, however, was not daunted, and the story of his last service to British botany may be prefaced by a further quotation from Smith:

If the fame and utility of Ray's great botanical work has neither of them been commensurate with the expectations that might have been formed, a little octavo volume that he gave to the world in 1690 amply supplied all such defects and proved the great corner-stone of his reputation in this department of science. . . . Of all the systematical and practical Floras of any country the second edition of Ray's synopsis is the most perfect that ever came under our observation. 'He examined every plant recorded in his work and even gathered most of them himself. He investigated their synonyms with consummate accuracy; and if the clearness and precision of other authors had equalled his, he would scarcely have committed an error. It is difficult to find him in a mistake or misconception respecting nature herself, though he sometimes misapprehends the bad figures or lame descriptions he was obliged to consult.'[2]

Yet the book which deserved such praise owes its occasion to a publisher's litigiousness.

We have already referred to the second edition of the *Catalogus Angliae* published in 1676 and to the list of new plants and localities contained in it. It was little more than a reprint and was sold out some ten years later. A third edition, to include many new species and some corrections, was planned. But Martyn, who had printed it, had died in 1680; and the rights in it had passed to his heirs and been transferred by them to another firm. This publisher, according to Ray's account, behaved very insolently to him, refused to offer any payment at all for the revision and enlargement of the book, saying that it was not worth his while to do so. Other publishers, apparently Walter,[3] offered to produce it, but being threatened with legal proceedings had to abandon the project.

So Ray had to content himself with the issue of a supplement of 32 pages, the *Fasciculus Stirpium Britannicarum post editum Plantarum Angliae Catalogum observatarum*, published by Faithorne with Carbery's imprimatur in 1688, under the authorship of 'John Ray and his friends'. His own additions are largely of Fungi, thirteen species, most of them already mentioned in the *History of Plants*, being listed. *Allium oleraceum* he had found 'in a corn-field in Black Notley belonging to the hall called Westfield adjoining to Leez lane plentifully'.[4] *Herniaria glabra*, now first

1 *F.C.* pp. 154–5. The book was sent by Captain Brooks in April 1688.
2 *Mem.* pp. 77–8: the latter portion is quoted from *Trans. of Linn. Soc.* IV, p. 277.
3 *F.C.* p. 291.   4 *Fasc.* p. 2.

given this name, he records from rocks near the Lizard: this he had previously confused with Gerard's Millegrana minima (i.e. *Radiola linoides*), being misled by G. Bauhin's use of the same name. A visit to the Essex coast in the summer of 1685[1] had given him several Fuci, *Zostera marina*, and *Atriplex sabulosa*, *Poa loliacea* ('I found it between Walton and St Osyth') and other grasses. His own neighbourhood added a few mosses and lichens. But for the most part the plants come from his friends. Thomas Lawson had sent him on 9 April—barely a month before the Imprimatur—from Great Strickland a long list of northern plants[2] found by him in the Lakes or on the coast from Witherslack to Whitehaven; many of these are inserted in the last three pages of the *Fasciculus* under a special heading: but many are given with Lawson's name in the general list, notably *Geranium lancastriense*, discovered at Walney Island on the spot where it still abounds: Ray notes that transferred to gardens it keeps its characteristic flower-colour and growth.[3] James Newton also sent him many novelties and from a wide range of localities, including *Hymenophyllum unilaterale*[4] from 'Buzzard Rough Crag near Wrenose', *Juncus filiformis* from Ambleside, *Allium ampeloprasum* from Steepholm (there is no mention of the Peony!) and *Arabis stricta* from St Vincent's rocks at Bristol. Edward Lhwyd is responsible for *Silene acaulis* and *Cerastium alpinum* from Snowdon, *Potentilla rupestris* and *Veronica hybrida*[5] from Breidden Hill ('Craig Wreidhin') in Montgomeryshire and *Thalictrum alpinum* from Cader Idris. Samuel Doody sent a number of plants, especially sedges and grasses, from London. Leonard Plukenet had discovered *Helianthemum polifolium* in its chief British locality, 'Brent Downs in Somersetshire near the Severn Sea'.[6] Peter Dent is represented by *Papaver dubium*[7] and *Silene anglica*, 'near the Devil's Ditch in Cambridgeshire',[8] and his son Pierce Dent, with Samuel Dale, found *Malaxis paludosa* 'in the fenny grounds near Gamlingay'.[9] Dale added several others. So did Matthew Dodsworth,[10] Rector of Sessay,[11] Yorkshire, who

1 Cf. letter to Sloane: *Corr.* p. 140.

2 Printed in *Corr.* pp. 197–210: cf. above, pp. 232–3.

3 Lawson states, *Corr.* p. 202: 'I have sent roots to Edinburgh, York, London, Oxford where they keep their distinction.'

4 This gave him some trouble: he was uncertain 'whether it be moss or fern' and whether it was distinct from *H. tunbridgense* shown him from Tunbridge by Plukenet and Doody: cf. *Fasc.* p. 1; *F.C.* pp. 190, 195.

5 Referred to by Ray in letter to Lhwyd, 2 Aug. 1689: *F.C.* pp. 191–2.

6 *Fasc.* p. 4.

7 This must have been a much rarer plant in Ray's time. He had reported all the other Poppies in the *Catalogus Cantab.*

8 *Fasc.* p. 15.   9 *Fasc.* p. 3.

10 He was nephew of Roger Dodsworth the antiquary: cf. Venn, *Alumni Cantab.* I, p. 53.

11 In the North Riding, midway between Ripon and Helmsley.

was working at the history of Ferns[1] and discovered *Lysimachia thyrsiflora* in the East Riding; and James Sutherland, the first Professor of Botany at Edinburgh and author in 1683 of the Catalogue of plants in the Physic Garden. Even Sloane is mentioned—in connection with *Salicornia perennis*,[2] which he found near King's Ferry (now Queen's Bridge) in the Isle of Sheppey.[3] If his own field-work was restricted, he had now a goodly band of helpers and correspondents.

One of these, Edward Lhwyd, he hardly yet knew; for the first of the letters that passed between them is dated 21 June 1689—Plot having suggested their correspondence on 10 June.[4] Lhwyd had gone up in 1682 to Jesus College, Oxford, but had not taken a degree, and became underkeeper at the Ashmolean in 1684. He left a Catalogue of North-Welsh plants for the guidance of visitors and herbalists at Llanberis in 1688 when he spent the summer at Snowdon;[5] and this was found by Robinson and passed on to Ray.[6] He was already known to Lister and keenly interested in natural history; and the early correspondence between him and Ray dealt mainly with plants, though later fossils and the archaeology of Wales became his chief interest. He succeeded Dr Plot as keeper of the Ashmolean Museum[7] in 1691 and apart from long journeys not only throughout Wales, but to Ireland, Scotland, Cornwall and Brittany, remained there until his death in 1709.

His letters, now collected and edited by the late Dr R. W. T. Gunther,[8] reveal him as an ardent, alert and indefatigable scholar, quick-tempered but generous, imaginative yet anxious to check his speculations by accurate knowledge. His work suffered from his lack of concentration, botany giving way to geology, and this to archaeology, and this to the Celtic language; from his credulity, as in the matter of the stack fires[9] and still more in his final theory of the origin of fossils; and from his poverty, though here his energy and ingenuity made up for the defect. But his enthusiasm vitalised every subject that he touched; and in geology and philology he deserves the honour due to a pioneer.

1 Cf. Ray's letter to Robinson May 1685: *F.C.* p. 147.

2 This is discussed by Ray in a letter to Sloane of 24 Aug. 1686: he agrees that it is a new species.

3 So Sloane, *Corr.* p. 186.

4 Gunther, *Life of Lhwyd*, pp. 87–8.   5 Gunther, l.c. p. 199.

6 Cf. Lhwyd to Lister: *F.C.* p. 189. Ray quotes it as sent to him by Robinson, *Syn. Brit.* p. 150, apropos of *Saxifraga oppositifolia*.

7 This was in Wren's building, now the Old Ashmolean, opened in 1683 by Prince James, who visited Oxford and saw some experiments in the laboratory. Plot was the first keeper, but was transferred to the laboratory in 1690.

8 It is singularly appropriate that the last gift of Dr Gunther to the studies which he had made his own should be this volume on his greatest predecessor at the Old Ashmolean.

9 Cf. letter to Lister (Gunther, *Life of Lhwyd*, pp. 218–20) and to Ray (*Corr.* pp. 281–2), and Ray's replies (*F.C.* pp. 243, 245): above pp. 69, 70.

His regard for Ray stands out from the beginning. 'I am lately entered in correspondence with Mr Ray who leads the solitary life he always affected at a place called Black Notley in Essex. I judge he is a man of the most agreeable temper imaginable.'[1] So he wrote on 12 November 1689 to his kinsman, the Rev. John Lloyd; and in May 1695 he repeated the same sentiments: 'Mr Ray is doubtless the best acquainted with Natural History of any now living....You may return me Mr Ray's letter when you have read it because I post up all his letters.'[2] If he did not quite repay Ray's manifest affection for him,[3] this may well be because of the difference of age and the fact that they never met.

But the *Fasciculus* was only a makeshift. In its Preface Ray had said: 'Next spring if God give me strength and health I have decided to publish a *Synopsis stirpium Britannicarum* with brief and characteristic notes not only of the genera but of all particular species: it will not exceed the size of a Catalogue but will give the careful reader, even without pictures, a clear and unmistakable knowledge of each.' In fact this was already in the press by the spring of 1689 but Mott, the printer, made difficulties,[4] so that the publication was transferred to Samuel Smith, the bookseller at the Princes Arms in St Paul's Churchyard.[5] In the following March[6] Ray was desperately ill with pneumonia, the illness of which Benjamin Allen, the young doctor who had lately come to Braintree, has left a record;[7] and Jane the baby was suffering with fits: this delayed his completion of the work.

The correspondence with Lhwyd which began in the summer of 1689 throws light on the work of preparation. Lhwyd, who was awaiting his appointment to the Ashmolean, spent much of it in Wales collecting plants and sending them to Ray. He was then as always enthusiastic, and packets of dried specimens were constantly arriving at Black Notley: but he was not an experienced botanist and some of the entries added on his authority to the *Synopsis* are mountain varieties or (in the case of ferns) immature examples. On 26 September,[8] thanking him for samples of plants forwarded by Mott the printer, Ray says that the *Fasciculus* contains all the new species of the *Synopsis* excepting Lhwyd's recent contribution, one or two from Plot's *Natural History of Staffordshire* and one from Lawson. He goes on to say that he has already been told about Sherard and would be glad to make his acquaintance or at least to correspond with him. On 3 November he thanked Lhywd for 'the draught you sent of the Subularia lacustris which is a plant to me altogether new and unknown' and added: 'I shall take care to get it engraven for the ornament and advantage of my

1 Gunther, l.c. p. 97. 2 Gunther, l.c. pp. 270–1.
3 Cf. his letter of July 1692: *F.C.* p. 230. 4 *F.C.* pp. 292–3.
5 *F.C.* p. 204. 6 Cf. *F.C.* p. 205 and Gunther, *Life of Lhwyd*, p. 99.
7 Cf. Miller Christy in *Essex Naturalist*, vol. XVII. 8 *F.C.* p. 198.

book'.[1] This he did, and the plate duly appeared; the plant, which grew under water in 'the little lake Phynonvrech near the summit of Snowdon , is listed at the end of the last or 'anomalous and aquatic' section:[2] it had no flowers or seeds, and by its picture is obviously *Isoetes lacustris* rather than *Subularia aquatica*.[3] He was at this time expecting lists and information from Bobart, Doody, Plukenet and Sherard;[4] and as the catalogue was being printed before their arrival grew anxious about their inclusion. In the spring his illness made it impossible for him to supervise the printing, and he was very anxious lest 'by reason of the multitude of loose papers' confusion might arise. The book, when it appeared, was 'without any errata, not because it is void of faults, but because I could by no importunity obtain the sheets of the printer, that so I might draw them out.' There are in fact more misprints than is usual in Ray's works, but no serious errors of order or printing. Some of the additions, especially Lhwyd's, are put in evidently at the last moment and without arrangement. But in the main the book has not suffered as much as the circumstances of its composition would justify.

On 22 January 1690 the Royal Society had given its imprimatur, Sir John Hoskyns, as Vice-President, being in the chair; and the book was printed, first copies appearing in May.[5] It was dedicated to Willughby's younger son Thomas,[6] who had succeeded to the baronetcy conferred on his brother when Francis died in his twentieth year in September 1688. There had been trouble between him and his step-father over the estates, but this had been settled in his favour, and Thomas was now in possession. He had gone up to St Catharine's Hall in 1683 as a fellow-commoner, had moved to Jesus College in 1685, but had not taken a degree.[7] Ray's devotion to his friend's memory and his affection for the child of whom he had had brief charge are revealed in the letter appended to the dedication. There is indeed here and in the Preface a note of personal feeling hardly to be found in any other of his published works. The intense reserve of the past is broken: his life's work has been accomplished: his anxieties for his family and for the country are relieved: he can now give expression, restrained indeed but evident, to his thankfulness and his hopes. The pages in which he does so are worth study.

1 *F.C.* pp. 201–2.   2 *S.B.* p. 210.
3 Both are found in the Welsh mountain-lakes, but this is certainly the Quillwort, *I. lacustris* or *I. echinospora*.
4 *F.C.* p. 205.
5 Cf. letter of 7 May to Lhwyd, *F.C.* p. 207, and to Robinson, *F.C.* p. 293. It was on sale first on 16 May, as is stated by Sherard to Richardson (Nichols, *Illustrations of Literature*, I, p. 340).
6 It is of interest to note that a fern, apparently a stunted *D. felix-mas*, is described as 'found on Lichfield Minster by Sir Thomas Willughby': *S.B.* II, p. 48.
7 Venn, *Alumni Cantab.* IV, p. 423.

He begins with a warm tribute to Ralph Johnson, the Vicar of Brignall, who had died in 1689;[1] he had been a trusted friend since Cambridge days, and a regular correspondent, contributing to the histories of birds and fishes as well as of plants; and to him Ray states that he owed the first suggestion, many years ago, that the *Catalogue* should be arranged in the order not of the alphabet but of nature. 'I knew that this was good advice but at the time dared not take it: I was not ready for such an attempt: it demanded a close study of the systems of others, and a thorough survey of all species of plants, their relationships and differences: and this was a work of years.' He goes on to tell of Wilkins, of the hasty and artificial, indeed almost extempore, tables drawn up for him; of his own *Method* published five years ago and embodied in the *History of Plants*; and of his conviction that the *Catalogue*, now that its second edition was sold out, could be remodelled according to Johnson's proposal. The present volume is in fact a succinct History of British Plants. It includes two hundred more species than the *Catalogue* and a third century in the appendix.

The work owes these additions and its completeness to the friends whose names appear in it, and especially Jacob Bobart, curator of the Medical Garden at Oxford; Samuel Dale; Matthew Dodsworth; Samuel Doody; Thomas Lawson; James Newton; Edward Lloyd (Ray spelled his name thus until he expostulated in November 1692[2]), 'whose friendship I value very highly although we have not yet met'; James Petiver; 'Robert Plot L.L.D. whose Histories of Oxford and Staffordshire have given me not a little and to whom I owe the friendship of Mr Lloyd';[3] Leonard Plukenet; Hans Sloane; William Sherard, fellow of St John's College, Oxford; and above all Tancred Robinson, who has read the manuscript, corrected mistakes and omissions, and done all in his power to improve it: 'with such a midwife my baby after a rather difficult birth has struggled out into the light'. All these names are familiar in Ray's writings, and in the botany of the time: but Plot and Sherard appear for the first time.[4]

Plot,[5] who had graduated from Magdalen Hall in 1661 and published his *History of Oxfordshire* in 1677, became Secretary to the Royal Society

1 Thoresby, *Diary*, I, p. 279, who speaks of him as 'a person of the greatest curiosity in botany, ornithology, antiquities etc.', planned to call on him in 1694—evidently not knowing of his death.

2 *F.C.* p. 232.

3 This is borne out by Plot's letter, Gunther, *Life of Lhwyd*, pp. 87–8.

4 The first reference to Sherard by Ray is in a letter to Lhwyd of 26 Sept. 1689, *F.C.* p. 198, from which it appears that Ray had heard of his botanical knowledge but wished Lhwyd to approach him for help.

5 A considerable number of his letters and some account of his life and writings are in Gunther, *Early Science in Oxford*, vol. XII.

along with Aston in 1682, keeper of the Ashmolean Museum on its foundation in 1683 and Professor of Chemistry in Oxford in the same year, owing his promotion to his patron Henry Howard, Duke of Norfolk, and to the visit of Prince James to the University. He was a man versatile but credulous,[1] witty and sociable, but a time-server and according to Lhwyd, his underkeeper at the Ashmolean, with 'as bad morals as ever disgraced a Master of Arts'.[2] Ray apparently never met him, but had formed an unflattering opinion of his character and capacity.[3] He had already criticised his geological work[4] and been sceptical about his descriptions of fishes:[5] and in botany, though he refers respectfully to his records, he does not accept them without question.

Sherard was a man of thirty who had gained his fellowship in 1683, and then been allowed to travel. He had studied with Tournefort at Paris and with Paul Hermann, author of the *Hortus Lugduno-Batavius*, at Leyden, had visited Geneva, Rome and Naples, and botanised, presumably on his return in 1689, in Jersey. A list of his discoveries there reached Ray after the *Synopsis* had gone to the press, and is included in the Appendix, pp. 238–9. It is full of interest as containing the first British records of *Helianthemum guttatum*, 'on the west side of the island near Grosnez Castle', *Echium plantagineum*, 'in the sandy grounds near St Hilary plentifully', *Scrophularia scorodonia*, 'by the rivulet betwixt the Port and St Hilary', and *Gnaphalium luteo-album*, 'on the walls and dry banks very common', although considering the number of plants peculiar to the Channel Islands it might easily have been more impressive.

The lists for which he had waited and which are inserted with it in the Appendix are not in fact very valuable. Plukenet's, although covering three full pages, contains nothing that is clearly new and specifically distinct. Doody's has a number of mosses and fungi, but the descriptions are too brief to make identification certain. Bobart reports *Hutchinsia petraea*,[6] 'brought by Richard Kayse from St Vincent's Rock near Goram's Chair in the parish of Henbury, three miles from Bristol', and *Centaurea solstitialis*,[7] 'not far from Cirencester'; but the rest consists of varieties,

1 Cf. the full prospectus of his proposed Journey through England and Wales (Gunther, l.c. pp. 335–45) which includes everything from spas to banshees and from ornithology to 'thaumaturgicks'.
2 Cf. *D.N.B.* XLV, p. 425, and Gunther, *Life of Lhwyd*, p. 131; Gunther, l.c. p. 64, denies the justice of this criticism.
3 Letter of Jan. 1692 to Lhwyd, *F.C.* pp. 224–5: 'I fear also he may be too much influenced by worldly advantages of honour and profit, for I have been told that he was inclinable to change his religion in the reign of the late King James.'
4 Cf. *Corr.* p. 154 and below, pp. 427–8.
5 *F.C.* pp. 142–3: of the Finscale of Dr Plot 'I suspect the Dr was therein mistaken'.
6 *S.B.* p. 236: the finder's name is spelled Kayle in *S.B.* II, p. 174.
7 *S.B.* p. 237.

some of them of Corn and Barley. Sloane's contribution is a list of plants growing in Jamaica and also in Britain.[1]

Having made acknowledgments, he continues:

I would close this Preface on a note of thankfulness. I am full of gratitude to God that it was His will for me to be born in this last age when the empty sophistry that usurped the title of philosophy and within my memory dominated the schools has fallen into contempt, and in its place has arisen a philosophy solidly built upon a foundation of experiment: against it elderly professors protest and struggle in vain; they are men who when fruit has been discovered prefer to live on acorns for fear they should be constrained 'to lose in age the lessons of their youth' and acknowledge that they have wasted their lives following the shadow of philosophy and embracing a wraithe instead of the Queen of Heaven. It is an age of noble discovery, the weight and elasticity of air, the telescope and microscope, the ceaseless circulation of the blood through veins and arteries, the lacteal glands and the bile duct, the structure of the organs of generation, and of many others—too many to mention: the secrets of Nature have been unsealed and explored: a new Physiology has been introduced. It is an age of daily progress in all the sciences, especially in the history of plants: to this end have been devoted the energies not only of private individuals but of princes and magnates, eager to find new flowers for their gardens and pleasances and to send plant-hunters to the further Indies: they have travelled over mountains and valley, forest and plains, exploring every corner of the earth, and bringing to light and to our view all that is hidden. Surely we may look for something splendid from those who have such abundant opportunities for the procuring, cultivating and depicting of plants, when we remember how often lack of material neutralises industry and frustrates effort.

There are those who condemn the study of Experimental Philosophy as a mere inquisitiveness and denounce the passion for knowledge as a pursuit unpleasing to God, and so quench the zeal of the philosopher. As if Almighty God were jealous of the knowledge of men. As if when He first formed us He did not clearly perceive how far the light of human intelligence could penetrate, or were it to His glory to do so, could not have confined it within narrower limits. As if He were unwilling that man should employ the intelligence which He had bestowed on the objects of which He had made it capable and which He had provided for its investigation. Those who scorn and decry knowledge should remember that it is knowledge that makes us men, superior to the animals and lower than the angels, that makes us capable of virtue and of happiness such as animals and the irrational cannot attain. Nevertheless I agree that those who pervert knowledge to evil arts, or allow themselves to be deluded by the instigation of demons into peering into the secrets of the future, ought to be constrained and condemned with severity.[2]

He goes on to add:

Nothing is universally blessed. Literary excellence, the study of languages and antiquities, Criticism, Grammar, carried by our immediate predecessors to a

1 S.B. pp. 250–2. Several of them are garden weeds: most of the rest mosses, fungi or aquatics whose identification is probably mistaken.

2 This is almost the only allusion in all Ray's writings to the subject of witchcraft, which absorbed the attention of so many of his contemporaries.

midday zenith, seem in our day to have passed their peak and to be sinking into a decline. Rarely indeed appears among us a Casaubon, a Grotius, a Salmasius, a Scaliger, a Selden, a Bochardt, a Vossius—those lights and ornaments of former days. Apparently the learned of to-day are concerned to study things, but care less about words.

It is a complaint that he had made before, but never so clearly. Latin was falling into disuse; and his younger scientific colleagues were too often guilty of solecisms in their learned writings. But in fact there was no lack of literary or linguistic attainment, as the records of archaeological and Anglo-Saxon scholarship prove.[1]

There follows almost the only passage in which he speaks his mind about contemporary events.

Above all I thank God that He allowed me to live long enough to see this dear land endowed by divine favour with princes such as in the recent stormy times I longed for but scarcely dared to expect, princes chaste and religious and distinguished in every form of virtue. Under their quiet rule, if only God grant us peace, we can rely upon prosperity and a real age of gold. If their subjects are ready to fashion their characters after their example, I see no reason why our people should not attain to perfect happiness. Superstition has been overthrown. Pure and reformed religion is honoured. Its profession and practice are not only freely permitted, but by our Sovereigns enjoined in word and encouraged by example. The yoke of slavery which our necks have never learnt to endure was beginning to oppress us; it has been broken: our heritage of freedom has been restored and secured: each order has received its proper privileges and immunities. The unbridled licentiousness of a wanton soldiery that insulted free-born citizens in their own houses with impunity and filled the land with violence, murder and outrage has been repressed: the guardianship of his estate has been made secure and unafraid for us all. The majesty of the country's laws, the very foundation of the realm, is inviolate. Philosophy and all sound learning, now that the favour of princes smiles upon the efforts and stimulates the industry of scholars, show promise of wonderful advances.

To secure the perpetuity of these benefits our first prayer should be that the august Sovereigns whose virtue has won them and whose counsels and foresight have preserved them for us may be as immortal as they are essential. That is indeed a condition not allowed to mankind: but valour begets valour, and we would humbly pray to God that when our Sovereigns full of years and glory, ripe for heaven, have passed to the abode of the blessed, they may be succeeded from the same exalted house by a continuing line of monarchs, filled with the same spirit, equal in all but years to their forebears, under whose happy rule we may enjoy all the blessings of peace and moral excellence.

It is easier to express such sentiments appropriately in Latin than in a translation; and from many of his contemporaries such an outburst would be nothing but conventionality and a bid for royal favour. Ray had never asked or accepted preferment. Charles Stuart's perjury had driven him

1 Cf. e.g. Douglas, *English Scholars*, for a recent account of Ray's contemporaries in this field.

from his fellowship, his lecture-room, his pupils, to a life of dependence and poverty. He had seen Josiah Child a millionaire, and Carbery the President of the Royal Society. Monmouth's rebellion and the Bloody Assize were fresh in his memory as symbols of the profligacy of the Court and the prostitution of justice. If the savageries of Claverhouse in Scotland or of Jeffreys in Somerset were remote, the hanging of Henry Cornish, the burning of Elizabeth Gaunt, the imprisonment of Richard Baxter were in London. He had lived so busy and secluded a life, deliberately cut off from the world of affairs, that we are apt to forget the turmoil of the times and its effect upon the naturalist of Dewlands and his young family. That he shared the disgust and anxiety of all decent Englishmen, that he had his own special reasons for indignation, hardly appears elsewhere in his letters or books. Now when the cloud had lifted, he could at least express his relief; and take up again those interests in education and religion which had been laid aside for thirty years. Since 1660 he had confined himself to the world of nature: now he was free to return to his profession. The *Wisdom of God* and the *Three Discourses* show that he had not forgotten the underlying motive and deeper significance of his work. Had he given to theology and philosophy the thirty years that he gave to botany and zoology the world would have lost perhaps as much as it would have gained. For himself he makes no complaint: for his country he cannot but rejoice that an evil thing has been overthrown. It is at least of interest to the historian to know how one of the best minds of the age regarded the Stuart régime and its ending.

The book, though it contains a large number of cryptogams not previously listed and some additional flowering plants from his friends, is naturally rather a rearrangement of the *Catalogue* and condensation of the *History* than a record of new discoveries. The growth of interest in botany, especially at Oxford, where Richard Dyer of Oriel[1] had like Sherard been elected to a fellowship and given leave to study abroad; John Wynne of Jesus was 'a diligent tutor and a good naturalist';[2] John Pointer of Merton was beginning his lectures and collections,[3] no doubt accounts for its success. In the summer of 1694 Smith the publisher called for a second edition,[4] and Ray set himself to collect further lists from his correspondents. He had himself done much work in preparation for it. Almost before the first edition was published he had written to Lhwyd: 'My Synopsis meth. Stirpium I am revising and enlarging with the addition of many species.

1 Cf. *Journal of Botany*, LXIV, pp. 43–4, and Gunther, *Life of Lhwyd*, pp. 197, etc. 'That excellent botanist', as Hearne calls him (III, p. 239). He had planted eight 'cedars of Virginia' in the College grove in 1685 (*Oriel Coll. Register*, pp. 117–18).

2 Cf. Gunther, *Life of Lhwyd*, p. 290: he was Fellow 1685 and Principal 1712–20.

3 These and the advertisement of his lectures are preserved in the Old Ashmolean.

4 *F.C.* p. 247.

I have this spring been diligent in the inquisition of Mosses whereof I have observed and distinguished above forty new sorts, all as truly and certainly distinguishable by characteristic notes as any other greater plants'.[1] He had also received from Dale a specimen of *Zostera marina* in seed, which he had carefully reported to Lhwyd and incorporated with a full description in the new edition.[2] But the pressure of other work, his new enthusiasm for insects, and the state of his health, which cut him off from any travel beyond Braintree and the immediate neighbourhood, prevented him from activity in the field; and the actual novelties in the second edition are mainly in its earlier portions. About these and especially the ferns he took great care to satisfy himself: for Lhwyd and others were pressing their discoveries upon him, and he had difficulty in maintaining his refusal to multiply species.[3]

He was now able to give their proper positions to the plants in the various lists in the Appendix, except those of Sloane from Jamaica; and made one other change in arrangement. In the first edition he had placed notes of the medicinal uses in the index: these he now put back into the text, following the description and localities: it is not clear that this is an improvement, but in any case the notes are much shorter than in the *English Catalogue*. The second edition was printed on a larger page, and this gives room for the letter of Rivinus and his reply to it at the end of the volume.

The Dedication and Preface of the first edition are reprinted, but after the list of authorities a second and very interesting Preface is added. In this Ray begins by declaring that he is now too old and ill to be able to look forward to any further editions of the work and therefore has done his best to make this as perfect and faultless as he can. Yet the number of plants is so great that in spite of the intensive researches of British botanists new species are constantly being discovered. Indeed, a new field of study has lately been opened up, the Mosses, Fungi, and Seaweeds: 'Our forebears and even ourselves once ignored these tiny plants and were content with three or four species of Capillary Mosses and a similar number of Terrestrial, whereas recent observation has disclosed and described at least thirty different species of the former and more than forty of the latter.' These are now included in the *Synopsis*.

I was inclined to think that I had played my part long enough and could rest content to classify the results obtained by others: but their industry stimulated

1 *F.C.* p. 220. Lhwyd passed on this information to Richardson (cf. Gunther, *Life of Lhwyd*, p. 151) and to Lister (l.c. p. 149).
2 *F.C.* p. 220 and *S.B.* II, p. 7. The type of seed-vessel was then entirely strange. Moyle and Stevens discovered it in seed in Cornwall shortly afterwards: cf. *S.B.* l.c.
3 Cf. his letter to Lhwyd of Aug. 1694 (*F.C.* pp. 250–1) discussing in detail the distinction between *Aspidium angulare* and *A. aculeatum*.

me. I was ashamed to think that my Catalogue should contain so much that they had seen and should lack what I might have seen myself. So I shook off my sloth, girded myself for the search, and urged my friends to collaborate with me. As a result this edition contains about a hundred fresh species in this family.[1]

There is a vast amount still to be done: 'Travelling over the ocean of plant-life none of us dare cry I see land ahead.' If plants were done, insects remain, and their number I should put at twice as many as the plants; and we know nothing of their origin and way of life, their habits, age or qualities. 'After five years continuous work within the narrow limits of a mile from my home I have not yet seen a tenth part of them all and yet I have described some three hundred in one family alone.'[2] One can guess how vast a number there must be in the whole country and in the world.

He goes on with a fine passage repeating his thanks to God for the gift of life in such an age of discovery and scientific enthusiasm, and praying that war and barbarian invasions may not again overwhelm Europe and that if peace is established there may not be a time of stagnation.

If this be granted then, to quote the great Boyle, I predict that our descendants will reach such heights in the sciences that our proudest discoveries will seem slight, obvious, almost worthless. They will be tempted to pity our ignorance and to wonder that truths easy and manifest were for so long hidden and were so esteemed by us, unless they are generous enough to remember that we broke the ice for them, and smoothed the first approach to the heights.

A passage follows which discusses a wholly new department of botany, links up with subjects that had been much in his mind, and reveals the dilemma in which he and his contemporaries found themselves. It deserves translation in full.

Before I bring this Preface to an end it will not perhaps be irrelevant to warn the reader that a good part of the plants which in former times existed in this island are missing in this *Synopsis*—that is if the opinion of a recent writer[3] to which I am not myself opposed is correct. He maintains that the elegant leaf-like appearances (*foliorum schematismi*) which we find so wonderfully impressed or engraved upon schist or other substances owe their origin to the very plants whose leaves they represent. This opinion I judge to be plainly preferable to that which I have studied in a work lately published[4]—namely that these imitations of plants are delineated by the hand or pen of sportive nature to display her

1 Second preface, p. 3.

2 He had been collecting and breeding moths and butterflies since 1690: see below, pp. 392–6.

3 He refers not to Lhwyd, who had sent him specimens in 1691 (cf. *F.C.* p. 213) and more in 1695 (cf. Gunther, *Life of Lhwyd*, p. 283) but to Woodward, who had declared that 'the substantial plants themselves left their impressions' (cf. Ray to Lhwyd in Oct. 1695: *F.C.* p. 259) and had published his *Essay towards a Natural History of the Earth* in the same year.

4 Plot's *History of Oxfordshire*, whose arguments against the organic origin of 'formed stones' Ray had criticised in a long letter to Robinson: *Corr.* pp. 151–5.

graphic art and to furnish a spectacle for mankind. This is a conclusion that seems inconsistent with the majesty of nature and with the foresight manifested in all her works. Yet there are certain considerations that compel me to hesitate to give my full and confident approval: and these I will briefly state.

There follow four objections, very typical of the perplexity in which honest men living between the old world of tradition and the new world of experiment found themselves.

First, it would seem to follow that antediluvian plants now lost were hardly less numerous than those that still survive; for it seems that in the class of Ferns the number of imprints of unknown species in this island or in Europe is greater than that of the species living and discovered by botanists. This involves belief in the destruction of species, that is in the mutilation of the universe, a conclusion which neither theologians nor the wiser kind of philosophers admit; for they maintain that God watches over the perfection of His creation and preserves each species from destruction as He did at the Deluge through Noah and the Ark. If we admit the loss of any species then we must accept the absurd conclusion that the earth might become wholly empty of inhabitants. Moreover, so far as we know no species described by ancient writers has ceased to exist.

Secondly, why are not other parts of plants besides the leaves preserved? It is incorrect to argue that the Deluge took place in May when only the leaves had unfolded: it began, according to the experts in exegesis, in November.[1]

Thirdly, how if the imprints are made by real leaves are they so flat and free from folds? Leaves, especially when young, can hardly be smoothed out by the skilled botanist: at the Deluge they were rolled up haphazard in the waters.

Fourthly, how were the leaves torn off from their roots in the universal dissolution that our author imagines? The pedicels and lowest fibres are too strong to be thus broken. Is it not more likely that they would sink sticking to their roots? Yet roots so far as I can hear do not appear.

Thus Ray reveals the conflict that was inevitably aroused. The Deluge had become the accepted explanation of the presence of fossils by those who believed them to be genuine organic remains. Ray had stood uncompromisingly for belief in their organic origin: they had once been living plants, molluscs, fishes. But he saw the difficulties that explanation in terms of the Deluge involved; and as we shall see elsewhere gives hints that he is dissatisfied with it. At that time he could hardly be expected to do more.[2]

1 Ray gives a long discussion of this point in *F.C.* p. 258: Nicolson, *Letters*, I, p. 85, says that most commentators place it in May and that Ray is mistaken.
2 He had discussed the whole matter very thoroughly with Lhwyd: cf. Ray to Lhwyd, 8 April 1695 (*F.C.* pp. 258–9, epitomised Gunther, *Life of Lhwyd*, pp. 171–2); Lhwyd to Ray, 28 Aug. (*Life*, pp. 281–2) and 12 Sept. (*Life*, pp. 283–4); Ray to Lhwyd, 8 Oct. (*F.C.* pp. 259–62).

He concludes with an alphabetical list of those who have helped him, Bobart, Dale, Doody, Lhwyd, Moyle, Petiver, Sherard, Sloane, Vernon and 'Amicorum Alpha' Robinson; and among them two new names. 'Walter Moyle, Esquire, a young man, scholarly, able, generous', appears in the body of the book[1] as of Cornwall and exploring that county in 1694 with Mr Stevens, 'a learned clergyman and skilful in Botanicks'. Moyle,[2] who was a member of the famous Cornish family of that name, was born in 1672 and went up to Exeter College, Oxford, in 1689. His versatility and brilliance gave him almost at once a footing in the literary world. He took no degree, came to London and was in Parliament from 1695 till 1698, but then retired and devoted himself to history and to the ornithology and botany[3] of Cornwall. Some time before January 1696 he visited Ray at Black Notley with Robinson and promised to send an account of his Cornish botanising.[4] His works, including letters to Robinson,[5] Sherard[6] and others, were published in 1721 after his death by Thomas Sergeant. Lewis Stevens,[7] who had taken his B.A. degree at Gonville and Caius College in 1675–6, had then transferred to Oxford and been ordained at Exeter in 1677, was Vicar of Treneglos and Warbstow from 1678 to 1685 and of Menheniot from 1685 till his death in 1724. His friendship with Sherard[8] at Oxford had perhaps given him his interest in plants. He was an expert on Sea-weeds[9] and devoted to the botany of Cornwall and left a Catalogue of Cornish plants in manuscript. William Vernon, who had gone up to Peterhouse in 1685 and gained a fellowship there in 1692, was already an enthusiastic botanist, and took also much interest in Lepidoptera. He had stayed with Ray at Black Notley,[10] in the autumn apparently of 1694 or 1695. In 1697 he was given leave of absence, as Sherard had been from Oxford, for the purpose of travel abroad, and went first with the German, David Krieg, to Maryland. He is responsible for a number of records, especially of mosses, and apparently had the help of one of his colleagues, 'Mr Davies of Peterhouse'.[11] Sherard, after his return from the Continent and Jersey, had gone to Ireland and been living with Sir Arthur Rawdon, the great gardener, at Moyra in Co.

1 *S.B.* II, pp. 4, 5.
2 For fuller details cf. *D.N.B.* or the 'Life' prefixed to the second edition of his *Works*.
3 He wrote to Sherard in 1719: 'I was once setting up for a botanist, but my ill state of health has obliged me to abandon that study' (*Works*, I, p. 412).
4 *F.C.* p. 262.
5 Containing a description of the Stormy Petrel, a bird then undescribed (*Works*, I, pp. 398–401).
6 Three letters from Sherard with Moyle's replies.
7 Cf. *Alumni Cantab.* IV, p. 156 and J. Venn, *Gonville and Caius College*, I, p. 445.
8 Cf. Moyle, *Works*, I, p. 411.
9 He is mentioned as the discoverer of a new species in *H.P.* I, p. 75.
10 *S.B.* II, p. 30.   11 *S.B.* II, p. 32: probably Richard, scholar 1693, fellow 1699.

Down.[1] He sent records of plants from the Mourne mountains[2] and of the true *Subularia aquatica* from Lough Neagh.[3] In 1694 he reported to Ray an interesting record of an Orchid, evidently *Cephalanthera longifolia*, which he had discovered nor far from Moyra 'on a rotten bog by a lough-side near the Dairy-house in Crevetenau Ballina-hinch',[4] that is, near Derry House on Derry Lough in the township of Creevy Tenant near the market town of Ballynahinch.[5] He also records a fungus as found 'in June on rotten oaks in Kilwarlin near Hilsborough'.[6] Petiver's notes and a long list of cryptogams and a few flowering plants from Doody, who does not seem ever to have found writing easy, must have arrived too late for incorporation. They follow Sloane's Jamaican plants in the Appendix. Many of them are notes criticising or amplifying the records in the *Synopsis*; to these Ray usually adds his comments.

This edition of the *Synopsis* is, as Ray anticipated, his last publication on the Flora of Britain, and as such deserves careful study by those who would estimate his quality as a botanist and his contribution to the department of science which he had specially made his own. To compare it with the *Cambridge Catalogue* is to realise the enormous development that he had achieved. He had seen in their living state most of the plants that he records. He had grown them whenever possible from seed and striven to base specific characters upon observation of their whole life-cycle. He had encouraged a band of helpers to explore the whole country and to send to him specimens and full particulars. As a result he had fulfilled his original intention and created a reliable and surprisingly complete *Phytologia Britannica*. Smith's praise of the work is testimony to its influence. It gave to any student of the country's plants a working guide to identification, locality and habit. When he began his researches there was no agreement as to names, specific characters or classification: large numbers of plants were unknown: many recorded species were non-existent or hopelessly confused: descriptions were defective and misleading: the beginner had to overcome almost insuperable obstacles. In the *Synopsis*, apart from the Linnean nomenclature, we have a modern hand-book. The survey of species is remarkably accurate, at least in the flowering plants. The country has been adequately explored. The names and brief descriptions make identification easy. The classification, if not scientifically perfect, follows a natural sequence and is as easy to use as the

1 Sloane, in the Preface to his *Voyage to Jamaica*, describes how Rawdon on seeing his Jamaican plants sent James Harlow to bring back a cargo for his gardens at Moyra.
2 *S.B.* II, p. 51.   3 *S.B.* II, p. 281.
4 *S.B.* II, p. 242: the discovery seems to have been actually made in June 1691: cf. letter of Sherard to Richardson in Nichols, *Illustrations of Literature*, I, p. 341.
5 I owe this interpretation to Dr R. Lloyd Praeger, who tells me that the Orchid has been gone from there since at least 1870.
6 *S.B.* II, p. 19.

modern scheme. British botany has been given a secure and intelligible foundation.

Where the *Synopsis* is in error, including species which are in fact either varieties or aberrant forms or duplicates, the mistake is almost always due to the failure of his correspondents to exercise a rigid criticism: they are eager to find novelties, and exaggerate points of difference; they mistake abnormalities of growth or flower for specific distinctions; they fail to allow for stunted or luxuriant development due to situation and soil; they have not Ray's experience or memory or caution or flair for essentials. Ray constantly admitted the difficulty of judging plants from dried specimens and lamented his inability to see them growing in the wild or even in gardens. Many of the observations reported to him he refused to accept as novel; many others he inserted with a note that they were probably varieties of previous species: but when enthusiastic contributors whose merits he was temperamentally inclined to overestimate sent him their plants and records it was difficult for him to reject them as ruthlessly as he had rejected the multiplied entries of Gerard or Parkinson. Sherard was an excellent and accurate scientist and Doody seems to have been generally reliable. But Lhwyd, who was a geologist rather than a botanist, Plukenet, who was jealous and difficult, and whose superficiality he learnt to recognise,[1] Plot, who was credulous and not really expert, the younger men whom he had inspired and encouraged, these he took at too high a valuation; and the book suffered in consequence. But when allowance is made for the addition of a small proportion of superfluous species, the list certainly gained from the mass of material supplied by his friends.

Nor was Ray's own contribution limited to the mosses. Black Notley is not a very rich neighbourhood, and there is evidence of only one expedition outside the area of a short walk. At some time between the appearance of the second edition of the *Catalogue* and the first publication of the *Synopsis* he went with Newton to Tollesbury by way of Braxted; for he reports *Trigonella purpurascens*—'Mr Newton in our company found it on sandy banks by the sea-side'[2]—at the former and *Cynoglossum montanum*, 'plentifully by the waysides',[3] at the latter. Otherwise Braintree was the limit of his travel. But there are several records of plants that he discovered himself. Of the 'Hieracium castorei odore' which he had recorded under a different name in the *Cambridge Catalogue*, and which Gibson identifies as *Crepis foetida*, 'I found it last summer (1690) plentifully in a field near my house, called Stanfield, which had lain a while since it was plowed';[4] and in the next year he has a note of two other species,

---

1 Thus he reported *Scilla autumnalis* as growing at Blackheath, Kent (*S.B.* II, p. 232).
2 *S.B.* II, p. 196.   3 *S.B.* II, p. 119.
4 *S.B.* II, p. 72: cf. Gibson, *Flora of Essex*, p. 193, and for these localities note at end of chapter.

one apparently *Arnoseris*. So too of Herb Paris (*P. quadrifolia*), 'here at Notley in Lampit Grove belonging to the Hall';[1] of *Herminium monorchis*, 'also in the parish where I live on the greens of a field belonging to the Hall, called Wairefield';[2] of *Allium oleraceum*, 'in a cornfield belonging to the Hall called Westfield adjoining to Leez Lane';[3] of Black Currants (*Ribes nigrum*), 'here in our neighbourhood by the Hoppet-bridge near Braintree';[4] and of *Mentha rotundifolia*, 'in a moist place of a little meadow adjoining Faulkburn Hall, the seat of my Honoured friend Edward Bullock Esquire'.[5]

To illustrate the quality of the list it will be well to take the Alsine group[6] for comparison with the entries under that name in the *Cambridge Catalogue*. Their identification is now made easy by accurate division and by the addition of short notes giving the characteristics of the more difficult species. We give the first name and the locality in the *Synopsis*[7] and the name now applied to the species, taking this in most cases from Bentham and Hooker.

Anthyllis maritima lentifolia. Sandy shores common. *Arenaria peploides*.
Alsines species anomalae, flore tetrapetalo.
Alsine tetrapetalos Caryophylloides. Sandy waste places in spring. *Moenchia erecta*.
Saxifraga graminea pusilla. Boggy and waste places and garden walks. *Sagina procumbens*.
Anglica alsinifolia annua. Garden walks at Balliol College and fields. *Sagina ciliaris* or *apetala*.
S. gram. pus. foliis crassioribus. Near Tynemouth. *Sagina maritima*.
Millegrana minima. Sandy heaths in summer. *Radiola linoides*.
Alsinae pentapetalae genuinae.
Caryophyllus holosteus arvensis glaber. Hedgerows, etc. in spring. *Stellaria holostea*.
C. hol. arv. glab. flore minore. Sandy fields. *Stellaria graminea*.
C. hol. arv. medius.[8] Isle of Ely and near Oxford. *Stellaria glauca*.
C. hol. arv. hirsutus flore majore. Dry banks and heaths. *Cerastium arvense*.
Alsine hirsutus myosotis latifolia praecocior. Fields in April. *Cerastium viscosum*.
A. hirs. myosotis. Later in fields. *Cerastium vulgatum*.
A. myos. facie Lychnis alpina flore amplo.[9] Streams on Snowdon. *Cerastium alpinum*.

---

1 *S.B.* II, p. 148.   2 *S.B.* II, p. 237.
3 *S.B.* II, p. 229. Leez is of course Leighs, the lane leading to the villages of that name.   4 *S.B.* II, p. 298.
5 *S.B.* II, p. 124.   6 *S.B.* II, pp. 206–11; and cf. above pp. 87–8.
7 L., P. and S. signify Lhwyd, Plukenet and Sherard and indicate that the entry depends solely on them.
8 A note adds that this was regarded in *Catalogus Angliae* (p. 168) as a variety of preceding.
9 Note two forms, green and hoary.

Alsine major repens perennis. Marshes and riversides. *Malachium aquaticum.*
A. vulgaris. Waysides and gardens. *Stellaria media.*
A. longifolia uliginosis proveniens locis. Boggy ground. *Stellaria uliginosa.*
A. myositis lanuginosa grandiflora. Snowdon. L. *Cerastium alpinum* var.
   lanatum.
Alsinae petalis integris, aliaeque affines, incertae sedis.
Alsine minor multicaulis. Waste places and walls. *Arenaria serpyllifolia.*
A. Plantaginis folio. Hedges and moist thickets. *Arenaria trinervia.*
A. tenuifolia. Cornfields in Cambs. *Arenaria tenuifolia.*
A. pusilla pulchro flore.[1] Hills around Settle. *Arenaria verna.*
A. palustris foliis tenuissimis. Marshes in July. *Sagina nodosa.*
A. montana minima rotundifolia. Hills in Herts. P. ? *Arenaria leptoclados.*
A. Polygonoides tenuifolia flosculis in spicam dispositis. Near Boston. P.
   Bufonia tenuifolia.[2]
A. Spergula dicta major. Cornfields. *Spergula arvensis.*
Spergula purpurea. Sandy places. *Spergularia rubra.*
Alsine Spergulae facie media. Salt-marshes. *Spergularia salina.*
A. Spergula dicta semine membran. fusco. Salt-marshes. S.[3] *Spergularia*
   marginata.
A. rotundifolia seu Portulaca aquatica. Marshes and stagnant water. *Peplis*
   portula.
A. parva palustris tricoccus. Marshes. *Montia fontana.*
A. spuria pusilla repens fol. sax. aureae. Cornwall. *Sibthorpia europaea.*

Recognising that he admitted uncertainty as to the position of the last
few species and that *Sibthorpia* has no business to be in this company, we
may yet feel that here is an admirable list of the British Stitchworts and
Chickweeds—a list almost complete except for *A. uliginosa* of Widdybank
Fell, for *A. ciliata* of Ben Bulben, *A. gothica* of Ribblehead (perhaps a later
arrival) and *C. cerastoides* of the Scottish mountains and a few others
whose specific distinctness or claim to British status is disputable.

Of the two other groups that we examined *Millefolium* has been split up
and *Geranium* divided into headings[4] so that the three (or four) species
now assigned to *Erodium* are separate; three new species are rather precarious: one rests on Morison, but may be perhaps *G. pyrenaicum*;[5]
another, a large form of *G. sanguineum* found by Dale; and a third called
'G. nodosum of C. Bauhin etc.', which a Mr Archer, probably John
Archer, nephew of William Nicolson and a friend of Lhwyd,[6] sent to

1 A note states that this is the plant called Auricula muris folio tenuissimo in the
Catalogus Angliae: it is in II, p. 35.
2 This famous plant figured in all British floras as found by Plukenet at Boston,
till Smith, *Flora Britannica*, 1800, I, p. 192, quoting Banks questioned its identification.
3 Sherard noted the membranaceous edge of the seed.
4 *S.B.* II, pp. 216–20.
5 Ray refers to his treatment of it in *H.P.* II, p. 1060.
6 Son of Miles Archer of Oxenholme, Queen's College, Oxford, 1690 (cf. Gunther,
Life of Lhwyd, p. 172), St John's College, Cambridge, 1694 (Venn, *Alumni Cantab.*).
Nicolson married the daughter of Miles's brother, also John.

Bobart and declared to be wild in Cumberland. In addition, there is *G. lancastriense* found by Lawson, as recorded in the *Fasciculus*, and *G. lucidum* found by Sherard in Jersey and at Swanage.

Of the new plants now recorded the greater number, as we have said, are cryptogams. Mosses and fungi are in fact now listed for the first time with recognisable descriptions and the beginnings of a classification; and the additions are mainly Ray's own. Among ferns the progress since his previous lists is remarkable: he is not wholly successful in distinguishing varieties from species, but this is hardly surprising; and at least twenty-eight species are identifiable. Lhwyd's description of *Woodsia ilvensis*, for example, 'Alpine fern with leaves like those of the Lousewort and hairy beneath', is sufficient to fix it: he adds: 'it is found on the wet rocks called Clogwyn y Garnedh juxta summitatem montis Gwydhvae totius Cambriae altissimi, a very rare plant even at Snowdon'.[1] He groups together the two Hymenophyllums, noting that their leaves are either bifid or trifid and that Plukenet insisted on separating the form found by Newton[2] and Lawson on 'Buzzard Rough Crag near Wrenose' from that found by George Dare near Tunbridge; and appends to them a note on the frond of a much larger specimen, collected in Ireland, 'all the leaves of which were bifid, trifid or some quadrifid',[3] which one would suppose to be *Trichomanes radicans*.

There is not much of importance among his third family, but in the Composites he has done a lot of work upon the Hawkweeds, adding a large number of new species, classifying them, and referring to his own investigations and raising of them from seed.[4] The Sow-thistles he now separates both from the Hawkweeds and the Lettuces, listing four forms of *Sonchus oleraceus—asper* of which he says 'I do not dare to assert that they are specifically different: let anyone with leisure and a mind to it grow them from seed'.[5] After *S. arvensis* he adds under a separate heading the record[6] of *S. palustris* 'near Greenwich and about Blackwall' as of a distinct species—with a note: 'Some people use this plant when green and tender as a vegetable and in salads: I leave it to be chewed by rabbits.' There is an interesting note attached to the record of Conyza caerulea acris (*Erigeron acre*): 'Conyza annua acris, alba, linariae foliis, Boccon. It occurs freely round London and elsewhere but is certainly not indigenous: this was observed by Dr Tancred Robinson'.[7] The reference is to *Erigeron*

1 *S.B.* II, p. 46.

2 He found it there among moss in July 1682: *H.P.* I, p. 141.

3 *S.B.* II, p. 47.   4 *S.B.* II, pp. 71–5.

5 It is curious that whereas *Hieracium* has been split up into innumerable species, *Sonchus*, which certainly varies almost as widely, has had no such attention paid to it.

6 Cf. *C.A.* p. 290: 'T. Willisel found it on the Thames near Greenwich; and I noted it in Sussex', but there he attaches it as a tall variety to *S. arvensis*.

7 *S.B.* II, p. 80.

canadense, which Robinson had reported in one of his earliest letters to Ray as establishing itself round Paris:[1] it is to-day an abundant weed, but usually regarded as of recent introduction. Among the spineless Thistles, which he now separates under the name Cirsium, in addition to two of his own observation (*C. pratensis* and *C. heterophyllus*) he now adds two which he suspects to form one species, found by Lhwyd at Clogwyn on Snowdon and on the Glyders and evidently *Saussurea alpina*.[2]

Of Umbellifers he adds Tordylium (*T. maximum*) as 'found by Mr Doody about Thistleworth' (Isleworth),[3] but suspects it to be an alien escape. He has a good note on the variety of leaf-form in *Apium inundatum*.[4] But the most important addition is what he names from J. Bauhin Apium petraeum seu montanum album, found on the Gogmagog Hills, and by the description of its leaves, stalk and seeds clearly *Seseli libanotis*.[5] Among Labiates the Thymes and the Mints are increased by varieties from fresh localities and more careful study; Dale contributing a series of the latter from 'Bocking river' (the Blackwater) and Great Yeldham on the Colne and including *Mentha piperita*. In the next family he has an admirably classified and complete account of *Ranunculus*, has brought *Dryas octopetala*, which appeared as Teucrium alpinum in the *Catalogue*[6] and was 'found by Mr Heaton[7] in Ireland on the mountains betwixt Gort and Galway', into close connection with the Geums, and has introduced *Potentilla sibbaldi* from Sibbald's *Prodromus* published in 1684. But it is only with the Gentians that a real problem arises. Here he starts with *Gentiana pneumonanthe*, the discovery of his journey with Skippon; then lists the two annuals (presumably *G. amarella* and *G. campestris*) which in the *Catalogue* he had refused to distinguish, but of which 'the taller' had now been found 'first by Dr Eales[8] near Welling [Welwyn] in Hertfordshire; then by Mr Dale in some barren layes at Belchamp St Paul, Essex: Mr Doody received it from Mr Stonestreet';[9] and then continues: 'Gentianella fugax verna seu praecox found by Mr Fitz-Roberts[10] on the back side

1 Cf. *Corr.* p. 135: in this letter Ray states that Morison denied that the plant was a Canadian alien, and regarded it as native to France and England.

2 *S.B.* II, p. 86. 3 *S.B.* II, p. 102. 4 *S.B.* II, p. 107.

5 Evans, *Flora of Cambs*, p. 85, is mistaken in saying that it was first reported by Relhan in 1783: Relhan himself refers to Ray, *Syn. Brit.*, and states that the plant had been long unnoticed until he found it. 6 *C.A.* p. 295.

7 Richard Heaton, D.D., of Dublin, Dean of Clonfert and a competent botanist: cf. *F.C.* p. 197.

8 Luke Eales is quoted for plants in Hertfordshire, e.g. *S.B.* II. pp. 124, 160, where he is called 'that learned and eminent physician'. He was probably known to Ray through John Hutchinson, an ejected fellow of Trinity who was practising medicine at Hitchin and was a close friend of his; Calamy, *Lives*, I, p. 218.

9 William Stonestreet, the collector of shells: cf. below p. 402.

10 John Fitz-Roberts supplied several plants from this neighbourhood; e.g. *Helianthemum canum*, *S.B.* II, p. 203; and the picture of *Andromeda polifolia*, *S.B.* II, p. 202.

of Halse-fell-nab[1] near Kendal; as also in the parks on the other side of
Kendal on the back of Birk-hags:[2] it flowers in April and on until June.'
As Gentianella fugax autumnalis is his name for *G. amarella*, this is probably
the vernal flowering form (the var. *uliginosa* Willd.) of that species;[3] for
at this time Ray knew nothing of English, or Irish, *G. verna*. The entry is
an example of the difficulty that arises when Ray has to depend upon correspondents and has not himself seen what they describe. There follows one
of the few cases in which Ray has introduced a plant on the sole authority
of a predecessor, the 'Gentiana altera dubia Anglica, brought to De l'Obel
out of Lancashire by Mr Hesketh' and described as a white five-petalled
flower with a yellow central boss and three black spots: his only comment
is 'no news of any such plant since'.

The Crucifers, his seventeenth family, are now very clearly classified
and are in fact an almost complete list. As a specimen of his success with
a homogeneous but not very easy group it may be worth while to set
out the modern names of his species in the order and 'genera' that
he gave.

Matthiola sinuata	Brassica rapa	Erysimum cheiranthoides
Cheiranthus cheiri	———	Sisymbrium irio
	Brassica campestris	Sisymbrium sophia
Draba incana		
	———	———
	Brassica nigra	Cardamine amara
Draba verna	Brassica sinapis	Cardamine pratensis
Draba incana ?[4]	Brassica alba	Cardamine impatiens
Draba muralis ?[5]		Cardamine hirsuta
	———	
	Raphanus raphanistrum	Arabis stricta
Alliaria officinalis	R. raph. luteum	Nasturtium officinale
Brassica oleracea		N. officinale[7]
	———	
Arabis glabra	Brassica muralis, or	———
	tenuifolia	Nasturtium amphibium
	Brassica monensis	Cochlearia armoracia
Arabis hirsuta		
	———	———
	Nasturtium sylvestre	Camelina sativa
Arabis thaliana	Barbarea vulgaris	
Draba muralis ?[6]	Sisymbrium officinale	Cochlearia officinalis[8]
		Cochlearia anglica
———	———	

1 Helsfell is some two miles north-west of Kendal.

2 Birks is on the neck between Hay Fell and Docker Fell, two miles due east.

3 Cf. F. Arnold Lees, *Flora of West Yorks*, p. 326 and Withering, *Arrangement*,
3rd edition, II, pp. 281–2.

4 Reported by Lhwyd, but without note of its seed-vessel.

5 So identified by F. Arnold Lees, *Flora of West Yorks*, p. 144: Lees is careful and
accurate in identifying Ray's species.

6 Reported by Plukenet from Axebridge, but without details.

7 Reported by Dale as having smaller leaves and earlier flowers.

8 He includes the mountain and coastal forms in the same species.

Cochlearia danica	Lepidium latifolium	———
C. danica[1]	———	Crambe maritima
———	Lepidium campestre	———
Lepidium ruderale	Lepidium smithii	Cakile maritima
Teesdalia nudicaulis	Thlaspi arvense	
Iberis amara?[2]	Thlaspi alpestre	Isatis tinctoria
Arabis petraea[3]	Lepidium smithii[4]	
Hutchinsia petraea	Thlaspi perfoliatum	Coronopus ruellii
———	Capsella bursa-pastoris	

In the next family he is very successful with *Veronica* but naturally less so with *Epilobium*. In the Papilionaceae his list is both well ordered and almost complete: he has separated the Medicks from the Clovers and added *Trigonella purpurascens*.[5] In the pentapetalous family we have already referred to the Alsine group and to Dodsworth's discovery of *Lysimachia thyrsiflora*: Lhwyd's discovery of *Saxifraga nivalis*[6] on 'Crib y Distilh' (Ddyssil) is new: on the strength of Dale's finding both forms true from seed he adopts Morison's mistake in dividing *Linum perenne* into two species.[7] Lhwyd's other great achievement is at the end of the next family: 'Bulbosa alpina juncifolia pericarpio unico erecto in summo cauliculo dodrantali: I have not seen the flower and cannot tell to what group of bulbous plants it belongs: on the stalk are two or three leaves rather narrow and short: on the highest rocks of Snowdon'—that is the first introduction of *Lloydia serotina* to the world of science.[8]

In the grasses as in the mosses he has made a vast advance: the classification, though not yet sufficiently detailed, is greatly improved; the chief groups are well defined; and in many of them the tale of British species is almost complete. Here again he suffers somewhat from the enthusiasm of his assistants and the defective descriptions of predecessors and correspondents: but with time and patience most of the species can be identified; and some of the newer finds are rarities.[9] The culmed grasses are separated

1 Reported by Lawson as more angular in leaf.

2 Reported by Lawson from Stonehenge, but without description.

3 Reported by Lhwyd from various parts of North Wales: no description but reference to Johnson, *Mercurius*, pt II, p. 8, where there is only the bare name, Nasturtium petraeum, which Johnson previously applied to *Teesdalia nudicaulis*: cf. *F.C.* p. 190 for Ray's comment.

4 Reported by Lawson as with glabrous leaves.   5 *S.B.* II, p. 195.

6 *S.B.* II, p. 213: cf. *F.C.* p. 201, where Ray writing to Lhwyd on 26 September 1689 speaks of it as new to him.   7 *S.B.* II, pp. 220–1.

8 In 1696 Ray received the plant in flower from Lhwyd, who had found it 'plentifully' (cf. letter to Robinson, *Life*, p. 308), and wrote (*F.C.* p. 277): 'The bulb with a single flower which you had seen in seed before, if it be not a plant *sui generis* but educible to any of the known kinds, I think it may be referred to Ornithogalum.'

9 Some of the localities are also of interest: thus of *Panicum verticillatum*, *S.B.* II, p. 249, 'in a field among rape beyond the village of Putney on the path leading to Rough-hampton: also beyond the neat-houses by the Thames-side going from the Horse-ferry above Westminster to Chelsey'.

from the sedges—these with the rushes forming a separate family. *Luzula* or Gramen hirsutum, of which he distinguishes four species, comes at the end of the grasses, *Triglochin*, *Sparganium*, *Acorus* and *Typha* after the rushes.

In the water plants he still lists the double-flowered Frogbit that he had found at 'Audrey Caussey' but has dropped l'Obel's small White and Yellow Waterlilies, for which he had searched in vain on the Thames. He groups together *Myriophyllum verticillatum* and *spicatum*, *Utricularia vulgaris* and *minor* and *Ceratophyllum demersum* under a single heading; follows them with *Hydrocotyle*, *Callitriche*, *Stratiotes*, *Lobelia dortmanna* and *Zannichellia*; and then puts in Sherard's *Subularia aquatica*, although describing its white four-petalled flower, and pods like those of a Cochlearia. There follows the 'Pepper-grass', apparently a mud-creeping form of *Callitriche*; *Impatiens noli-tangere*; *Myosurus*; *Polygala*; *Cuscuta*; and Lhwyd's Subularia, which is probably *Isoetes lacustris*. These are of course the Herbae anomalae for which no family affinities can be appropriately found elsewhere.

The Trees and Shrubs which complete the list contain little that is new and their arrangement is inevitably unsatisfactory. It is curious that Ray, who had worked out so thorough a classification for the rest of his subject, taking the whole life and structure of the plant into account even if paying special attention to its flowers, should have accepted almost unaltered the old scheme for the final class—'trees with the flower separate from the fruit', including nut, cone, berry, and catkin bearers—'trees with flower contiguous to the fruit', with drupes, berries or 'dry fruit'. He must surely have realised that to conclude his book with a group *Staphylodendron* (which he records from Pontefract but as 'not certainly wild'), *Buxus*, *Ulmus*, *Fraxinus*, *Acer*, *Myrica gale*, *Erica* (including *Frankenia*), *Genista* (including *G. tinctoria* and *Ulex*), and *Tilia* was to stultify his own principles. Apparently, though he was still prepared to defend the traditional distinction between herbs and trees, he did so more for the sake of avoiding innovation than from any conviction of its natural value; and then having made this concession to tradition, he felt it best to be thoroughgoing in his traditionalism. If he had wished to provide a contrast between his own scientific 'method' and the sort of arrangement that he had elsewhere rejected, he could hardly have done so more effectively.

To complete the record of Ray's work on the British flora it remains to mention his notes on the 'more rare plants growing wild' in the counties, contributed to Edmund Gibson's[1] edition of Camden's *Britannia* published in 1695. Ray first mentions the project in November 1693, when he writes to Lhwyd:

I perceive the undertakers for the reprinting of Camden's Britannia intend to make a glorious work of it: for they have addressed to most of the learned

1 Fellow of Queen's College, Oxford, 1694; afterwards Bishop of London.

men in England to assist them in it and to communicate what they know....
I hear Mr Churchill[1] intends to write to me about it. But truly I have neither
skill nor leisure to contribute anything considerable.[2]

In June 1694 he says that in reply to Mr Churchill's request

I told them that I had indeed by me some observations made in my simpling
voyages.... If they thought it might be acceptable to the readers or advantageous
to the sale of the book, I would collect and send them Catalogues of the local
plants in each county.... This offer I found them willing to accept: so I drew up
and sent them such Catalogues of the particular counties excepting Middlesex
for which I referred them to the London Herbarists, and Wales for which I must
refer them to you.[3]

On 14 March[4] he had written to Robinson that he had 'already sent up
those from Cornwall to Kent and received a letter of thanks from Mr
Gibson, who manages the whole work for them and seems by his writing
to be a good scholar and ingenious person'; and adds, 'so that I perceive
they have a great opinion of my contributions and better I think than they
do deserve'.[5] He says that he had also sent lists for Cornwall of the sea-
fish, the sea-fowl and various other details: but these were not printed.
In the Preface Gibson acknowledges Ray's help in a paragraph:

The Catalogues of Plants at the end of each county were contributed by the
Great Botanist of our age, Mr Ray. They are the effect of many years' observations: and as that excellent person was willing to take this opportunity of hand-
ing them to the public, so were the undertakers very ready to close with such a
considerable improvement, though it exceedingly enhanced the expenses of
printing, and they were in no way tied to it by their proposal.[6]

The Catalogues differ greatly in quality. Cornwall, the first county, is
good: it contains Ray's own discoveries with references where possible
to the figures in Plukenet's *Phytographia*. Devon is poorer; Dorset
poorer still, being largely a criticism of reports by Parkinson and l'Obel.
Somerset has a better list, including Willisel's record of Ornithogalum
'on a hill three miles on this side Bristol in the way to Bath'.[7] Wiltshire is
only notable for the 'Long trailing Dog's grass with which they fat hogs
and which is four and twenty foot long' reported in the *Phytologia
Britannica*, of which Ray remarks: 'we are not yet satisfied what sort of
grass this might be'.[8] Hampshire has several records by Goodyer, including *Pulmonaria longifolia* 'near Holbury House in the New Forest'.[9]
Berkshire has only four plants; Surrey six; and Sussex seven, all of his own
observation. Kent has much the longest list, running to five columns,

1 A. and J. Churchill at the Black Swan in Paternoster Row with A. Swalle were
the publishers.
2 *F.C.* p. 239.   3 *F.C.* p. 244.
4 So epitome of letter in *F.C.* p. 299: the letter as printed by Derham was undated.
5 *Corr.* p. 273.   6 Pref. p. 3.   7 *C.B.* c. 83.
8 *C.B.* c. 113.   9 *C.B.* c. 135.

many being Ray's own records from his stay with the Barnhams, others from De l'Ecluse on his visit to England, and others, many of which Ray questions, from Gerard and Parkinson, including Parkinson's 'very pleasant story' of the Roman nettle (*Urtica pilulifera*) which Julius Caesar's soldiers sowed when they landed at Romney, so-called from Romania, to chafe their limbs when cold—to which Ray objects that the Nettle is found elsewhere and that Caesar did not land at Romney.[1] Gloucester includes the Bristol and Holm plants mostly reported already in Somerset. Oxfordshire has several plants taken from Plot's History of the County[2] and includes *Digraphis arundinacea* 'foliis ex luteo variegatis' 'found by Mr Bobart in the Thames...though it be but an accidental variety, it deserves to be mentioned being very ornamental in gardens'[3]— a sentiment endorsed by its universal popularity. Buckinghamshire he confesses that he has had no opportunity of searching and only knows one rare plant found by Plukenet, who at this time had a farm on Horn Hill near Chalfont St Peters.[4] Bedfordshire is remarkable for the full description of the preparation of Woad—the grinding, 'balling', second grinding, 'couching', 'silvering' and bagging: 'the best Woad is usually worth £18 per ton': he notes the Latin name Vitrum and connects this with British and Welsh glass, suggesting that both are due to the blue colour of much glass.[5] Hertfordshire has nothing of importance: but Essex has nearly four columns, most of them from his own neighbourhood and one, *Euphorbia platyphyllus*, 'in mine own orchard here at Black Notley',[6] others from his exploration of Maldon and the coast, and others found by his friend Samuel Dale. Suffolk has several of the rare Breckland plants originally found by Willisel and a long note on the Aldeburgh peas.[7] Norfolk has *Verbascum pulverulentum* 'about the walls of Norwich', where it is still abundant, and a note on the colour-variation of *Medicago sylvestris*, 'usually with a yellow flower but varying to become purplish'.[8] Cambridgeshire has naturally a long list, nearly five columns, but containing nothing new to the *Catalogue* and its supplements. Of Huntingdon he has nothing, 'the more rare being common to it with Cambridgeshire'.[9] Northamptonshire has a very short list, with one strange plant, the 'Gentiana concava Gerardi';[10] Leicestershire and Rutland none; Lincoln-

---

1 *C.B.* c. 228. 2 Published in 1677. 3 *C.B.* c. 236.
4 Cf. *S.B.* II, p. 22. 5 *C.B.* c. 292. 6 *C.B.* c. 362.
7 Reproducing his letter from Friston: see above, p. 130.
8 *C.B.* c. 402. 9 *C.B.* c. 428.
10 *C.B.* c. 442: writing later to his friend J. Morton, author of the *Natural History of Northamptonshire*, Ray said: 'I was well assured by those that searched the spinney Gerard mentions that it was there....All the plants of that kind cultivated at this day owe their original to the plants that Gerard and other herbalists dug up till they had quite extirpated it' (*History of Northants*, p. 365). He did not include it in *Synopsis Britannicarum*.

shire and Nottinghamshire nothing of importance. Derbyshire has his own records from the Peak district. Of Warwickshire he says: 'though I have lived some years in this county, yet have I met with no peculiar local plants';[1] and his list is small and almost entirely from Middleton. Worcestershire has very little; Staffordshire a few from Plot's History;[2] Shropshire three only; Cheshire one; and Herefordshire none. Yorkshire has a long list, mostly from the Ingleborough district.[3] Durham has only four plants, all previously recorded. Lancashire has more, thanks largely to Lawson's Walney Island discoveries. Westmorland is richly supplied, thanks to Lawson and Fitz-Roberts and to Ray's own exploration round Shap and Orton. Cumberland is less rich, but has appended to it a list supplied by Archdeacon Nicolson of Carlisle,[4] including *Tragopogon porrifolius* from 'fields about Rose Castle' and several varieties of familiar species. Northumberland has nothing except the plants found by Willisel or by Ray in his journey in 1671.

The lists only profess to give notes on the rare and local plants, and are not very adequate. The criticisms of the *Phytologia Britannica*, of Gerard and Parkinson, remind us of Ray's earliest days when he compiled from these sources records of the plants of every county as a guide to his own and his friend's explorations. His comments show how much trouble these careless and unverifiable statements had caused; and by repudiating them he saved others from being similarly misled. That he should have something to say at first hand about almost every county, and that in many his list should be of real value, is proof of the extent and accuracy of his knowledge of the British flora: but the notes are generally too scrappy to give any real idea of a county's plant-life. They testify rather to the extent of his own observations than to any full survey of the botanical features of England. They helped to stimulate enthusiasm for county lists and led to the production of a large volume of evidence. But a careful study of his *Synopsis* would in fact be found equally adequate.

One final record completes the tale of his services to the study of our Flora. In 1700 Lhwyd made an expedition to the west of Ireland[5] and next spring sent a packet of his discoveries to Ray. Ray's letter of 11 June 1701

1 *C.B.* c. 516. 2 Published 1686.
3 A list of Yorkshire specialities is also given in his letter to Robinson, *Corr.* pp. 273–4, to prove that there do occur plants peculiar to a single area.
4 Who contributed to the second edition of *English Words*: he was for a time a keen botanist: cf. *Letters of W. N.* (London, 1809), I, pp. 36, 52, etc.
5 He visited Dublin and the north in 1699: but the tour from Sligo (Ben Bulben) through Mayo and Galway into Kerry on which he found these plants seems to have been a year later. A short report of it in a letter to Robinson dated Penzance, 25 Aug. 1700, was printed in *Phil. Trans.* XXVII, no. 336, pp. 524–6. If so, it must have preceded his visit to Snowdon with Richardson in June 1700: cf. *Journal of Botany*, LXI, p. 226. For the tours of these years cf. Gunther, *Life of Lhwyd*, p. 332.

comments upon some of these [1]—*Saxifraga umbrosa*, 'in the woods under the mountain of Cruach inhii and by Kil Arni';[2] *Dabeocia polifolia* or, as he calls it, Erica S. Dabeoci, of which he expresses doubt as to its being a genuine Erica as its corolla falls off and the fruit seems to be different; *Adiantum capillus-veneris*; *Gentiana verna*, which he believes to be new to science; and a 'Sedum serratum foliis pediculis oblongis insidentibus', which must surely be *Saxifraga geum*.[3] It is pleasant to think of the old naturalist receiving St Dabeoc's Heath and the Spring Gentian before he died.

NOTE. *Localities at Black Notley*

The precise identification of the fields and other localities at Black Notley mentioned in the *Synopsis Britannicarum* is not without interest for the student of Ray's life and of the changeless countryside. By the kindness of Mr Alfred Hills of 'the Buck', Black Notley, in Ray's time the residence of his friend James Coker, and of the Old House, Bocking, then the home of Samuel Dale, I have received a copy of the bill of sale of Black Notley Hall estate, dated October 1938, which gives full details of the area.

From this and from Mr Hills's knowledge of the district the following details can be determined:

'Westfield adjoining to Leez Lane' is still known by that name: it lies along the north side of the lane half a mile from the Braintree-Witham road, and is crossed by a footpath from the Hall.

'Wairefield', now corrupted into Wirefield, is at the south-west corner of the estate, adjoining the north side of Leighs Lane a mile from the main road: its name 'waire' or ware is the Essex version of weir or dam; a ware-pond is one formed by embanking and closing a stream: there is such a pond on the edge of the lane just opposite the field.

'Stanfield' is the local abbreviation of Stanwells Field—this being the name given it in the bill: it lies directly to the west of the Dewlands orchard.

'Lampit Grove' seems to have been destroyed: but there are memories of a copse on the site of what is now the sanatorium, and this formerly belonged to the Hall.

'Hoppet-bridge' is still Hoppit Bridge, and crosses the Brain, carrying the main road from Braintree to Black Notley.

1 *F.C.* pp. 280–1.
2 'The Tories of Kil Arni obliged us to quit those mountains much sooner than we intended': Gunther, *Life of Lhwyd*, p. 457.
3 Lhwyd also found *Potentilla fruticosa* 'plentifully amongst limestone rocks on the banks of Loch Crib' (Lough Corrib): Gunther, *Life of Lhwyd*, p. 433.

CHAPTER XI

LAST WORK IN BOTANY

Vir pius et modestus V.D.M. maximus ab hominum memoria botanicus.... Omnium botanicorum plurima opera edidit, uno Linnaeo excepto.

ALBRECHT VON HALLER, *Bibliotheca Botanica*, I, pp. 500, 506.

Ray's work on the British Flora was only one comparatively small part of the output of the years after the publication of the *Historia Plantarum*. Busy as his life had been, no period of it was so productive as the years between 1688 and 1698; and hardly any writers can ever have produced so many and such varied books in a single decade.

Most of these publications deal with subjects other than botany. For as soon as his magnum opus was finished, he took up the task which Willughby had suggested and Robinson in his letter of 18 April 1684 had revived.[1] The volumes in which he surveyed the Mammals, Reptiles, Birds and Fishes—those hand-books which he regarded as journeyman's work[2] but which actually laid the foundations of zoological study—were finished in 1692 and 1694. They will be discussed in detail later. Here we need only remark that though the second of them, the *Synopsis Avium et Piscium*, was little more than a condensation of his two larger volumes, the first broke fresh ground and involved not only researches for which he had had little previous training, but the handling of general problems in biology and taxonomy in which he made new and very important contributions to learning.

In addition, in 1691[3] he prepared a second and much enlarged edition of his *Collection of English Words*—the edition which is the starting-point for the study of dialect and folk-speech. Of the first edition in 1674 we have already described the main features; and the second maintains the division into northern and southern. In other respects it is almost a new book. He had received, as he tells in the Preface, 'a large catalogue of Northern words, their significations and etymologies' from his friend and former colleague Francis Brokesby,[4] Rector of Rowley near Brough in

1 *Corr.* p. 141: cf. above, p. 212.

2 So to Robinson Aug. 1693: *F.C.* p. 298.

3 Cf. letters to Lhwyd of 7 November 1690 (*F.C.* p. 210) and of 25 Nov. 1691 (*F.C.* p. 224): 'My Collection of Local Words is published as you may learn by the *Gazette*. But the Bookseller concerned in that copy is so stingy and sordid as not to allow me copies to present to my friends.' The publisher was Christopher Wilkinson 'at the Black Boy over against St Dunstan's Church in Fleetstreet'.

4 Cf. above, p. 168: a letter from him to T. Hearne dated 12 December 1708 and referring in detail to his contributions to Ray's *Collection* is in *Letters by Eminent Persons*, ed. by Bliss and Walker, I, pp. 184–5 (London, 1813).

the East Riding from 1668 to 1682, who also sent observations on their pronunciation.

This had been supplemented by 'my honoured and dear friend' Tancred Robinson. In the Southern collection he had been helped by contributions from Nicholas Jekyll of Sible Hedingham, grandson of William Jekyll the antiquary, who had died in his home at Bocking in 1653 leaving forty volumes of manuscript material for the history of Essex, Norfolk and Suffolk, many of which passed to Nicholas; and from Mansell Courtman, minister of Castle Hedingham, from whom Ray got help also in insects.[1] In addition, while the new edition was being printed he received lists from Edward Lhwyd,[2] whose name he still spells Lloyd, and through Lhwyd from Robert Tomlinson,[3] 'of Edmund Hall and a Cumberland gentleman'; and a third 'from Mr Wilkinson a bookseller in Fleet Street...sent him from Mr William Nicholson[4] an ingenious minister living in Cumberland', who is of course Archdeacon, afterward Bishop, Nicolson, from whom he got later a list of Northumbrian plants. These lists arrived too late for inclusion in the alphabetical series, and being printed separately seriously impaired the form and utility of the book.

In addition, he got help in etymology from the earliest work of the greatest of all historical scholars of the period, George Hickes, fellow of Lincoln, Dean of Worcester, and one of the first to be consecrated a non-juring bishop, who had published in 1689 his *Grammatica Anglo-Saxonica et Moeso-Gothica* and also an 'Islandish Dictionary' which was appended to his edition of R. Jonas's *Grammatica Islandica*.[5] The Preface discusses and corrects Hickes's views of some of these Icelandic words; cites Sir Thomas Browne's eighth tract,[6] and refers to Sir Henry Spelman's *Glossarium*.[7] Ray also adds a Postscript and an interesting 'Account of some Errors and Defects in our English Alphabet, Orthography and Manner of Spelling', in which he draws largely upon Wilkins's *Essay*

---

1 Cf. *Hist. Ins.* pp. 127, 337.

2 *Coll.* II, pp. 122–30. He had promised in Nov. 1690 to send a collection of Welsh words which 'bear affinity both in sound and signification with some amongst the north-country words': *Corr.* p. 226 and Gunther, *Life of Lhwyd*, p. 110.

3 *Coll.* pp. 131–8. Skeat, l.c. p. xiii, calls him Edward; Ray merely 'Mr.' Reference to Foster, *Alumni Oxon.*, shows that he must be Robert, son of Richard of Aikhead, Cumberland, who graduated from St Edmund Hall in 1689.

4 *Coll.* II, pp. 139–52. It is in Latin and headed Glossarium Northanhymbricum. Nicolson alludes to it in a letter to Thoresby dated 8 February 1692 (*Letters of W. N.* I, p. 24).

5 It is in his great *Thesaurus*, Pt III, pp. 73–92 (published 1703): but he makes no reference to Ray's criticism.

6 Wilkin, *Browne's Works*, IV, pp. 195–213.

7 That is to the edition of 1687 entitled *Glossarium Archaiologicum, continens Latino-barbara, peregrina, obsoleta et novatae significationis vocabula.*

towards a Real Character, points out the confusion caused by the difference between spelling and pronunciation and makes many concrete proposals for a simplified spelling. The whole essay throws light not only upon the efforts then beginning to be made to establish an agreed orthography,[1] but also upon the changes in pronunciation since the seventeenth century; and many of the points of criticism arise directly out of his experience in teaching his children to spell. Thus discussing the vowel *i* he says:

Children take notice of this difference between *i* when pronounced as a diphthong and when as *iota*. One of my children in all words where it is to be pronounced as a diphthong pronounced it as a simple iota or *ee*. As for mine, thine, like, bile it pronounced meen, theen, leek, beel and so in all others of that nature; the child, it should seem, finding it more facile to pronounce the single vowel, not being able to frame its mouth to pronounce the diphthong.[2]

Putting forward some twenty proposals, he affirms: 'If all these faults were amended, viz. the superfluous letters cut off, the wanting supplied, and to every letter his proper power attributed, spelling would be much more regular, uniform and easy.' The lists of birds and fishes printed in the first edition are omitted, but the account of manufactures is preserved and revised.

Skeat, although regarding this edition as 'by far the best' and stating that 'it has long been considered a standard work', rightly complains of the haste and errors of its arrangement, and expresses the wish that it had reached a third edition in Ray's lifetime. Towards such an edition he received in April 1703 a further long list of Yorkshire words from Ralph Thoresby of Leeds,[3] the antiquary who contributed to Gibson's *Camden* and was a friend of Nicolson. Ray's reply on 12 June[4] holds out hopes of another edition and asks for a further part of the list.

That the second edition was produced hurriedly is hardly surprising. In November 1690, before it was printed, he had begun on the first of the two books which brought him popular fame, *The Wisdom of God in the Works of Creation*, and in July 1691 almost before its publication was planning the second, which at first he refers to as *The Dissolution of the Earth*.[5] These were the expression alike of his abiding conviction that 'Divinity is my profession'[6] and of his satisfaction that with the Revolution the religious outlook was more stable and his own work would be

1 Thus he condemns what he describes as the recent introduction of 'scituate' for situate and 'scent for sent signifying a smell': both of which are false to the derivation, the Latin situs and sentio.

2 P. 161.

3 *Corr.* pp. 418–30: Thoresby refers to this, *Diary*, I, p. 421.

4 Printed in *Letters of Eminent Men to Ralph Thoresby*, II, pp. 22–4 and by Skeat, l.c. p. xxvii.

5 So to Robinson: *F.C.* p. 294.   6 Cf. letter to Aubrey: *F.C.* p. 163.

appreciated. We have seen his joyful relief at the passing of the Stuarts. Now his friend John Tillotson, stepson by marriage of Wilkins, a man sensitive, generous, conciliatory, was constrained against his will to accept the Archbishopric of Canterbury, where one of his first acts was to offer Ray preferment.[1] Ray knew his own happiness and work too well to consider accepting: but at least he could support Tillotson's endeavours by his writings. To do so he took up themes which he had had no possibility of developing since he left the University.

Each of his two books was in its original form an expansion of Commonplaces delivered in Cambridge: each was sold out in a few months; and was then reprinted in a much enlarged form. *The Wisdom of God* became in fact a treatise upon the adaptation of form to function in nature, the forerunner of Paley's *Natural Theology* and of the series of Bridgewater treatises. The *Discourses* were expanded into a survey of cosmology and geology. They will be discussed at length in later chapters.[2] Here it is enough to note that these new ventures were unexpectedly and enormously influential; that they involved him in a mass of work; and that they were produced while the *Synopsis Quadrupedum* was being written, and without any break in his other activities.

Before a second edition of the *Discourses* was published a further task was laid upon him. On 25 October 1692 Charles Hatton, to whom he had dedicated the *Historia Plantarum*, wrote to him[3] about the *Itinerary* of Leonhart Rauwolf,[4] claiming for it a high value but urging that 'being printed about a hundred years ago[5] it is very rare and being never translated out of High Dutch it is unintelligible to those who do not understand the German tongue'. He had proposed to Sloane and Robinson[6] that the book be translated; Sloane had borrowed it out of the library of the Royal Society; and Nicholas Staphorst, a German, whom Ray describes as 'chemical operator to the Company of Apothecaries',[7] was at work on an English version. This would need correction and revision 'by a master of the English language': it would also be the better for notes and comment, since some of Rauwolf's statements, particularly one about unicorns in Abyssinia, had been criticised, and changes had taken place since Rauwolf's day.

It was highly to be wished that some person duly qualified would where requisite make some brief animadversions; and if any person would, in a short preface, give some account of the author and by a favourable character of the

1 *F.C.* p. 294.   2 See below, chs. XVI, XVII.   3 *Corr.* pp. 255–7.
4 He set out from Augsburg in May 1573, went to Tripoli, Aleppo, Babylon, Bagdad, Mesopotamia and Palestine, returning in Feb. 1576.
5 Actually in 1583 at Frankfort.
6 Apparently Robinson had already written: for in Derham's epitomes Rauwolf's name occurs on 12, 21 and 26 Oct.
7 *F.C.* p. 232.

work give it a recommendation, it would be an invitation to all ingenious persons to peruse it: for which achievement there is no person on earth so duly qualified as the justly-renowned Ray. Therefore, pardon me, sir, if I join my humble desires to those of our afore-mentioned worthy friends...if you will please so far to condescend Mr Smith engages to return you, in a fitting manner, his thanks for the benefit he shall receive...and all persons of learning or ingenuity will, I doubt not, acknowledge it as an obligation from you to them.

No doubt, despite its pomposity of style, the request was kindly meant. Ray was competent, and poor; and Smith's unspecified acknowledgment might be welcome. But he was busy and contentedly independent; and the suggestion seems to have annoyed him. His reply has not been preserved. To Lhwyd he wrote, on 7 November,[1] 'some of my friends and acquaintances at London would fain put another task upon me, that is revising and making notes upon Rauwolfius his *Itinerary*....But I am not inclinable to undertake work of that nature.' Nevertheless, on 28 December[2] he states: 'Rauwolfius is an author of good account among the learned. Clusius and both the Bauhins mention him always not without great respect.[3] No more than the first part of his *Itinerary* is as yet sent to me which I have perused and well approve of: he seems to me to have been very diligent in observing and faithful in setting down his observations.' He must then have been considering acceptance.

On 16 February 1693[4] Sloane wrote on lines similar to Hatton, speaking favourably of Staphorst's translation, discussing the existence of a fourth part of the book, and stressing the interest of the work to scientists. Robinson seems also to have returned to the subject.[5] Ray then assented and on 10 April wrote to Sloane:[6] 'The translator has done his work as well as could be expected from a foreigner: I have revised it and altered the phrase and language....Annotations...I have had no leisure to add but have referred that back to Mr Robinson.[7]...A catalogue of more rare Oriental plants I have drawn up to be added to the end of the work referring such as are found therein to the page where they are mentioned.' He says nothing here or, apparently, elsewhere as to the series of extracts reprinted from the records of other travellers, Pierre Belon, Francis Vernon,[8]

1 *F.C.* p. 322.   2 *F.C.* p. 233.
3 Ray had himself frequently quoted him, no doubt from them, in his *Historia Plantarum*.
4 *Corr.* p. 260.   5 *F.C.* p. 298.
6 *Corr.* pp. 262–3: he explained that he did not answer earlier because of pain of his 'exulcerated pernios'.
7 Apparently on 7 April: *F.C.* l.c.
8 In a letter of 10 Jan. 1676 to Oldenburg describing his journey to Greece and Smyrna: cf. *Phil. Trans.* XI, no. 124, p. 575. On this journey he was murdered in Persia.

Sir George Wheeler,[1] Thomas Smith,[2] John Greaves,[3] Robert Huntington,[4] Sir Henry Middleton[5] and many others. These are mostly drawn
from *Philosophical Transactions* and may have been Robinson's work.[6]
The list of plants at the end of the book, lists from Egypt and
from Crete being added to that from the Levant, must certainly be
Ray's.

The two 'tomes', *A Collection of Curious Travels and Voyages*, were a
valuable contribution to the development of science in England. Rauwolf's
work—the first volume—deserves Ray's commendation: indeed the rarity
and character of his comments are sufficient to confirm his high opinion
of its accuracy. To Rauwolf's description of birds 'black with long necks
and feeding upon fish' as Sea Eagles, he notes: 'I guess them rather to have
been Cormorants';[7] to his report that at the site of ancient Babylon there
are insects bigger than lizards with three heads, 'Rauwolf was here too
credulous: that there neither are nor ever were any animals with more
heads than one naturally, I do confidently affirm';[8] to his story of unicorn
and 'Griffins' with the beaks and wings of eagles, the tails of lions and the
feet of dragons sent by Prester John to the king of Persia—a story which
Rauwolf says 'I did believe the sooner because he could also tell me what
trees grow there'—'too soon; for that there are no such creatures in the
world as either Unicorn or Griffin I am as sure as I can be of a negative'.[9]
In commenting upon the argument that Jerusalem is the highest site in
Palestine because Scripture always speaks of 'going up' to it he points out
that though it stands on a hill, yet 'whosoever travels up to London,
goes up thither; and whosoever travels from thence, goes down into the
country let his habitation be never so much higher'.[10] So too he corrects
the statement that 'James the Lesser was first Bishop of Jerusalem' by
affirming that James the son of Alphaeus was James the Less, whereas the
first bishop was James the Just called the Lord's brother.[11]

---

1 Cf. above, p. 221: here is printed a list of plants from Greece and Asia Minor: to
this is added a note: 'For these the readers may consult Mr Ray's Collection of Exotic
Catalogues, especially the Oriental one where the Synonymous names are added.'

2 Described as 'Fellow of Magdalene College Oxford and of the Royal Society':
his observations on Constantinople and Prusa in Bithynia were made in 1668–71 and
printed in *Phil. Trans.* XII, no. 152, pp. 335–46.

3 Professor of Astronomy at Oxford till 1648: travelled to Egypt 1638–9: published
his *Pyramidographia* in 1646.

4 Fellow of Merton College, 1659–83: then Provost of Trinity College, Dublin and
Bishop of Raphoe: chaplain at Aleppo, 1671–81. His paper here printed was in *Phil.
Trans.* XIII, no. 161, pp. 623–9.

5 The merchant captain under the East India Company: the paper here printed
deals with his imprisonment at Mocha in the winter of 1610–11.

6 I agree with L. C. Miall, *Early Naturalists*, pp. 128–9, that they are not by Ray.

7 P. 157.   8 P. 176.   9 P. 196.

10 P. 275.   11 P. 286.

How much alteration he made in the text there is no means of discovering: the style of the translation is not very distinguished, but is
generally lucid and readable: probably the time spent upon it was not
great; and as botany was Rauwolf's main object, and he gives full accounts
of the plants at the various places with references to Dodoens and De
l'Ecluse, Ray must have found the work interesting, valuable and relevant
to his own studies. It helped to prepare him for his final tasks—the
emendation of his *Methodus* and the compilation of the third volume of
the *Historia*. But it was hack-work: and in his state of health and business
he had to husband his strength, and ought certainly to have been spared
efforts which gave no real scope to his powers.

For in addition to all this literary output he was still actively engaged
in field-work. We have seen that he had returned to botany and discovered, described and arranged about a hundred new species of cryptogams. Since 1690 he had also been enthusiastically collecting and describing Lepidoptera with a view to completing his debt to Willughby by
a History of Insects. The unfinished records of achievements in this field,
published in 1710 without editing or supervision from the mass of material
that he left, will be considered in a later chapter: but the diaries there
printed containing careful descriptions of some three hundred species,
caterpillars, pupae and perfect insects, as he received them from his children and neighbours or secured them in his house and garden, are proof of
the time and care, the thoroughness and insight expended upon this
entirely unexplored research. That in his cramped surroundings and
crowded life he managed the mere routine of providing cages, foodstuff
and apparatus for the breeding and storing of his specimens; that at his
age and without any reliable books he succeeded in discriminating and
identifying forty-seven British butterflies and a large number of moths;
that he grasped, as no entomologist for more than a century after him
succeeded in grasping, the necessity of studying not merely the imago but
the whole metamorphosis of a species; and that even when he realised the
magnitude of the task and the certainty that he could not live to complete
it he still went on collecting, describing, classifying: these are facts which
more than any other reveal the heroic quality of his character. Any account
of his last botanical work must set it against a background of incessant and
almost overwhelming activities in other directions. He accomplished the
Historia Plantarum and the *Synopsis Britannicarum* in continuous concentration: the final books were produced in days filled with a variety of
other and exacting pursuits.

The *Historia Plantarum*, whatever its permanent influence, had given
him a unique status in the world of science. When in 1691 *Philosophical
Transactions* spoke of him as 'that incomparable botanist'[1] it was only

1 XVII, no. 193, p. 528.

anticipating what Rivinus of Leipsic said in 1694: 'I asked your criticism because I knew you to be the best skilled in this science and easily the first of all botanists who have ever lived'.[1] As such he was in correspondence with scientists of all sorts: Wilhelm Ernst Tentzel wrote to him about his Mammoth bones found in Thuringia, on which the Royal Society would not risk an opinion;[2] Francis Vaughan, the Irish doctor, wrote from Clonmel about the baneful effects of *Oenanthe crocata*;[3] Andrew Paschall, Rector of Chedsey, expounded in many letters 'long, not to say tedious by reason the hypothesis is abstruse'[4] his fancies of a 'mutual contranitency between parts central and circumferential'[5] and of tidal senaries and their influence upon health;[6] John Morton, now of Oxendon, invited his views upon Manna[7] or suggested a history of scriptural plants.[8] In addition, he wrote constantly to Sloane, Robinson and Lhwyd and less regularly to Aubrey and Petiver and several others.

He had neither secretary nor assistant; and in his small home the presence of four young children, even if their enthusiasm helped him with caterpillars, did not make for ease of work. When Lhwyd in 1695 sent him a box of fossils he has to report that 'one of them, and that too which you value most, is to my grief broken asunder crossways by the negligence, I suppose, of some child or servant who, in my absence taking it up to look upon it, let it fall'.[9] And the same tale has to be told to Sloane in 1699 of some books: 'I must beg your pardon for having in some measure defaced them by sullying them myself, being forced to use them by the fireside, and partly by a child's unluckily scattering ink upon them. I intend', he adds, 'to have the books myself and if you please to send me back these again, I will order Mr Smith to get them bound for you in the same manner these are.'[10] Considering the number of parcels of plants and insects and books that came to him it is surprising that such accidents in his crowded life were so rare.

Plainly he was happy: in the house that he had built and the work that he had chosen, he was free to fulfil his own evening prayer and 'to grow daily that so our last days may be our best days'.[11] It was a quiet ordered life with a regular routine and simple standards seldom altered except on the rare occasions when one of his friends from the great world paid him

---

1 *Epistola De Meth.* p. 3.

2 *Corr.* pp. 322–3 and 472; and *Phil. Trans.* XIX, no. 234, pp. 757–76. Tentzel in this letter *De Sceleto Elephantino* Nov. 1697 quotes Ray to prove that the fossil remains are of an elephant and not a 'lusus naturae'.

3 *Corr.* pp. 304, 314.   4 So Derham, *Corr.* p. 273.

5 *Corr.* p. 271.   6 *Corr.* p. 280.   7 *Corr.* p. 369.

8 *Corr.* p. 379.   9 *F.C.* p. 261.

10 *Corr.* p. 365: Smith is of course his publisher and bookseller.

11 Cf. *Mem.* p. 62; one of the four prayers written by Ray which Scott added to Derham's *Life.*

a brief visit or when some unusually exciting contribution from the carrier made him put aside his immediate task and deal with something new. Until the end of the century, at least, his output remained exceptionally large. Unless he had been not only persistently diligent but methodical and a rapid worker, he could not have coped with it: and his family and household must have combined to make his wishes and life's work their common purpose.

For it was now becoming a life of broken health and almost incessant pain. We have seen the beginning of the trouble with his attacks of diarrhoea and the inflamed sores on his legs during the writing of the *Historia*. When it was finished, with rest and exercise he might have shaken off ailments which at first were evidently not serious. But early in 1690, when the *Synopsis* was being printed, he was attacked by pneumonia. Jane, his youngest, had been having teething-fits which had given him much anxiety. On 6 March he was taken 'by a violent cough and fever attendant thereupon'[1] and for some days was desperately ill. Benjamin Allen, who was treating Jane, has left a full account of 'Mr Ray's case of Peripneumonia' in his Case-book discovered by Mr Miller Christy and described in the *Essex Naturalist*.[2] He recovered and by May, though 'not yet quite freed from the dregs of the distemper',[3] was able to get back to work. But his general health was impaired, and his powers of resistance weakened. For the next two years he seems to have been well, and was certainly intensely active: but the long and bitter winter of 1692–3 brought on a fresh outbreak of sores. The first notice of them seems to be in a letter of 22 March 1693[4] to Lhwyd:

I have been this winter so afflicted with exulcerated pernios[5] upon the back side of the small of both my legs, occasioned at first by inadvertency and neglect (for they might have been prevented or at first easily healed) that writing or indeed any other business is become troublesome.... You would not imagine that ulcers of which so little account is made should be so painful and vexatious, they giving me very little respite from pain night or day.

On 26 November[6] he referred to them again:

I know not whether I acquainted you with the ulcuscula...which broke out last winter and have continued ever since upon one of my legs. At first I looked upon them only as exulcerated chilblains or pernios, and so I believe they were in a great measure. They are very stubborn and ill-natured and have hitherto resisted and frustrated all means and methods of cure; so that now I have little hopes of getting them healed and dried up this winter, if ever. The cold

---

1 So his letter on 28 March to Lhwyd: *F.C.* p. 204.

2 Vol. XVII.   3 *F.C.* p. 206.

4 *F.C.* p. 234. He had mentioned them to Robinson on 3 Feb., but the letter only exists in Derham's epitome.

5 In his *Dictionariolum*, p. 91, he gives Pernio as meaning 'Kibe or chilblain'.

6 *F.C.* pp. 238–9.

weather doth affect and greatly exasperate them, causing them to run abundantly.

He had mentioned his sores to Sloane in April[1] but with the hope that the warm weather would heal them. Apparently during the summer he consulted Dr William Briggs, fellow of Corpus Christi College, Cambridge, 1668, and of the College of Physicians, 1682, well known for his *Ophthalmographia* published in 1676: for in an undated letter also to Sloane, which must clearly have been written in September or early October,[2] he writes: 'I have lately received a letter from Dr Briggs wherein he tells me that he had consulted with Dr Lister and yourself concerning the ulcers on my legs wherewith I have been troubled now the best part of a twelvemonth, and that you expressed a great concern for my condition; for which kindness I return you hearty thanks': he then gives details of the effect of the 'Calomelanos'[3] which he had taken and proposes in spite of its mixed benefit to repeat. On 16 October[4] he wrote again to report that the calomel though 'well prepared being done by our physician himself' had had very violent results not yet abated without any 'remission of pain or abatement of the running of my sores'. He continues:

Yesterday I, by the advice of Dr Robinson, applied the emplastrum e ranis cum mercurio,[5] which is so far from giving any present ease that it aggravates the pain especially in the night.... What good it may do in the future I know not; but in other plasters that I have used at first they have been most easy and afterwards more troublesome; only this which I have used all this summer and part of the spring gave me present ease the first night that I applied it.... I have this summer made use of a decoction of sassafras, sarsa [sarsaparilla] and china[6] with some sage and hypericon,[7] and shavings of hartshorn,[8] but without guaiacum, for a short time: I found that it heated and dried my body but gave me no sensible relief, so I gave it over because it was nauseous and ungrateful to my stomach; but upon your recommendation I will make another trial of it.

A postscript, added after missing the post, reported that the plaster *e ranis* had had disastrous results. On 7 November[9] he acknowledged 'woods

[1] *Corr.* p. 262.

[2] *Corr.* p. 475. It is certainly earlier, though from the references to Martens' Voyage to Spitzbergen not much earlier, than the letter of 16 Oct.

[3] Protochloride of mercury: he concluded later that any preparation of mercury was dangerous for him: 'calomelanos' is of course the genitive of 'calomelas', calomel.

[4] *Corr.* pp. 264–5.

[5] Lest a plaster of frogs and mercury seem impossible as a remedy, we may note that Dale in his *Pharmacologia* (ed. 1693) gives ointment of mercury as curing every skin-disease (p. 68) and says of the Common Frog 'sperma...manuum scabiem curat, herpetes interficit' (p. 607).

[6] The root of *Smilax china*, a plant closely allied to Sarsaparilla (*Smilax officinalis*) and at this time equally valued as a drug.

[7] Oil prepared from St John's Wort.

[8] I.e. ammonia prepared from shavings of horn.

[9] *Corr.* p. 267.

and roots' sent by Sloane for his malady and promised to use them: he
continues:

Lime-water I have made use of outwardly to wash the sores...and have
received some benefit....I have lately applied a plaster in form of a cerecloth
which I had from a neighbour who knew not the ingredients of it: but by the
scent and consistency of it I plainly perceive that there is Burgundy pitch [1] and
rosin in it, which at first agreed marvellously well with the ulcers...so that had
the weather favoured I was in some hopes...but it happening lately to be cold
and frosty they fell off itching and spread again and are now come to be as bad
as before.

On 29 November [2] he reported that he had 'for some time made use of a
diet drink made by decoction of those materials you were pleased to send
me....I think for the present while I am taking it it doth somewhat abate
the running of the sores (which yet after a day or two's intermission returns as copious as before), but makes them more painful'. The rest of the
letter is a long discussion of the nature of the sores, which he is now convinced are chilblains.

So it went on month after month for the rest of his life. Sloane took an
infinity of pains, trying many prescriptions and receiving reports which
begin with gratitude and end with the admission of no change. Sometimes,
as for a few months early in 1694, the sores dry up; only to break out
afresh: 'They puzzle my philosophy and I am at a loss how to order
them.' [3] Sometimes, as in the spring of 1697, he reports a new remedy—
'a salve made up of two parts of diapalma and one of basilicon'. [4] Always
the spell of bad weather—there was a succession of cold and wet summers
for almost the whole of the last decade of the century—with lack of
exercise and constant sedentary work prevented any permanent cure. But
he seems to have grown accustomed to the irritation and until the summer
of 1697, if no better, was, 'considering my infirmities and craziness ad-
monishing me of the near approach of death', [5] able to keep his pain under
control and to do a full life's work. In the winter of 1694–5 he had a re-
currence of the 'long-continuing diarrhoea', which though he 'could stop
it at any time for a while by the taking of a liquid laudanum, after a few
days would return again'. But this he eventually cured 'by a very easy
and pleasant medicine, Naples biscuit boiled in milk'. [6]

1 The resin of the Spruce (*Abies excelsa*), so called because obtained from Neufchâtel.
2 *Corr.* p. 268.   3 *Corr.* p. 279.
4 *Corr.* p. 294: diapalma was an ointment composed of palm-oil, litharge and
sulphate of zinc; basilicon another ointment apparently prepared from Sweet Basil
(*Ocymum basilicum*).
5 *Corr.* p. 296: he had had a bad cough and fever a few weeks earlier: cf. *Corr.*
p. 295.
6 So to Lhwyd, April 1695: *F.C.* p. 256: Evelyn, *Acetaria*, p. 53, speaks of 'Cakes
cut long-wise in shape of Naples biscuits'.

During this period two of his last botanical tasks were discharged, the cataloguing of European plants, and the controversy with Augustus Quirinus Rivinus (A. Q. Bachmann) and Tournefort over classification.

The revision of the Catalogue of foreign plants appended to his *Observationes* was apparently urged upon him by Robinson in the autumn of 1692.[1] Robinson, who had probably used Ray's list on his own travels in Provence and Italy and certainly realised the value of such small hand-books, may well have urged that to compile from the *Historia Plantarum* a companion volume to his *Synopsis Britannicarum* dealing with European plants would be a great service to botany and a small task for its author. Ray worked at it while finishing the *Synopsis Quadrupedum* during the winter of 1692-3 and in April gave an account of it to Sloane.[2] To the original Catalogue

of such plants not native of England as I myself found growing spontaneously in my travels beyond the sea, I have added what escaped my notice out of Magnol's Catalogue and Appendix of Montpellier plants; M. Hoffmann's Catalogue of Altdorfine; G. Bauhin's of Basilian; Commelin's Holland Catalogue; P. Boccone's Sicilian plants; moreover all Clusius' Pannonic plants; and all the Helvetic and Rhaetic besides, mentioned and described in G. Bauhin's *Prodromus*. These are all put in one alphabetic catalogue wherein I have still room left to receive what you shall please to contribute or procure from Mr Sherard: for this book will not be printed before September next, my bookseller having his hands so full that he cannot attend to it.

The *Sylloge* received the imprimatur of Robert Southwell, President of the Royal Society, on 27 March 1694, but apparently was not published until the end of the year.[3] It was dedicated to Edward Bullock of Faulkbourne Hall, whose father, Ray's friend, had lately died, a young man who seems to have acknowledged the compliment and the good wishes of the inscription by neglect of the naturalist and his family.[4] In its Preface he recalled his three years' travels and the Catalogue published in 1673, described how when this was sold out he planned a revised edition on the lines of his letter to Sloane, and added that his 'taskmaster and very dear friend Dr Tancred Robinson' had urged him to enlarge it still further both by including all the species recorded in De l'Ecluse's *History of rare Plants*, Gaspard Bauhin's *Prodromus*, and Colonna's *Ecphrasis*, and also by attaching the separate lists of Spanish, Portuguese, Greek, Cretan, Baldensian, Alpine, Pyrenean, Roman and Parisian rarities: so that in fact he had listed all except Indian and American. What is the use of it? To help students to know what plants to expect; to identify their finds by the help of synonyms and localities and to detect novelties; and to remind experienced botanists of species that they have forgotten. The lists of regional plants will serve a similar purpose.

1 *F.C.* p. 297.   2 *Corr.* p. 263.
3 Apparently Ray sent a copy to Robinson on 1 Jan. 1695: *F.C.* p. 300.
4 Cf. *F.C.* pp. 182, 274 and above p. 204.

The book itself is divided into four main sections; first, an alphabetical Catalogue, similar to and often verbally identical with that of 1673 but with the details revised and enlarged and the medical uses removed, and with twice as many species included; those which he had not seen are marked with a letter to indicate the source of the record: secondly, a series of lists of plants seen at all the places visited by him in Italy, Sicily, Malta, Switzerland and Provence, beginning with the journey to Genoa on 18 March 1664;[1] these are copied exactly from the lists inserted in the text of *Observations*: thirdly, a series of Catalogues, Sicilian from Paolo Boccone, Epist. 21;[2] Swiss from Jean Jacques Wagner;[3] of the Campagna from Gian Giacomo Roggeri of Rome;[4] of the Lido at Venice from Antonio Donati;[5] Parisian from Jacques Cornut, *Enchiridion*;[6] of Monte Baldo from Giovanni Pona,[7] the Veronese pharmacist; Spanish and Portuguese from De l'Ecluse and others (this is a longer list); of the Pyrenees from Tournefort's *Catalogue of the Royal Garden in Paris*;[8] European plants extracted from the *Viridarium Lusitanicum* of Gabriel Grisley;[9] and finally, a short list from William Sherard, who being in Ireland had sent it too late for insertion in the general Catalogue: fourthly, a reprint of the three lists, Greek, Egyptian and Cretan, appended to the edition of Rauwolf.[10] As will be seen, the book is a compilation, a hand-list of European plants for the English traveller.

For the history of botany and of Ray's work the importance of the book lies in the subject to which he devoted the greater part of the Preface. Augustus Quirinus Rivinus, whose native name was Bachmann,[11] appointed Professor of Physiology in the University of Leipsic in 1691, had then just published his *Introductio generalis ad Rem Herbariam* and had sent a copy of it to Ray. This gave opportunity for a discussion of classification.

I emphatically commend Rivinus' work for its careful notes on floral differences, for its ingenious analysis of them as a basis for classification, for its

1 Recorded in *Obs.* p. 250.
2 Included in his *Museo di Piante rare della Sicilia, Malta* etc., Venice, 1697.
3 *Historia Naturalis Helvetiae curiosa*, Zürich, 1680.
4 *Catalogo della Piante del Suolo Romano*, Rome, 1677.
5 *Trattato de Semplici nel lito di Venezia*, Venice, 1631.
6 *Enchiridion botanicum Parisiense*, Paris, 1635.
7 *Plantae quae in Baldo monte et in via a Verona etc.*, Venice, 1595.
8 Known to Ray in *Schola Botanica*, i.e. Sherard's edition of Tournefort's list.
9 There is a pleasant note at the end of this list: 'If the author of this Viridarium had not professed himself a Chemist [Chymiatrus] his inflated style betrays the typical Chemist's quality which one rarely finds untouched by a trace of madness; either because half-baked persons take to Chemistry or because the fumes of the metals, minerals and other muck [res sordidae] which they handle derange their wits.'
10 This section though bound up and included in the contents at the end of the volume is paged 1–45, whereas the first Catalogue begins on p. 45.
11 He was third son of Andreas Bachmann, who translated this name as Rivinus.

brevity and clarity, its purity of Latin, its elegant plates. But the learned author will forgive me if, as is the way of mankind, I am too partial to my own system and think it more congruous to the nature of plant-life than his. When I say 'mine', I mean the general classification of plants into those whose seed-plants have two characteristically distinct leaves and those which have no such seed-leaves: and of the former a threefold grouping, those whose flowers have stamens but no petals; those with petals but no seed-vessel except the perianth; those with petals and a distinct seed-vessel.[1]

The details of Bachmann's system being wholly derived from the shape of the flower need not concern us. He was interested solely in nomen-clature, not in structure or relationships, and his system was simply a con-venient means to identification.[2] Ray, who had been unable to accept Cesalpino's insistence upon the sole significance of the seed, was not likely to be impressed by a scheme which was equally narrow and was based upon a much less reliable character. But Bachmann had evidently asked for Ray's comments upon his proposals; and the request provided an opportunity for surveying and explaining the principles of 'Method'. In doing so he divides his answer into four sections, two dealing with his objections to Rivinus's system, the third discussing and rejecting the classification by seeds and seed-vessels alone, and the fourth expounding several points in his own method.

First he treats of the division between trees and herbs. Bachmann, following logically his conviction that flowers alone are decisive, rejected this division: 'I have no scruple', he wrote, 'in bringing trees, shrubs and herbs into the same group provided they are united by the natural cri-terion.' Ray admits that he once rejected the division before he had discovered sufficient grounds for retaining it in the fact that trees produce buds while herbs can hardly be said to do so. 'In a sense all trees can be called annual plants since each year they are invested in a new ligneous covering.' A yearly growth of new wood and of buds is the characteristic of a tree.

If you urge that flowers and fruit are alone significant and that if these are ignored Elder and Dane-wort, Tree Mallow and herbaceous, Shrubby Potentil and Cinqfoil are wrenched apart, I reply that I admit the importance of flowers and fruit for determining subordinate groups, but the primary divisions must be derived from the whole habit and constitution or from one outstanding peculiarity like a woody substance: I do not think it absurd or inconvenient if in each main class there are found allied groups—Tree Mallow under trees, Herbaceous under herbs. Even in subordinate groups flowers and fruit are not always decisive: thus Anemone nemorosa, though its fruit agrees with Ranun-culus and it is placed there by C. Bauhin and others, is to be retained with Anemone.

1 Preface, p. 10 (unnumbered). This threefold division is not clearly marked in his own previous book.
2 He supplied the basis for Linnaeus, Critica botanica.

Secondly, he points out the difficulties of Bachmann's system. Thus the Bulbosi are plainly a homogeneous group: yet Bachmann separates the six-petalled Tulips from the monopetalous Narcissus, and from the monopetalous but irregular Gladiolus. So too he detaches the Viper's Bugloss from the Borages, the Tormentil from the Cinqfoils, the four-petalled Alsines from the five-petalled, the Gentianella from five-petalled Gentians, and the Skull-cap, because its calyx is operculate, from all other Labiates.

Thirdly, he admits that in spite of these defects if he had to choose he would prefer Bachmann's system to that which divides all Enangiosperms (that is, plants with seed-vessels) according to the number of divisions in each seed-vessel—Unicapsulares, Bicapsulares, etc. This, as he demonstrates, is just as full of anomalies; it separates the five-capsuled Rose of Sharon (*Hypericum calycinum*) from the other St John's-worts, and the Horned Poppies from other Papaveraceae, etc.; so too the Leguminosae, an obviously natural group, are split up on this system quite arbitrarily; as also are the Crucifers.

Finally, he defends the broad principles indicated in his own *Methodus*. The division into monocotyledons and dicotyledons there indicated has been confirmed by further reflection. In particular the grouping of the Grasses next to the Bulbs agrees with the findings of famous botanists, De l'Obel and the Bauhins, although they had not perceived the characteristic point of the resemblance. The Cyclamen, Peony and Asparagus, which seem to be monocotyledonous, differ from monocotyledons in other respects so widely that they must be grouped as Anomalous. The tuberous Irises should be kept with the bulbous on account of the obvious resemblance in flowers and fruit. The three groups, Apetala, Petalodea nudo semine and Petalodea vasculifera, are obviously natural. 'I call a flower petalled when it falls or withers before the seed ripens: thus the Plantains about which I hesitated I regard as petalled: the Bistort (*Polygonum bistorta*) and others of its kind although the flowers simulate the colour of petals belong to the Apetala.' He then states five reasons for his belief that the stamens are of vital significance, 'an essential part of the flower as containing the pollen analogous to the male seed of animals, endowed with life-giving power and serving to fertilise the seed'. Then after admitting that no Method can be without anomalies he urges that the ancient distinction between Gymnosperms and Enangiosperms is valid and defines what he means by 'naked seeds'.

He concludes by saying that 'if God prolong my life and give me vigour and health, I shall have more to say about Method in the future' and by an apology for imputing error to G. Bauhin: 'I was hallucinated by Robert Morison whom I followed too carelessly.'

The intention thus announced and the clear indications here given that his mind had been moving towards ideas which had been barely hinted in his earlier *Methodus* involved him in almost the only controversy of his career. But work on the subject did not begin till the next year. In the summer of 1694, while the *Sylloge* was at the printer's, he was very busy with insects, and in the autumn he wrote a Tract on Respiration[1] which he proposed to prefix to his *Synopsis Avium*.[2] This was in part a criticism of ideas put forward by Lister in his book on Molluscs in 1685; but, as we gather from hints in his letters, contained an attempt to deal with the physiology of oxygenation and the blood-stream, on the lines laid down by Richard Lower in his *Tractatus de Corde* in 1669 and by John Mayow.[3] The essay was sent to Robinson in December and emended in January; for Ray was anxious lest it should offend Lister. As Smith did nothing with the *Synopsis* the Tract was further submitted to Sloane in 1696[4] and by him to Bernard Connor (or O'Connor), whom William Vernon had commended and whose 'medico-physical dissertations published at Oxford'[5] Ray had seen and approved. Connor's criticism gives us an indication of the contents of the Tract and is in itself interesting as an example of physiological thought at the time.[6] But his letter may well have helped to decide Ray against publication; for in October 1697 he told Robinson that he intended to suppress it, giving as his reason that it would offend Lister, who had apparently never forgotten the matter of the gossamer-spiders[7] and who though not backward in criticising others hotly resented any criticism of himself. The story is typical of Ray's unwillingness to enter upon any sort of controversy: but we know so little of his ideas of physiology that the disappearance of the Tract is a real loss.

During 1695 he was also very busy reading and commenting upon Sloane's History of Jamaican plants. This was arranged on the plan of his *Methodus*, and parcels containing the various tribes were constantly sent to and fro between London and Black Notley. Ray comments upon the style, which he finds sometimes obscure—'I must impute it to my own dulness if I do not sometimes apprehend your meaning',[8] and the Latinity —'the language as far as I am able to judge is proper and good only some typographical errata...none of them except in the Greek words con-

1 So letter to Robinson, 28 Nov., *F.C.* p. 299: the subject was one in which he had doubtless been interested ever since his association with Needham, whose *De Formato Foetu*, ch. VI, deals with it.

2 This had been sent to Smith in February, but its publication was interminably delayed.

3 For an account of their work cf. M. Foster, *Lectures on History of Physiology*, pp. 181–99.

4 *Corr.* p. 308.

5 *Corr.* p. 302; *Dissertationes medico-physicae*, published 1695: he had been elected F.R.S. in Nov. of that year.

6 Printed in full, *Corr.* pp. 308–10: Connor was a very learned and able doctor.

7 See above, p. 139.   8 *Corr.* p. 289.

siderable':[1] but in the main expresses admiration for the thoroughness and accuracy of the work. The book, a preliminary hand-list, was published in 1696 and Ray received his copy of it early in June:[2] but the parcels continued to arrive for use in his own Supplement to the *History of Plants*, and for Sloane's larger work.

It was in May 1695[3] that he reported to Robinson that he had received Bachmann's Epistle[4] in reply to his Preface and had sent off an answer. He proposed to publish them in the second edition of his *Synopsis Britannicarum*, then in the press:[5] but this was delayed owing to Doody's dilatoriness in sending his promised contribution[6] and was not published until 1696. The *Epistola de Methodo Plantarum Augusti Quirini Rivini ad Johannem Raium cum ejusdem Responsoria* was printed with a separate title-page and pagination but bound up with the *Synopsis*.

Bachmann begins with a courteous and generous acknowledgment of Ray's outstanding eminence as a botanist and of his complimentary references to his own book; and then, with suitable apologies, gives his reply. In dealing with the distinction between trees and herbs he argues that the traditional fourfold division has much to be said for it and that Ray ought either to accept or reject it; that the compromise is not sound since all plants produce buds, as Ray himself quoting Grew has recognised in his *History*; and the existence of a woody envelope is not a primary characteristic: he quotes a large number of plants, Thyme, Salvia, Lavender, Restharrow, Cistus, Spiraea, Ivy, etc., in which the division between tree and herb breaks down. The rest of his letter is a long but much less successful attempt to defend his own system—an attempt which amounts to little more than saying how inconvenient it is for a beginner to look for a four-petalled plant among the five-petalled.

Ray's reply begins with a similar courtesy. He fears that his criticism has given a wrong impression. He had no desire to wound, or to repay Bachmann's kindness by injury. Let each maintain his own opinion until he is convinced of his mistake: only let discussion be conducted without acidity or any savour of contempt. He then proceeds to defend his distinction of trees from herbs, though he now admits that all alike bear buds, by repeating that their woody substance and buds on the branches sufficiently define them; and to deal once more with the specific examples by which he has criticised Bachmann. This part of his reply is not important.

1 *Corr.* p. 286.

2 *Corr.* p. 295: *Catalogus Plantarum in Insula Jamaica etc.* He reviewed it in *Phil. Trans.* XIX, no. 221, pp. 293–6.

3 *F.C.* p. 300.

4 It is dated 5 Nov. 1694: no doubt he received Ray's comments and perhaps the *Sylloge* before its publication.

5 So letter of 7 May to Aubrey: *F.C.* p. 183.

6 *F.C.* p. 261.

The latter part of it is concerned with a vindication of the principles on which his Method is based, and of his use of them in the *Historia Plantarum*. He first explains his procedure: well-known groups like Umbelliferae, Verticillatae and Papilionaceae he retained, though he tried to define their characteristics more clearly: such characteristics should be plain and obvious, since the purpose of a Method is to assist identification: flower and fruit supply the most appropriate marks of distinction, but in some cases other parts serve more suitably and can be used either alone or in association with flowers. Then he expounds the various classes in his *History*, giving short paragraphs to each main section. He touches on many points of novelty and interest: thus of the Fungi and Algae he says that he now has good evidence, partly first-hand, that they are seed-bearing, a fact only suspected in 1685; of the Ferns he outlines a new scheme necessitated by Sloane's and other tropical specimens, and based on the position of the seed either on special fronds as in *Osmunda*, or on the back of ordinary fronds and either on the margin or on the ribs or in circular spots; of the Apetala he expounds his previous definition; of the later groups he expounds the relationship and deals with points on which he has been criticised.

Finally, he returns to Bachmann's scheme. To show the force of his two objections that it is too meticulous and thus obscures the distinction of kind, he compares its treatment of the Labiates with Tournefort's, and argues that where two schemes both based upon flower-structure differ so widely their basis cannot be adequate. So with a wealth of good wishes —which nevertheless ring true—he closes his letter.

The delay in the publication of the *Synopsis* gave him the opportunity to mention the new matter raised in his last section and to add a postscript to his *Responsoria*. In the summer of 1695 there had been sent to him a copy of Joseph Pitton de Tournefort's *Élémens de Botanique*, written in French and published in the previous year at Paris. This, as he confessed to Lhwyd, 'exercised him for some weeks'[1]—probably from its language rather than its contents. But as he mastered it—and in French he was 'but a smatterer'[2]—he discovered not only that he was himself hardly ever mentioned except for condemnation, but that a Method was being propounded to which he could not give assent. Writing at a later date to Sloane[3] he confessed that if it had not been for Tournefort's incivility he would have taken no public notice of his Method although he thought it 'faulty and liable to many exceptions'. As it was he had time to write a note on it for insertion in his *Synopsis*; and resolved that he would deal more fully with the matter later. Bachmann, though a man of good ability, was not one whose views concerned Ray closely. Tournefort was the great professor in Paris, the head of the Royal Garden, the teacher of

1 *F.C.* p. 261. 2 Cf. *Corr.* p. 123. 3 In April 1698: *Corr.* p. 339.

Sloane and Robinson; and Tournefort had in his book constantly criticised Ray for including too many and often superfluous characteristics in his classification. Like Bachmann he was advocating the use of a single feature of the plant, a proposal suitable enough for an elementary hand-book but wholly incapable of providing a scientific taxonomy. Unless a protest were made, systematic botany might receive a fatal set-back.

In the Postscript[1] he defends his action in selecting a number of generic marks on four grounds:

1. Although the perfect definition would give us the particular group and the essential individual differentiation, we seldom know the essential but have to rely on a number of accidentals: we must collect as many of these as we can.

2. Since the notes reported by him apply to all the known species of each group they may be accepted as characteristic pending further evidence.

3. Other notes, perhaps unnecessary, have been added to help identification and serve as a description.

4. These notes are all easily visible, many of them during the whole life of the plant; they therefore help identification even when the plant is not in flower.

He then comments briefly on six points on which Tournefort has criticised him but declines to answer fully. Finally, he notes certain points in Tournefort's work, its accurate observation and delineation of the form and structure of flowers and seed-vessels; its resemblance to Bachmann in Method, and its consequent separation of cognate species—of this he mentions examples; and finally its peculiarity in classifying Veronica as a 'regular' flower, while Bachmann, surely correctly, holds it to be 'irregular'.

This reply was admittedly temporary and incomplete. He had in fact begun to plan an essay on the subject as early as February 1696, when he told Robinson[2] that he would deal with three points: the origin and development of 'method'; his own principles and system; his answer to Tournefort's criticism, and his criticism of Tournefort's 'method'. He seems to have worked at the book during the spring and had perhaps finished it by June. It was published before the end of the year, a tract of 48 pages.

In the Preface, after explaining the scope and circumstances of the book, he declares that though nature often affects an air of 'method', yet she constantly produces species wholly peculiar and species of dubious affinity; that therefore there cannot be a general and complete system; and that he sees no reason why any characteristic, if it is clear and gives results free from anomaly, should not be used as distinctive, though no doubt

1 *Epist. et Respons.* pp. 52–5.   2 *F.C.* p. 300.

generally the flower is the most convenient. 'Universals are always figments of the human mind: nature always particularises.... Species are essentially distinct and immutable; for a seed is nothing but a small plant cognate with its mother.' Nevertheless, if Methods are all unsatisfactory, yet Method is essential to the study of botany: 'A young man with its help will learn more in two years than in ten years without it'—so Petrus Hotton of the University of Leyden said in his inaugural lecture. 'I should like a method which would unite all such groups as are naturally cognate and have many points in common and would not disjoin any kindred species, and which would reduce to classes and groups the species that are not really akin but can be conveniently classified by the shape of the flower or the divisions of the fruit.' Such a method cannot be derived from the flowers alone, as both Tournefort and Bachmann have shown.

The Dissertation itself begins with a section on the Origin and Development of Method.

Method like language aims at grouping a number of individuals under a common and appropriate category. Man when he began to speak distinguished between Animate and Inanimate, between Animal and Vegetable, between Trees and Plants: and this last distinction seems sound and scientific; for Trees are woody, long-lived, bud-bearing and Plants were more difficult to classify: locality, use, structure of the principal parts have all been tried: structure of root, stalk, leaves, flowers, fruit is the best and yields a large number of natural groups. Some of the most learned botanists, Cesalpino, Gesner, Colonna, have fastened upon flowers and fruit only; and with them I partially agree.

The next section on the Marks characteristic of Groups is an exposition of this partial agreement. In effect he shows by a number of examples that reliance upon flowers and fruit alone is unsatisfactory and sometimes impossible. Sometimes indeed one peculiar characteristic can be found common to a whole group: where this is not the case, as many attributes as possible common to the group should be taken. The familiar and easily recognised groups do not include all plants: there remain certain anomalous and unclassifiable—some which need a group to themselves.

Turning to his own Method he describes the fourfold division of Herbs —those without flowers, the apetalous, and the two others—and the difficulty of discriminating between apetalous and petalled, and between naked seeds and capsules, in almost the same words as he had used in his earlier writings.[1]

This is followed by a long list of the points which Tournefort had criticised as superfluous in Ray's Method and descriptions. He gives the references to the *Elementa Botanica*, quotes and answers its objections in detail: as there are about a hundred of them the section is somewhat long.

---

1 Cf. *Diss.* p. 17 top to p. 18 bottom and *Resp.* pp. 41–2, verbally identical; *Diss.* pp. 21, l. 12–22, l. 4 and *Sylloge*, Preface, pp. 12–13, very nearly so.

Finally, he deals with Tournefort's own Method, based like Bachmann's upon the structure of the flower or rather the number and symmetry of the petals;[1] shows that its author is frequently unable to follow it and when he does is often led into manifest absurdity, as when Caryophyllaceae and Umbelliferae are grouped together as 'rose-flowered', or when Rose, Poppy, Rue, Buttercup and Cistus or Arum, Aristolochia, Foxglove, Butterwort, Milkwort and Broomrape are forced into alliance. He concludes by promising an emended edition of his own system.

Tournefort's reply, in a letter addressed to Sherard and entitled 'De optima methodo instituenda in re herbaria', was published in 1697:[2] it merely reiterates the position of the *Élémens*. But in 1698, when he sent his colleague André Gundelsheimer to England to see Sloane's Jamaican plants, he paid a special visit to Ray at Black Notley to express Tournefort's apologies and admiration;[3] and when the Latin version, the *Institutiones Rei Herbariae*, appeared in 1700 it was evident that the criticism of Ray had been largely removed and the preface contained a glowing tribute to him.[4] Whatever the merits of the controversy it was conducted with good feeling on both sides.

Ray had apparently completed the *Methodus Emendata* before seeing Tournefort's letter: he had begun it early in 1696;[5] had nearly got it ready for the press in April 1697;[6] and began to negotiate for its publication in 1698.[7] Unfortunately Smith, who had not yet printed his *Synopsis Avium*, showed no desire to publish any more Latin hand-books, and was not very accommodating about the Supplement to the *Historia Plantarum*. In 1700, when it was still held up, Ray received Tournefort's Latin version of the *Élémens* and revised his script accordingly.[8] A year later it was still unprinted. Indeed it might have been delayed indefinitely but for the generosity of Petrus Hotton, the Dutch botanist, first mentioned in the *Dissertation*. Ray had been in correspondence with him since 1698[9] and was much impressed by him; so that when early in 1701 he offered to get the book printed in Holland Ray readily agreed.[10] The script was sent over

1 Tournefort was primarily interested in classification, not in the structure or natural relationship of plants. Thus he ignored the work of Malpighi and Ray on the seed-plant and of Grew on structure. His chief contribution to botany was in developing a binomial nomenclature.

2 He says that he only received Ray's *Dissertatio* in June 1697, l.c. p. 27.

3 *Corr.* p. 339: cf. Pulteney, *Sketches*, II, p. 81.

4 *Instit.* I, p. 53.   5 *F.C.* p. 301.

6 *Corr.* p. 318.   7 *F.C.* p. 303 and *Mem.* p. 50.   8 *Corr.* p. 372.

9 By Hotton's special request to Mrs Ray shortly after Ray's death none of his letters were handed over to Derham or printed. For an account of this cf. Dale's letter of 20 Aug. 1712 to Sloane (Sloane MSS. 4043, f. 82 and Gunawardena, l.c. pp. 146–7; and *Mem.* p. 50).

10 On 19 Feb. 1701 he had written: 'My Methodus Emendata lies by me finished... no bookseller will undertake it': *Corr.* p. 370. Hotton offered to print it on 26 Feb.

to Leyden by Petiver[1] and the printing was supervised by Hotton, 1100 copies being issued and the portrait of Ray in the *Sylloge* being re-engraved. Although the publishers were Jansson and Waasberg of Amsterdam, the title-page described the book as of London and published by Smith and Walford, a misstatement at which Ray is said to have protested vigorously.[2] It appeared in 1703; and on 20 January he sent two copies to Sloane, describing them as 'magni affectus exiguum effectum'.[3]

Not unnaturally it was dedicated to Hotton; without his help it might never have been published, and between him and Ray there had sprung up a close affection. The Preface gives a brief account of Ray's work on classification—the Tables and Morison's attack upon them, the *Methodus Nova* of 1682, the *Historia Plantarum*, and then the letter of Bachmann and the subsequent discussions: and of the systems of Bachmann, Tournefort and Hermann, this last being based upon the divisions of the seed-vessel. A further Statement of Preliminaries ('Praecognoscenda') sets out the gist of his arguments in the Responsoria and Dissertatio, and leads up to Six Rules for Classification. These are:

1. Names must be changed as little as possible: the multiplication of names produces mistakes and confusion.

2. Notes characteristic of the group must be clearly defined: those that depend upon comparison are to be avoided: if used, the difference between the groups compared must be large.

3. Characteristics must be obvious and easily observable, the use of a Method being principally to help beginners.

4. Groups approved by nearly all botanists should be preserved: they are in fact natural, and have many common attributes besides that from which they are named.

5. Care must be taken that cognate plants are not separated, nor strangers united; valuable as is Rivinus's method for teaching, it fails in this regard.

6. Characteristics must not be multiplied unnecessarily, and should not be more than is needed to determine the group.

The Method itself is not a second edition of its predecessor but a new book. It follows the sequence of the *Historia*, beginning with Sea-weeds and Cryptogams and uniting the Shrubs with the Trees. The main divisions are no longer 'imperfect' and 'perfect flowers', but 'plants without flowers' and 'flowering', the former including the submarine and subaquatic and the Fungi, Mosses and Ferns, the latter being divided into Dicotyledons and Monocotyledons—both words being now regularly used.[4] The tables in which the sections are worked out are similar in

1 On 15 April Ray sent it to Petiver for transmission: *Corr.* p. 393.
2 *Mem.* pp. 50–1.   3 *Corr.* p. 409.
4 *M.E.* p. 1, etc.

construction and often in content to those previously published, but instead of the casual notes of the *Methodus Nova* there is now a systematic list of what would to-day be called the genera included under each family: these are defined in a few clear lines describing one or two salient features. In these definitions he frequently acknowledges the work of Tournefort, and also makes frequent mention of Rivinus, Hermann, Abraham Munting,[1] Bobart, editor of Morison's *Historia*, and Banister, of Charles Plumier,[2] Johann Georg Volckamer[3] and of Charles Preston, with whom he had come into touch apparently in 1697,[4] and whom he described to Sloane as 'very knowing and curious in the method of plants'.[5] Preston, an M.D. of Edinburgh and later the successor of James Sutherland as professor at the Botanical Gardens there,[6] had apparently written and sent specimens to Ray before sending the long letter dated 13 January 1701,[7] which comments upon Tournefort's *Institutiones* (the Latin version of the *Élémens*) and sends a list of notes, which Ray used in some cases by direct quotation[8] in the *Methodus*. At the end of the Trees—that is, when the whole list has been finished—there is added a Special Classification of Grasses, Rushes and Sedges, which Ray compiled at Hotton's request and sent to him in October 1701.[9] There follow eight pages of additions and emendations; a glossary of terms, a very full index, and a list of synonyms for many of the plants mentioned.

The system thus propounded is a very real improvement. Though his unwillingness to reject traditional views made him slow to admit that his distinction between trees and herbs was not valid, he had now based all his other major divisions upon structural and structurally important considerations. While refusing to adopt the artificial simplifications of Bachmann and Tournefort he had learnt much from them, and on their evidence had cleared up many points on which he had previously been uncertain or mistaken. In consequence his treatment of the smaller divisions—genera as we should call them—is masterly; short descriptions fastening

1 'Last week Mr Smith sent me a large Dutch Herbal of A.M. [the 2nd edition, published Leyden, 1696]...it will be of little use to me, being written in Dutch.' Letter of April 1697: *Corr.* p. 316.

2 *Description des Plantes de l'Amérique*, Paris, 1695: Ray first heard of this from Robinson in 1693: cf. *Corr.* p. 270.

3 *Flora Noribergensis*, 1662.   4 So *F.C.* p. 302.

5 *Corr.* p. 345.

6 He contributed freely to *Phil. Trans.* on medical matters in 1696–7, and Ray speaks of him as 'Carolus Preston, M.D. Edinburgensis Scotus.' He and his brother George were successively in charge of the Edinburgh garden: cf. Pulteney, *Sketches*, II, p. 9.

7 *Corr.* pp. 380–8.

8 E.g. on *Herniaria*, *Corr.* p. 384, *M.E.* p. 24; on *Cakile maritima*, *Corr.* p. 385, *M.E.* p. 99.

9 *Corr.* p. 398.

upon two or three characteristics, a sequence which reveals the natural relationships, and a plain statement on points or species which cause difficulty.

As before, so still more now, he laments his inability to see and study the living plants imported into the gardens of London and Oxford or even to consult a botanical library: new plants have been discovered in vast numbers; dried specimens almost never afford the essential details; classification cannot be accurate under such conditions. Thus in the Ferns, where he is convinced that the character and position of the seeds are the true basis for systematic arrangement, the vast number of newly discovered species in which seed-structure has not been recorded and ought to be studied, cannot be arranged that way: so he falls back upon an older and more superficial method. Yet his constant study of plants and revision of his records; his appreciation of the importance of structural differences and flair for fastening upon them; and his modesty and readiness to learn from others; these give him a power to see the individual plant and the vegetable kingdom as a whole, an insight into its natural affinities, and a capacity, very rare in men of his age, for reviewing previous conclusions and reshaping cherished ideas. 'Creature of his time', says Gunawardena,[1] 'he has created the model on which our systems are based.... The foundations of plant taxonomy were laid in his *Methodus Emendata*.'

The work thus begun in 1696 was in fact virtually finished before the summer of 1697. It was in that year that his health became definitely worse and began to interfere seriously with his undertakings.[2]

On 7 July he wrote to Sloane:[3]

One thing I have to acquaint you with and to beg your advice. In the beginning of May last, if you remember, there was about a week of extraordinary hot weather which had such influence upon the sores of my legs (which were then almost wholly dried up and healed) that it altered the nature of them and turned them into a kind of herpes or tetter which hath spread very much and encompassed my legs; it was and is still attended with an extraordinary heat and itching. I have used by the advice of our physician at Braintree a decoction of litharge of his own preparing to bathe them...and have taken flowers of brimstone inwardly, and applied an unguent to the soles of my feet, which though they mitigate and put a little check to the spreading of the herpes yet do not, as they say, kill or cure it. At first it issued out a thin humour out of the small pimples, but now there is no visible humour but only a scurf upon the eruptions. But enough of this.

1 L.c. pp. 9, 10: he claims that in spite of the temporary triumph of artificiality with Linnaeus 'Ray's system formed the basis of the classification of De Jussieu'; this was reformed by De Candolle; and was then enlarged and improved by Bentham and Hooker in their *Genera Plantarum*: cf. note at end of chapter.

2 'Mr Ray comes seldom among us', wrote Thomas Kirk from London in 1699 to Thoresby: *Letters to R. Thoresby*, p. 80.

3 *Corr.* pp. 324–5.

Five days later, deeply touched by Sloane's speedy reply, he described his medical treatment in more detail:[1]

I take inwardly flowers of sulphur, half a drachm at once, which keeps my body soluble....Outwardly I use a decoction of elecampane, dock-root and chalk in whey, twice a day bathing the affected places therewith....Mercury I dare not be bold with....I am now come to a suspicion that these tumours are owing to insects making their burrows under the cuticula, their juice mixing with the serum of the blood causes an ebullition, and excites the tumours, pustules, inflammation and itching. But this I propose only as a conjecture.

Then having discussed bleeding, which Sloane had suggested, he explains his 'reverie in relation to sulphur' that 'taken into the blood it may be heated to that degree as to emit a fume sufficient to kill the insects lodged in the tumours'.

The medicines proved useless, and for the rest of his life the sores seem never to have been healed. In the winter his digestion also gave him trouble; 'it is a distemper that usually attends me in very cold weather, proceeding I guess from the relaxation of the tonus of the bowels'.[2] Unfortunately Sloane's prescription though it checked his diarrhoea also 'rendered his sores very painful in the nights and causes them rather to spread'.[3]

The letter which reported this told also of a graver anxiety.

I have now another case to beg your advice in. My daughter Mary, one of the twins, after a long trouble with the chlorosis, is fallen into the jaundice, all the symptoms whereof she hath in a high manner. We have made use of our neighbouring physician Mr Allen who first gave her some powders which taking no effect he gave her, I suppose, Riverius, his first medicine for the jaundice, which she hath now taken five days, half a quarter of a pint thrice a day; which notwithstanding all the symptoms continue or rather increase, and she grows faint and feeble. Now, sir, myself and wife (who tenders her very humble service to you) earnestly entreat your counsel and direction how we are to order her and what remedies you think most proper and effectual for her....Be pleased to write a word or two in answer by the next post, for we are very much concerned for the child.

So he wrote on 22 January. On 1 February[4] a brief note announced:

My dear child, for whom I begged your advice, within a day after it was received became delirious and at the end of three days died apoplectic, which was to myself and wife a most sore blow. I doubt not but you will commiserate our sad condition. Nothing afflicts me so much as that I did not in time make use of that remedy which I had proved so effectual to my own relief and cure in the same disease....I am not in case to write much.

1 *Corr.* p. 326.	2 *Corr.* p. 333.
3 *Corr.* p. 311.	4 *Corr.* p. 312.

The blow was very heavy. On 2 March[1] he asked Sloane for further sections of his Jamaican work:

For I intend to apply myself wholly to it being desirous to get the work off my hands. For upon this sad accident and by reason of my growing infirmities, I am well mortified as to natural studies and inquiries, though I shall not, so long as life and strength last, wholly desert them, but make them some part of my parergon and diversion as I should only have done before.... My wife is full of grief having not yet been able fully to concoct her passion.

On 12 March[2] he broke through his hitherto invariable reticence about personal matters in writing to Lhwyd, and after describing his own ailments said:

A sadder calamity than these hath lately befallen me. You may possibly have heard, though I do not remember I ever told you, that I had four daughters, the two eldest of which were twins under the age of fourteen years. Of these the younger was cut off by a disease that is rarely mortal, the yellow jaundice, for want of using convenient medicines, we putting too much confidence in a young physician who thought himself to have an infallible cure. This makes the loss more grievous; though the child was very ingenious, and helpful to me, and upon that account the more dear too, which considerations added to that natural στοργή in parents, render this death a sore blow to me.

We have quoted these letters at length because here for the only time in all the mass of his correspondence he breaks through his reserve. The poignancy of his grief stands out from the awkwardly stilted phrases, as it does from his harsh judgment of Allen,[3] and his regret at not having tried his own remedy. Study of the *Historia Insectorum* and the references in its notes to his 'filiola Maria' reveals her helpfulness and his affection. His desire to concentrate on his work and his doubt as to that work's value disclose his pain. But most significant is the lament that he had not used the concoction with which twenty-seven years before he had dosed himself.[4] The scientist in him would criticise Allen and appeal to Sloane, trusting to the expert and his latest therapeutics: but the man was the herb-woman's son of Black Notley, brought up in an atmosphere of folklore and white magic and living in an age of transition. In the shock of his child's illness he returns to the primitive, to the very traditions against which his thought and work were a protest. The remedy which he would have used was 'an infusion of stone-horse dung steeped in ale for a night with a little saffron added and in the morning strained and the liquor sweetened with a little sugar... about half a pint at a time'. It is against this horrid background thus displayed that we can best judge the greatness of Ray's work.

1 *Corr.* p. 313.   2 *F.C.* pp. 267–8.
3 Allen's own story of the case is in his Commonplace book, p. 203. He regarded green sickness rather than jaundice as the cause of death and treated it with 'yellow electuary with chalybeate': Miller Christy in *Essex Naturalist*, XVII, p. 10.
4 *Corr.* p. 86: cf. above p. 153; and for the same drug *Diary of A. De la Pryme*, p. 38.

During the summer his anxieties were continued. His wife broke down and for some weeks was very ill; and Margaret, the elder twin, was on 1 June 'inclining to the jaundice'.[1] On 28 June he wrote to Sloane that she had been very ill and was not yet out of danger: 'we have plied her with chalybeate medicines judging her disease to be complicated of the jaundice and chlorosis'.[2] In spite of swellings of the feet and legs, 'especially in the afternoon and at night', she made a complete recovery before the autumn. But his own sores were 'spreading and increasing and growing very deep and running extremely, being also so painful that they do very much hinder my rest; sometimes the heat and itching is so violent that they force me to quit my bed'[3]...'round every sore there are small red tumusculi, flat and bigger near the sore which I conjecture to be the nests or swarms of those insects'[4]....'I intend to have the issues cut which I hope will deliver me from this misery, else *vita minime vitalis esset*'.[5] So it went on: perpetual irritation from his ulcers and recurrent attacks of diarrhoea giving him no rest nor opportunity to gain strength. In November 1699 he sent a further appeal to Sloane:[6] 'I am in sad pain and have little heart to write or to do anything else: the days are so short that the forenoon is almost wholly spent in dressing my sores which are now more troublesome than ever, notwithstanding I have used fomentations, traumatic drinks, mercurial purges etc., I pray you what think you of vomiting with Turbith mineral [i.e. basic sulphate of mercury] which is recommended to me as a medicine which will certainly give ease.'

Nevertheless, his preparation of a Supplement to his *Historia Plantarum*, and in the summers his collection of insects, went forward steadily. In addition, in September 1700 he wrote to Sloane:[7] 'I have a small present for you, a little tractate of about half a dozen sheets of paper which I drew up at the request of a friend last winter, entitled a *Persuasive to a Holy Life*. It hath been finished a pretty while and I wonder Mr Smith is so slow in putting it to sale.'

The *Persuasive* is the outcome partly of the change in the temper of the Church, which as we have seen led Ray to publish his *Wisdom of God* and *Discourses*, and partly of the shock of his daughter's death and his own ill-health. He had never felt quite at ease in conscience that being an ordained minister of religion he had devoted his life to science; and now that he felt his days numbered the desire to discharge the obligation of his ministry evidently weighed upon him. It only needed the suggestion of a friend to encourage him to this further authorship.

Edmund Elys, whom he names in the margin as the friend in question, is a man of whose connection with him we have no knowledge except a

---

1 *Corr.* p. 341.   2 *Corr.* p. 342.   3 *Corr.* p. 343.
4 *Corr.* p. 344.   5 *Corr.* p. 347.   6 *Corr.* p. 369.
7 *Corr.* p. 376: he sent a copy to Robinson at the same time: *F.C.* p. 304.

single sentence in the epitome of a letter of March 1692 to Robinson.[1] We know so little about Ray's private life, indeed our whole knowledge of him except as a scientist is so fragmentary, that this need not surprise us. But it is probably explained by Elys's own character. This and his contact with Ray deserve discussion.

At first sight it is tempting to assume a long-standing acquaintance, indeed to see in Elys a link between Ray and his Cambridge contemporaries. For Elys, though an Oxford man (a fellow of Balliol from 1655 to 1659), became an enthusiastic disciple of Henry More, submitting to him his first thesis, sharing his taste for religious and mystical verse,[2] and exchanging with him a number of letters during the years 1667 to 1684.[3] If these letters contained any reference to Ray they might throw light on his connection with the Cambridge Platonists and especially on the origin of his *Wisdom of God*. But in this volume, and so far as can be discovered in Elys's other publications, there is no mention of him or his books.

Rather it is clear that Elys was a man of a familiar kind, studious, devout, widely read, especially in devotional and doctrinal literature, and deeply concerned with the discussion and development of religion, meticulous in his scholarship, but muddle-headed in his general outlook. Having been in trouble in his benefice and eventually deprived of it as a non-juror, and finding little outlet for his scholarly interests, he entered into correspondence with any men whose writing attracted or annoyed him. Along with the letters of More are printed several from Thomas Pierce, President of Magdalen College and Dean of Salisbury,[4] and one from Robert Sharrock,[5] replying (in the latter case with ill-concealed amusement) to Elys's eulogies; and in addition letters from him to 'the author of an Enquiry into the Constitution, Discipline, Unity and Worship of the Primitive Church' expostulating with him over certain of his contentions; to the author of a pamphlet on the Doctrine of the Trinity—very violent; to William Sherlock, Dean of St Paul's,[6] protesting against a tritheistic phrase in his 'Vindication', and to Richard Bentley, similarly protesting

1 *F.C.* p. 296. 'Lr of Mr Elys about Dr Sherlock's *Prin.*' and 'Dr S's notion of *Prin*'—which is surely a misreading for Person: cf. Elys, *Letters*, pp. 47–8.

2 The *D.N.B.* describes his poems as 'a series of tiresome conceits strung together in execrable rhythm'—which is too severe.

3 More's letters together with other papers to and from Elys were published in London in 1694 under title *Letters on Several Subjects*: Elys's letters to More, covering the years 1668–77 and 1685, are preserved in the Library of Christ's College.

4 Described in *D.N.B.* as 'controversialist', a man of violent speech and somewhat unsavoury reputation.

5 Of New College and Archdeacon of Winchester, a man of some learning and with botanical interests.

6 Sherlock had been a non-juror who recanted and succeeded Tillotson at St Paul's in 1691: his 'Vindication' of the Trinitarian theology against the Socinians gave rise to the 'Battle Royal': cf. Robert South's *Animadversions* and *Tritheism charged*.

against a sentence in one of his Boyle Lectures. All these are the work of a conscientious and earnest man, even if, then as now, the recipients of such letters might regard them as an unwarrantable nuisance.

It is reasonable to suppose that his only contact with Ray was through reading the *Wisdom of God*; that its references to More gave him occasion to write, and that in doing so he urged Ray to develop his thoughts on their religious side. To such invitations Ray's self-depreciation made him pay attention out of proportion to their real value.

The *Persuasive* is a modest little volume of 133 pages to which, as he explains in the Preface, 'finding myself at leisure from other business and considering that it was suitable to my profession and present condition as being conducive to my preparation for that change which the pains and infirmities I laboured under seemed to threaten...I was easily induced to employ those intervals I had of ease and remission of pain'. He adds: 'I do not pretend to anything new, or not delivered by others.... To do every man right I must acknowledge myself to have borrowed a good part of my matter out of the Right Reverend Father in God Dr John Wilkins late Lord Bishop of Chester his Treatise of Natural Religion.[1]...This Tractate may possibly fall into the hands of some who never saw or would else have seen that; and recommend to them the reading of the whole.'

From Wilkins he quotes repeatedly; indeed both arrangement and contents are largely drawn from the second part of his work. Hammond's *Paraphrase*[2] and Owtram's *De Sacrificiis*,[3] both already quoted in his *Discourses*, Tillotson and Bishop Lloyd[4] are also quoted. So are Simon Patrick, his exact contemporary at Cambridge,[5] now Bishop of Ely,[6] Cockburn[7] quoted in the third edition of the *Wisdom of God*, and Riverius[8] the medical writer, whose treatment Allen had followed for his daughters. The thesis, that honesty is the best policy, is not one that lends itself to attractive treatment; and when it appears in the form that religion is a sound investment it easily becomes nauseous. In Wilkins's case the praise of religion as the source of prosperity is almost indecent—for even if he was not a Vicar of Bray he had certainly made the best of both worlds. Ray's treatment of the theme is at least free from this reproach: religion had brought him no preferment, nothing indeed except worldly loss: his book gains greatly by his ability to separate happiness from money or

1 A posthumous work published in London in 1675 with a preface by Tillotson and a funeral sermon by William Lloyd.

2 *Pers.* pp. 8, 9.   3 *Pers.* p. 3.   4 *Pers.* p. 74.

5 Cf. *F.C.* p. 29, where mentioning to Courthope Patrick's election at Queens' College in 1662 Ray calls him 'a deserving person and one that wants nothing but years to qualify him for such preferments'.

6 *Pers.* p. 104, quoting from *The Witnesses to Christianity*, pt. II.

7 In the Addenda printed after the Preface.

8 *Pers.* pp. 23–6, 35–6, quoting his *Institutiones* published in 1655.

prestige, and to see the effects of religion rather in health, contentment, peace and good work than in prosperity, acquisitiveness and security. But even so the theme is not one on which he has anything of much importance to say; and its style is not sufficiently distinguished to cover the dullness of its contents.

In the letter of 30 April 1701 in which Ray apologised to Lhwyd, who had been in Ireland, for not having sent him a copy of the *Persuasive* he also describes his own activities. He was then hoping, as he had been for some time, that the Supplement of his *Historia Plantarum* was being printed. He had been working at it with energy even after his renewed illness in 1697; in 1698 he had promised to let Smith have the copy on the same terms as those which Faithorne had offered for the previous volumes, £30 in money and twenty copies of the book; and to this the publishers agreed in March 1699.[1] Negotiations seem to have dragged on all through the years 1698–1700.[2] In 1701 proposals for printing by subscription were advertised by Smith and Walford;[3] but Robinson wrote to Lister that he doubted whether 200 subscribers could be procured.[4] In 1702 Compton took up the matter and proposed to get a large contribution from the queen both for publishing and for producing copper-plate pictures for the whole three volumes.[5] Robinson's view was that the expense would be much larger than was supposed and that artists and supervisors could not be obtained; and to this view Ray reluctantly agreed.[6] But the proposal seems to have encouraged Smith to start printing in earnest. Even then there were delays—the casting of a new fount of type,[7] the preparation of the appendix, the final collection of delayed material,[8] and perhaps the death of Smith, which seems to have been in 1703:[9] it was not until 8 June 1704 that he could tell Sloane 'the third volume of my History of Plants is now finished at the press'.[10] It was published at the cost of '5/- in hand and 15/- more on delivery'.[11]

The letter to Lhwyd gives a brief résumé of what the Preface states more fully, the sources of the book. He first gives a list of the many volumes that have been published since his own *Historia Plantarum*—six more volumes of the *Hortus Malabaricus*, Breyne's second *Prodromus*, Hermann's *Flora* and *Paradisus*,[12] Tournefort's *Schola*,[13] *Élémens* and *Institu-*

---

1 Cf. *F.C.* p. 303 and *Corr.* p. 365.

2 Cf. epitomes to Robinson, 31 May, 11 Nov., 17 Feb., 14 March, 18 April, 31 May, 16 Aug., etc.: *F.C.* pp. 303–4.

3 *F.C.* p. 305.   4 *F.C.* p. 307.   5 Cf. *Corr.* p. 406.

6 *Corr.* p. 410.   7 So to Lhwyd: *F.C.* p. 281.   8 *Corr.* p. 433, etc.

9 So *Dictionary of Booksellers*, 1668–1725: Walford carried on the business for some six years.   10 *Corr.* p. 445.

11 So Sherard to Richardson: Nichols, *Illustrations of Literature*, I, p. 345.

12 Ray reviewed this in *Phil. Trans.* XXI, no. 249, pp. 63–7: cf. *Corr.* pp. 349–52.

13 *Schola Botanica* edited by S. W. A. (W. Sherard), Amsterdam, 1689.

tiones, and Parisian *Histoire,* Charles Plumier's *Description,* Plukenet's *Phytographia, Almagest* and *Mantissa,* Boccone's *Museum,*[1] Commelin's *Hortus Medicus Amstelodamensis,* Jacob Bobart's third volume of Morison's *Historia Plantarum,* Francesco Cupani's *Hortus Catholicus,* Johann Georg Volckamer's *Flora Noribergensis,* and Bachmann's *Methodus.* The plants recorded in these books fall into three classes, those which have been studied and described in their wild state by the authors, those so described in gardens, those known only from dried specimens often defective. Of these last Ray declares, as he had repeatedly done in his letters, that such evidence is not sufficient for accurate classification: but that in dealing with plants thus recorded he has not ventured to do more than quote the author accurately. Out of all these books he has collected all the species not previously described or included in his *History.*

In addition, a large number of new plants have been added as a result of his own and his friends' efforts, including some 300 fuci, fungi or mosses, of which the mosses have been split up into groups. Hans Sloane has allowed him to extract the details of all the new species from his manuscript History of Jamaican plants,[2] and has also sent the collection of Maryland plants made by Vernon and Krieg; Petiver has supplied many novelties from the collections sent to him by his many correspondents; Sherard has contributed more than a thousand of his own discoveries and has revised and corrected the script. His three other friends, Robinson, Lhwyd and Dale, have also given much help.

Then after a tribute to the labours of the older botanists, Gesner, De l'Ecluse, the Bauhins and Colonna, he points out that in their time the vegetation of the world outside Europe was almost entirely unknown and depended upon random specimens brought home by seafarers as curiosities. 'Now in our day botanists fully trained not only in the European flora but in the contents of public and private gardens have themselves travelled to America, Africa, the East Indies, Japan and China and have studied, collected and described the indigenous plants, thus adding an incredible number of new species.' These explorers he then recites: Piso and Marcgraf; John Banister, mentioned in his former volumes, now lately killed by a fall from rocks; Sloane, with his sixteen months' exploration of Jamaica; William Vernon, elected fellow of Peterhouse in 1692, who had been given leave of absence from Cambridge in 1697 in order to visit Maryland with his friend, the German doctor David Krieg—these two being both keenly interested in insects as well as in plants, and having visited Ray at Black Notley; Paul Hermann; Hendrik van Rheede van Draakenstein; and, finally, Georg Joseph Kamel ('Camelli'), the Jesuit

1 Ray reviewed it in *Phil. Trans.* xx, no. 247, pp. 462–8: cf. *Corr.* pp. 352–8.
2 His *Catalogue* had been published in 1696 but the *Natural History* which Ray was allowed to see and use did not appear till 1707.

doctor and missionary at Manila who had collected, drawn and sent to him in 1698[1] the plants of Luzon in the Philippines. To these he adds a quotation from his friend Petrus Hotton of Leyden.

Vernon, whom Ray came to know in the last years of his life, first worked with him in 1694 in the matter of mosses. He 'hath been more industrious in searching out and more successful in finding the species of that tribe than any man I know...he hath communicated to me dried samples of many sorts which I had not discovered and shown me more'—so wrote Ray to Lhwyd.[2] In 1695 he seems to have stayed with Ray,[3] and Ray hoped for a visit from him and Krieg in 1699 to discuss their Maryland plants. He came in the winter of 1704 and wrote on 20 January to Richardson: 'I likewise waited on Mr Ray who is very old and infirm in body though his parts are very vivid'.[4]

Camelli, whom Ray describes as 'well skilled in botanics...a German by nation, native of Brin [Brno] in Moravia',[5] wrote to him in 1698 and sent some drawings. One of his letters, that of 28 October 1700, with Ray's reply, has been printed by Derham.[6] From it we learn that he had dispatched pictures in January 1698, but the ship had been intercepted by pirates; other pictures a year later were sent to Samuel Brown,[7] Petiver's correspondent in Madras; but he having died they seem to have been transmitted by Edward Bulkley,[8] chief surgeon at Fort St George. Ray in reply says how opportunely the records and pictures have come, since his *Supplement* is about to be printed, and assures him that he will try to get the pictures engraved—though this, he adds, will not be easy. Apparently the difficulties of transport, and Camelli's expressed desire not to burden an aged invalid, gave rise to some misunderstanding; for the pictures and descriptions of trees, presumably the last section of his work, were sent to some other correspondent than Petiver, and never reached Ray.[9] Camelli's contribution occupies the first 96 pages of the Appendix to the *Supplement*, and though the last sections of it are a bare list the plants and some of the trees are described in short paragraphs. Ray apologises in a note for the omission of plates[10] on the ground of expense.

1 *F.C.* p. 303: 14 Oct., 'recd. Rev. Camelli's designs': cf. *Corr.* p. 345.
2 *F.C.* p. 250: there is a curious error in the epitome of the letter of 17 Oct. 1694 to Robinson (*F.C.* p. 299): 'Vernon's Hist. of Moths and Doodie's', which must of course be 'Vernon's list of mosses'.
3 Cf. *Corr.* p. 302. 4 *F.C.* p. 308.
5 *Corr.* p. 347. 6 *Corr.* pp. 377–9.
7 From whom Petiver received the plants listed in *Supplement*, Appendix, pp. 234–8.
8 *Corr.* p. 448: Ray asked Petiver to have a copy of the *Supplement* sent to him: he afterwards supplied the list of birds of Madras printed in the *Synopsis Avium*.
9 Cf. *Corr.* p. 404, where Ray complains to Petiver that 'Camelli hath not dealt ingenuously'.
10 They passed into the collection of De Jussieu but were in the main too inaccurate to help in identification.

Sherard is in fact, apart from Sloane, the most important contributor, and he evidently took a great interest and a considerable share in the production of the work. Since returning from Ireland he had been tutor to Viscount Townsend in 1694, and to the Marquis of Tavistock,[1] touring France and Italy in 1695–9: then he had gone to Badminton[2] as tutor to the second Duke of Beaufort and in uncongenial surroundings 'found consolation in botanical work for Ray'.[3] In 1702 he was appointed a commissioner for sick and wounded soldiers and sent to Portsmouth;[4] and in 1703 he went out as consul to Smyrna, where he lived for many years and became rich.

Ray concludes his Preface with a discussion of the benefits which this exploration of the tropical flora has conferred upon mankind. First, he commends the sheer variety of vegetable life—'parasitic plants' (he uses the term) of which the Mistletoe has been our sole example; leafless plants, like many of the Cacti; trees which increase by branches drooping to the earth and rooting, like the Banyan;[5] plants which root from their leaf-tips, like the five or six ferns observed by Banister and Sloane; the Sensitive plants; the trees which shed their leaves and then flower and set fruit; the 'pitcher-plants'—plantae cisternis quibusdam naturalibus; huge fruits like the Coconut or the 'Shaddocks', the largest and sweetest of Oranges;[6] timber as hard as iron, like that called Iron-wood (probably *Sloanea jamaicensis*). Then he points to its use for food, for medicine, and for crafts; for food, Coffee, Chocolate, Tea, Spices and Fruits like the Banana and the Pineapple; for medicine, Peruvian bark (Quinine) in agues, 'Ipepochoanna'[7] in dysentery, the Chinese root Nisi or Ginseng,[8] a universal panacea, and the St Ignatius bean, a kind of Nux Vomica;[9] for crafts, in dyeing Logwood, Fustick wood (*Cladrastis tinctoria*) and Indigo (*Indigofera anil*), in furniture 'Spanish Elm or Princes wood' (*Cordia gerascanthoides*), in candle-making the Candleberry tree (*Myrica cerifera*). Finally, after drawing attention to two specially useful trees and to the

1 *F.C.* p. 271.

2 Hence the 'Badminton plants' sent to Ray: they were from Sherard's collection, then in Sloane's care: cf. *Corr.* pp. 413, 436–7.

3 *D.N.B.* LII, 67. Ray wrote to Lhwyd, 'he takes as much pains in the work as if it were his own': *F.C.* p. 279.

4 Cf. *Corr.* p. 406: letter of Dec. 1702.

5 Ray never uses this name: it is to him Ficus Indica.

6 *Suppl. Dendr.* p. 80: it is *Citrus decumana*, the Grape-fruit: Sloane, *Jamaica*, I, p. 41, says that Shaddock is derived from an East Indiaman sea-captain of that name who called at Barbados and left the seed there.

7 See above, p. 227.

8 From an Aralia, *Panax ginseng*; the medicinal qualities appear to be solely psychic! Cf. *H.P.* II, p. 1338.

9 *Strychnos ignatii*, still used as a source of strychnine; first made known in *Phil. Trans.* XXI, p. 87.

very large number of Brooms and other leguminous shrubs, he concludes by warning the reader that the arrangement of many of the plants now included is uncertain and possibly mistaken.

The book itself falls into three parts—the Supplement to the *History of Plants*, occupying pages numbered 1–666; the Supplement to the *Dendrology*, pages 1–135; and the Appendix consisting of Camelli's list, pages numbered 1–96, an alphabetical list of new plants chiefly oriental from Tournefort's *Institutiones*, pp. 97–112, and a series of short lists from William Dampier's *Voyages*,[1] pp. 225–6, Friderich Martens of Spitzbergen, pp. 226–7, Kaspar Commelin, mainly African, pp. 228–32, several catalogues from Petiver, pp. 233–49, a 'farrago plantarum incertae sedis', pp. 251–5, and a short index of five pages, three columns to a page. On the back of the last page of the index is the note: 'I am glad to insert here a description of one or two plants omitted in the History': the last of them is a Sedge found at Notley by Samuel Dale, an appropriate ending to his last botanical treatise.[2]

In form it is an extension of the Appendix in vol. II. Each entry is referred to the appropriate place in the *Historia* at which it should be inserted; and the classification is consequently the same. It is thus simply a compilation, bringing up to date the record of all known plants, and assigning to their proper grouping the discoveries which had so vastly extended the number of identified species. It is a compilation—as Linnaeus's work was a compilation; and it included more than 10,000 new entries. Many of them, like those from Sloane or the *Hortus Malabaricus*, are adequate and recognisable: others, and the majority, are little more than brief descriptive names with a reference to the discoverer or to the book in which it is recorded, unfortunately often without page-reference.

During the latter part of its composition he was in fact a dying man. In 1700, when he was finishing the long task of making extracts from Tournefort, he had written to Sloane: 'I thank God I am able to go on with this work, though I have little or no absolute intermission of pain',[3] and three months later: 'I am now much worse, being by the sharpness of my pains reduced to that weakness that I can scarce stand alone, so that I lay aside all thoughts of the History of Insects and despair even of life itself. I remember that in one of your former letters you told me that my condition might lead to a dropsy.... I now find your prediction true.'[4] In 1701: 'Sir Thomas Millington coming to see me, discovering my condition told me that he believed no outward application would do me any good and advised me to use a plain anti-scorbutic diet-drink, made of

1 Published in 1699.
2 This Sedge had been entered in *Syn. Brit.* II, p. 265 as no. 9 of the group Gramen cyperoides polystachion: if genuinely distinct, it appears to be *Carex strigosa*.
3 *Corr.* p. 376.   4 *Corr.* p. 380.

dock roots, water-cress, brooklime, plantain and alder-leaves which I have
done now this fortnight but as yet have received no sensible benefit by it,
my sores running as bad and being as painful as ever.'[1] At the end of the
year Sloane wrote of him to Thoresby: 'he is become of late sickly and is
pretty well in years, but a very extraordinary person'.[2]

Nevertheless, in 1702 and 1703 he managed to carry on with his work,
getting together his observations of insects and producing the *Methodus*,
which was to be preliminary to his *History*. He was almost entirely confined to the house, as his legs were bandaged or kept in 'strait stockings',[3]
but this did not prevent him maintaining his correspondence and even
adding to it letters to the Rev. William Derham,[4] preparing fresh editions
of his *Wisdom of God* and *Discourses*, and planning to survey the insects in
the collections of his friends and the London virtuosi.[5]

Then in February 1704 he was 'brought even to death's door'.
Following on

the sharp cold brunt which happened in January...there befel a very strange
accident; one of my small sores began to run with that extraordinary rage as no
man could believe...and so continued night and day for five days together, till
it had reduced me to that weakness that I was unable to rise up from my chair.
...All this while a fever attended me which at last abated in a plentiful sweat.[6]
...Since this another sad accident has befallen me. A part of the skin of one of
my insteps by degrees turned black and now is with the flesh under it rotted and
corrupted...yet runs a copious gleet.[7]

In June, when 'the sores seemed to be in a fair way of healing', there was
another outbreak and the discharge continued 'notwithstanding all my
physic':[8] but despite it he could add: 'I have drawn up a little Method of
insects which may take up two sheets. It is very lame and imperfect,
especially in the tribe of Muscae. I did intend to have sent it up this day
but fear it is now too late and the carrier gone'; and in August he could
say: 'I have now begun the History of Insects',[9] and enclose a draft proposal for subscriptions to plates for it. Later that month Sloane came
down to see him,[10] and until the coming of winter he was full of hope that
the book might yet be produced. Smith's delays ('of a long time I cannot
extort a letter from him'[11]) and the difficulty of getting support before the
work was begun discouraged him: but though on 1 November he wrote:
'The History of Insects must rest if I continue thus ill, and I see no likelie-

1 *Corr.* p. 397: Millington was his contemporary at Trinity.
2 *Letters to R. Thoresby*, p. 108. 3 *Corr.* p. 446.
4 Cf. *Corr.* p. 399, in reply to Derham's first letter (of 24 April 1702); p. 401, in
reply to Derham as printed in *Hist. Ins.* pp. 262–6; pp. 414, 455: Derham was then at
Upminster and visited Dewlands in the summer of 1704.
5 *Corr.* pp. 410, 430, 432. 6 *Corr.* p. 412. 7 *Corr.* p. 413.
8 *Corr.* p. 445. 9 *Corr.* p. 448. 10 *Corr.* p. 452.
11 *Corr.* p. 456: he seems to have died in 1703, cf. p. 300.

hood of amendment unless I overlive this winter',[1] he seems to have struggled on until the end of the year.

Then on 7 January 1705 he sent his last letter to Sloane—

Dear Sir, the best of friends,

   These are to take a final leave of you as of this world. I look upon myself as a dying man. God requite your kindness expressed anyways towards me an hundredfold, bless you with a confluence of all good things in this world and eternal life and happiness hereafter, and grant us a happy meeting in heaven.

I am, Sir, Eternally yours, JOHN RAY.

   When you happen to write to my singular friend Dr Hotton I pray tell him I received his most obliging and affectionate letter for which I return thanks and acquaint that I am not able to answer it.[2]

   Ten days later, at 10 o'clock in the morning of Wednesday, 17 January, he died.[3]

NOTE. *Ray and Linnaeus*

To discuss in detail the influence of Ray upon Linnaeus (which was in fact much larger than the Swedish naturalist was ready to allow) or to comment fully upon the merits of their respective contributions to botany lies outside the scope of this book and the competence of its author. But because the fame of Linnaeus and his jealous and critical attitude towards Ray have obscured the value of Ray's work, it is necessary to make certain brief statements about the two men.

   There can be no sort of doubt that Ray had the truer appreciation of the real task of a scientist: for Linnaeus revealed his lifelong conviction when he defined the fundamentals of botany as classification and nomenclature[4] and thus limited science to systematology. He could neither accept Ray's suggestion that classification was mainly a matter of convenience for the benefit of the beginner,[5] nor could he fully realise the importance of relating it to structural conditions or natural affinities. He wanted a clear, symmetrical and easily recognisable system; and naturally found it by fastening upon a single character, the sexual structure of the flower. Tournefort was his model; and the results, though admirably lucid and workable, bear no relation to any system which a modern student could accept. When in his Classis III of Triandria monogynia he groups together Valeriana, Iris and Scirpus, or in Classis VI, Hexandria monogynia, Narcissus, Lilium, Erythronium, Acorus, Juncus, Berberis, and Frankenia

---

1 *Corr.* p. 457.
2 *Corr.* p. 459. Derham added a note: 'his strength failing him he was forced to break off abruptly'.
3 So Dale to Sloane.   4 *Philosophia Botanica*, p. 97.
5 *Phil. Bot.* p. 131, quoting Ray, *Methodus Emendata*, Reg. iii, p. 26.

(and these two examples are typical of the chaos which his system produced), we can only be amazed that students who knew Ray's *Methodus*
could tolerate so lamentable a substitute. Linnaeus gave us binomial
nomenclature and a convenient starting-point for the fixing of the earliest
Latin names; he not only attached an exaggerated importance to the
business of identification, but because he had little interest in the life or the
structure of organisms even his taxonomy was artificial and ill-founded.
He violates most of the six rules which Ray laid down and is exposed to
all the criticism which Ray addressed to Rivinus and Tournefort. It is one
of the misfortunes in the history of botany and to a less extent also of
zoology that the sound principles laid down by Ray should have been so
largely abandoned or ignored by his great successor.

Fortunately Ray's lead was not lost. His principles were embodied in
the more natural systems proposed by de Jussieu and de Candolle;[1] and
these made possible the development of systematics in accordance with
evolutionary ideas. Ray's studies of the seed-leaves and his use of the
cotyledons and the categories of gymnosperms and angiosperms have
proved of permanent value. It is interesting to note that even his distinction between arboreal and herbaceous habit (for which he has been so
generally ridiculed) has lately been reaffirmed by a competent authority.[2]

1 So Lindley and Boulger (*Transactions of Essex Field Club*, IV, p. 182, etc.).
2 Cf. J. Hutchinson, *Families of Flowering Plants*, where the *Archichlamydeae* are
divided into two groups according to arborescent or herbaceous habit.

CHAPTER XII

THE *ORNITHOLOGY*

The foundation of scientific Ornithology was laid by the joint labours of
Francis Willughby and John Ray.
ALFRED NEWTON, *Dictionary of Birds*, Introduction, p. 7.

If botany was the field of Ray's greatest and lifelong interest, it is by no
means his only or even perhaps his chief claim to the gratitude of posterity.
In it, as we have seen, he accomplished work which would of itself make a
fine record: to it, as he constantly claims, he returned from other excursions as to his proper task. But the astonishing feature of his career is not
his mastery of a single subject, but the range of his knowledge and the
value of his parerga. In these days of specialisation it is difficult to believe
that a man could make himself expert in the whole of zoology literally as
a sideshow and in the intervals of his main study; and Ray himself never
claimed to have done so. But the fact remains that after Willughby's death
he set himself to produce books on birds, fishes, mammals and reptiles, and
insects; and that these books, even more than his botanical writings, laid
the foundation for serious scientific progress in each subject.

It has been customary to regard this aspect of his work as little more
than the editing of his friend's material and to give Willughby the credit
for the result. This is obviously a mistaken estimate. Not only had he been
interested in them before Willughby and in the years of their partnership had
done the major part of the observations and records, but, while accepting
the task as a debt to the dead and making full use of Willughby's researches,
he created out of a few incomplete memoranda a series of books, each of
which marked a new epoch in its special field. They had travelled together,
kept notes and made collections: they had compiled together the Tables
in Bishop Wilkins's *Real Character*: Willughby had left many descriptions
of the plumage and anatomy of birds, a few of fishes, and a number of
observations of insects. To these preliminary studies Ray brought the
trained mind and technical knowledge of an expert already disciplined and
tested by his work on plants. He knew all the writings of previous students
and could estimate the value of their evidence, collate their records, and
identify the species described by them. He knew the principles of classification, the significance of structural differences, the criteria of specific
distinctness. Thus on the basis of the material gathered by Willughby and
himself—material which even in insects is hardly ever Willughby's alone
—he was able to achieve results analogous to, if less complete than, his
History of Plants.

Willughby's death gave him the incentive and the opportunity; and he took up the task almost at once. In his last letter to Courthope, dated 17 January 1673–4, he wrote:

I am at present, and have been a twelvemonth, almost wholly engaged in reviewing and preparing for the press Mr Willughby's *Ornithology* for which his relict is content to be at present at the charge of engraving brass figures, though I doubt not that the work, when published, will reimburse her. I believe we shall hardly get it abroad this twelvemonth yet. The death of Mr Willughby hath cast more business upon me than I would willingly have undertook.[1]

He had in fact resolved to carry through their plan directly after Willughby's death. His own story of it should be quoted.

Viewing his MSS. after his death I found the several animals in every kind, both birds and beasts and fishes and insects, digested into a method of his own contriving, but few of their descriptions or histories so full and perfect as he intended them; which he was so sensible of, that when I asked him upon his deathbed whether it was his pleasure they should be published, he answered that he did not desire it nor thought them so considerable as to deserve it, or somewhat to that purpose, though he confest there were some new and pretty observations of insects. But considering that the publication of them might conduce somewhat to, first, the illustration of God's glory by exciting men to take notice of and admire his infinite power and wisdom, displaying themselves in the creation of so many species and animals; and secondly, to the assistance of those who addict themselves to this most pleasant and no less useful part of philosophy; and, thirdly, also the honour of our nation in making it appear that no part of real knowledge is wholly balked and neglected by us, (he not contradicting) I resolved to publish them, and first took in hand the *Ornithology*.[2]

There was indeed every reason to do so. In botany the practical interest both in 'simples' for the physician and in fruit and flowers for the gardener had created a great enthusiasm and a modicum of knowledge. In zoology there was no such stimulus; and ignorance was still almost abysmal. If it was no longer universally agreed that Aristotle had spoken the last word on the subject, yet the 'pandects of birds, comprising whatever had before been written by others, whether true or fabulous'[2] were the only literature: Gesner, Belon and Aldrovandi[3] represent the best—Gesner, learned, lengthy, unsystematic; Belon, with less learning but more first-hand knowledge and some grasp of arrangement; Aldrovandi, a compiler, jealous of Gesner and plagiarising Belon. The subject was entangled, as Ray put it,[4] with 'hieroglyphics, emblems, morals, fables, presages'; and even the exploration of the New World and the curiosity of its explorers, though

1 *F.C.* p. 38. 2 Preface, *Ornithology*.

3 His three great folios are a strange mixture of fact, fable and fancy: thus e.g. Book IX contains 'Birds of mixed nature', the Bat and the Ostrich; Book X, 'Birds of fable', the Gryphon (which lives in the mountains of Arimaspia), the Harpy and the Sphinx.

4 Preface.

it produced a vast mass of material, only 'led up to what may be deemed the foundation of scientific Ornithology'—'the foundation laid by the joint labours of Willughby and Ray'.[1]

In Britain almost nothing had been done. Merret's *Pinax* had contained a bare catalogue, neither full nor accurate, and only of interest because it contained four or five notes sent in by Willughby after his return from the Continent:[2] these refer to birds seen in Warwickshire, Shrikes, a Merganser, a Sandpiper, and one, a Dipper, in Cumberland. Walter Charleton's *Onomasticon Zoicum*, published in 1668 with the imprimatur of the Royal College of Physicians, was not very much better. Its purpose as described in the Preface was to produce a list of the names of quadrupeds, serpents, insects, birds, fishes and minerals derived from Gesner, Wotton, Rondelet, Mouffet, Salviani, Belon, Aldrovandi and Johnstone, and to check them where possible by reference to specimens in the royal menagerie in St James's Park or in the Museum of the Royal Society. The birds are slightly better than the rest: there are plates of the Hawfinch which Charleton had exhibited to the Royal Society, of a Bee-eater brought to England by Sir Thomas Crew, of whose pictures of birds Ray says that they 'are humoured by the painter to make them more beautiful', and of a Hoopoe 'killed last winter ten miles from London':[3] but most of the book is simply a digest of previous writers. Ray mentioned the book to Lister on 10 September 1668,[4] but seems never to have used it. Writing in 1693 to Robinson[5] he acknowledges receipt of its second edition in folio,[6] and says:

Whatever he may boast of his performance, he is (as you say) no better than Johnson[7] and did not understand animals nor had any comprehensive knowledge of them...to me it is a great argument of learned men's ignorance in that History that such a book...should come to a second impression. As for his figures of birds they were taken most from Sir Tho. Crew's designs drawn at Montpellier which I saw there.[8]

He adds: 'I expect to find little in him.' There was plainly need for something more scientific.

Ray was in fact well qualified to take up the task; for if birds had never been a principal object of his studies, he had shared with his friend in all their field-work and been eager to use what opportunities he could. The

1 Newton, *Dictionary of Birds*, Introd. p. 7. This gives what is still the best short account of the history of Ornithology.

2 Cf. *Pinax*, p. 171, where Merret acknowledges Willughby's notes.

3 P. 92. 4 *F.C.* p. 119. 5 *F.C.* p. 101.

6 The first edition was a small quarto.

7 John Johnstone, author of *Historia Naturalis De Avibus*, etc., published in 1652, and of *Dendrographia*, published 1662: Ray had condemned him as 'a mere plagiary and compiler of other men's labours': *F.C.* p. 160. His books had no originality, but attained some influence. The *De Avibus* groups the Bats with the Owls, and says that the Bird of Paradise, being footless, can only rest by tying its streamer-feathers to a branch and hanging (p. 171). 8 Cf. above p. 137.

Cambridge Catalogue by its record of the seeds found in the crop of a
Great Bustard (*Otis tarda*)[1] proves that even as early as 1659 Ray was in
the habit of dissecting and studying such birds as he could obtain. He had
already on the journey of 1658 shot, dissected and described a Common
Gull (*Larus canus*) 'on the bank of the Lake of Bala':[2] and a letter to
Courthope[3] describes how he dissected four other birds, a Bittern, a
Curlew, a Yarwhelp and one other, with Nidd. In 1660, when as we have
seen the two made a journey to the Isle of Man, though Willughby's
description of the young 'Puffin of the Isle of Man' (*Puffinus puffinus*)
obtained on the Calf is quoted, the account of the cliffs off the southern
end of the Isle and of the nocturnal habits and daytime solitude of the
Shearwater colony are plainly Ray's work. Nor is it fair to accuse him of
confusing the birds then seen with those now called Puffins.[4] He found
the Shearwaters on the Calf of Man, named by the Manx-men 'puffins';
but when he visited Prestholm he distinctly separates 'the puffins, so called
there, which I take to be *Anas arctica clusii*',[5] and does the same at Bardsey
('there builds the Prestholm puffin').[6] In the *Ornithology* the two species
are widely and rightly separated, the Puffin of to-day being grouped with
the Auks, the Puffin of the Isle of Man with the Shearwater supplied to
him by Sir Thomas Browne.[7]

On the 1661 journey the visit to the Bass gave opportunity for further
study of sea-birds,[8] 'the Scout which is double-ribbed [the Razorbill,
Alca torda], the Cattiwake [*Rissa tridactyla*], the Scart [the Shag, *Phalacrocorax aristotelis*] and a bird called the Turtle-dove' (presumably a
version of Dovekey, the local name in parts of Scotland for the Black
Guillemot, *Uria grylle*), 'whole footed and the feet red'. He adds:

There are verses which contain the names of these birds among the vulgar,
two whereof are,

'The scout, the scart, the cattiwake,
The soland goose sits on the lack,
Yearly in the spring.'

The 'Soland Geese' (or Gannets, *Sula bassana*), then as now 'innumerable
on the rocks', were carefully described and he noted their boldness—
'they sit in great multitudes till one comes close up to them'. It is noticeable that in Chapter VII of the first Book of the *Ornithology* which deals
with 'some remarkable isles about England', mentioning the Bass he says:
'When we were there near mid-August all the other birds were departed,

---

1 *C.C.* p. 34.   2 *Orn.* p. 346.
3 *Early Science in Cambridge*, p. 347: cf. above p. 47.
4 As Newton, under Shearwater, in *Dict. of Birds*, pp. 830 and 752.
5 *Mem.* p. 168.   6 *Mem.* p. 171.
7 Pp. 333–4 and the excellent picture on Plate LXXVIII; *Fratercula arctica* is on
pp. 325–6.   8 *Mem.* pp. 154–5.

only the Soland Geese remained upon the island, their young being not
yet fully grown and fledged.'¹

Next year, with Willughby, the first record of importance is of a visit
to the Black-headed Gull, 'Puit', colony in a mere at Norbury near Nantwich. 'They build altogether in an islet in the middle of a pool. Each hen
layeth three or four eggs (of a dirty blue or sea-green, spotted with black.
...They come the beginning of March, and are all gone by the latter end
of July or before.' Later on in the same tour, at Prestholm (Puffin Island
in the Menai Straits), he saw 'two sorts of sea-gulls, Cormorants, Puffins,
so called there,...Razorbills and Guillems, Scrays, two sorts, which are
a kind of Gull' (Terns); at Bardsey 'Prestholm Puffins, Sea-pies'
(Oyster-catchers, *Haematopus ostralegus*) and some other birds; and at St
Margarets, near Caldey, Puits (Black-headed Gulls, *Larus ridibundus*) and
Gulls, and Sea-swallows (Common Terns, *Sterna hirundo*)—'the nests lie
so thick that a man can scarce walk but he must needs set his foot upon
them'.² At Aberavon, 'on the sandy meadows by the sea-side', a bird
breeds believed to be the 'Hamantopus magnitudine inter vanellum et
gallinaginem minorem media', and there follows an admirably brief but
decisive description of a Redshank (*Tringa totanus*), ending with the
words 'the legs long and red, the ungues black; it hath the posticus
[hallux or fourth toe]; these she stretches backwards in flying which make
amends for the shortness of the tail; it makes a piping noise'.³ Near
Padstow 'we saw great flocks of Cornish Choughs' (*Pyrrhocorax pyrrhocorax*) and at St Ives and Godrevy more sea-birds, including 'those which
they call Gannets which prey upon pilchards' and which he never identified with the Solan Goose of the Bass.⁴ He adds: 'Another bird they told
us of here, called wagell, which pursues and strikes at the small gull so long
till out of fear it mutes, and what it voids the wagell follows and greedily
devours, catching it sometimes before it is fallen down to the water'—
a familiar habit of the Skuas, which however Ray only reports on the
authority of 'several seamen'.⁵

The only other evidence of activity in this period is in the second of his
letters to Courthope during his stay at Friston. This letter as printed in
the volume of *Further Correspondence*⁶ is mutilated, omits an interesting
sentence about John Nidd's dissections, and makes it appear that four

1 *Orn.* p. 19.   2 *Mem.* p. 175.
3 *Mem.* p. 177. The editor makes the egregious suggestion that the bird is the
Black-winged Stilt. For 'haemantopus or redshank' cf. *Sir T. Browne's Works*, IV,
p. 184.
4 Even in the *Synopsis Avium* he lists the 'Anser Bassanus' with the Cormorant on
p. 122 and the 'Catarractes or Gannet' with the Gulls on p. 128: Moyle pointed out the
identity of the two to Sherard (*Works*, I, p. 424).
5 In *Orn.* pp. 348–9 he confuses the Cornish Gannet with the 'Skua of Hoier' and
the Wagell with the Burgomaster Gull.   6 *F.C.* p. 35.

birds were obtained by Ray at Friston, whereas they were really specimens obtained by Nidd and preserved in his chamber in Trinity College.[1] But it does record that at Friston Ray busied himself in enquiring out and describing 'such birds as frequent the channel near us. I have gotten some and cased them', among them a bird which seems to have been the Yarwhelp, though he did not then recognise it as such. 'The yarwhelp is a name that I never read or heard of before or since.' It is in fact the East Anglian name for the Godwit, and by the description in the *Ornithology* Ray's bird was the Bar-tailed Godwit, *Limosa lapponica*.[2]

On the continental tour 'fowls' were among the objects collected; the *Ornithology* is full of records of birds found in the markets of Germany, Italy or Sicily and described and dissected; and Willughby bought certain books of pictures. Yet the notes on birds are few; and in view of the claim that Willughby was the real ornithologist it is interesting to observe that in the journal of his solitary tour in Spain,[3] though he records the amethyst mines, the Arbutus, the crops grown, and the habits of mules, and in his letter to Ray describes the 'Oleum petrol',[4] there is no word of natural history, of birds or fishes or insects. Ray in the *Observations* records in Van der Meer's museum at Delft 'a cassowaries' or Emeu's egg' (presumably the former) and 'the tail of an Indian peacock',[5] and in the Grand Duke's gardens at Florence 'in an enclosed place two male ostriches and one female':[6] he repudiates as 'without all doubt false and frivolous' the opinion that birds are hatched from barnacles such as he saw growing on the back of a turtle on the voyage to Malta:[7] and he gives an attractive description of the 'remarkable grove' at Sevenhuys[8] 'where in time of year several sorts of wild-fowl build and breed'. The passage deserves quoting: 'Schoffers, i.e. Graculi palmipedes, in England we call them Shags; they are very like to Cormorants only less' (they must certainly have been Cormorants, *Phalacrocorax carbo*, for these still breed in trees at Lekke-

1 The recovery of the whole of this letter along with the rest of the series to Courthope is recorded and the letter itself published in Gunther, *Early Science in Cambridge*.
2 Yarwhelp is commonly stated (e.g. by Newton, *Dict. of Birds*, p. 1055; Patterson, *Nature in Eastern Norfolk*, p. 238, etc.) to have been the name of the Black-tailed Godwit (*L. belgica*), which at this time bred in Norfolk, as it still does in Holland. But Browne, 'Norfolk Birds' (*Works*, IV, pp. 319, 321), clearly distinguishes the Godwit, a marshland bird and 'the daintiest dish in England', from the Yarwhelp, 'a grey bird' with a shorter bill; and *Orn.* pp. 292–3 describes the Yarwhelp as having 'the tail-feathers crossed alternately with black and white lines' and the 'other Godwit' with 'the middlemost tail-feathers almost wholly black'.
3 Ray, *Obs.* pp. 466–99.   4 *Corr.* p. 7.   5 *Obs.* pp. 28–9.
6 *Obs.* p. 335.   7 *Obs.* p. 291: cf. above p. 135.
8 Now Zevenhoven: the marshland described by Ray has now been drained: in his time it must have resembled the drier parts of the Naardermeer.

kirke near Rotterdam, whereas the Shag is exclusively maritime).[1] 'We were much surprised to see them, being a whole-footed bird, alight and build upon trees. Lepelaers called by Gesner Plateae...we may term them in English Spoon-bills' (*Platalea leucorodia*, they nest freely in the Naardermeer though not there in trees). 'Quacks or Ardeae cinereae minores, the Germans call this bird the Night-raven because it makes a noise in the night...' (*Nycticorax nycticorax*—a regular tree-nesting species still breeding in Holland). 'Reyers or Herons' (*Ardea cinerea*, still common and nesting both in trees and, in Texel for example, in reed-beds). 'Besides the forementioned birds there build also in this wood Ravens, Wood-pigeons and Turtle-doves.'[2] It is probable that the *Observations* afford a poor estimate of the work that he actually carried out; for his dissections at Padua recorded in a posthumous paper published in the *Philosophical Transactions*[3] are wholly unnoticed. He worked there with Marchetti, probably Pietro, lecturer in anatomy in the University, in December and January 1663–4, and dissected at least one bird, a 'Gallina montana' or Capercailzie.[4] He certainly obtained a number of birds in the markets at Rome; and in Switzerland noted 'Urogallus or Cock of the Wood and Lagopus, a milk-white bird somewhat bigger than a Partridge, feathered down to the toes' (Ptarmigan), as well as 'Merulae torquatae and aquaticae'.[5]

Of the later journeys the second with Willughby to the West is represented by only two surviving records from Land's End, that of 'the Haliaetus or Bald Buzzard' (the Osprey, *Pandion haliaetus*)—'of a dark colour on the back, and white under the belly, having strong legs and talons so placed that she can bend two backward and two forward...this bird was shot having a mullet in her talon'—and of 'the Godwit which they call a Stone Curlew,[6] the same I suppose which is common upon our coasts of Suffolk: the colour differs somewhat from a bird of the same kind I was shown at Venice, which was more cinereous or dun and had not that variety of colours in the wings and tail'.[7]

On his return on 1 October he wrote a long letter to Lister,[8] in which more plainly than anywhere at this early stage of his work he revealed not only his extraordinary flair for natural history, but his knowledge of

1 In *Orn.* p. 330 Ray speaks of the Sevenhuys birds as Cormorants.
2 *Obs.* p. 38.
3 XXV, no. 307, pp. 2302–3.
4 Cf. *F.C.* pp. 87–8. Seward (*John Ray*, pp. 16–18) has shown that Gunther's dating (1683) is due to a misprint.
5 *Obs.* p. 419.
6 This confusion led him to classify Stone Curlew with long-billed waders in *Cat. Birds*, p. 90: he corrected this in *Orn.* p. 306.
7 *Mem.* p. 190 note.
8 *F.C.* pp. 114–15, a letter written in Latin.

anatomy and points of structural importance. Lister had evidently sub-mitted certain questions to him. He replies:

You needn't hesitate about the Teal; but I am not sure about the Knott—never yet having happened to see it alive.[1] In the hawk family and all our birds of prey nature seems to make fun of us, so that it is sometimes difficult, indeed almost impossible, to distinguish species accurately. The same bird varies in the colour of its plumage in accordance with age; in older birds the colour becomes less bright and often shades down towards whiteness. But perhaps you will find this Wood Kite of yours among the eagles. The Bald Buzzard of the English is certainly no other than the Osprey.... About your long-billed Sheldrake, I have not got books at hand and do not remember very clearly Gesner's description or picture of the Merganser.... Your observation of the Green Woodpecker corresponds with my own of the Black and both the Spotted Woodpeckers and the Wryneck. The use of that structure is quite clear, to extract larvae from rotten wood. I once got out from the crops of these birds on dissection larvae as big as my small finger. The muscles and tendons by which they shoot out and retract their tongues deserve curious study.[2] So too what you note in the Grey Heron is true of the whole family, the Bittern, the Small Grey,[3] the Great White and the Lesser White:[4] that is, they have the claw of the middle toe serrated on the left edge,[5] the use being to hold eels and slippery fishes.

The tour with Willisel in 1671 included a visit to Bamborough, and a list is given of the birds building on the Farne Islands:[6] 'Guillemots, Scouts or Razorbills, Counternebs' (Puffins), 'which build in holes and I guess them to be Anates arcticae clusii. Scarfs, i.e. Shags cornub.' (meaning that Shag is the Cornish name), 'Cuthbert Duck' (Eider, *Somateria mollissima*), 'bigger than a Wild Duck, of a brown colour; the drake is white on the back, the tail and feathers of the wings black, the legs also black. It hath a posticus digitus; the bill is scarce so long as a duck's; the superior mandible is a little crooked at the end and overhangs the inferior; but that which is most remarkable is that on both sides the bill the feathers come down in an acute angle as far as the middle of the nostril below'—if this note were by Willughby it would be hailed as a striking example of the skill in picking out a characteristic feature which Ray ascribes to him. He continues: 'Puffinet' (Black Guillemot, *Uria grylle*), 'which is as big as a pigeon, hath a small sharp-pointed bill, black in the summer having only a white spot in each wing, but white in the winter, lays two eggs, and builds in an hole under a rock'. This is again an admirable account, and of

---

1 He saw it 'on the coast of Lancashire near Liverpool, many flying in company' in Feb. 1672 (*Orn.* p. 302).

2 The hyoid bones of the Picine birds are unique: cf. *Orn.* p. 136.

3 I.e. Night-heron, *N. nycticorax*.

4 Little Egret (*E. garzetta*), which Ray 'bought in the market at Venice'; *Orn.* p. 282.

5 Here again he has noted a salient feature: he draws attention to it in *Orn.* p. 282.

6 *Mem.* p. 152 note.

interest because the bird in question no longer nests either on the Farnes or where he first mentioned it, the Bass. 'Annet' (Kittiwake, *Rissa tridactyla*), 'a gull, but small and white, only the tips of the wings black, yellow bill: Cattiwike' (the same). 'Mire Crow' (Black-headed Gull, *Larus ridibundus*), 'all white-bodied, only hath a black head, a little bigger than a pigeon. Puets. Pickmire, i.e. a Sea-swallow *forte*' (*Sterna sandvicensis* or *macrura*). 'Tern, a small gull, the least of all, having a forked tail' (probably *Sterna albifrons*). 'Sea-piots, i.e. Sea-pies, as far as I could guess'. He concludes with 'Kir-bird, a sort of Colymbus, less than a Magpie, black and white, and stands strait upright' (probably the Guillemot, *Uria aalge*). 'Gorges, a fowl bigger and redder than a Partridge'—presumably the Red Grouse.

So far the *Itineraries*. The letter to Lister[1] reporting on his solitary tour of 1668 has an interesting reference to the same game-bird.

Of birds [he writes] only four or five species were found by me that I had not seen before, to wit the Grygallus major of Gesner which the Italians call a Francolin; it is common on heather-clad mountains; sportsmen and countryfolk call it Red Moorgame. I am well aware that Gesner thinks the Italians' Francolin is called the Hazel-hen. I think this bird is the same as that whose picture Mr Thomas Crew showed us[2] at Montpellier whose French name I have forgotten.

He continues with a list; 'Merula saxatilis seu montana quite different as it seems to me from Torquata' (probably the bird is the immature or perhaps female, *Turdus torquatus*: that is, of course, the Ring-ousel[3]); 'Merula aquatica' (the Dipper, *Cinclus cinclus*); 'Caprimulgus' (the Goatsucker); 'and of small birds two or three species, but I do not know whether they have been described, or under what names'. On his return the friend, Francis Jessop, with whom he had stayed at Broomhall, wrote to him to say that he had stuffed the skins of a Moor-cock and Moor-hen,[4] and to Willughby[5] that he had packed them. In this letter he states his certainty that the birds are not the Hazel-hen and his doubts as to their being the Grygallus of Gesner: 'the feet', he says, 'are not like those of the Urogallus minor, but nearer resembling those of the Lagopus, being feathered all over'. The birds were in fact Red Grouse (*Lagopus scoticus*) and not Blackcock and Greyhen.

In March 1669 he sent a message of thanks from himself and Willughby to Lister[6] 'for saving us from a notable error':

We had supposed the Snipe and the smaller birds that you call Gids and we Jack Snipes to be different sexes of the same bird—partly from vulgar prejudice,

---

1 *Corr.* p. 27.
2 Lister was with Ray and Skippon at Montpellier in 1665: cf. *Corr.* p. 62.
3 So he suggests in *S.A.* p. 66 and *F.C.* p. 229.
4 *Corr.* p. 33.   5 Derham, *Phil. Letters*, p. 367.
6 *Corr.* p. 39: he repeats this acknowledgment in *Orn.* p. 291.

partly because our specimens of one happened to be all males, of the other all females: since your letter we have looked into the matter and found males and females of both kinds.

In the same year a gift of the dried leg of a Buzzard came from Lister on 22 December[1] with the note that upon comparing them with the Kite, the Bald Buzzard (Osprey) and Wood Buzzard (probably *Buteo buteo*) you will find them exceedingly different: 'but Mr Willughby did almost persuade me it was the Milvus aeruginosus Aldr.' Ray replied that he was confidently persuaded of this last identification—that is, that the bird was the Marsh Harrier (*Circus aeruginosus*).

In 1672, while staying at Chester, he obtained a male Scoter 'killed on the coast thereabouts and bought in the market by my Lord Bishop Wilkins his steward'.[2] So too his friend Dent, the Cambridge apothecary, in February 1675, ransacking the market for a Thornback Skate, put up in a box with it 'some waterfowl, viz. a Pocker (*Aythya ferina*), a Smew[3] (*Mergus albellus*), three Sheldins (*Tadorna tadorna*), a Widgeon and a Whewer, which last two are male and female of the same kind' (*Anas penelope*)....'We could not meet as yet with a Pintail.' He adds a point of interest:

I have put up some hollow bones which are annexed to the windpipe of each male; for in females I can find none, otherwise than you will find in the paper writ upon Whewer. The difference of shape of these bones doubtless causes their different tones. If you steep one of the Sheldin's windpipes a while in warm water to make it lax, you may observe the pretty motion to be found in the middle protuberance and pick out a little philosophy from it.[4]

This matter of the structure of the trachea in the male sex of most though not all the ducks is a nice point to find noticed so early.[5] Dent is right about its sex-limitation and also about the peculiarity of the Sheld-duck in which the drake has two bony ampullae, one on each side, as is specially noted, no doubt from this specimen, in the *Ornithology*.[6] Ray dealt at some length with the subject in a letter of 13 August 1684 to Robinson[7]— the letter in which he identified the Macreuse with the Scoter (*Melanitta nigra*). He there noted

in this bird, and in some others of the sea-ducks, which are much under water, that they want that vessel, or ampulla, situate in the very angle of the divarication of the windpipe, which, for want of a better and fitter name, we are wont to call the labyrinth of the trachea.[8]...I am somewhat to seek about the use of this

1 *Corr.* p. 50. 2 *Orn.* p. 366.
3 Cf. *Orn.* p. 338: it was a female; the drake being called White Nun.
4 *Corr.* pp. 16–17.
5 It was mentioned by Walter Charleton (author of the *Onomasticon*) at a meeting of the Royal Society on 15 Feb. 1665 (Birch, *History of the R.S.* II, p. 13).
6 *Orn.* p. 363. 7 *Corr.* p. 150.
8 Ray's name is still used.

vessel, and I think it were worth the while to examine what sort of birds have it, whether the males have it or in some the females also. I observed it in the Mergus cirratus longiroster noster or the Dun-diver [*Mergus serrator*] and that very large and extended by very strong bones; and yet I thought myself to have sufficient reason to judge that bird to be the female of the Merganser;[1] but I dare not be confident that it is the female because of this labyrinth.

That is a remarkable note: the purpose of the labyrinth is still unknown: the Scoters are quite without it: in the Mergansers it reaches its fullest development.

As regards foreign birds the continental tour had introduced a certain number of species in the wild state, and still more in museums and collections. But the 'Royal Aviary in St James's Park near Westminster',[2] where M. de Saint-Évremond was appointed 'keeper of the ducks', was a richer source of material. Here there were Pelicans—as there have been ever since—two of them specially presented to Charles II by the Ambassador of the Emperor of Russia;[3] and two kinds of Geese, the Spur-winged[4] (*Plectopterus niger*) and the Canada (*Branta canadensis*); also a 'Cassowary or Emeu' (*Casuarius casuarius*), one of four in London that Ray knew and described;[5] an Eagle,[6] two Eagle Owls,[7] a Vulture,[8] a brown-coloured Magpie[9] and Carrier Pigeons.[10] There were also some novelties to be found among the small birds imported from the East Indies or the New World in cages and kept by fanciers. Ray reports,[11] for example, that in 1673 he saw many 'Anadavads'[12] (generally called Avaduvats, *Estrilda amandava*) in the house of a certain citizen of London to whom he had been introduced by his friend Thomas Allen; these 'had been brought out of the East Indies kept all together in the same cage'; and that the Virginian Nightingale (*C. cardinalis*) was 'brought into England out of Virginia'.[13] Parrots were naturally favourites, and Gesner's story[14] of Henry VIII's parrot which fell out of a window in the palace of Westminster into the Thames and cried out 'A boat, a boat for twenty pounds', is typical of the interest taken in them.[15]

1 Males in their first plumage and in eclipse plumage are similar to the female and very unlike the drake in full dress: Ray repeats his uncertainty in *Orn.* p. 336.

2 Hence Birdcage Walk, 'so named from the cages of an aviary disposed among the trees which bordered it', Knight, *London*, I, p. 195.

3 *Orn.* p. 327: for these and other birds in the Park, cf. Evelyn, *Diary*, 22 Feb. 1665.

4 *Orn.* p. 360. The plate, T. 71, supposed to represent this is an Egyptian Goose, *Chenalopex aegyptiacus*.

5 *Orn.* p. 151.   6 *Orn.* p. 59.   7 *Orn.* p. 100.
8 *Orn.* p. 67.   9 *Orn.* p. 128.   10 *Orn.* p. 181.
11 *Orn.* p. 266.

12 Newton, *Dict. of Birds*, p. 11, derives the name from Ahmadabad, 'whence according to Fryer (*New Account of East India*, 1698) examples were brought to Surat'.

13 *Orn.* p. 245.   14 Quoted in *Orn.* p. 109.
15 Cf. Skelton's poem 'Speak, Parrot'.

Finally there were the museums. It was an age of collecting; and though at this time the great accumulations gathered by Sloane or Petiver were not begun, Tradescant had a collection of 'cases' and the Royal Society had its own specimens at Gresham House. Nehemiah Grew, whose Catalogue of them was published in 1681, devotes twenty-eight pages to birds, and although he includes bats among them takes most of his descriptions from the *Ornithology*. No doubt Willughby, who had given much to the Society's Museum, drew from it some of his material; and Ray certainly used it, as he declares in a letter to Lister: 'The Tropic-bird dried I have seen in the Repository of the Royal Society, and have described as well as I can.'[1] With such preparation Ray was fully competent for the work: and in spite of obstacles and other duties it went forward. Two of his friends, Francis Jessop of Broomhall near Sheffield, who was one of Willughby's executors, and Ralph Johnson, Vicar of Brignall near Greta Bridge,[2] contributed very largely both by the supply of specimens and by their own observations. Johnson is described in the Preface as 'a person of singular skill in zoology, especially the history of birds, who, besides the descriptions and pictures of divers uncommon and some undescribed Land and Water-fowl, communicated to us his Method of birds whereby we were in some particulars informed, in many others confirmed, his judgment concurring with ours in the divisions and characteristic notes of the Genera'.[3] Sir Philip Skippon and Sir Thomas Browne contributed pictures, and Browne's notes published in his paper 'On Norfolk Birds'[4] were an additional help. Martin Lister, who wrote on 20 June 1673[5] to express his delight that Ray intended to publish 'the Natural History designed by Mr Willughby', and realising 'the great pains it will ask to perfect any one part of it', urged him to publish one part before undertaking the next, confessed that his own notes on birds were slender, but nevertheless sent them.

But the largest part of the task was not the arranging of Willughby's scripts or the enlarging it by personal study, but the collation of the writings of other students. 'I am going on as fast as I can with the Ornithology', wrote Ray to Lister in November 1673.[6] 'That the work may not be defective, I intend to take in all the kinds I find in books which Mr Willughby described not, and to have a figure for all the descriptions I can procure for them. I have sent this week to Mr Martyn to begin to get some figures engraved.' This was in fact a very arduous and difficult undertaking. It involved collecting, criticising and arranging a great mass of

---

1 *Corr.* p. 112; *Orn.* p. 331; Grew, *Mus. R.S.* p. 74; the species is *Phaëthon aetherius*.
2 Johnson went up to St John's College as a sizar in 1648 from Sedbergh School: he was born at Newsham, a farmer's son, and kept a successful school at Brignall.
3 *Orn.* Preface.   4 *Works*, IV, pp. 313–24: cf. below p. 337.
5 *Corr.* p. 103.   6 *Corr.* p. 105.

material provided not only by Gesner, Belon and Aldrovandi, but by more
recent students and travellers:[1] Clusius[2] or De l'Ecluse, Marcgraf,[3]
Hernandez,[4] Bondt,[5] Worm,[6] and Piso.[7] Moreover, it meant plunging into
writings of widely differing value, where some genuine observations and
accurate descriptions were mixed up with ignorant, fanciful and legendary
elements almost inextricably. If the book was to be an Ornithology and
not simply a record of such species as he and Willughby had personally
had opportunity to describe, the task was inevitable. But for Ray, who
had always insisted on the need for personal knowledge of the living
subject, this dealing with inadequate and often unreliable stories at
second-hand, with little means save his own insight for discriminating between them, must have been arduous and distasteful. With
what skill he rejected beliefs almost universally held, weighed up the
reliability of his authorities, checked their evidence, and compared birds
described by them with specimens known to himself can be seen from
the *Ornithology* on almost every page. There is proof enough here,
if he had written nothing more, to vindicate him against the foolish
charge of ignorance of birds and to attest his claim to outstanding
scientific ability.

In the Preface, more explicitly than in his letters, he defines the purpose
of the book and the steps taken to fulfil it. 'Our main design was to
illustrate the History of Birds, which is in many particulars confused and
obscure, by so accurately describing each kind and observing their
characteristic and distinctive notes, that the Reader might be sure of our
meaning and upon comparing any bird with our description not fail of
discerning whether it be the described or no.' He aimed, here as in his
botanical works, at making identification easy. To do so 'we ourselves did
carefully describe each bird from the view and inspection of it lying before
us'. 'That this diligence was not superfluous will appear in that we have
thereby cleared many difficulties and rectified many mistakes in the
writings of Gesner and Aldrovandi...mistakes in the multiplying of
species and making two or three sorts of one....Many such mistakes occur
in Aldrovandi...and some also in Gesner notwithstanding his great skill
and circumspection.' He goes on to explain that everything not strictly
relating to the natural history of birds has been omitted; that what has

---

1 The list is given in *Orn.* Preface.

2 *Exoticorum*, lib. x, Leyden, 1605.

3 *Historia Rerum Natural. Brasiliens.*, Leyden, 1648: he went out with
Piso as physician to Prince Moritz of Nassau, who was governor of Brazil (1637–
44): cf. above p. 219.

4 *Plantas y Animales de la Nueva Espana*, Mexico, 1615.

5 *Hist. Natur. et Med. Indiae Orientalis*, Amsterdam, 1656.

6 *Museum Wormianum*, Leyden, 1655.

7 *De Indiae utriusque Re Naturali*, Amsterdam, 1658.

been included has been carefully selected as reliable; and that such inclusions, drawn as they are from all the available sources, have been inserted 'each kind, as near as I could, in its proper place'. Anyone who
compares the *Ornithology* with its predecessors carefully will realise that
here as in his previous field of work he has brought order out of chaos;
that the species which he and Willughby have seen and described are
almost in every single case identifiable; that considering the scantiness of
his material he is exceedingly acute in detecting differences of sex and
plumage; and that though he could not avoid using the work of others for
species outside his own knowledge and though their descriptions are
always less accurate and often quite indeterminate, he is remarkably
successful in his comments and criticisms upon them.

There remain two further parts of the task to be considered, the plates
and the classification. Mrs Willughby had offered to pay for the illustrations. Willughby had bought 'of one Leonard Baltner, a fisherman of
Strasburg, a volume containing the pictures of all the Waterfowl fre-
quenting the Rhine near that city': 'at Nuremburg he bought a large
volume of pictures of birds drawn in colours': 'he caused divers species
to be drawn by good artists'. Sir Thomas Browne of Norwich had lent
drawings of several rarities.[1] Aldrovandi, Giovanni Pietro Olina,[2] Marcgraf and others provided pictures from which a selection could be made.
Yet, as Ray admits, the result was not satisfactory: 'though the gravers
were very good workmen, yet in many sculps they have not satisfied me'.
He ascribes this to his 'own absence from London': he could not supervise
the work except by letter; and in letters his advice was sometimes mistaken
or obscure, and sometimes neglected or misunderstood. Yet 'the figures
such as they are, take them all together, are the best and truest, that is most
like the live birds, of any hitherto engraven in brass'.[3] They are in fact, as
might be expected, very uneven in quality. Those borrowed from Marcgraf are frankly grotesque: those from Aldrovandi, which in the large
original cuts have an archaic vigour and dignity, do not preserve their
quality when copied and reduced—the Golden Eagle on the first plate is
typical of them.[4] Browne's figures, easily recognised because they usually
represent the bird lying dead, are in the main good: his Shearwater, taken
from his own living specimen, is excellent; his 'Mergulus melanoleucus
rostro acuto brevi', for which Ray could find no English name nor any
previous record, is immediately identifiable as a Little Auk, and his Razor-

1 Sent while Ray's *Collection of English Words* was in the press and mentioned in
the Preface to it. For these and others cf. note at end of chapter.
2 Author of *Uccelliera della Natura*, Rome, 1622, the engravings of small birds in
which are often very attractive.
3 Preface.
4 This picture appeared first in Gesner, *Icones*, p. 3: Aldrovandi copied it.

bill is very satisfactory: but the Stone Curlew, if the distorted horror on Plate 58 is his, deserved to be replaced by the much better picture on Plate 77. Baltner's, some of which have his name attached, are stiff and inexact. The small birds are on the whole bad, though most of them are recognisable. The ducks, evidently most of them from the same source, are all in the same position but distinct and not unpleasing. The 'Penguin' or Great Auk and the Dodo are both adequate. There is an excellent plate of the various breeds of domestic pigeon; and an admirable Sheldrake looks as if it were by the same artist. But these, it must be admitted, show up the exceeding badness of many of the others. No wonder Ray was disappointed.

The classification is by common consent one of the outstanding merits of the book. Of his predecessors since Aristotle only Belon[1] had made any serious attempt at an ordered arrangement, and his was not profound: for he took habitat as his basis and grouped the birds according to the type of country in which they were found. First of course he puts the birds of prey, diurnal and nocturnal, Eagles, Owls and the Bats. Then the web-footed birds of the rivers—Swan, Pelican, Goose, Chenalopex, Ducks, Cormorant being the first six and a mixture of Ducks, Gulls, Grebes and Plovers making up the section: then those without webbed feet, Crane, Herons and Waders, the Kingfisher and Bee-eater. The next section deals with birds of the plain, and begins with the Ostrich, Peacock and Bustard, and so to the Game-birds, the Larks and the Woodcock. The two final sections—'birds which live in all kinds of places'—include roughly the larger and the smaller picine and passerine species. The book is a charming volume: its woodcuts and some of its descriptions are excellent; and the author knew much of the life and something of the structure of his subjects. But it is not a scientific treatise; and to read it is to realise how great a pioneer was the *Ornithology*.

Belon's emphasis upon habitat was largely followed by Willughby in the Tables which he and Ray drew up in the winter of 1666–7 for Wilkins, and which are printed on pp. 144–56 of the *Essay towards a Real Character*. He begins these with the statement that 'birds may be distinguished by their usual place of living, their food, bigness, shape, use and other qualities', though actually he often employs the structure of beaks and feet as characteristics. It is difficult to judge the Tables because he, like Ray in those of Plants, was working under conditions imposed by Wilkins and had to fit his classification into nine divisions. But it is interesting to compare them with the 'Catalogue of English Birds' which Ray appended in 1673 to the first edition of his *Collection of English Words*. This, though retaining the broad distinction between Land and Water birds and following generally the same sequence as the Tables, is classified otherwise upon

1 *L'Histoire de la nature des Oiseaux*, Paris, 1555.

structural lines and very closely resembles the 'method' adopted in the *Ornithology*.[1] Comparing the three systems, the transition from Belon becomes plain: we can trace the gradual replacing of habitat by structure, and admire the skill with which anomalies are removed and a scientific grouping attained.

It would be easy on the evidence of these lists to maintain the view that Willughby was content with superficial characters—he groups his passerines solely by their size, pays great heed to colour and song, and thus classes Oriole, Hoopoe, Bee-eater and Kingfisher as 'more beautiful' thrushes,[2] and Nightingale, Lark and Robin as 'canorous' and thus distinguished from the 'not canorous' Wheatear and Wagtails[3]—and that Ray with clearer perception of scientific principles insisted upon the importance of anatomy, studied the relation of form to function, and succeeded in working out a true arrangement of the main families of small birds, of the ducks, the mergansers, the auks, and the group that includes Gannet, Cormorant and Shag. But it is unfair to Willughby to emphasise the contrast; and there is plain evidence that they were working together,[4] and that even in the *Ornithology* there are still traces of the influence of the Tables. It is of course still more unfair to Ray to suggest that Willughby was alone responsible for the classification in the *Ornithology*[5] or that his was the master mind.[6] Ray in the Preface to the *Ornithology* states that Willughby took an eager interest in examining, dissecting and describing the different species—a fact to which the book bears clear witness—that he had great skill in discovering the characteristics of each kind, but that when he failed to do so his descriptions were so minute as to ignore the fact of individual variation. The only allusion to classification is in the sentence: 'Viewing his manuscripts after his death I found the several animals in every kind both Birds, Beasts, Fishes and Insects, digested into

1 The only differences, beyond the increase of subdivisions, are these: In *Orn.* Part I, Sea-eagle or Osprey appears as well as Bald-buzzard; Woodchat, Brown Owl, Hoopoe, Cock of the Wood (i.e. Capercailzie), Rock-pigeon are added: the heading 'Birds having a hard protuberance in the upper chap' (to include the Buntings) disappears: the Rail takes the place of the Bustard (omitted by oversight). In Part II, White Heron, Totanus (i.e. Bar-tailed Godwit), Dotterel, Ash-coloured Diver, Sea-turtle (Black Guillemot), Lesser Sea-swallow, two Black Terns, Larus fidipes (Red-necked Phalarope), Swan-goose, Brent, Scaup Duck are inserted: the Stone Curlew (wrongly described in the *Catalogue* as long-billed) is transferred to its proper class: the Greatest Diver is put with the Grebes with a note that its feet are different: the Auks and the Steganopodes change places: the Kingfisher placed in the *Catalogue* among the water-birds is moved up next to the Woodpeckers.
2 *Real Character*, p. 149.
3 P. 151.
4 As is stated in Willughby's letter to Ray, *Corr.* p. 63.
5 As is argued by Neville Wood in the *Ornithologist's Text-book*, pp. 3–4.
6 As Newton states, *Dict. of Birds*, Introd. p. 7.

a method of his own contriving.' He does not discuss this method, nor claim to have adopted it. Instead on the title-page of the Latin version and in a letter to Aubrey[1] Ray expressly claims responsibility for the classification; and a comparison with the *Catalogue* proves this claim to be correct.

The system is in fact simple and, because based upon real structural differences, in the main successful in bringing together kindred forms. All birds are divided into Land and Water birds: Land birds into (I) those with crooked beak and claws and (II) those with straight beak; (I) into (*a*) carnivorous, (1) diurnal birds of prey, greater, the Eagles and Vultures; lesser, the Falcons and Hawks; the Shrikes and Birds of Paradise[2] and (2) nocturnal, the Horned Owls, Owls and Goatsucker; and (*b*) frugivorous, Macaws, Parrots, Parakeets; (II) includes (*a*) large and flightless, Ostrich, Cassowary and Dodo; (*b*) middle-sized, having (1) large bills feeding on flesh, the Crows; fish, the Kingfisher; insects, Woodpeckers, or (2) smaller bills, having white flesh, Poultry kind; dark flesh, Pigeons or Thrushes; (*c*) small birds, either (1) soft-beaked, insect-feeders or (2) hard-beaked, seed-eaters: Water birds into (I) those that frequent watery places, (II) those that swim; (I) into (*a*) piscivorous, Crane, Heron, Stork; (*b*) insectivorous, (1) long-billed, crooked, Curlew; or straight, Woodcock, Snipe, (2) middle-sized, Sea-pie, Redshank, Stints, (3) short, Lapwing, Plovers; (II) into (*a*) cloven-footed, Moorhens, Coot; (*b*) long-legged, Flamingo, Avocet; (*c*) three-toed, Auks; (*d*) four toes webbed together, Soland Goose, Cormorant; (*e*) back toe loose, (1) blunt-billed, either Mergansers or Shearwater, (2) sharp-billed, either short-winged, Divers and Grebes, or long-winged, Gulls and Terns, (3) broad-billed, either large Swans and Geese, or small, Sea-ducks or River-ducks. It will be seen that he fastened upon some really significant alliances, and in the main, though his subordination is unsatisfactory, the resulting groups are homogeneous and natural. If he attached too much importance to the feet, and so grouped the webbed-footed long-legged Avocet with the Flamingo, he put the Kingfisher next to the Woodpecker,[3] and the Gannet with the Cormorant, and in one case, the Divers, waived the whole-webbed foot and placed it near the 'fin-toed' Grebes. That the Eagle and the birds of prey should head the list, and that therefore the large and predatory Crow tribe should be taken out of alliance with the Passerines, are natural errors; that the two Reed Warblers (*Acrocephalus arundinaceus* and *A. scirpaceus*[4])

1 *F.C.* p. 165: 'There are in Mr Willughby's *Ornithology* two tables of birds as exact as I can contrive.'

2 Before Ray's time it had generally been believed that these were footless: he credits them with strong feet and claws fit for taking prey.

3 Not as in the *Catalogue* after the Sandpipers.

4 The Lesser Reed-Sparrow is surely this species and not, as Gilbert White (Letter XXV, to Pennant) believed, the Sedge Bird (*A. schoenobaenus*).

should be grouped near the Woodpeckers is less excusable;[1] that Game-birds, Pigeons and Rails should be hard to place is a familiar fact; that he did not recognise the affinity between Waders and Gulls, or between Herons and Pelicans, is hardly surprising. No one will claim that he has produced a fully satisfactory classification: no one, who realises how small were his data and how great the chaos of tradition, will refuse a real admiration for his achievement. He had based his system upon accurate study and dissection; he had tried to take the whole life and structure of each species into account; he has given us the first 'scientific' classification.

On 19 December 1674 Ray reported to Lister[2] that the work on the Birds was finished; on 17 June 1675[3] the Council of the Royal Society ordered that the treatise entitled *Francisci Willughbeii de Middleton Armigeri quondam e Societate Regia Ornithologia* be printed by John Martyn their printer; on 9 December a copy was exhibited by Oldenburg;[4] on 8 February 1676 Lister wrote to express his gratitude for 'the kind token' and to say: 'Certainly never man was so happy in a friend as he has been in you who have been so just to his memory and labours.'[5] But the Latin version had only a limited appeal, and if the expense of the plates were to be recovered a more popular circulation was desirable. Ray was invited to produce an English rendering of the work; and Lister (and no doubt others also) suggested that this should be augmented by chapters on falconry, on the care of small birds in cages, and on wild-fowling. Ray agreed to the proposal,[6] though he roundly accused the source proposed to him of mere plagiarism.[7] When, in 1678, the English version was published, it contained along with other corrections and alterations a long treatise on the Art of Fowling inserted in Book I;[8] three pages on the care of singing birds in Book II;[9] a Summary of Falconry[10] after the Appendix (which dealt with birds suspected to be fabulous); a paper on sea-fowl[11] taken from the *Foeroe Islands* by Lucas Jacobson Debes;[12] and in the Preface certain notes, supplied along with reference to this book on the Faeroes, by Lister.[13] Ray also took occasion to notify his readers of the mistakes which he had now recognised in the Latin work and to emend them in the new text.

1 They are grouped with Nuthatch, Wall-creeper, Tree-creeper and Hoopoe (*Orn.* pp. 142–5)—this is one of the few points in which the *Ornithology* is identical with Willughby's list in *Real Character*.
2 *Corr.* p. 112.   3 Birch, *History of the R.S.* III, p. 222.
4 *F.C.* p. 83.   5 *Corr.* p. 116.   6 *Corr.* p. 121.
7 Lister had recommended an English book, *The Gentleman's Recreation*: Ray demonstrated both in his reply and in the Preface to the English *Ornithology* that this was entirely derived from Turberville and Latham: 'I do not blame him for epitomising, but for suppressing his authors' names and publishing their works as his own.'
8 Pp. 29–53.   9 Pp. 200–2.   10 Pp. 397–437.   11 Pp. 438–41.
12 Written in Danish 1670, "Englished by John Sterpin" London 1676.
13 On 8 Feb. 1676: cf. *Corr.* pp. 116–18.

The book itself will always remain an object of reverence to the ornithologist and of admiration to the historian. When we consider the confusion of its predecessors, the short time and scanty material available to its authors, and the difficulties of the subject in days when collecting and observation had none of their modern instruments, the quality of the achievement stands out above any record of defects or shortcomings. They had seen and drawn up first-hand descriptions[1] of at least two hundred and thirty species, often adding accounts of differences of sex or plumage, and in many cases dissecting them and noticing peculiarities of structure and diet. They (or more probably Ray) had collated their specimens with those recorded by all previous writers and done great service in eliminating superfluous species or suggesting that birds previously described might be identical with or the other sex of those they had seen. They had included all previous records that seemed clearly authentic or for which the evidence was reasonably good: but had rejected those vaguely described, or reported on hearsay, as well as those which they regarded as fabulous. Though there are naturally some cases of duplication, as for example the black-breasted summer Dunlin and the same bird in winter plumage,[2] or the insertion of the 'Coddy-moddy' or young Black-headed Gull[3] or of the young drake and duck Golden-eye along with the adult drake,[4] or of the caged Linnet and the wild red cock-bird,[5] there is plenty of evidence of real insight in the recognition of sex differences, as in the case of the Ring-tail and Hen Harrier,[6] the cock and hen 'Coldfinch' or Pied Flycatcher (*Muscicapa hypoleuca*),[7] the White Nun and the Smew (*Mergus albellus*).[8] In the birds that they have themselves handled and described there is seldom or never any serious difficulty in identification. Thus their accounts of the Citril (*Carduelis citrinella*), Siskin (*C. spinus*) and Serin (*Serinus canarius*) not only prove that they knew the three as distinct but that they had fastened upon the very points of difference which the most modern handbook stresses.[9] Dealing with the Rock-thrush (*Monticola saxatilis*), though in deference to the authorities Ray lists the Solitary Sparrow[10] and the Greater Redstart[11] separately, and describes specimens of his own obtained in Florence in August, yet he insists that they are probably the same, and his own descriptions exactly tally with the plumages of an adult hen and of young birds; and he expresses the opinion, 'to speak freely what I think', that 'the Blauvogel of

---

1 It is obvious that Ray had made his own descriptions even of birds that they had both seen; for he often inserts in the middle of Willughby's account a note in brackets ('In the specimen I, J. R., described' etc.); and in others, after what is presumably his description, 'Mr Willughby observed etc.'

2 *Orn.* p. 305.	3 P. 350.	4 Pp. 368–9.
5 Pp. 258, 260.	6 Pp. 72–3.	7 P. 236.
8 Pp. 337–8.	9 *Orn.* p. 256: cf. *Handbook of British Birds*, I, pp. 65, 83.
10 *Orn.* p. 191.	11 P. 197.

Gesner' which he lists separately and had found 'at Chur in the Grissons country' is also the same species.[1] So too the 'Small bird without a name like to the Stopparola of Aldrovandi' is plainly the Spotted Flycatcher (*Muscicapa striata*);[2] the 'Great pied Mountain-finch' is the Snow Bunting (*Plectrophenax nivalis*);[3] the 'Greater Plover of Aldrovandi' is the Greenshank (*Tringa nebularia*);[4] and 'Mr Johnson's small cloven-footed Gull', which 'from the make of its feet I judged to be of the Coot-kind', is the Red-necked Phalarope (*Phalaropus lobatus*).[5]

As has been already noticed, the book is mainly concerned with descriptions of plumage and structure. Occasionally there are admirable accounts of habits. In the case of the Honey-Buzzard (*Pernis apivorus*), which 'hath not as yet (that we know of) been described by any writer', there is not only a full picture of the bird, including the finding of 'a huge number of green Caterpillars of that sort called Geometrae' in its stomach, but it is added:

It builds its nest of small twigs, laying upon them wool and upon the wool its eggs. We saw one that made use of an old Kite's nest to breed in, and that fed its young with the nymphae of wasps: for in the nest we found the combs of wasps' nests and in the stomachs of the young the limbs and fragments of wasp-maggots. There were in the nest only two young ones, covered with a white down, spotted with black. Their feet were of a pale yellow, their bills between the nostrils and the head white. Their craws large, in which were Lizards, Frogs etc....This bird runs very swiftly like a Hen. The female as in the rest of the Rapacious kind is in all dimensions greater than the male.[6]

So too of the Nuthatch (*Sitta europaea*):

It builds in the holes of trees, and if the entrance be too big, it doth artificially stop up part of it with clay, leaving only a small hole for it to pass in and out by. It feeds not only upon insects but upon nut-kernels. It is a pretty spectacle to see her fetch a nut out of her hoard, place it fast in a chink, and then standing above it, with its head downwards, striking it with all its force, breaks the shell and catches up the kernel.[7]

Notes of locality are rare, though it is usually stated where the specimen described was obtained: but of the Dipper (*Cinclus cinclus*) it is said:

It frequents stony rivers and water-courses in the mountainous parts of Wales, Northumberland, Westmorland, Yorkshire etc. That I (J.R.) described was shot beside the river Rivelin near Sheffield in Yorkshire: that Mr Willughby described near Pentambeth in Denbighshire in North Wales. It is common in the Alps in Switzerland, where they call it Wasser-Amzel.[8] It feeds upon fish, yet refuseth not insects. Sitting on the banks of rivers it now and then flirts up its tail. Although it be not web-footed, yet will it sometimes dive or dart itself

1 Pp. 192–3.   2 P. 217.   3 P. 255.   4 P. 298.
5 P. 355.   6 *Orn.* p. 72.   7 P. 143.
8 Cf. *Obs.* p. 419 of it in Glarus.

quite under water. It is a solitary bird, companying only with its mate in breeding time.[1]

Of the Swallow (*Hirundo rustica*) and the problem that so exercised Gilbert White[2] a century later, Willughby writes:

About the end of September we saw great numbers of them to be sold in the market at Valentia in Spain when we travelled through that country anno 1664. What becomes of Swallows in winter time, whether they fly into other countries or lie torpid in hollow trees and the like places, neither are natural historians agreed nor indeed can we certainly determine. To us it seems more probable that they fly away into hot countries, viz. Egypt, Aethiopia etc.[3]

Similarly, Ray writes of the Crane (*Grus grus*):

We saw many to be sold in the poulterers shops at Rome in the winter time, which I suppose had been shot on the sea-coast. They come often to us in England: and in the fen-countries in Lincolnshire and Cambridgeshire there are great flocks of them, but whether or no they breed in England (as Aldrovandi writes...) I cannot certainly determine.[4]

Of the Cormorant (*Phalacrocorax carbo*):

This bird builds not only on the sea-rocks but also upon trees....For on the rocks of Prestholm Island near Beaumaris we saw a Cormorant's nest, and on the high trees near Sevenhuys in Holland abundance. Which thing is worthy the notice-taking: for besides this we have not known or heard of any whole-footed bird that is wont to sit upon trees, much less build its nest upon them.[5]

Of the Puffin of the Isle of Man (Manx Shearwater, *Puffinus puffinus*):

They feed their young ones wondrous fat. The old ones early in the morning, at break of day, leave their nests and young and the island itself and spend the whole day in fishing in the sea...so that all day the island is so quiet and still from all noise as if there were not a bird about it. Whatever fish or other food they have gotten and swallowed in the day-time, by the innate heat or proper ferment of the stomach is (as they say) changed into a certain oily substance (or rather chyle) a good part whereof in the night-time they vomit up into the mouths of their young, which being therewith nourished grow extraordinarily fat.[6]

Of the 'Wild Swan, called also an Elk and in some places a Hooper' (*Cygnus cygnus*), a bird 'not as yet described by any author', Ray writes:

The bill towards the tip and as far as the nostrils is black: thence to the head covered with a yellow membrane (Mr Willughby describes the bill a little differently, thus...). The wind-pipe after a strange and wonderful manner enters the breast-bone in a cavity prepared for it and is therein reflected and after its egress at the divarication is contracted into a narrow compass by a broad and bony cartilage, then being divided into two branches goes on to the lungs....

1 *Orn.* p. 149.
2 Cf. Letter 38 to Pennant and Letters 9, 12 and 18 to Barrington.
3 *Orn.* p. 212.   4 *Orn.* p. 274.
5 *Orn.* p. 330.   6 *Orn.* p. 333.

We have observed in the wind-pipe of the Crane the like ingress into the cavity of the breast-bone....If you ask me to what purpose...I must ingenuously confess I do not certainly and fully know.[1]

In dealing with birds that they had not personally seen, Ray is on the whole remarkably judicious. He laid down in the Preface the principles on which he acted—to omit 'as being no part of the subject' birds admittedly fabulous: 'Phoenixes, Griffins, Harpies, Ruk and the like': but in cases where he was doubtful to put their descriptions in an Appendix. Thus he includes in the body of the book the Dodo[2] and the Penguin or Great Auk,[3] both of which he had studied in 'Tradescant's cabinet at Lambeth near London',[4] the Cariama (*Cariama cristata*) and the Ankima (Screamer, *Palamedea cornuta*) described by Marcgraf, which Ray places near the Crane,[5] and Marcgraf's Anhinga (Snake bird, *Plotus anhinga*), which he rightly groups with the Pelican.[6] But he relegates the Hoatzin[7] (*Opisthocomus hoazin*), Hernandez' description of which is very vague, to the Appendix: and as he confessed later[8] rejected the Condor altogether on the ground that the stories of its size were incredible. He accepts the Curucui[9] (one of the Trogons) and places it among the Picarian birds: but the Quetzaltototl[10] (*Pharomacrus mocinno*), described fully by Hernandez and the most gorgeous of all Trogons, he puts in the Appendix: the Jacana (*Parra jacana*) appears as the Brazilian Water-hen,[11] while the Colins (*Ortyx*) or 'Indian Quails'[12] are in the Appendix, probably because in this case the former is much more clearly described. The fact is that most of the species included in the text can be identified with tolerable certainty; and that Ray chose the quality of the descriptions as a criterion. Certainly the book would be the poorer if the notices of the Waxwing, Wall-creeper, Rhea, Jabiru, Glossy Ibis, Stilt, Purple Gallinule, Black-throated Diver and Little Auk had been omitted: it would have suffered still more if he had not been strict in his exclusions or critical in his scrutiny of those accepted.

Of the book's influence enough has perhaps been said. Newton, after surveying the earlier history, declared that it laid the foundations of

---

1 Pp. 356–8.

2 Pp. 153–4. The accounts are taken from De l'Ecluse, who saw a foot at Peter Pauw's house at Leyden in 1605, and from Bondt, *Hist. Nat. et Med. Ind. Orient.*

3 Pp. 322–3. There is some confusion here: the plate, Worm's description and Tradescant's specimen are *Alca impennis*: but De l'Ecluse's description of the 'Magellanic Goose' and that in Terry's voyage refer to some species of Penguin.

4 In the *Catalogue* published in 1656 is the record, 'Dodar from the island Mauritius: it is not able to fly being so big': this may have been the specimen exhibited alive in London in 1638 and seen there by Sir Hamon Lestrange: it passed with much of Tradescant's museum to the Ashmolean in 1684, but was destroyed by order of the Vice-Chancellor in 1755, its head and right foot being preserved.

5 *Orn.* pp. 276–7.   6 P. 332.   7 P. 389.   8 *S.A.* p. 11.

9 *Orn.* p. 140.   10 Pp. 391–2.   11 P. 317.   12 P. 393.

scientific ornithology.[1] Linnaeus 'in his classification of birds for the most part followed Ray, and where he departed from his model seldom improved upon it'.[2] Buffon 'is indebted to it for almost the whole of the anatomical portion of his History of Birds'.[3] Pennant and Gilbert White and the whole school of British ornithologists drew their inspiration from it and from its successor. This volume, though largely an epitome, became the popular hand-book on the subject, and deserves separate consideration.

The two versions of the *Ornithology* represent Ray's chief contribution to the study of birds. But when he had discharged his obligation over Willughby's Fishes and finished his own greatest work on Plants, and the invitation came from Tancred Robinson to produce a complete *Systema Naturae*, he turned back for a short space to them. In 1692 he wrote to Lhwyd that his taskmasters had put him on to this new work;[4] and by the next summer the *Synopsis Quadrupedum* was finished.[5] He went straight on from it to the sequel, and working at it with his usual energy sent it to Robinson on 29 February 1694 and was able to tell Lhwyd on 10 June:

The copy of my *Synopsis Methodica Avium et Piscium* hath been already some while in the bookseller's hand, but he makes no haste to print it and perchance he may have reason: such kind of books he knows in these strait times will go off but heavily, which in the best are but few men's money, as well because of the subject as because of the language in which written.[6]

Two months later he wrote:

I am not troubled that Mr Smith makes no more haste to print my Synopsis of Birds and Fishes. It contains nothing new or unpublished in other books, except what Dr Sloane has communicated. The method is the same with that of Mr Willughby's Histories of the same subjects. No anatomical observations, those would have swelled the book to too great a bulk for an enchiridion [hand-book]. It may be of use to poor students who have not ability to purchase great volumes and desire some insight into those Histories. I have given short descriptions of each species containing the chief characteristic notes, whereby they may be distinguished from others of the same genus.[7]

He was by this time too old and ill, and too busily engaged on the *History of Insects*, to care much for a piece of work which he evidently regarded in the light of a task and an epitome—certainly he was not prepared to deal with the excuses and subterfuges of his publishers. So the manuscript lay neglected and forgotten, 'a prey to dust and moths', until long after its author's death, when the firm's affairs were cleared up and the stock sold. Derham at last unearthed it through Innys and insisted on its publication early in 1713.[8]

1 *Dict. of Birds*, p. 7.   2 L.c. p. 8.   3 Cuvier in *Mem.* p. 106.
4 *F.C.* p. 227.   5 *F.C.* p. 175.   6 *F.C.* p. 249.
7 *F.C.* p. 252.
8 *Mem.* p. 46. Smith died in 1703, Walford sometime in 1709–11: Innys then bought the stock and found the script.

In the main Ray's estimate of the work, allowing for his modesty, is not unjust. It is a compilation, summarising the *Ornithology* and adding to it a large number of the new species from Asia and America which in those days of exploration were pouring in to the cabinets of the virtuosi. In a few points the classification is improved. He has taken the Creepers and Nuthatch away from the Woodpeckers and made new groups for them and for the Kingfisher and Bee-eaters. The Land-rail is still with the Quail and Game-birds: but the Bustard and Little Bustard are separated from them. The Bohemian Chatterer has now become the Silk-tail (the Waxwing, *Ampelis garrulus*) and has been moved from the Jays to the hard-billed Finches; and reference is made to Lister's report of it.[1] But the Goldfinch,[2] by the error of following the sequence of the *Ornithology*, is placed among the small Finches from Brazil. Thus also small improvements in order and in the definition of species can be found in the small birds, and in the Waders and Ducks. Thus he brings the Mergansers next to the Duck tribe; and definitely identifies the 'Swallow-tailed Sheldrake of Mr Johnson' (the Long-tailed Duck, *Clangula hiemalis*, in winter plumage) with Worm's Havelda (the same in its dark breeding dress);[3] and the 'Smaller reddish-headed Duck' as the hen Golden-eye (*Bucephala clangula*)—though he still leaves the 'Larger' (or young drake) as a separate species.[4] But the greatest improvement is undoubtedly in the condensation of the descriptions. In the *Ornithology* there was, as Ray himself had admitted, a tendency to multiply details, so that it was often difficult to determine what was crucial. For example, in the two Godwits, though the barring of the tail in one and the black central feathers in the other are mentioned, they occur in very long accounts of plumage and do not catch the eye. In the *Synopsis* the Godwit, *Limosa belgica*, is explicitly said to differ from the Yarwhelp, *L. limosa*, 'by the colour of its tail'; and this is almost the only point described. Ray's whole training and system of nomenclature had fitted him to produce concise and distinctive definitions; and the excellence of this hand-book lies in the ease with which it enables identification.

There is, in addition, some remarkable new material mainly American and derived from Ray's friends, but partly due to the plan to publish an account of several late Voyages and Discoveries by Sir John Narborough and others. This was to include Friderich Martens's Voyage to Spitzbergen, published in German at Hamburg in 1675—a book which owed much to certain questions submitted to its author from Oldenburg of the

1 Cf. *Phil. Trans.* XIV, no. 175, p. 1161 and Johnson's letter, *Corr.* p. 183.

2 *S.A.* p. 89.

3 *Orn.* p. 364, where he says: 'I am almost persuaded that these are specifically the same', and *S.A.* p. 145.

4 In *Orn.* pp. 369–70 he had inserted the three, but indicated that they were old and young drake and duck of the same. Cf. pp. 142–4.

Royal Society. It deals largely with whaling, but has accurate and interesting lists of plants, birds and fishes. Ray received the English version of Martens from Sloane in October 1693[1] before its publication and drew upon it, as he did from other material supplied to him. He records the nesting of the Golden Eagle (*Aquila chrysaetus*) in 1668 in wooded country near the Derwent in the Peak Country of Derbyshire.[2] The Condor, which he formerly rejected as fabulous, he now on evidence supplied by Sloane inserts among the Vultures.[3] The Toucan is removed from the Magpies[4] to the Birds akin to Woodpeckers and put next to the Aracari.[5] To the Turkey, which here as in the *Ornithology*[6] he describes as African in origin, he now adds the New England Wild Turkey,[7] which quoting John Josselyn[8] he declares to be larger, reaching forty or even sixty pounds in weight.[9] The 'American Mock-bird' (*Mimus polyglottus*) is placed among the Thrushes and described in the Appendix.[10] The American habitat of the Flamingo is added, but not the specific difference of the New-World bird. Martens's account of Spitzbergen[11] gave him three new species of Gull—the Ivory (*Pagophila eburnea*), clearly identified by white plumage and black feet, the Arctic Skua (*Stercorarius parasiticus*), of whose persecution of Kittiwakes he tells the story that Ray told of his Cornish Wagell, and the Burgomaster (*Larus hyperboreus*).[12] Lister, in a letter of September 1685, has provided him with the 'Anser palustris noster Grey Lagg dictus',[13] breeding in the Yorkshire marshland and distinguished by the colour of beak[14] and feet from his Wild Goose, which from its yellow feet and black-barred bill seems to have been the Bean.

Most of the new matter is included in the Appendices, in which he summarises the work of recent explorers and writers. The very accurate account of the 'Fregata' (Frigate Bird) from Du Tertre—'it opens and shuts its forked tail like a tailor's scissors: "the plumage of the males is all black: its wings are immensely wide: it nests in trees" ';[15] a full and classified list of Hernandez' birds, in many cases identifiable; and a series of Jamaican birds obtained or observed by Sloane during his stay in the

1 It was published in 1694 and reprinted by the Hakluyt Society in 1855.
2 *S.A.* p. 6.   3 P. 11, and *Corr.* p. 275.
4 Where Ray put it in *Ornithology* out of deference to Aldrovandi.
5 *S.A.* p. 44.   6 *Orn.* pp. 157–8.   7 *S.A.* p. 51.
8 *New-England's Rarities discovered*, London, 1672, pp. 8, 9.
9 The problem of the earliest introduction and origin of the Turkey is very obscure: the farmyard bird (*Meleagris gallopavo*) came from Mexico, where it was probably already domesticated; the northern species is similar but distinct (*M. americana*).
10 *S.A.* pp. 64, 159.
11 Cf. *Corr.* pp. 266 and 474: Sloane sent a copy of it to Black Notley in Oct. 1693.
12 *S.A.* pp. 126–7.   13 P. 138.
14 This Lister describes wrongly as black at base and tip.
15 P. 153, quoting *Histoire générale des Antilles* (Paris, 1667), II, pp. 269–71.

island,[1] and containing the 'Old Man' (*Hyetornis pluvialis*), the 'Logger-head' (probably *Pitangus caudifasciatus*) and the 'Bonana Bird', so named because 'it frequents trees called bonanas'.[2] Sloane's list is certainly part of Ray's original script, for it is mentioned in his letters:[3] but the final section of the Appendix, a list of birds of Madras supplied by 'Edward Buckley'[4] to Petiver, was added after Ray's death when the manuscript had been recovered. The *Synopsis Avium* ends with two indices, one of names, 'English, Welsh, Cornish, Northern and others, British and Irish'—a regular glossary of local and dialect titles—and the second of 'Latin, French, Dutch, Brasilian, German, Indian, Mexican, Italian, Portuguese etc.', which may be Derham's or more probably Ray's with additions from the Petiver section.

The character of the book gives little scope for notes on habits: but under the 'Lesser Spotted Woodspite or Hickwall'[5] (*Dryobates minor*) he summarises a record which he had sent in a letter to Dr Robinson on 13 August 1684 from Black Notley. As this is one of the last of his observations and deals with a matter still hotly debated, it deserves to be quoted in the original form:[6]

I propose to your consideration whether that sort of bird mentioned by Dr Plot to be often heard in Woodstock-park, from the noise it makes commonly called the Woodcracker, be not the lesser sort of Picus martius varius? For since the publishing of Mr Willughby's 'Ornithology' I have observed that bird sitting on the top of an oaken tree, making with her bill such a cracking or snapping noise as we heard a long way off, the several snaps or cracks succeeding one another with that extraordinary swiftness that we could but wonder at it; but how she made the noise, whether by the nimble agitation of her bill to and fro in the rift of the bough, or by the swift striking of the mandibles one against another as the stork doth, I cannot clearly discern; but an intelligent gentleman, who was very diligent in observing the same bird, said it was the former way.

This same letter brought to an end a discussion which had been conducted for some time and which reveals the need and value of the work which Willughby and Ray had done. A question had arisen as to the nature and origin of the bird called Macreuse which Catholics were allowed to eat as fish in Lent and which had a ready sale in Paris. Stories were told connecting it with the barnacle legend to justify its fishlike affinities. Ray was apparently asked to look into the matter and wrote early in 1683 to Robinson, who after leaving Cambridge four years

1 Sloane went as physician to the Duke of Albemarle and was there from 1687 till 1689: cf. *Corr.* pp. 189, 192, 210.
2 Pp. 187–8: for the 'Bonana-tree' cf. *H.P.* II, p. 1375.
3 E.g. *Corr.* pp. 270, 276.
4 No doubt the Edward Bulkley to whom Camelli sent his pictures of plants for transmission to Petiver: cf. *Corr.* pp. 377, 448 and above, p. 302.
5 *S.A.* p. 43.   6 *Corr.* p. 150.

previously had been studying abroad. Robinson replied in July that the Macreuse was out of season but that 'everyone says that it was originally a fish':[1] he enclosed a description taken from a book by Graindorge which had been suppressed by the priests. Ray, who had originally assumed that it might be a Puffin, found the description 'scarce sufficient to determine to what genus it belongs', but thought it must be a Colymbus.[2] Robinson, who had moved on to Montpellier, arranged that his friend Charleton,[3] staying on in Paris, should procure and send specimens. On 1 August 1684 Robinson, then returned to England, sent off the skins of two Macreuses to Black Notley, and the riddle was solved.

> I had no sooner opened the box [wrote Ray] but instantly I found the Macreuse was no stranger to me, though unknown by that name....A particular description of the cock you may find in Mr Willughby's 'Ornithology', p. 366 of the English edition, among the sea-ducks to which kind this bird belongs and not to the Divers or Douckers as I falsely fancied to myself....It is so rare and uncommon that I take it not to have been described by any that have written the history of birds before.[4]

A note appeared from Robinson in *Philosophical Transactions*[5] stating that the Macreuse was the Scoter (*Melanitta nigra*) and including a large part of Ray's letter. Ray himself adds 'Gallis Macreuse audit' and the reference in his notice of the Scoter in the *Synopsis*.[6]

NOTE. *Ray and Willughby*

It is not necessary to discuss the details of the unpleasant controversy which raged a century ago, at about the date of the foundation of the Ray Society in 1844, as to the respective share which he and Willughby had taken in their joint enterprise. Certain remarks of Sir J. E. Smith in his Introductory Discourse to the Linnean Society[7] were taken as a reflection upon Willughby: they were that 'certainly it is by no means a fair statement of the case to say, with Dr Derham, that Mr Willughby had taken the animal kingdom for his task, as Mr Ray had the vegetable one' and 'indeed Ray was so partial to the fame of his departed friend and has cherished his memory with such affectionate care, that we are in danger of attributing too much to Mr Willughby and too little to himself'. This, judicious as most students of the evidence will think it, provoked a storm: the ornithologists rose in wrath to denounce this impertinent botanist and to defend their patron saint. 'It is our duty to say that the amiable and gentle Ray, whatever he might be in botany, had very little merit as an ornithologist....We are sorry to observe that the credit of Willughby's

1 *Corr.* p. 132. 2 *Corr.* p. 134.
3 I.e. William Courten, the collector: cf. above, p. 209.
4 *Corr.* pp. 148–9. 5 XV, no. 172, pp. 1036–40.
6 P. 142. 7 April 1788, cf. *Transactions*, I, p. 18.

system, and also of his names, is generally, most unjustly, awarded to Ray in works on natural history in the present day.' So wrote Neville Wood in the *Ornithologists' Text-Book*:[1] and the author of the 'Memoir of Francis Willughby' in *Jardine's Naturalists' Library*[2] devotes twenty pages[3] to an attempt to prove by long and oft-repeated quotations from the Preface to the *Ornithology* that, if Willughby was not solely responsible for all the important part of the book, then Ray was a liar and a sycophant, and to sneering very ungenerously at the great naturalist.

It is of course impossible, when two men have not only collaborated on a long piece of research but also shared a close friendship, to imagine that the work of either can have been uninfluenced or that the results can be accurately apportioned between them. But certain points in this connection are indisputable.

(1) Ray was, as Smith was well qualified to judge, in his opinion 'the most accurate in observation, the most philosophical in contemplation, the most faithful in description of all botanists of his own or of any other time'.[4] His University record bears out this testimony—that he was a man of outstanding ability amounting to genius, in the Galtonian sense of that word.

(2) Ray was responsible for the collections made abroad, as the independent evidence of Edward Browne proves;[5] for the recording of observations on birds and fishes made in England; for independent study of birds, fishes and insects as shown in his letters;[6] for the editing and amplifying of the *Ornithology*.

(3) Ray was scrupulously punctilious, indeed generous to a fault, in acknowledging the help of others: witness his insistence in his botanical works on stating that this plant was first discovered by Mr M. Lister and shown to me, this was found by Mr Johnson and shown to me, this was shown to me by T. Willisel—and so forth; and above all his insistence that a contribution to *Philosophical Transactions* be corrected because the caption to it seemed to ascribe credit to him instead of to Lister,[7] and his letters[8] establishing the claims of Hulse and Lister to priority in the discovery of the projection of spiders' threads.

(4) Ray wrote with enthusiasm of all his friends, as for example in the dedication of the *Synopsis Quadrupedum* to Courthope and Burrell: he never utters a word of criticism—we could never gather from his letters that

1 P. 4 (1836), following Swainson, *Study of Nat. Hist.* (1834), pp. 23–30.
2 Vol. XVI, published 1843.
3 Pp. 101–21: he returns to the subject apropos the Fishes and Insects on pp. 124–6.
4 *Mem.* p. 87.
5 Letter of 16 Jan. 1665 to his father: *Sir T. Browne's Works* (ed. S. Wilkin), I, p. 86.   6 To Dent, Lister, Jessop, etc. in Willughby's lifetime.
7 Cf. *F.C.* pp. 47–8: the references to *Phil. Trans.* are IV, no. 50, pp. 1011–16; V, no. 65, pp. 2103–5.   8 *Corr.* pp. 62, 83–4; *F.C.* pp. 48–9.

Lister was one of the most self-satisfied and obstinate of men, or that Sloane was a snob and a dilettante. The Preface to the *Ornithology* is at once a tribute to a devoted friend and an obituary notice of an eminent naturalist; and in the seventeenth century eulogy, as a hundred epitaphs demonstrate, was habitually in superlatives.

Remembering these points we may add the following:

(1) Willughby was manifestly a man of high principles, great fertility of mind and energy of disposition, keen, interested, enthusiastic. He was one of the circle of men of means and position who, sickened by the political, moral and religious extravagances of the time, turned to science and the exploration of nature. He was in the language of the time a 'virtuoso'.

(2) Willughby has left very little certainly original work; and the only letters to him that bear on this question are those from Barrow dealing with mathematics,[1] and his own to Ray urging him to note particulars of the government of French towns on his way home.[2]

(3) Willughby had a real gift for speculative ideas, which Ray did not always think sound or probable—though he does not explicitly condemn them;[3] and a flair for discriminating points of interest, for noting specific differences and inventing characteristic marks, as Ray rightly insists.

(4) Willughby contributed almost nothing to Ray's botanical work except in sharing in the experiments on the flow of sap: whereas a dozen of Ray's friends are named in his Catalogues for help with plants, Willughby is not among them.

(5) Willughby's best work, by his own statement, was done in the study of insects: the records of this are reproduced under Willughby's initials in Ray's *Historia Insectorum*: we can here judge his achievement in its original state: it shows keenness, ability and a wide range of observations, but is elementary as compared with the best of Ray's work.

We may sum up the matter in the following conclusions:

(1) There was a whole-hearted partnership between the two men; and to decry one or the other is to do an injustice to their friendship.

(2) Willughby not only made possible the common task by his generosity and enthusiasm, but stimulated the older man by his alertness, fertility of ideas and boldness of vision. He had less respect for tradition, but also less knowledge, patience and judgment.

(3) If it is necessary to describe them, the evidence makes it certain that Ray was a scientist of genius and probable that Willughby was a brilliantly talented amateur. There is no evidence at present that they were in the same class as naturalists.

1 Derham, *Phil. Letters*, pp. 360, 362–5.   2 *Corr.* p. 9.
3 E.g. his theory of the moult in birds, *Orn.* p. 373.

NOTE. *Sir Thomas Browne on Birds*

A question of some interest to students of early Ornithology arises in connection with the paper 'An Account of Birds found in Norfolk' written by Sir Thomas Browne, preserved among the Sloane MSS., and published in Wilkin's edition of Browne's *Works*, IV, pp. 313–24. The paper on fishes in Norfolk, l.c. pp. 325–36, was evidently written at the same time and originally as part of the same document. We know from his correspondence that Browne supplied much material to Merret for a revised edition of his *Pinax*,[1] but this seems to have been in the form of occasional notes and was never used. The opening sentence of the paper: 'I willingly obey your command; in setting down such birds, fishes and other animals which for many years I have observed in Norfolk', plainly indicates that he was replying to a specific request from someone whom he knew and respected. Internal evidence suggests that the account was drawn up in the year 1664: for the notes on birds end with a record of a Roller (*Coracias garrulus*) killed 'about Crostwick' and sent to him on 16 May of that year: and this record was originally inserted in the script in the middle of the fishes. The only other date given is 22 May 1663, when a Pelican which he suggests was one that had escaped from St James's Park was shot at Horsey. The notes contain no mention of the Little Auk, which he obtained in February 1669, reported to Merret[2] and afterwards to Ray.[3] There is no sign that he was in touch with Merret before 1668, and in his first letter[4] to him he states that 'three years ago a learned gentleman of this country' desired him to give account of animals in these parts, but that this friend died while he was writing it. It seems obvious that the Account is the document then prepared.

Browne must have sent his 'draughts of birds in colours' to Ray shortly after Willughby's death; for they arrived in time to be mentioned in the prefatory pages of the *Collection of English Words*, which was printed late in 1673.[5] There seem to have been no notes with them; but these were certainly sent later, since some of the records in the Account, notably those of the Grey Loon (*Podiceps cristatus* in juvenile or winter plumage) and of the Shearwater (this latter a story of how two of these birds were kept for six weeks by Browne 'cramming them with fish which they would not feed on of themselves'[6]), are reproduced in the *Ornithology*.[7] It is surprising to the modern student to observe that Browne's most

---

1 *Works*, I, pp. 393, 401.
3 *Collection of English Words*, To the Reader, p. 7.
5 Cf. letter of 17 Jan. to Courthope: *F.C.* p. 37.
6 *Works*, IV, p. 316 and letter to Merret, I, p. 402.

2 *Works*, I, p. 406.
4 *Works*, I, p. 393.

7 Pp. 340, 334.

interesting records (of the Cranes seen in hard winters, the Spoonbills[1] which formerly built in the heronry at Claxton and Reedham and 'now at Trimley in Suffolk', the Cormorants nesting in trees at Reedham, the Godwit (*Limosa belgica*) and Ruff 'taken in Marshland', the latter abundantly, the Avocet as 'a summer marshbird'[2]) are not cited by Ray. But this is due to the fact that he and Willughby had already described them and that in editing he inserted only new species and paid little attention to localities: he was writing a History of Birds not an Avifauna of England.

Browne, who complained in 1682[3] that the pictures sent to Ray had not been returned, must have recovered them from Martyn. They are certainly not in the large volume of original paintings and of engravings which Willughby had collected and Lord Middleton still possesses. These paintings come mainly from two sources. Either they are named in Latin and French and are largely passerines, sitting on conventional branches and in stilted and often ill-drawn postures: or they have names in Latin and German with occasional notes in broken English and dates in 1644–5; several are drawn as dead birds hanging up, a Starling, a Pochard and a Great Bustard among them; these may belong to the Nürnberg collection. Nearly all have full names added in Willughby's hand. One, a Sand-grouse (probably *Pterocletus pyrenaicus*), is headed 'Le Jangle de Languedoc A Birde of Passage', and under it in Ray's neat script 'This bird is I suppose the same that is figured and described by Olina under the title of Francolino,[4] though the colours differ being corrupted by the painter to make the bird show beautiful': it is dated on the back 'anno domini 1670'. It is noteworthy that most of these pictures were not reproduced in the *Ornithology*, and of those which are in it the engravings fully deserve Ray's censure.

1 'Platea or Shovelard', *Works*, IV, p. 315: cf. Ticehurst, *Birds of Suffolk*, p. 318.
2 *Orn.* p. 322 only records it for winter.
3 *Works*, I, p. 337.
4 The Francolin (*F. francolinus*) occurred then in Italy, but almost certainly not in France: Ray is mistaken in identifying this picture with it.

CHAPTER XIII

THE *HISTORY OF FISHES*

Le caractère particulier des travaux de Ray consiste dans des méthodes plus claires, plus rigoureuses que celles d'aucun de ses prédecesseurs et appliquées avec plus de constance et de précision.

CUVIER, *Biographie Universelle*, XXXV, p. 256.

As soon as the *Ornithology* was off his hands Ray turned to his second task, and in December 1674 wrote to Lister:[1]

Having finished the History of Birds I am now beginning that of Fishes, wherein I shall crave your assistance, especially as to the flat cartilagineous kind and the several sorts of Aselli; especially I desire information about the Coal-fish of Turner, which I suppose may sometimes come to York. When I was in Northumberland I saw of them salted and dried, but could not procure any of them new taken. Besides the common Cod-fish, the Haddock, Whiting and Ling I have in Cornwall seen and described three other sorts of Aselli from which I would gladly know whether the Coal-fish be specifically distinct. I am also at a loss about the Codling of Turner what manner of fish it should be, and how certainly differenced from the Cod-fish. Of the flat cartilagineous I have seen and described four or five sorts, but I am to seek what our fishermen mean by the Skate, and what by the Flair, and what by the Maid. By the affinity of name one would think that the Skate should be *Squatina*, which yet I believe it is not.[2] The sorts of Raia that I have seen and described are the Thornback or *Raia clavata*, a certain and characteristic note of which is want of teeth. 2. The *Raia laevis vulg.* [*R. batis*]. 3. *Raia laevis oculata*, with only two black spots on the back, one on each side [probably *R. circularis*]. 4. The *Raia Oxyrhynchos*. 5. The *Rhinobatos* or *Squatinoraia*. Rondeletius, and the following authors out of him, have many more sorts.

At the same time he evidently enlisted the help of other friends; for a few weeks later Peter Dent, the Cambridge physician, wrote[3] to say that James Mayfield the fishmonger, who went up to the London market every Friday and looked out for rare fishes,[4] had not been able to procure any 'dried Maids or Thornback at the mart' but had helped him to a fresh Thornback, a female weighing ten pounds. This Dent 'dried a little with

---

1 *Corr.* pp. 112–13.

2 In *Historia Piscium*, p. 70 he explains that Skate is the English for the Laeviraia (apparently *Raia oxyrhynchus*—so the Index to *Historia Piscium* and the Plate C 5, though on p. 69 he makes Skate and Flair synonyms), while *Squatina* is the Monk-fish.

3 *Corr.* pp. 15–16.

4 So *F.C.* p. 113 note. On the occasion of the visit of Charles II to Cambridge, 4 Oct. 1671, the following is an item for his entertainment: 'To James Mayfield, fishmonger, in full of his bill for fish £12. 15s. 0d.' (Cooper, *Annals of Camb.* III, p. 550).

salt, yet not so much but that it will recover its form if soaked in hot water' and sent off to Middleton in a box with some wildfowl.[1] He added a few sentences about Maids and Flairs, and the information that Mayfield had once sold 'a Flair of two hundred weight and upwards' to the cook of St John's College, which had served all the scholars of the College at that time, being 'thirty mess for commons'.[2] In his next letter[3] Dent amplified his account after obtaining a Flair with eggs; and in June 1675 sent another long description, of the sex organs of the male.[4]

These letters form a good starting-point for consideration of Ray's work on fishes. They make clear that he had in fact done a good deal of study and description; that he was engaged in the task of identifying the specimens obtained from markets and fishermen and the names employed by them, and found it none too easy; and that Willughby, whose work he was purporting to edit, had not left any large amount of material.

Ray's own interest in the subject had plainly been considerable, and had apparently begun early in his career—earlier than the publication of his *Cambridge Catalogue*, and long before the famous occasion on which he and Willughby decided to divide the world of nature between them; and our first knowledge of it concerns the discovery of a species then new to science. It is a matter of some obscurity; for the evidence is not contained in any letter or Itinerary; and in his books he gives no date for it. But the fact appears evident. In the *Historia Piscium*, in the second section of Chapter IV, 'On the Trout family', he gives a full description of the 'Guiniad[5] Wallis, piscis lacus Balensis, Ferrae (ut puto) idem'[6]—this is of course *Coregonus pennantii*. Of the locality he says: 'It is found in the lake called Pimblemere near the town of Bala in the county of Merioneth but not in the river Dee which flows through it.' He continues: 'I think that the fish called "Schelley" [*C. stigmaticus*] in Cumberland which is found in Lake Ullswater not far from the town Pereth [Penrith] is altogether the same as the Lavaretus'—that is, as *C. lavaretus*, which he had described in the previous section as found in Lake Constance and in the Ammerzee in Bavaria. The 'Ferra' of Lake Leman he describes later and says that it differs only in size from the 'Lavaretus'. After inserting three others, two from Gesner, he adds a paragraph on the whole group and says that in

1 This identical specimen is described, *H. Pisc.* p. 77: for the birds sent with it, cf. above p. 317.

2 This is repeated in *H. Pisc.* p. 69, but there it is said to have given dinner to all the alumni, 120 men.

3 *Corr.* p. 119: this is undated, but refers to his last letter of the '15th inst.', which is obviously that in *Corr.* p. 15.

4 *Corr.* p. 118.

5 The Guiniad is mentioned as caught in Pimblemere (Bala Lake), but not in the Dee, by Camden, *Britannia*, Merionethshire, c. 656 (Gibson's edition).

6 *H. Pisc.* p. 183.

spite of the many forms described by various authors he does not believe that there are more than two or perhaps three species. 'I have myself seen and described as accurately as possible the Albele of Zurich, the Lavaret of Constance, the Ferra of Geneva, the Guiniad of Wales, the Schelley of Cumberland: the chief difference was in size.'[1] Whether he was right or wrong in refusing to multiply specific distinctions may be open to question: the Whitefish like the Chars are a problem for systematists. But the fact that he had seen the Guiniad is significant: for he was only once at Bala and that in the year 1658. It must have been then—on 3 to 5 September, when he found the plant *Corydalis claviculata* 'in the hedges as you go from Bala down to Pimblemere'[2]—that he discovered the fish.[3]

On the journey to the north in 1661 with Philip Skippon the *Itinerary* records a number of fish seen at Scarborough:[4]

We saw Ling [*Molva vulgaris*], Codfish [*Gadus morrhua*], Skate [*Raia batis*], Thornback [*R. clavata*], Turbot [that is Halibut (*Hippoglossus vulgaris*)],[5] Whiting [*Gadus merlangus*] and Herring [*Clupea harengus*]; they take also Conger [*Conger vulgaris*], Bret [or Turbot (*Rhombus maximus*)], Haddock [*Gadus aeglefinus*] and Mackerel [*Scomber scombrus*]: we saw there among others a long, large, cartilagineous fish, which they call a Hay [*Acanthias vulgaris*], not unlike (they say) to a Dog-fish.

It also describes his dissection of Ling and Turbot, and the distinction between Whiting and Haddock, 'the latter hath a great head *pro ratione corporis* and is marked about the middle of the body with a black spot on each side about the bigness of a three-pence which (they say) is the print of St Peter's thumb and forefinger'.[6] At Whitby there were the same sorts; and others, 'dabs [*Pleuronectes limanda*], billiards,[7] white maws, sword-fishes [*Xiphias gladius*]', were named to them. At Dunbar he records the 'great confluence of people to the herring-fishing'. But neither there nor later are there any special notes.

On the western journey in the next year his first note is of 'a fish called torgoch, blackish upon the back, red under the belly (unde nomen)',[8]

1 *H. Pisc.* p. 186: he had declared the Guiniad, Schelley, Ferra and 'Weissfisch or Alberlin' of Zug to be the same in *Obs.* p. 430.

2 *C.A.* p. 102: see above, p. 113.

3 The *Itinerary* is very brief—mentioning Bala but with no details: a short record appears in the Catalogue appended to the *Collection of English Words*: see below, p. 346.

4 *Mem.* p. 145.

5 In *H. Pisc.* pp. 94 and 99 he explains that in North England they call a Halibut a Turbot, and a Turbot a Bret: cf. *Corr.* p. 94, where he explains this confusion of names to Lister.

6 *Mem.* p. 145.

7 This he describes in *H. Pisc.* p. 169 as the young Coal-fish (*Gadus virens*).

8 This he identifies with 'the Red Charr of Lake Winandermere' (*Salvelinus willughbeii*) and Gesner's *Umbla minor*: *H. Pisc.* p. 196.

taken in the lakes at Llanberis, Bettws and Festiniog[1]—'at Llanberis they say that they are there taken only in the night and that when it is not moonshine': the fish is one of the Chars, *Salvelinus perisii*. Then at Tenby he recounts a very long list—all those mentioned at Scarborough and in addition:

Soles [*Solea vulgaris*], Plaice [*Pleuronectes platessa*], Hake [*Merluccius vulgaris*], Dog or Hound-fish [*Scyllium canicula*], Hornfish or Sea-needles [*Rhamphistoma belone*], Gurnards, red [*Trigla pini*] and white [*T. gurnardus*], Sprats [*Clupea sprattus*], Mullet [*Mugil capito*], and Bass [*Morone labrax*], Suins [? *Salmo cambricus*], Sharks [*Carcharias* ?], Dunhounds [? *Mustelus vulgaris*], Bream [*Pagellus centrodontus*], Flukes, grey and white [*Pleuronectes flesus*], Cowes [? Cuckoo, *Labrus mixtus*], Bleaks or Pollacks [*Gadus pollachius*], Ballon [*Labrus bergylta*], Smelts [*Osmerus eperlanus*], Bull-heads [*Agonus cataphractus*], Butterfish [*Pholis gunnellus*]

—and many crustaceans and molluscs.

After Willughby had left, when Ray went on with Skippon into Cornwall, there are reports of many more fish at St Ives, including Pilchards (*Clupea pilchardus*), 'for this fish it is the best place in Cornwall', 'Tomlins, which are nothing but a young Cod-fish', Shads or Schads (apparently the Allis Shad, *Clupea alosa*),[2] Dories (*Zeus faber*), Sand-eels (*Ammodytes tobianus*) and Launces (*A. lanceolatus*). At Penzance 'we saw and described several sorts of fish, to wit, Mullus major[3] [*M. surmuletus*], Trachurus [*Caranx trachurus*], Pagrus [*Pagellus centrodontus*], Erythrinus [*P. erythrinus*]...Whistling-fish [Three-bearded Rockling, *Motella tricirrata*], Rawlin[4]...and Tub-fish, which is no other than a Red Gurnard'. He also described the simple press by which Pilchards were crushed for their oil. On his second visit to Penzance in 1667 he added to these 'a large Tunny (*Thynnus thynnus*) which was taken in the pilchard nets: they call them Spanish mackerel: it was about seven feet long, and of a great bigness; his stomach was full of pilchards'.[5]

On the continental tour he and Willughby evidently paid much attention to fish, although the *Observations* contain few references to them. At Strasburg Willughby bought pictures of waterfowl and fish from the fisherman Leonard Baltner[6] and apparently also his manuscript record of

1 He names the lakes 'Llanberis, Llynunber and Travennyn', presumably Llyn Peris, L. Munbyr (near Capel Curig) and L. Cwellyn (near Bettws Garnon).

2 In *H. Pisc.* p. 227 he describes the Shad or *Clupea* as running up the Severn.

3 Recorded in *H. Pisc.* p. 286: 'we saw it for sale in the market at Penzance'.

4 The Cornish name for Coal-fish (*Gadus virens*).

5 *Mem.* pp. 190–1 note: so too *H. Pisc.* p. 176, where he notes their migratory habits.

6 *Orn.* Preface and above p. 321.

the fishes of the Rhine.[1] There are many references in the *History* to fishes of the Rhine and Danube, to the fish-markets at Venice, Genoa and Naples, in Sicily and Malta. *Observations* contains the story of the catching of Sword-fish with 'harping irons' in the straits of Messina,[2] which Ray and Skippon watched from a boat. But it was the market in Rome during the winter after Willughby's departure that produced the richest results— 'scarce any fish to be found anywhere on the coast of Italy but some time or other it may be met withal here'.[3] The list that follows is not itself very complete: but taken along with the constant references in the *History* it is evidence that Ray spent much of his time on the subject. A list from Zug[4] and allusions to Geneva and to Marseilles which he visited from Montpellier bear out the same conviction. He was deeply interested, and collected material with enthusiasm.

The results of these expeditions were first used in the Tables of Fishes drawn up in 1666 by Willughby and Ray for Bishop Wilkins and printed in his *Real Character*, pp. 132–43. The obligation to produce nine divisions and group the names in pairs is here as in the other lists a handicap. It prevents the Cetaceans from being given a division to themselves; for Willughby only included two, Whale and Porpoise, equating the latter with the Dolphin, and grouping them with the Sharks, the Saw- and Sword-fish and the Sturgeon. But the 'Viviparous Cartilagineous', occupying the first two divisions, are distinguished from the Oviparous; these are divided into salt-water and fresh-water; and the former into those 'with back fins soft and flexible', with 'two back fins, one spinous and the other soft', and 'with one back fin', 'with oblong figure having slimy skins', and 'plain or flat fish'.

That Ray contributed to the composition of this list is certain; for a number of the entries are of fish which he alone had seen. Thus in the first division the Hog-fish or Centrina (*C. salviani*) is included, and this according to the *History*[5] was only seen in the market at Rome, where it was called the Pesce Porco: Willughby had left before they visited Rome, and this record must therefore be Ray's. One pair of fishes in this division, the Zygaena (*Z. malleus*) and the Fox (*Alopias vulpes*), they had not seen, though they had a dried specimen of the former[6] and had heard of the latter in the Mediterranean.[7] Of the last pair Willughby had seen and described the Sturgeon (*Acipenser sturio*) at Venice[8] and had heard and left notes of the 'Huso' (*A. huso*) from the fishmongers at Vienna.[9] The next

1 *H. Pisc.* Preface contains an acknowledgment to Dr Frederick Slare for translating this from the German: the difficulties of this list are described by Ray, *F.C.* p. 143.

2 *Obs.* pp. 316–17.	3 *Obs.* p. 362.
4 *Obs.* p. 430.	5 *H. Pisc.* pp. 58–9.
6 *H. Pisc.* p. 55.	7 *H. Pisc.* p. 54.
8 *H. Pisc.* p. 239.	9 *H. Pisc.* p. 243.

division begins with a pair that are derived from Rondelet and Belon, the Aquila (*Myliobatis aquila*) and the Pastinaca (*Trygon pastinaca*). It includes the 'Crampfish' (the Electric Ray, *Torpedo nobiliana*), which they had seen at Genoa, Leghorn, Rome and elsewhere in Italy. The third division contains the Cod-family, the Tunny (*Thynnus thynnus*) and Pelamys (*P. sarda*), which Ray distinguishes clearly by its banded spots,[1] the Kite-fish (Flying Gurnard, *Dactylopterus volitans*), which Ray records as on sale in Rome, Sicily and Malta, and the 'Swallow-fish or Hirundo Plinii' (Flying-fish, *Exocoetus volitans*), which he obtained at Reggio.[2] Three fishes in this division are derived from Marcgraf. Of another, the Lupus marinus (Cat-fish, *Anarrhichas lupus*), there is a note that the tops of its teeth are commonly sold for toad-stones: this Ray records as stated by Merret in his *Pinax*: the fish itself, he says, is taken near Heligoland ('insula terrae sanctae'), but his friend Johnson had obtained and described one caught near Hartlepool.[3] In the fourth division, Mullets and Gurnards, there are at least two fish which Ray had only seen at Rome, the Saurus, of which he says that although for the five months of his visit he constantly frequented the fishmarket he only saw a single specimen,[4] and the Trumpet-fish (*Centriscus scolopax*), of which he found several there among other tiny fishes.[5] Several of the others were seen and described by Willughby at Venice or Genoa. The fifth division contains the Sea-Breams and Wrasses—a large number of species having been noted in Italy, *Pagrus auratus*, *Dentex vulgaris* and *Box vulgaris* being easily identifiable. The sixth contains the Eels, Lampreys and other long-bodied fishes, including the 'Sheat-fish or River Whale' (*Silurus glanis*), which they had seen at Vienna:[6] the seventh is the Flat-fish—all British: the eighth the Globe-fishes and Sea-horses, of which latter it is remarked that some reckon them amongst insects:[7] and the ninth the 'Squamous River-fish', from the Pike to the Bansticle or Stickleback (*Gasterosteus aculeatus*).

We have examined these Tables at some length not so much for the sake of their classification, which is much inferior to that of the *History*, but because they demonstrate, what is less evident in the Tables of Birds and Animals, that although Wilkins ascribes them to Willughby they are in fact the joint work of the two friends, who composed them together at Middleton on the basis of the notes of their tour.

Ray's first independent publication on the subject was of the two Catalogues appended with that of English Birds to his *Collection of English*

---

[1] *H. Pisc.* p. 180: he lists both the Pelamys of Belon and the Sarda of Rondelet but refuses to regard them as really distinct.

[2] *H. Pisc.* p. 283.   [3] *H. Pisc.* p. 130.
[4] *H. Pisc.* App. p. 29.   [5] *H. Pisc.* p. 160.
[6] *Obs.* p. 141; *H. Pisc.* p. 128.   [7] *Real Character*, p. 142.

Words in 1673. The former of these is of 'Fishes taken about Penzance and St Ives, given us by one of the ancientest and most experienced fishermen, the most whereof we saw during our stay there':[1] the second is of 'Fresh water Fish found in England'.[2] He divides the Cornish into the Cetaceous kind, the Cartilagineous kind, Long (the Sharks) or Broad (the Rays) and Spinous or Bony Fishes; these last are split up into 'Flat-fishes that swim upright, which at present we will distinguish into 1. Long and narrow, 2. Broad or deep'; the 'Long and narrow' are further split into the Cod kind, the Gurnard kind, 'having as it were fingers before the fins on their bellies', the Herring kind, the Mackerel kind and 'Miscellaneous of several kinds'. The list contains the names of sixty-three fishes, five of which he specially notes that he did not see. The Butter-fish (*Pholis gunnellus*) and the Sea-adder (probably the Viviparous Blenny, *Zoarces viviparus*) he describes as 'very small fishes and not described or mentioned by any author I know of',[3] and of the latter he says in the *History*: 'this little fish the Cornish boys at St Ives pulled out of the sea and brought to us',[4] and its size he compares to 'the quill of a crow or less often of a goose'.

In detail the Catalogue has many points of interest. Of the Sharks he lists the Blue Shark (*Carcharias glaucus*) and the 'Tope' (*Galeus vulgaris*), both of which in the *History* he claims to have seen and described at Penzance.[5] The 'Picked Dog' (*Acanthias vulgaris*), the Rough Hound (*Scyllium catulus* though he calls it Mustelus laevis),[6] the Morgay (*Scyllium canicula*) and the White Shark are the others. With the two Rays he groups the Monk-fish (*Rhina squatina*), 'not rare in the Cornish sea',[7] the Piper or Raio-squatina of Rondelet (*Rhinobatus granulatus*), of which in the *History* he gives only Naples as the locality,[8] and the 'Frog-fish or Sea Divel' (*Lophius piscatorius*), from Genoa.[9] Of 'Flat-fishes that swim side-ways and lie most part grovelling at the bottom' he has eight, but two, 'Lanterns' and 'Queens', he has never seen.[10] Of the Cod tribe he claims to have seen and described all except the 'Bulcard', a fish which as he afterwards realised does not belong here: the 'Bib or Blinds' (*Gadus luscus*) is the only one not found in his earlier records. In addition

---

1 *Coll.* p. 97: presumably this is the 'Catalogus Piscium' which he was finishing in Oct. 1667 at Black Notley: cf. letter to Lister, *F.C.* p. 115, written immediately after his second visit to Cornwall. He alludes to it also in a later letter, *Corr.* p. 95.

2 *Coll.* p. 108.   3 *Coll.* p. 104.

4 *H. Pisc.* p. 160.   5 *H. Pisc.* pp. 50, 51.

6 Cf. *H. Pisc.* pp. 62 and 64 where he identifies Rough Hound with Morgay but claims to have seen both Dogfishes in Cornwall.

7 *H. Pisc.* p. 80.   8 *H. Pisc.* p. 79.   9 *H. Pisc.* p. 84.

10 The flat-fish gave him much trouble, cf. letter to Lister, 2 March 1672 (*Corr.* pp. 94–5): but on the whole he is skilful and successful in disentangling local and dialect names. The Lantern ('Lug aleth Cornubiensibus' as he adds) is *Arnoglossus megastoma*, and the Queen *Pleuronectes microcephalus*.

to the two Gurnards he now mentions the 'Tub-fish or Piper' (*Trigla lyra*); and with the Herrings the 'Alose called in other places Shads which are of the Anadromi coming up rivers, commonly taken in the rivers of Thames and Severn, called in Latin Clupeae and Alosae. They are the biggest of this kind, growing to be far greater than a Herring'[1]—*Clupea alosa* is its modern name, but it no longer spawns in the Thames: it is now in fact rare with us. Of the miscellaneous group two are Wrasses, the first probably *Labrus bergylta*, which at Tenby he had called the Ballon, the other the Cuckoo (*L. mixtus*): 'Old wives which I saw not' may be the Sea-Bream called by that name, *Cantharus lineatus*;[2] the 'Girrock or Horn-fish' is *Rhamphistoma belone*, and the 'Skipper' its kinsman, *Scombresox saurus*; 'Calken, i.e. Weever', whose poison he comments upon in the *History*,[3] is *Trachinus vipera*; the 'Mulgronock', which in the *History* he identifies with the Bulcard,[4] is probably the Shanny (*Blennius pholis*); and 'the Father-lasher, so called by the Cornish boys', is *Cottus scorpius*. The list ends with the 'Gilt-head' (*Pagrus auratus*), 'Sea-bream or Chad' (*Pagellus centrodontus*), 'Dory' (*Zeus faber*) and 'Sunfish or Mola' (*Orthagoriscus mola*), whose name he declares in the *History* to be due to its shining at night and of which he says that there is a stuffed skin in the Royal Society's collection and that he saw and described a specimen taken at sea near Penzance.[5]

The second list, 'Catalogue of Freshwater Fish found in England',[6] is divided into four headings: Anadromi, Lacustres, Fluviatiles squamosi and Fluviatiles laeves. The first consists of four bare names, Sturgeon, Salmon, Shad or Alose, and Smelt. The second has only two; the Char of 'Winandermere', 'this I take to be the same with the Welsh Torgoch...the same I saw and described at Zug in Switzerland...there are two sorts taken in Winandermere; the greater having a red belly they call the Red Char and the lesser having a white belly the Gilt or Gelt Char'—the Windermere Char, *Salvelinus willughbeii*, varies greatly in colour according to sex and age, the deepest red belonging to the males at spawning time; and the Guiniad 'in the lake of Bala...the same with the Ferra in the lake of Geneva and the Alberlin of the lake of Zurich...found also in a lake in Cumberland called Huls water where they call it the Schelley'. Of Scaly River Fishes he first lists the Trout, 'of these there are said to be several sorts: as the Lincolnshire Shard, the Salmon-trout, the Bull-trout, Grey-trout or Skurf: but to me these differences are not well known'. In the *History* he separates the Salmon-trout but applies the name only to the

---

1 *Coll.* p. 102.

2 In *Syn. Pisc.* p. 136 he identifies the Wrasse with the Old Wife on the strength of the Welsh name for Wrasse which in *H. Pisc.* p. 320 he gives as 'Gwrach id est Vetula'.   3 *H. Pisc.* p. 289.   4 *H. Pisc.* p. 133.

5 *H. Pisc.* p. 152.   6 *Coll.* pp. 108–12.

great fish of Lake Leman, where 'I saw them often of 30 pounds weight while staying at Geneva'.[1] He makes one species of the River Trout, borrowing most of his comments upon it, including an account of 'tick-ling', from Gesner; but includes two pages[2] on the various forms of Salmon and Trout from Ralph Johnson, who wrote to him at length on the subject. The list continues with the Samlet, 'of the trout-kind taken in Hereford River' and presumably Salmon Par;[3] the Grayling—'I take this to be the same Fish which in some places of the North they call the Umber' (*Thymallus vulgaris*); the Pike (*Esox lucius*); the Carp, 'this fish though now there is none more common with us was but lately brought over to England:[4] Leonard Mascall[5] in his book of fishing saith that himself was the first that brought in Carps and Pippins' (*Cyprinus carpio*); Bream (*Abramis brama*); Perch, 'called at Huls water the Bass' (*Perca fluviatilis*); Ruffe (*Acerina vulgaris*); Tench (*Tinca vulgaris*); Barbel (*Barbus fluviatilis*); Chub or Chevin (*Leuciscus cephalus*); Dace or Dare (*L. leuciscus*); Bleak or Bley (*Alburnus lucidus*); Roach (*Leuciscus rutilus*); and Gudgeon (*Gobio fluviatilis*). Of River Fish without scales, the Eel (*Anguilla vulgaris*); Eel-pout or Burbot (*Lota vulgaris*); Lampern (*Lampetra fluviatilis*); Minnow, 'Minim or Pink' (*Phoxinus laevis*); Loach (*Cobitis barbatula*); Stickleback, 'of this there are two kinds, one that hath only three prickles on the ridge of the back' (*Gasterosteus aculeatus*), 'another that hath six or more' (presumably *G. pungitius*); the Bull-head or Miller's thumb (*Cottus gobio*).

The Catalogue appeared at the time when Ray began serious work upon the *History*. It was a heavy task; for though the previous lists show some acquaintance with the literature of the subject, they did not involve the thorough compilation and comparison of authorities necessary if a full record of all known fishes was to be prepared. Books were more numerous than in the case of the Birds. Gesner, whose vigorous but fanciful picture of a Shark[6] is reproduced in Ray's *History*;[7] Belon, whose *De Aquatilibus* published in 1553 in Paris Ray certainly knew,[8] and by whose classifica-tion, defective as it is, he was in a few instances[9] influenced; Aldrovandi,

1 *H. Pisc.* p. 198. 2 *H. Pisc.* pp. 192–4.

3 He still gives this a separate entry in *H. Pisc.* p. 192.

4 It is in fact an eastern fish and introduced into this country: in *H. Pisc.* p. 246 Ray dates Mascall's introduction of it as 'about a century ago': Izaak Walton, *Compleat Angler*, ch. IX had mentioned Mascall and the carp, but there is no sign that Ray knew Walton's book: Albert Günther (*Study of Fishes*, p. 590) gives 1614 as the date of its introduction.

5 Clerk of the kitchen to Matthew Parker; died 1589: *A Booke of Fishing with Hooke and Line*, by L. M., was published in London, 1590.

6 *Anim. Marin.* p. 152, published 1555. 7 B 7.

8 He probably did not know his other book, *La Nature et Diversité des Poissons*, published in 1555.

9 E.g. in grouping together all 'fishes of a snake-like form'.

whose *De Piscibus* was published in Bologna in 1638 and who combines extensive borrowings from all his predecessors with fabulous accounts of sea-apes,[1] dragons[2] and unicorn-fish;[3] these three authorities are familiar from their treatises on Birds. Clusius—De l'Ecluse—in the sixth book of the *Exotica* published in 1605 has fourteen pages[4] devoted to fish; and the picture of the Whale in the *History*[5] was drawn from one of his woodcuts. But the two chief authorities, exact contemporaries of Gesner and Belon, are Guillaume Rondelet of Montpellier, whose *De Piscibus Marinis* was published at Lyons in 1554, and Hippolyto Salviani of Rome, whose *Aquatilium Animalium Historiae* appeared in the same year.[6] Rondelet, who had studied medicine and anatomy in Paris, classifies fish from the places—sea, lakes, rivers and out of water—in which they live; and includes among them squids, crustacea, sea-urchins and a few prodigies, including a pleasant wood-cut of a Monk-fish drawn as a Monk.[7] But although Ray rightly complained of his multiplying of species, often with very inadequate descriptions, his work is on the whole sensible, his notes on anatomy and structure useful, and his woodcuts sufficiently accurate for identification. Salviani's[8] is in some respects superior and is certainly a beautiful piece of printing: the plates are engravings and very finely done: the larger part of Ray's best plates, if not original, is copied from him— most of the Sharks and Rays and very many others—and has not been improved in the process. The fish are very badly classified, but there is an alphabetical table at the beginning in which are given the Latin, Greek and Common names, a brief but useful list of characteristics and then the references in Aristotle, Oppian, Pliny, Athenaeus, Aelian and others! The letter-press is in fact almost entirely a mosaic of quotations from these authors, and admirably summarises all that had been written about the name, description, habits and culinary uses of each species. Since these books, published simultaneously in the middle of the sixteenth century, there had been no work of similar importance. Apart from Aldrovandi the only other book regularly quoted by Ray is Stephanus à Schonevelde's *Ichthyologia*, an alphabetical record of marine, river and lake creatures in Schleswig-Holstein and Hamburg, published in that city in 1624. This is a slim quarto, with seven engraved plates of fishes, but including entries of all aquatic life from the Beaver to the Cuttle-fish and the Crab. Some few

1 *De Pisc.* pp. 405–6, copied from Aelian with an imaginative picture.
2 E.g. the *Reversus Indicus*, l.c. pp. 367–8, copied from Peter Martyr.
3 From Ambrosius Paraeus, pp. 692 and 695.
4 Pp. 130–43.
5 Plate A 1, from Clusius, *Exot.* p. 131.
6 It is a curious fact that the four chief books should have appeared within the three years 1553–5.
7 P. 492.
8 Salviani was a man of wealth, physician to three successive popes.

of them are new and clearly identifiable, but the descriptions are generally brief: Ray drew from it all that is of any value. In addition there are the fishes described by Marcgraf and Piso,[1] Bondt, Worm and others, including Nieuhof's East Indians, which Ray relegated to the Appendix. To digest and correlate all these books was of itself a large undertaking: to carry it through was to initiate a new era in the subject.

That Ray made good progress with the mastery of this literature and in pursuing enquiries from his friends is shown not only by the correspondence with Dent about Thornbacks already quoted and by a letter from Johnson about Salmon and Eels,[2] but by the references to printing and to Child's meanness in letters to Lister.[3] The strained relations with Mrs Willughby which led to his removal to Sutton Coldfield must have made his work difficult; and in 1676 he wrote that the *History* 'will take up still a year or two's time'.[4] It is unlikely that he was able to do much continuous work at it until in 1679 he settled down at Black Notley: but from the remark in the dedication that the script had 'long lain in his bookcase a prey to beetles and moths' it is certain that he finished it some time before Tancred Robinson brought its existence to the notice of the Royal Society early in 1685. No doubt when he began the work he was confident that his friend's family would arrange for its publication: their treatment of him and of the book must have been a bitter disappointment. He carried through the self-imposed task thoroughly but without zest; and his every allusion to it in the later letters reveals how sore had been the blow to his affections. He had done his share of the work: it was not his business and he was not in a position to undertake its publication: if Mrs Child had no interest in her first husband's researches, then the book could be forgotten and he could get back to his own proper concerns.

Nevertheless, when Robinson pressed him for the script he was ready to hand it over.[5] On 25 February the Royal Society was informed that it was fitted for the press: on 11 March it was presented to them for publication. Martyn, the Society's printer, had recently died and his stock had been sold.[6] So after discussion it was resolved to approach Dr Fell, the Bishop of Oxford, and the University Press in which he was deeply interested.[7] He agreed to undertake it if the Society took a hundred copies;

---

1 The fourth book of the *Hist. Nat. Brasiliens.* treats of fishes: the descriptions are generally accurate but the pictures are poor—though better than the birds.

2 *Corr.* pp. 127–8.   3 *F.C.* p. 136.   4 *Corr.* p. 126.

5 An account of the negotiations for its publication abstracted from Birch, *History of the R.S.*, has been compiled by Gunther, *F.C.* pp. 88–97.

6 For the difficulties caused by Martyn's death cf. Preface to *Fascic. Stirp.* and above, p. 244. Henry Faithorne was not appointed printer to the Royal Society till 18 Sept. 1685.

7 'Except in the capital and at the two Universities there was scarcely a printer in the kingdom' (Macaulay, *History*, I, p. 191).

and Lister, who had given up his practice in York and moved to London in 1684, was asked to go into the question of illustrations for it.

On 13 March Ray wrote a formal letter to Robinson saying that he had realised that no ordinary bookseller would deal with so costly a work unless encouraged by subscriptions, thanking the Society 'for their forwardness to promote the publishing',[1] answering certain suggestions as to the Narwhal, and suggesting that if delay is necessary 'Mr Willughby's son when he comes of age will be at the charge of plates rather than it should be suppressed'.[2] He adds: 'As to designs for the cuts I have several drawn by hand from the life and have already for every species made a reference to the place where the best figure of it is extant in Gesner, Aldrovandi, Rondeletius, Salvianus etc. (I mean in my judgment) in a paper I have by me which you may command.'

This letter was duly read to the Society on 18 March and a week later Plot, who was apparently acting for the Bishop of Oxford and the press of the University, wrote criticising the proposal to copy plates from other books and saying that nothing could be done until the question of their accuracy had been investigated.[3] Pepys, who was at this time President, was not often at the meetings and Lister, who was eager to see the book in print, took his place; the Society agreed to pay the expense of publication; and a committee was appointed to deal with the matter. Subscriptions were obtained for the pictures, Pepys contributing fifty pounds, which provided seventy-nine of them:[4] type and paper were chosen; printing both of script and plates was begun. But difficulties arose; for the letterpress was printed in Oxford, the illustrations and appendix in London. Negotiations dragged on for more than a year and it was not till 5 May 1686 that a letter was read from Ray thanking the Society for the twenty unbound copies of the book presented to him.[5]

Ray was at this time so busy with his *History of Plants* that the turmoil about the Fishes hardly affected him. But the letters to Robinson printed in full by Gunther[6] deal with a number of points that arose in connection with its contents. That of 1 April discusses two species, a Lampetra and the Finscale, mentioned by Plot in his *Natural History of Oxfordshire* and also by Baltner, the latter of which he thinks may be new. On 29 April he writes that the Finscale of Plot or Rotele of Baltner are the same as the Rudd of which Lister has given him information, and asks Robinson to

---

1 *F.C.* p. 142.   2 *Corr.* p. 164.
3 Cf. Gunther, *Early Science in Oxford*, XII, p. 86.
4 He gave £50 to be laid out as the Council should decide: the Council with Lister in the chair decided to put it to the plates of the *Historia Piscium*: Birch, *History of the R.S.* IV, p. 428.
5 These and the thanks of the Royal Society were apparently all the remuneration that he received. The book seems to have been ready in March, cf. Evelyn, *Diary*, 10 March 1686.   6 *F.C.* pp. 142–54.

add a description of it, using that already given of the Orfus unless any-thing better can be found.[1] This same letter[2] contains a very illuminating remark: 'if I had Mr Willughby's notes, I doubt not but I could find out a more exact description of the Orfus than will be met with in authors; for that fish, I am sure, was more than once described by us. But it is almost impossible to procure a sight of them.' He goes on to add the sentence already quoted:[3] 'I did describe most of the animals we met with in our travels; but all my notes of high and low Germany were unfortunately lost.' He also acknowledges receipt of a book of designs but found nothing suitable for the plates. On 12 May he wrote that Baltner's descriptions were as he had feared of little use; that the East Indian fishes had best be put in the Appendix; that Lister's figure of the Rudd confirms its identity with the Rotele but that 'ours of the red Orfus (Golden Orfe) taken at Augsburg from the life is much better'. On 5 June he discusses the 'Misgurn' (*Misgurnus fossilis*), which he had seen at Nürnberg and Ratisbon, insisting that it is specifically different from the Loach[5] (*Cobitis barbatula*); and says that he has written to Skippon for his designs of fishes. On 8 September Robinson sent him a question about a fish figured by Gesner, a method of distinguishing Pilchard from Herring[6] and copies of Bondt[7] and Piso[8] and assured him that 'the sculps are now much better done than at first, for I complained early: three parts of them will be better than Salvian'.[9] Ray replied on the 14th that he feared that he had omitted through oversight Gesner's fish, that he was delighted that the plates would be so beautiful, and that 'looking over Bontius I make no account of his fishes: he has heathenish figures and no descriptions that are worth anything...this paper is not big enough to contain a censure of the rest: in general I reject them all'.[10]

The important passage about Willughby's notes raises a question which can hardly be avoided. Willughby had undoubtedly left a number of careful and detailed descriptions of many species; and in the *History* his name and notes appear with some regularity, particularly in connection with fishes seen at Venice, Genoa, Leghorn and Naples—that is, on the

1 *H. Pisc.* pp. 252–3 contains the Rudd (*Leuciscus erythrophthalmus*) with a description and discussion of its relation with the Orfus: it is said to be 'broader than the Roach and thicker than the Bream': he does not distinguish it from the Orfe or Nerfling (*L. idus*). He ought to have identified it with his German 'Red-eye', which he describes *H. Pisc.* p. 249.

2 In the part of it printed in *Corr.* p. 165.

3 Cf. above, p. 133 note.   4 *F.C.* p. 146.

5 *F.C.* p. 148 has 'leche', which must be a misprint for 'loche'—Ray's usual spelling.

6 *Corr.* p. 174. This is inserted in *H. Pisc.* p. 224 at the end of the section on Pilchards.

7 *Hist. Nat. Indiae Orient.*   8 *De Indiae utriusque Re Nat.*

9 *F.C.* p. 150.   10 *F.C.* p. 153.

part of the Italian journey on which he and Ray were together. Presumably his notes of German fishes shared the fate that befell Ray's; for his name nowhere occurs in the accounts of them. There are no notes at all of fishes seen on his Spanish journey; and almost all the British records are Ray's or from some of his other friends. A large number of the accounts are taken from the books of others; and the classification, though not uninfluenced by the Tables in *Real Character*, is clearly Ray's own.[1] What then did Willughby contribute?

So far as it is possible to judge the answer can only be that almost all the book is Ray's work. No doubt during the last year that he was living at Middleton he digested Willughby's notes and records and wrote some of the book. But the *Ornithology*, though completed in December 1674, had still to be published and then translated and amplified in the English version. He did not begin the Fishes until the Birds was finished; and the break with Mrs Willughby came before the end of 1675. After that, as appears from this letter, it was difficult for him to get Willughby's papers. Loyalty to his friend and the memory of his enthusiasm and help no doubt justified him in allowing the book to be published as a posthumous work: but its actual contents, as well as all the evidence of Itineraries and letters, show that in fact it was, far more than the *Ornithology*, his own. No doubt as the Tables prove they had collaborated: no doubt Willughby's interest and contribution had been greater than our limited knowledge would suggest. But such facts as are available demonstrate beyond question that, though Willughby had gathered some of the material, the studying, arranging and producing of it were almost entirely Ray's. Here too Willughby was the Maecenas, but Ray wrote the poems.

In justice to the man and to the facts so much must be said. There is no sign that he ever regarded it as a pious fraud or ever resented the decisions of the Royal Society to embellish Willughby's book with plates and to refuse such embellishment to his own *History of Plants*. Rather, as he had written in one of his earliest letters to Courthope,[2] he was a man deeply touched by any act of kindness or consideration and eager to show gratitude and acknowledge obligation. Willughby had given him a home and friendship when Cambridge and the Church rejected him. Willughby at his death had provided him with the means to carry on his researches and to devote the rest of his days to science. Whatever memorial he could erect in honour of his friend would be offered eagerly and without hesitation. If as a result he has made Willughby's name famous in the annals of biology, he was not the man to be jealous or to regret that credit due to himself should be given to his benefactor.

1 As is shown by his letter of June 1675 to Oldenburg; *F.C.* p. 84.
2 *F.C.* p. 29.

The book when it appeared was indeed a sumptuous production.[1] It was published in two volumes; for the plates, 187 pages and an engraved title-page, were printed in London on larger paper and bound up separately. The letterpress, with an elaborate dedication to Samuel Pepys as President and principal contributor to the illustrations and to the Council and Fellows of the Royal Society by 'one who thinks it an honour to hold the lowest place in so illustrious a company', with three pages of Preface, 343 of text, 30 of Appendix, and a full index, was published as from the Sheldonian Theatre at Oxford.

The Preface consists of a note on the Willughby family and of a list of those who deserve thanks for their help—John Fell, the Bishop of Oxford; Pepys, who had contributed more than seventy plates; Lister, who had supervised the engraving and supplied a classification of the Orbes (Porcupine fishes or *Plectognathi*); Robinson, who had dealt with the script, corrected and supplemented it; Ralph Johnson, especially for help with the *Salmonidae*;[2] W.C., that is William Courten or Charleton of the Middle Temple, for the supply of fresh and dried specimens for illustration from his museum; James Frasier, for pictures of Indian fish;[3] Francis Aston, at that time Secretary of the Royal Society, for translating Nieuhof's Dutch work on East Indian fish, and arranging the publication; Frederick Slare, for translating Baltner's German notes; Edward Tyson, the anatomist, for the anatomy of a Shark; Sir Philip Skippon, for particulars of the curing of Herrings and a picture of the Rudd; Peter Dent, for observations of the eggs of Skates; Samuel Langley, Rector of Tamworth, for information about 'Sprats'; and the Museum of the Royal Society for many dried fishes for the plates.

Of these names most are already familiar. Fell, who as Dean of Christ Church was the victim of Tom Brown's epigram 'I do not like thee Dr Fell', had become Bishop of Oxford in 1675, but still continued his energetic interest in the University Press. His enthusiasm for the Church of England and for the Stuart dynasty did not prevent him from co-operation with the Royal Society, though it might not predispose him in Ray's favour. He died in 1686, worn out by his immense labours in the service and disciplining of the University and his efforts to enlist its members against Monmouth. Aston, an exact contemporary of Newton at Trinity

1 The Society has been criticised for this, cf. More, *Isaac Newton*, pp. 307–8: but it is unfair to blame them for not having foreseen and saved up for the *Principia*.

2 His brother Jo. Johnson (cf. *Corr.* p. 105) obtained at Middlesbrough on 18 Sept. 1681 the most interesting of the new species, the 'Brama marina cauda forcipata', *H. Pisc.* App. p. 17 and Plate V 12, now called Ray's Bream (*Brama raii*).

3 Plates G 10, M 5.1, 0.5, V 9, 10, 11, X 9 are marked 'e picturis archetypis D. Jac. Frasier': they were apparently obtained through Lister: cf. *H. Pisc.* App. p. 24. He presented a Capercailzie 'found in the fir-wood in the Highlands of Scotland' to the Royal Society in 1687 (Birch, *History of the R.S.* IV, p. 549).

and a fellow there in 1667, was one of the least distinguished of office-bearers in the Society during the difficult period between Oldenburg's death and the Revolution: he seems to have been concerned rather with organisation than with scientific interests and resigned the secretaryship rather offensively in November 1685.[1] Slare, though born in North-amptonshire, is called a Palatino-German by Foster in the *Alumni Oxonienses*: it is perhaps significant of his origin that he should have trans-lated Baltner's not very literate notes. Little is in fact known of him until 1680, when he became an M.D. of Oxford and a Fellow of the Royal Society: but he was a zealous experimenter, and Evelyn[2] records an evening at Pepys's house when he demonstrated with phosphorus. Edward Tyson is much more definitely linked both with biology and with Ray. He had graduated in 1670 from Magdalen Hall, but had taken his M.D. at Cam-bridge, and soon after his admission to the College of Physicians had begun to give lectures on Anatomy, and to publish elaborate monographs in the *Philosophical Transactions*. The manuscript of his lectures still exists in the Sloane collection, and Ray made much use of his publications both in this book—the anatomy of an embryo Shark[3] and of Lumpus Anglorum (*Cyclopterus lumpus*)[4]—and in his *Synopsis Quadrupedum*.

The contributions of Skippon and of Langley are of sufficient interest in themselves and for the study of Ray to be worth fuller consideration. They appear consecutively on pp. 220–2 of the *History*.

Skippon deals with the Herring-fishery and curing at Yarmouth, Lestoffe (Lowestoft) and Southwold.[5] The season is mid-September to mid-October. Nets two and a half fathoms deep are joined together, so as to stretch if need be a full mile, by a stout cable called the Wallop. The fishermen watch the sea-birds and so locate the shoals. They shoot the nets if possible across the flow of the tide. As soon as a boat is full it goes ashore and hands over its catch to the 'tower', who manages the curing. For this they use a vat holding a last or half a last and put a quantity of salt into it. Then they pour in a 'swilling' of Herrings—about 500, stir them once or twice, sprinkling salt over them; then another swilling, salting and stirring; and so on till the vat is full. The Herrings at the bottom soon become stiff. After sixteen or at most twenty-four hours they pack them into wicker crates, loosely woven so that when they are washed in running water the scales and dirt run out easily. Then they string them on thin wooden spits and hang them up on a wooden framework suspended from the roof at a height within reach of a middle-sized man. The spits are fitted five inches apart to poles called 'loves', and these again to beams

---

1 Birch, l.c. IV, pp. 449–50 and above p. 208.   2 *Diary*, 13 Nov. 1685.
3 *H. Pisc.* App. pp. 13–16.   4 *H. Pisc.* App. pp. 25–7.
5 Sent by him to Ray in Sept. 1671: cf. *Corr.* p. 87.

called 'bawks' fixed upright to the roof, so that the spits hang in tiers ten fingers lengths apart. On the floor below they pile up logs cut into billets, which they light every four hours. When the doors are shut the fire smokes; for there is no chimney and the smoke oozes out through the cracks in the tiles. Five hundred billets are wanted to dry each last. The fish for sale in the home market are cured in a month: those that are exported overseas require six weeks. If the weather is wet or windy, the Herrings on the windward side of the house take longer to dry. So if possible they erect these houses in places sheltered from the wind by other buildings or by trees, mounds or hedges. They use Spanish salt as the best for curing: a barrel and a half of salt is enough for a last. A barrel holds about seven hundred fat Herrings; of the other kinds a thousand, more or less. Ten barrels go to make a last. The fishermen, if they get only a few fish, salt them at sea; but these are thought inferior to those cured on shore. White Herrings, called spring Herrings because caught at the summer solstice, are large and fat: they are gutted and cured in brine and so are called 'pickled Herrings'. The Dutch know how to prepare these more skilfully than our people. Three barrels of Spanish salt are needed for one last of these.

Ray's Latin adds a peculiar flavour to the narrative: but even a translation reveals his keen interest in all processes of manufacture, and is a worthy supplement to the records of mining and smelting which he had published in his *Collection of English Words*.[1] He adds a similar but less detailed account of the capture and salting of Pilchards for export to Spain.[2] It is this natural enthusiasm rather than his friendship for Skippon that explains what is strictly an irrelevant insertion. He liked to know how things were made, to ask like Clerk Maxwell how does it go? If he had not been a botanist and biologist, he might have been a chemist or an engineer.

The second contribution comes from a man of considerable interest, Samuel Langley, Vicar of Tamworth. He was the son of 'the eminent Thomas Langley of Middlewich, Cheshire',[3] and after being educated apparently[4] at Oxford was admitted at Emmanuel College, Cambridge, in 1642–3 and proceeded at once to his B.A. In 1644 he was one of six fellows 'intruded' at Christ's College, but in 1649 married and left Cambridge for the rectory of Swettenham in Cheshire. He removed to Tamworth in 1657, thus becoming a near neighbour of Willughby at Middleton. In 1663

1 Pp. 113–78: he describes the smelting of silver, tin and iron, operations in farming and wire-work, the making of vitriol, red-lead, alum and salt.
2 *H. Pisc.* pp. 223–4: this he had seen and partly recorded in the Itinerary of his second Western tour.
3 Urwick, *Cheshire*, p. 200.
4 His name is not in *Alumni Oxonienses*.

and from that time till his death in 1694 he lived at the village of Bole Hall.[1] There seems to be no other mention of him in Ray's letters or books, but the record here inserted is plainly based upon his personal knowledge and Ray judged it reliable. The Robert Thorley of Knutsford to whom he refers was a member of a strongly puritan family,[2] long established there; and Langley's connection with the puritan cause must have brought him into contact with his neighbour. He describes the Eastertide catches of small fish at Rostherne mere in Cheshire. These lake fish, as Langley had been assured by Thorley and others who had eaten them, were the same as the Sprats caught at sea. 'Ten or even twenty fish are caught at one haul of the seine. At the same time as the catches in the lake, or a little earlier, these fish are caught in the salt water below the bridge at Warrington in the river Mersey, which is tidal, seven or eight miles below the lake. Although those who buy a licence fish the lake whenever the weather is suitable, they never catch these Sparlings except at this particular date'. He then describes the relation of Rostherne to the two rivers Birkin and Bollen which join and flow into the Mersey; notes that Sparlings are caught between Warrington and Runcorn on the latter, but never in either of the two tributaries, even in the weirs that obstruct their course. Have the fish come at flood-time straight from the Mersey to Rostherne avoiding the Bollen and Birkin, as Salmon run up the neighbouring Dane but avoid the Weever which joins it at Northwich? If so, do they live and perhaps breed in the deepest part of the mere and only come to the surface at Easter? So far Langley. Ray adds that he feels sure that they run up from the sea as the Shad does; for if they lived in the mere they would grow into Herrings— a Sprat being as he was convinced a young Herring. 'Yet, if they are never caught in the weirs, Langley's argument is strong. I make no rash statements but leave the matter for further examination.'[3]

It is of course plain from the term Sparling, which Ray gives as a synonym for Sprat but which is in fact a regular name for the Smelt, that the fish were not Sprats (*Clupea sprattus*) as Langley had stated, but Smelts (*Osmerus eperlanus*) coming to the shallows to spawn. Ray knew the Smelt and its habits—'piscis est anadromus in Thamesi aliisque fluviis majoribus frequens'[4]—and had described it in its proper place among the *Salmonidae*: but did not realise that it occasionally became resident in fresh-water lakes. As to the false identification he was here as elsewhere

1 Cf. Venn, *Alumni Cantab.* III, p. 45, and Peile, *Christ's College Biographical Register*, I, p. 464: for Bole Hall cf. Dugdale, *Antiquities of Warwickshire*, p. 824.

2 His son William was baptised in the parish church 25 Feb. 1662: but John, apparently an older son and afterwards a 'minister of the gospel', signed the Allestock Agreement as a non-conformist in 1689 and attended a meeting of Cheshire ministers at Knutsford in 1706.

3 For the problem of the Rostherne Smelts see note at end of chapter.

4 *H. Pisc.* p. 202.

too trustful of his correspondents, ready to accept the evidence of Langley and Thorley as in botany he accepted that of Lhwyd or Plukenet.

The tribute to the Royal Society's Museum is explained in a number of the plates, obviously drawn from dried specimens and marked M.S.R. in the corner: presumably Courten supplied the other and unmarked mummies. They are on the whole not very successful, shrunken and eyeless; but even so a vast improvement upon the plates in the *Ornithology*: and except for a Shark from Gesner and a few of the Marcgraf drawings they are the worst in the volume. In general the standard of merit is high, and some of the commoner species, engraved apparently from fresh examples obtained in the London fish-market, are quite excellent. Engraving, as the portraits of the period and the title-page of this volume prove, had reached a high level; and the artists employed for the original pictures have fulfilled Ray's expectations. Both in technical skill and in accuracy of detail plates like those of the Plaice, Dab, Flounder and Halibut,[1] Coalfish,[2] Char,[3] Pilchard and Herring,[4] Carp,[5] Mullet[6] and Perch[7] leave no room for criticism. Many of those drawn from pictures in other books are hardly less satisfactory: Salviani supplied most of these, and his Sharks, Rays and Monkfish come out well. Baltner's pictures, though small, are well drawn, and a plate like Q 10 on which seven, including Chub, Dace, Bream and Roach, appear together gives a very good impression of their characteristics.

The list of donors whose names appear at the foot of the plates presented by them shows that Pepys's example and influence were effective. Fell, the Bishop of Oxford,[8] Sir John Hoskyns, who had been President of the Society,[9] Abraham Hill, its Treasurer,[10] Francis Aston, its Secretary,[11] and Dr Thomas Gale, the famous Master of St Paul's School[12]—these four having taken an active part in its publication; John Lord Vaughan, who in 1686 became Earl of Carbery and President of the Society,[13] Richard Waller, afterwards its Secretary;[14] men of affairs like Sir John Lowther, afterwards first Viscount Lonsdale and Lord Privy Seal[15] and Richard Baron Coote of Coloony, afterwards Earl of Bellamont, Governor of New York and famous from his connection with Kidd, the pirate;[16] men of science like Sir Christopher Wren,[17] Robert Boyle,[18] John Evelyn[19] and Elias Ashmole;[20] friends of Ray like Edward Browne, son of Sir Thomas,[21]

1 *H. Pisc.* F 3–6.　　2 L 1.3.　　3 N 7.
4 P 1.　　5 Q 1.　　6 R 3.
7 S 13.　　8 X 12, Scorpaena.　　9 N 2, Salmon.
10 P 1, Herring.　　11 G 5, Eel.
12 I 27, Swordfish; Q 3, Rudd.　　13 P 7, Sturgeon.
14 L 1.1, Cod and others.　　15 N 5, Gilt Char.
16 L 2.2, Ling.　　17 E 1, Angler.
18 F 3, Plaice; F 4, Dab.　　19 G 6, Conger.
20 F 2, Turbot.　　21 C 4, Skate.

Martin Lister,[1] Walter Needham,[2] Robert Plot,[3] Leonard Plukenet,[4] Tancred Robinson,[5] Hans Sloane,[6] Frederick Slare[7] and Edward Tyson;[8] these and many others, doctors and members of the Society, contributed. Few of the plates except those illustrating the Appendix were unsubscribed.

Of the plates themselves only one, the elaborate title-page depicting fishermen emptying a net against a sea full of sea-monsters and a sky travelled by flying-fish with Britannia accompanied by nymphs and tritons drawing one of the catch in the foreground, is signed—by Paul van Somer. Some few of them had been drawn for the Latin edition of Bishop Wilkins's book; and these were in the hands of Hunt and were engraved and printed at Oxford—Hunt being paid two sums of £7. 10s. 0d. for his work. The rest were produced in London apparently under Hunt's direction, Halley and Lister being given supervision, and £232. 11s. 7d. being paid for them by the Society.

The book itself begins with introductory essays. In the first Ray defines a fish, rejecting the popular view endorsed by most of his predecessors which included squids, crustaceans and molluscs,[9] indeed everything that lives in water, seals and turtles. He argues that cetaceans, although manifestly viviparous quadrupeds in structure, yet because they are hairless, footless and wholly aquatic ought to be grouped with fish. But he accepts none of the other classes. In the second he surveys the general anatomy, noting the structure of the eyes and of the lateral line and paying special attention to the number and position of the fins: on this he based his classification. The third is a brief treatment of the brain derived from Thomas Willis's[10] book on its anatomy published in 1664. The fourth deals with the problem of hearing, demonstrating that fishes can hear although only the cetaceans have external ears, but admitting ignorance of the auditory mechanism. The fifth deals with the branchial clefts and respiratory system, recognising the analogy between gills and lungs and the necessity for oxygen—air as he calls it—but hesitating to consider the physiology of the process except by quotations from Rondelet, Needham and Willis. The sixth consists of notes on the digestive organs, the heart and kidneys, and here he speaks of his own dissections. The seventh is much

---

1 L 1, 2, Haddock.   2 C 2, 3, Sting-rays.
3 Q 9, Orfe.   4 X 6, Sea-perch (*Serranus cabrilla*).
5 D 4, Electric Ray.   6 V 3, Dentex.
7 L 2.1, Hake.   8 N 11, Lumpsucker.
9 Ray left the study of Molluscs to Lister, and his own references to them are few. Lister's letter of Oct. 1674 (*Corr.* pp. 110–11) shows that he was interested and observant.
10 Willis was Sedleian Professor of Natural Philosophy in Oxford 1660–6, but then came to London as a physician. He was a member of the group who founded the Royal Society. He had died in 1675.

longer and discusses the air-bladder, a subject on which he had written a letter to Oldenburg in 1675,[1] read to the Royal Society[2] and reported in its *Transactions*.[3] In this he had supported Boyle's belief that its function was to 'keep them up in any depth of water', since if it is pricked as Willughby and others had shown they sink to the bottom. He noted that flat-fish 'have no swimming-bladders that I could find'; that the cartilagineous fishes, both Sharks and Skates, have no bladders; that in 'most fishes there is a manifest channel leading from the gullet which without doubt serves for conveying air thereinto, as may easily be tried by anyone that pleases'; that it appears to be closed by a 'valve or some other contrivance' and is coated with strong muscular tissue; that it varies greatly in shape; and that 'in some e.g. the Hake, called in Latin Merlucius, it is inwardly covered with a red carneous substance'. In the *History* he reproduces most of this letter with certain quotations from Needham. The eighth chapter is derived from Borelli's book on the Motion of Animals.[4] The ninth treats of the reproductive organs: he discusses the pairing of the cartilagineous fish, but unlike Rondelet and most of his predecessors rejects the belief that other fishes copulate and maintains that the eggs are laid and then fertilised by the male: he quotes the evidence for spontaneous generation and admits the difficulty of explaining the presence of fish in isolated pools and lakes, but protests his own conviction against it. The tenth chapter deals with the food, age and growth of fish, and refuting the Aristotelian view that Tunnies live only two years quotes the evidence of salmon fishers who 'distinguish each year's fish; calling the yearlings Smelts, then Sprods, then Morts, then Forktails, then Half-fish and only in the sixth year when they are adult Salmon'[5]; and states that Tunnies also are called by different names each year until they are six years old. In Chapter XI he turns to classification, rejecting Rondelet's arrangement by locality on the ground that it separates obvious kindred, such as Eel and Conger, ignores the fish that pass from salt water to fresh, and fails to distinguish between pelagic fish. He refers favourably to Aristotle's division, Cetacei, Cartilaginei, Spinosi; divides the Cartilaginei into the long or Sharks and the broad or Skates, and the Spinosi into flat-fish and fish that swim with the back erect. These last, comprising the vast majority, he divides into the 'long and slippery' or 'anguilliformes', those that lack ventral fins, 'Orbes et congeneres', and those which have two pairs of fins, branchial and ventral. This last group he divides into those with 'soft' and spineless fins on the back and those with spiny rays; the former

---

1 *F.C.* pp. 84–6.
2 Birch, *History of the R.S.* III, p. 227.   3 X, no. 115, pp. 349–51.
4 Giovanni Alphonso Borelli (1608–79), philosopher, mathematician and correspondent of Willis and Boyle: his *De Motu Animalium* was published in Rome in 1680.
5 For a discussion of these names cf. F. Day, *British and Irish Salmonidae*, p. 58.

into classes with three or two or one dorsal fin, those with two being sub-divided into Truttaceae and Gobionaceae, those with one into those 'with a long continuous fin like the Hippurus',[1] the Harengiformes or Herrings and the Fluviatiles or Leathermouthed Fish; the spiny-rayed fall into two sections, those with two and those with one dorsal fin. Finally, in Chapter XII he gives a list of British fish classified on these principles, a list which is much fuller and rather better arranged than that in his *Collection of English Words*, and reveals the value and defects of the system. Four of his main divisions correspond to the modern orders Selachii, Batoidei, Physostomi and Acanthopterygi: but his Eel tribe is a strange medley and his classification by fins removes the Hake and Ling from the Cods and groups the Mackerel with the Salmon. Here as in the Birds he has done good service in instituting a classification by structural characteristics: much more study of physiology was needed before the essential structures could be determined.

Of the Whales Ray admits that his account is very defective. The Dolphin (*Delphinus delphis*) Rondelet had described at length; the Por-poise (*Phocaena communis*) Ray had obtained in April 1669 at Chester;[2] the Whale, of which he recognises only two species,[3] he describes from Faber and Rondelet, though he records one cast up on the shore at Tynemouth in August 1532; the Orca (presumably *O. gladiator*), whose ferocity he notes; the Physeter and the Cete, which he separates though both apparently refer to the Cachalot (*Physeter macrocephalus*), whose picture taken from a specimen stranded in Holland in 1598 he borrows from Clusius; and the Monoceros (or Narwhal, *Monodon monoceros*), of which he had seen many horns and read many fantasies, but which he rightly assumes to be a cetacean.

When he gets to the Sharks and Skates his knowledge is much more exact. He knew the Lamia or White Shark and argues that this is the sea-monster that swallowed Jonah, the Blue Shark (*Carcharias glaucus*), the Hammer-head (*Zygaena malleus*), the Humantin (*Centrina salviani*), which he saw for sale in the fish-market at Rome, and all the common British species: he includes the Saw-fish (*Pristis antiquorum*) among the Sharks. Of the Skates he describes the Eagle Ray (*Myliobatis aquila*); Sting Ray (*Trygon pastinaca*); several Rays (*Raia batis, oxyrhynchus, clavata* and *radiata*—this last from Lister); *Rhinobatus granulatus*; the Monk or Angel-fish (*Rhina squatina*): and the Cramp-fish (*Torpedo*

---

1 Ray did not himself know this fish, *H. Pisc.* pp. 213–14. It appears to be a kind of Blenny.

2 Cf. *Phil. Trans.* VI, no. 76, pp. 2274–9.

3 It is not clear whether he intends to discriminate between them: he describes one as less obese than the other, but both as lacking the back fin: probably both are *Balaena australis*.

nobiliana [1]), whose 'narcotic power' he discusses very fully, quoting Redi, Lorenzini and Borelli, but himself showing some reserve: 'I handled, skinned and dissected several with my own hands and got no sort of harm: so it is evidently confined to the living fish.'[2] Then in a separate book headed 'Cartilagineous but oviparous' he deals with the 'Toad-fish, Frog-fish or Sea-Divel' (the Angler, *Lophius piscatorius*), which he had seen and described at Genoa and of which he gives anatomical details from Sir George Ent[3] and Walter Charleton, the author of the *Onomasticon*; and with it the 'Guaparua or American Toadfish' (File-fish, *Balistes*), of which six species are depicted in his plates.[4]

These species are all dealt with accurately and so as to be easily identifiable, but in this section as elsewhere he adds a number of other headings and brief records, some from Rondelet and some from Marcgraf, which are much less valuable. His love of thoroughness and perhaps of cataloguing led him to give space to insertions which he often recognised as superfluous and which in the case of the American species were too few in relation to the whole fauna to be of any real importance. Thus in the next section dealing with Flat-fish he lists seven species of Sole on the authority of Rondelet and Gesner, but adds a note: 'we cannot but be surprised that Rondelet observed so many species of Sole, for although I sought for fish with all possible care throughout the whole of Italy I could only note two species, no more than Belon';[5] and among the Turbots he includes two on Rondelet's authority but states that in his own opinion one is identical with the Brill (*Rhombus laevis*) already recorded.

In the next section he divides the Eel tribe into those without fins like the Muraena (*M. helena*), which is in fact a true Eel but lacking pectoral fins, and the Lamprey (*Lampetra fluviatilis*); those with pectoral but without ventral fins, the Sea Serpent very fully described from Italy and perhaps *Lepidopus caudatus*,[6] the Eel (*Anguilla vulgaris*), Ophidion (*O. barbatum*), which he had seen and dissected at Venice, and the Sand-eel (*Ammodytes lanceolatus*); and those with both pairs, the Taenis or Bandfish (*Cepola rubescens*), which he saw at Rome, the Misgurn (*Misgurnus fossilis*) at Nürnberg,[7] the Remora (*Echeneis naucrates*) from Marcgraf, the 'Mustela or Sea Loach' (*Motella tricirrata*), which he got from a fish-

---

1 He knew at least two species, but regarded the differences as of coloration only.
2 *H. Pisc.* p. 82.
3 The famous anatomist knighted after a lecture in 1665 by Charles II, and an original member of the Royal Society.
4 I, 19–24, taken from dried specimens in the Museum of the Royal Society.
5 *H. Pisc.* p. 102.
6 Yet Ray describes its tail as 'smooth and ending in a sharp point with no fin' (*H. Pisc.* p. 108), which applies better to *Trichiurus lepturus*.
7 In *H. Pisc.* p. 118 as Misgurn with Ray's description: on p. 124 as Mustela fossilis Gesner and Schonevelde.

monger in Chester in 1671, the 'Mustela fluviatilis, Eel-pout or Burbot'
(*Lota vulgaris*) of the river Trent, the Silurus of Vienna, and the Alaudae
or Blennies, of which he names the Gattorugine (*Blennius gattorugine*) of
Venice and the Mulgranoc (*B. pholis*) of Cornwall. He adds a note that
he would naturally classify this third group with the fish that have one or
two soft fins on the back, but places it here in deference to Willughby's
arrangement: he has transferred to it two or three spiny fish on the ground
of resemblance—the Dragonet (*Callionymus lyra*), seen at Genoa and
Rome, the Miller's Thumb (*Cottus gobio*) and Father-lasher (*C. scorpius*).

He then turns to the Fishes with shorter bodies and without ventral
fins: these have only a single pair of fins. Most of them are of an unusual
shape with a round and globose body like the Orbes (Globe-fishes,
Tetrodontidae, etc.), which can inflate their bodies like a bladder according
to Marcgraf, and whose skin is hard and spinous: or triangular like
Marcgraf's Guaimaiacu (apparently a species of Trunk-fish, *Ostracion*);
or almost circular, with its tail (so to speak) torn off, like Salviani's Mola
(Sun-fish, *Orthagoriscus mola*). These have only four fins besides the tail,
a pair at the gills, one behind the vent, a fourth opposite far down the
back. 'To this class we have referred the Sword-fish [*Xiphias gladius*],
which others place among the cetaceans—deservedly if you regard size
alone, but it neither breathes with lungs nor bears its young alive as the
true cetaceans do: perhaps it should be put among the cartilaginous.'
In fact the chapter consists mainly of the various plectognathous species
so beloved by museums, which Ray did his best to describe and of which
Lister produced an excellent classification into three groups depending on
the number of their 'teeth', two, four or more.[1] To these are added the
Pipe-fishes (*Hippocampus antiquorum* and *Syngnathus acus*), the Sea-adder
(*Zoarces viviparus*), the Trumpet-fish (*Centriscus scolopax*) and the Sword-
fish.

The next chapter is concerned with the Cod family or those of them
with three separate dorsal fins—mostly his British species with a few from
the Mediterranean. These lead on to the two-finned, the Hake and Ling
at their head; the Tunnies and Mackerels; the Trout family—with a section on the various forms of salmon and trout contributed by Ralph
Johnson; the Gobies and the Pogge (*Agonus cataphractus*), a name also
derived from Johnson. The one-finned follow, divided into three, those
with a dorsal fin extending the length of the body, those with it central,
and those with it near the tail. To the first belong a number of species
which Ray had not seen and which are hard to identify: thus the first,
Rondelet's Hippurus, is from the description *Trachypterus iris*; but the
plate if correctly numbered is more like a Blenny. The second begins with

1 Cf. the modern classification of the Gymnodonts into *Diodontidae* and *Tetra-
dontidae*: his third class is the *Ostraciontidae*.

foreigners, including *Naseus unicornis* described by Grew from the Royal Society's Museum,[1] and contains the Herrings—the Sardine, seen at Rome, which he declares to be a young Pilchard; the Anchovy (*Engraulis encrasicholus*) at Venice, Genoa and Rome; the 'Agonus' (*Clupea finta*), which Willughby had described at Verona and which Ray rightly supposes to run up the Po to the lakes; the 'Argentina' (*A. sphyraena*), which he got at Rome; and the 'Horn-fish or Gar-fish' (*Rhamphistoma belone*). The third contains the 'Mugil alatus' (the true Flying-fish, or *Exocoetus volitans*[2]), the Pike (*Esox lucius*) and the Sturgeon. He then turns to the fresh-water fishes of similar structure, and records thirty-four species, three of them from Marcgraf, the rest British or continental. He has included all the British species except the Crucian Carp (*Carassius carassius*), which he only knew from Germany, and one of the Loaches, not then specifically distinguished.

There follows his other main division, the 'spinous fishes whose dorsal fins have aculeate rays'. The first section is of those with two dorsal fins, the anterior spined. He begins with the Bass (*Morone labrax*), rejecting Rondelet's two species and saying that in Italy he noticed much variation in colour, the young being spotted, the adults spotless. The Mullet (apparently *Mugil capito*) he describes at length, noting that it feeds on sea-weeds and runs up into estuaries, and giving an account of its capture at Martigues in Provence; then many species of Gurnard, both British and Mediterranean; the Red Mullet (*Mullus surmuletus*); the 'Draco or Weever' (*Trachinus draco*), the 'Trachurus or Scad' (*Caranx trachurus*); the Perch (*Perca fluviatilis*); 'Lucioperca' (the Sander), which he had seen caught in the Danube; the 'Faber or Doree' (*Zeus faber*) and 'Aper' (Boar-fish), which he says is either not found in Britain or is thrown back into the sea as worthless; several fish called Glaucus described by him and Willughby in Italy, the first being *Lichia glauca*; and the 'Umbra or Corvo' (*Umbrina cirrhosa*), also described from Venice and Rome.

The final section is of those 'with a single fin on the back, the rays being spinous in front', and most of the fifty-six species are described from Marcgraf and obscure. Ray tries to disentangle the names given by his authorities to the various forms called Scarus (the Sea-breams) and to identify them with those which he and Willughby have seen: he fixes upon the teeth as a good characteristic and three species at least can be recognised with reasonable probability. Of the 'Pagrus' (*Pagellus centrodontus*) he tells that 'at Genoa in 1664 we bought a large specimen and kept it overnight in the room meaning to dissect it next morning: lo, in the dark

1 Plate O 4: Ray calls it Monoceros piscis (*H. Pisc.* p. 216).
2 From the description this and not his other entries under Milvus etc. is the *Exocoetus*.

the whole fish shone marvellously like a glowing coal or red-hot iron: never before had we seen such a phenomenon'.[1] Of the 'Turdi' or Wrasses he first describes the species that he had seen in Cornwall (*Labrus bergylta*) and then several from his Italian journey: the 'Pavo' (*Labrus mixtus*) and the 'Julis' (*Coris julis*) are easily identified by the descriptions and plates. The section ends with the Ruffe (*Acerina vulgaris*), 'found in our larger English rivers: the Yare at Norwich, the Cam at Cambridge, the Isis at Oxford'; and with the two Sticklebacks (*Gasterosteus aculeatus* and *G. pungitius*).

The Appendix consists of a series of short notes on the East Indian fishes described by J. Nieuhof, plates of which were included; of extracts from Schwenckfeld on the nature and cookery of fish; of a summary of the available evidence on the Narwhal by Robinson; of a long account of the anatomy of an embryo Shark (*Mustelus vulgaris*) by Edward Tyson the great anatomist and physician[2]; and of a description of the Albacore (*Thynnus germo*). There follow two notes by Jo. Johnson, brother of the Vicar of Brignall. The first, already mentioned, contains the earliest account of the fish now known as Ray's Bream (*Brama raii*), from a specimen washed up in the Middlesbrough marshes near the mouth of the Tees on 18 September 1681: with it he also sent a sketch of the fish elaborated into Plate V 12. The second is of the 'Liparis called in Durham and Yorkshire the Sea Snail' (*Cyclogaster liparis*), caught on 15 September 1685 and very accurately described—especial attention being given to the ventral sucker. Other notes follow; and then a long list of descriptions by Lister of dried specimens of various Globe-fishes and their allies found in the museums; and finally a record of the anatomy of a Lump-sucker by Tyson and a few brief notes by Robinson. Two further pages of 'additions' contain an account of the digestive juices of fish, Ray observing that in dissecting a Pike he noticed that the tail of a fish in its gullet was untouched while the body in its stomach was dissolved as if by an acid: he adds a note by Dr Townes[3] about similarly rapid digestion in a shark. Another paragraph deals with the hearing of fish and with the question why their undersides and bellies are so commonly white while the upper are richly coloured. Finally, there

1 *H. Pisc.* p. 312.
2 He had taken his M.D. at Cambridge in 1680 and published his *Phocaena* in the same year. He contributed records of several dissections to *Phil. Trans.* and these appear in Ray's *S.Q.*: cf. below pp. 382, 387, 475. For his work as a doctor cf. F. G. O'Donoghue, *Story of Bethlehem Hospital*, pp. 223–33.
3 Thomas Townes, 'a learned and ingenious physician' in Barbados (*Corr.* p. 111), was a correspondent of Lister's. He had sent the letter here quoted to Dent of Cambridge with a stuffed shark-skin by his London agent Samuel Penn: cf. *Corr.* p. 115. He had been born and educated in Barbados, graduated from Christ's College, Cambridge, B.A. 1667–8, M.D. 1674 (Venn, *Alumni Cantab.* IV, p. 257).

is a description of a Saurus found at Rome, obviously omitted by over-sight.

The book closes with an Epilogue.

We have not attempted to produce Pandects of Fishes—Conrad Gesner has done that—but only to record the observations of ourselves or our friends or of reliable authorities. We shrink from unnecessary multiplication of species, and to avoid it have visited almost all the chief fishing ports of England, and the markets of Belgium, Germany, Italy, and France; have bought all the species new to us and described them so that the reader can easily recognise them. Rondelet has, we fear, multiplied species unnecessarily: we and Belon and other authors have failed to discover very many of his; and yet have found most of the species pictured by other students of fish. Yet his reputation led us to include all his records—the more readily because the number of Globe-fish discovered by Mr Martin Lister has proved to be much larger than we expected. In view of the many famous scholars who have worked in this field we cannot claim to have discovered many new species; we have found some and can claim to have de-scribed, discriminated and classified more accurately.

That is a modest but just estimate of his work.

The success of the book did not come up to the expectations of the Royal Society or to the merits of the production. The times were singu-larly ill-suited to such work. Monmouth's rebellion and the Revolution, if they had little disturbing effect upon the sage of Black Notley, threw the world of learning and leisure into chaos, and made scientific studies seem almost an irrelevance. The Society had carefully estimated the price at which the volumes should be sold; and on an edition of five hundred copies calculated that they must charge to members of the Society and contributors of plates £1 for a copy on the cheaper paper and £1. 3s. 0d. for a copy on the best, and to others £1. 5s. 0d. and £1. 8s. 0d.[1] But this, though only sufficient to cover expenses, was too large a sum for the general public. Interest in fishes, even if it had led to closer study than in the case of birds, was confined to the few: it is not even to-day wide-spread. Those who possessed libraries or enjoyed beautiful books were distracted by political concerns: the clergy and the growing number of students of natural history could ill afford so large an expense: the conti-nental market, which Hooke was instructed to approach through a book-seller in Amsterdam, was not very responsive: the Society itself was in a precarious condition, unable to pay the salaries of its officers and obliged to suspend publication of its own *Transactions*. For a few years science was under a cloud.[2] This and Ray's greatest work, the *History of Plants*, could hardly have been published at a more unfortunate time.

1 Reports submitted to the Council 21 April 1686 and summarised in *F.C.* pp. 95–6.
2 It was only by Halley's personal efforts that Newton's *Principia* was printed: cf. Weld, *History of the R.S.* I, pp. 307–13.

But as we have seen Ray himself was comparatively uninterested. He had done the work, and was absorbed in botany. His isolation and increasing ill-health made study of fish impossible; and Ralph Johnson's death deprived him of his most enthusiastic correspondent. He only returned to the subject when Robinson's plan for a Systema Naturae led him to produce the *Synopsis Avium et Piscium*, the story of which has been already told.

The second part of this is, like the first, an epitome, with such additions only as the published work of the intervening ten years provided, and as Sloane and Petiver supplied. The brief Preface shows that his conviction that whales are not fish has grown stronger:

Some people refuse to recognise the cetaceans as entitled to be called fishes and insist that the name fish be limited to creatures that breathe through gills and have only one ventricle: I entirely agree that if 'fish' is used strictly and philosophically it must be thus restricted: there are no important characteristics in which cetaceans agree with fishes; they have nothing in common with fishes except the place in which they live, the external shape, the hairless skin, and the motion of swimming; in everything else they agree with viviparous quadrupeds. But to avoid too violent a break with convention or the appearance of holding paradoxical opinions for the present we will treat such creatures as fishes.[1]

It is in fact in the whales that the *Synopsis* shows its greatest advance upon the *History*. Not only does he now clearly distinguish the Right Whale from the 'Fin-fish' or Rorqual, and add the 'White Fish' (*Delphinapterus catodon*) of Friderich Martens[2] and a note on a two-horned Narwhal from Edward Lhwyd, but before the book was finished he had been given by Sir Robert Sibbald a copy of his *Phalaenologia nova*[3] and inserts an epitome of it and of its references to recently stranded cetaceans. The rest of the *Synopsis* shows hardly any evidence of new knowledge. It is an excellent summary of the *History*: the brief descriptions are generally adequate for identification, and the material added so confusedly in the Appendix of the larger work is here inserted in its proper position. The book as a guide to ichthyology gains by compression. But there is much less fresh information than in the Birds. The Indian fishes are still listed separately at the end; and are followed by an abstract from Rochefort and Du Tertre dealing with the fishes of the Antilles and by a list of fishes from Jamaica supplied by Sloane.

1 *S.P.* pp. 3, 4.

2 The *Voyage to Spitzbergen* contains an account of several species of Whale, and this Ray was able to use: see above, pp. 331–2.

3 Published in Edinburgh 1692: Sibbald having become a Catholic and fled from Scotland to London in 1686, but returned to Protestantism and Edinburgh before the Revolution.

Another and final addition, probably added after Ray's death, a Catalogue of rarer Cornish fishes with 'elegant pictures' by George Jago,[1] parish priest of Looe, sent by him to Petiver, has a special interest for comparison with the similar list published by Ray in his *Collection of English Words*. Jago had added two Flat-fish, the first which he calls a 'Kitt', possibly *Zeugopterus punctatus*, the second, a 'Whiff', drawn with eyes on the right side and a strongly bent lateral line, can only be the Megrim (*Arnoglossus megastoma*)—which the description strongly suggests—if the position of the eyes in the picture is wrong; three Wrasses, the Goldsinny (*Ctenolabrus rupestris*) being recognisable; the 'Power' (*Gadus minutus*), which Ray had reported from the Mediterranean;[2] the 'Greater Forked-beard' (*Phycis blennoides*) and a smaller form of it— these he claims to be undescribed; the 'Smooth Shan' (*Blennius pholis*);[3] and the rare Streaked Gurnard (*Trigla lineata*): notes of other species beginning with the Burton Skate (*Raia marginata*) are also given.

Ray's work upon fishes, confined as it was to a short period and even then never regarded as a primary interest, is better fitted to give us an appreciation of his ability as a scientist than his work on birds or animals or insects. As we have seen, there is less reason here than in the *Ornithology* for regarding Willughby as in any real degree responsible for it. Of animals and reptiles Ray had little opportunity for first-hand study: insects he did not live to complete. At fishes he obviously worked hard during his travels, especially in the months in Rome when he was alone with Skippon, visited the fish-market daily, and acquired what was till Cuvier a unique knowledge of Mediterranean ichthyology. Evidently he did not regard himself as an expert on the subject: he knew from his botanical work the difference between the expert and the amateur. Hence he felt an obligation to his predecessors, especially to the revered figure of Gesner and the elaborate researches of Rondelet, which he did not so easily acknowledge in his own proper subject. He is indeed not prepared to accept their findings without criticism. He remains the constant enemy of the legendary and the fanciful: he knows that age and sex and locality are responsible for variation, and maintains his objection to the reckless multiplication of species. But he was aware of the limits of his own study and was slow to reject what creditable witnesses recorded. In consequence

1 Possibly the George Jago, son of John of Llanteglos, who matriculated at Oxford in 1663 (Foster, *Alumni Oxon.* p. 798): he had planned a full history of Cornish fish, but at his death his notes for this were lost: cf. Borlase, *Natural History of Cornwall* (1758), p. 264.

2 *H. Pisc.* p. 171.

3 He claims that this differs from Ray's Mulgranoc (*H. Pisc.* p. 133) because of its jugular spines, which Ray's description had omitted: Ray had in fact said that its shape and fins resemble those of the Gattorugine and had described the spines fully and as 'fins though they resemble barbules'.

his books are loaded with names that remain of dubious value and uncertain application: they need precisely that scrutiny and expurgation which Artedi was able to give.

Moreover they suffer, as did the Birds and his early books on plants, from the defects of the current nomenclature. He adopted the names already in use, some of them gravely misleading, all of them unsystematic. Sparus, Sargus, Rubellio, Pagrus, Erythrinus—there is nothing to suggest that the fishes thus designated have anything in common, or to convey, as his brief descriptive titles of plants so often convey, any idea of the particular specimen. For a list of the British freshwater fishes Rudd, Roach, Bream or their Latin equivalents may suffice: for a larger and more complex fauna system is essential. He was not content with the 'Alsine 6 of Gerard' in botany: if he had been able to work more thoroughly at fishes he would have rejected or replaced the 'Turdus alter Rondeletii' in ichthyology. He seems indeed all his life to have supposed that his readers were as familiar with the previous literature as he was himself, or that they would refer to the authorities quoted as a matter of course. But it is strange that the value of a better system of nomenclature never influenced him to alter established usage.

Even so, the work where it is based on first-hand knowledge is astonishingly accurate. When he has seen and described a fish, it is hardly ever difficult to identify it. He picks out salient points, chooses his language exactly, does not obscure the picture by over-elaboration. He knows what characters are specific, and is seldom misled into basing differences upon minor features: indeed he is more ready to unite than to divide unless he is clear that distinctions are demonstrably of structure. Among fish, more easily than among birds, because good illustrations in earlier books were available, he has sufficient material to discover what are the essential marks of the species and of the group. In consequence his descriptions are not only vastly superior to those of his predecessors but, as Cuvier declared,[1] often more accurate and intelligible than those of Linnaeus. If his books are too largely a compilation to escape the need of revision, he provided in them records of so large a number of clearly recognised species as to make the task of his great successors relatively simple.

Considering his lack of time and of material, the ignorance of anatomy, habits, and classification in his day, the difficulty of procuring subjects and the pressure of other work, the *History of Fishes* is a remarkable achievement; and the *Synopsis*, even if it is not more than an epitome, at least provided indispensable help to the naturalists of the next century. Gilbert White and his correspondents might argue the merits of the *Raii nomina* and the *nomina Linnaei*: but Ray's influence is manifest in almost every

1 In *Biographie Universelle* (*Mem.* p. 105).

subject discussed in the *Natural History of Selborne*, and though the *Methodus Plantarum*, the *Ornithology*, the *Observations*, the *Wisdom of God* and the *Letters* were all treasured, it was the *Synopses* which were White's starting-point and his text-books.[1] The debt which natural history in Britain, and indeed world over, owes to them can hardly be overstated.[2]

NOTE. *The Smelts of Rostherne Mere*

Langley's contribution to the *History*, though mistaken in its confusion of the Sparling with the Sprat, raises a point of some interest. Rostherne Mere is nearly three-quarters of a mile in length by half a mile in breadth: it is very deep, in places about a hundred feet; and the water in the deeper parts is heavily impregnated with salt from the deposits which are found in that area. The mere is connected by the Rostherne brook with the Birkin, and this runs into the Bollen and so into the Mersey, exactly as described by Ray. There are local rumours of a subterranean outlet connecting it directly with the Mersey:[3] but there is no evidence of this. Günther, who investigated the matter, secured two small specimens of the Smelts in 1895 and these are registered in the British Museum of Natural History as 'Rostherne Mere (introduced)': but in his *Study of Fishes*[4] he indicates his belief that they are a resident race. This is supported by Regan, who in his *British Freshwater Fishes*[5] (1911) says: 'In many Swedish lakes the Smelt is a freshwater resident throughout the year, and in Britain this seems to be the case in Rostherne mere in Cheshire': in his *Guide to the British Freshwater Fishes in the British Museum* (1932) he says: 'In Rostherne mere the Smelt is a permanent resident.'[6] This is consonant with Ray's evidence that no Smelts are caught in the Bollen or Birkin, though he himself inclined to believe that they were migratory, partly because they were caught annually in the Mersey between Runcorn and Warrington, but chiefly because he thought that they were Sprats and if resident would grow up into Herrings.

It seems clear that the Smelts are and have been since Ray's time[7] resident in Rostherne; that they live in the deeper waters and are of small

1 White's own copy of the *Synopsis Avium et Piscium*, obtained by him in 1742 when he was twenty-two and containing his marginal notes, is preserved in the Newton Library, Cambridge.

2 Cf. above, p. 31.

3 So Green, *Knutsford, its Traditions and History*, p. 92.

4 P. 647. 5 P. 130. 6 P. 18.

7 His statement destroys that in Methuen's *Little Guide to Cheshire* that 'the Smelts sometimes called Sparlings were introduced there about a hundred years ago'; though this tradition may have misled Günther.

size; and that in the spring they visit the shallows for spawning. At Easter dead Smelts are still frequently found on the edge of the mere.[1]

How the Smelts first came to Rostherne is a question that can only be settled when the origin of these Cheshire meres is more clearly explained. At present opinion is apparently divided as to whether they go back to a time when this area was covered by sea or are due, as is the case with some, to subsidence.

NOTE. *The Pictures of Fishes*

Lord Middleton still possesses the collection of pictures begun by Willughby and completed, in 1685, by Ray for the illustration of the *History*. They are in a folio volume (originally of a work by Luther) the pages of which have been cut off, leaving a margin of two inches to which the pictures are attached. They are much more varied in source and style than the similar collection of birds, and include water-colours, pencil and ink drawings, two or three oil-paintings on canvas, a number of engravings from books, and a large selection of proofs of the plates actually found in *H. Pisc.* with Ray's notes on them.

Among them are a series by a 'workman' employed by Willughby, illustrating stages of his dissection of a male Flair (*Raia batis*), and the two drawings by Jo: Johnson of Ray's Bream and of the 'Sea Snail' with his description and dating and their names in Ray's hand. The water-colours, including a lovely picture of a Perch (Plate S 13), have the names in Italian and like some exquisite studies of Lizards inserted loose in the same book seem to have come from Rome. Baltner's work is not included: no doubt it and his birds were in a book or portfolio of their own. There are a few cephalopods and crustaceans.

The engravings come from three chief sources—*Libellus varia genera Piscium continens*, thirteen plates by Nicolaes de Bruyn, apparently published in Amsterdam in 1630; *Piscium vivae Icones*, signed 'Nicholaus Joannes Vischer' (Claes Jansz Visscher of Amsterdam) and dated 1634; and the older collection with quaint landscapes in the background by the Fleming Adrian Collaert, published by De Witt in Amsterdam.

1 This information is due to the kindness of Mr R. T. Wigglesworth, agent to Lord Egerton. I am also indebted to Miss W. Comber, the Rev. G. A. Payne and Dr E. Trewavas.

CHAPTER XIV

OF MAMMALS AND REPTILES

So correct was his genius that we view a systematic arrangement arise even from the chaos of Aldrovandus and Gesner. Under his hand the undigested matter of these able and copious writers assumes a new form, and the whole is made clear and perspicuous.

THOMAS PENNANT of RAY, *Synopsis of Quadrupeds* (1771), Preface, p. iii.

It was, as we have seen, on 18 April 1684 and after the completion of the *Historia Piscium* that Tancred Robinson, writing from Geneva, revived Willughby's project for a General History of Nature.[1] The proposal in its main outline was clear—that he should produce a series of hand-books or Synopses surveying the whole order of Nature, and anticipating the work carried out in the next century by Linnaeus. According to the classification then in vogue this would have meant surveys of Animals, Reptiles, Birds, and Fishes; Exanguia, that is Cephalopods, Crustaceans, Shell-fish, and Insects; Plants; and Fossils—these including Minerals as well as 'formed stones'. Ray had got the material for some of these: to condense the *Ornithology* and the *Ichthyology* and even the *Plants* would not be a very serious business. But to deal with Animals and Reptiles was to break new ground; and the rest, except for some small notes of Insects, were wholly untouched.[2] We have seen that he very rapidly put together the *Synopsis Stirpium Britannicarum* and followed it up with the *Sylloge Europeanarum*. We shall see with what energy he took up the collection of Insects, finding the task vastly larger than he expected but struggling along with it literally until his death-bed, and producing a preliminary *Methodus* as a foretaste of the complete survey. But of the second volume, the *Synopsis Animalium Quadrupedum et Serpentini Generis* published in 1693, the year in which the *Synopsis Avium et Piscium* was written, we have unfortunately very little information. We know that he was planning to write it in June 1690;[3] that in 1691 he was busy with the second edition of his *Collection of English Words*, with the publication of the *Wisdom of God* and the preparation of the *Miscellaneous Discourses*; that in 1692 he prepared his *Collection of Curious Travels*, a task rather cruelly laid upon him by Hatton; produced the much enlarged second edition of the *Wisdom of*

---

1 *Corr.* p. 141; cf. above, p. 212.
2 In a letter of March 1693 (n.s.) to Lhwyd Ray states his intention to omit 'crustaceous and testaceous animals', these having been already dealt with by Lister, *F.C.* p. 235.　　　　　　　　　　　　　　　　　　　3 *F.C.* p. 293.

God; was constantly corresponding with Robinson about the Animals [1] and by November could write to Lhwyd that it was finished; [2] and that in March 1693 it was being printed 'slowly by reason of Mr Smith's late illness and indisposition'. [3] It was dedicated to his old pupils and Sussex friends, Peter Courthope and Timothy Burrell, [4] in a letter which shows that the warmth of his affection and the beauty of his Latin had not been impaired by time, a letter the dignity of which hardly conceals its pathos. It appears that he received five guineas for it. [5]

The lack of information about the book is the more regrettable because it not only deals with subjects hardly touched in his other writings, but contains a treatment of classification which marked a great advance and had a great influence, and a record of anatomical and structural particulars which reveals the extent and accuracy of his zoological knowledge. To those who know Ray only from the published record of his life and as primarily a botanist it is something of a surprise to find him not only familiar with physiological literature but obviously trained in dissection. Looking over his career and wondering at the magnitude and mastery of his achievements one cannot but ask when and under what circumstances he acquired his knowledge of zoology and of comparative anatomy. His Cambridge days would seem to have been devoted to University duties, to Hebrew and Greek and Latin, to disputations, sermons and 'Commonplaces', to College business and tutoring. The University took no great interest in physiology; and in botany, which was essential for a medical training, had on Ray's own showing no one with even the slightest claim to proficiency. Yet it must be in this period that he gained the groundwork of his knowledge; and it is probable that his friend John Nidd was responsible for it; for almost the only references to him apart from the tribute in the *Cambridge Catalogue* [6] are an account of his study of the breeding of frogs in the *Wisdom of God*, [7] and a reminiscence of dissecting birds in his chamber in Trinity College, contained in a letter to Courthope. [8] From the *Catalogue* it is clear that though he was Ray's friend and colleague yet his interests were not primarily botanical; from the later references it is evident that he was concerned with experiments in physiology and anatomy and had something of a collection, if not a laboratory. Walter Needham after taking his degree seems to have joined Nidd and

---

1 *F.C.* p. 296.   2 *F.C.* p. 230.   3 *F.C.* p. 236.

4 Courthope lived till 1725; Burrell till 1717: holograph letters of Ray to Burrell (dated July 1690 and March 1695) are preserved in the Library of Trinity College, Cambridge, and are printed in Gunther, *Early Science in Cambridge*.

5 Letter to Robinson, Jan. 1693: *F.C.* p. 297. If these letters had been preserved, we should know more of the book's production.

6 Cf. above, p. 82.   7 Part II, p. 84 (2nd ed.).

8 Printed in Gunther, *Early Science in Cambridge*, p. 347.

Ray; for the researches described in his *De Formato Foetu* were done in Trinity College in 1654 and cover subjects of which Ray shows first-hand knowledge. Ray's work in anatomy was not that of an amateur, but had been undertaken seriously and in company with experts.[1]

It is of course well known that among Ray's contributions to the Royal Society is a long account of the anatomy of a Porpoise (*Phocaena communis*) dissected by him during a visit to Wilkins at Chester in April 1669; communicated to Oldenburg on 12 September 1671; and published in *Philosophical Transactions*, VI, no. 76, p. 2274.[2] He claimed to have 'observed some things omitted by Rondeletius in his description of the Dolphin', and these he pointed out. It is a thorough and careful piece of work, evidently that of an experienced student. Further, during his stay at Padua he had attended the lectures and taken records of the dissections of Marchetti. From the lecture list for the year of his visit, 1663–4, it appears that this was not Dominico di Marchetti,[3] who was Reader in Surgery, but the more famous Pietro, whose subject was Anatomy,[4] and whose son and successor Antonio seems also to have taken part in the work of the laboratory.[5] Ray's 'Anatomical Observations' show that he worked there from 10 December and spent the next ten days on the dissection of a human subject: he reports Marchetti's technique and deductions; and comments upon them briefly but so as to demonstrate his own knowledge of the subject: 'he said that he had noticed...I do not believe'; 'he never finds...my experience convinces me of the opposite'; 'he has seen...this seems to me true'. Then on 25 December he saw a Caesarean operation. After Christmas he went to comparative anatomy, and wrote notes on a hare and a 'Gallina montana' (Capercailzie),[6] recording here exactly as in the *Ornithology* the salient points of interest in the structure of the organs and the contents of the stomach. These reports were found by Dale at Black Notley among Ray's papers after his death and were sent by him to Sloane, who passed them on to the Royal Society for publication.

But though in the description of the Porpoise he constantly compares its organs with those of other quadrupeds which he had obviously studied by dissection and thus justifies Cuvier's claim that he was the first zoologist to make use of comparative anatomy, there is little in what we know of his earlier work to prepare us for the thoroughness and grasp displayed in the *Synopsis*, still less for the very important introductory essays attached to it. These are among the most interesting of all Ray's papers;

1 Cf. above, pp. 44–8.
2 *F.C.* pp. 58–62.   3 So Seward, *John Ray*, p. 17.
4 The list is printed in *Observations*, pp. 209–13; cf. above, p. 133.
5 Cf. letter of Dale, *F.C.* p. 108; *Phil. Trans.* XXV, no. 307, pp. 2282–2303 and, for the young Marchetti, Edward Browne in *Sir T. Browne's Works*, I, p. 91.
6 *Orn.* p. 173: 'At Padua we saw many sold in the poulterers' shops.'

for they deal with subjects of contemporary controversy and so give us an insight into his general biological theories which we can only find elsewhere alluded to in his religious writings: they indicate his attitude towards the larger problems of science, and prove that he was much more than a collector and compiler.

The first deals with the plain question 'What is an animal?'[1] and raises the issue at stake between Descartes[2] and More with which he had dealt more briefly in his *Wisdom of God*.[3] Aristotle had taken sense-perception and especially touch as the characteristic: the Cartesians contended that in the strict meaning the brutes had no sense but were automata without trace of knowledge or intelligence—that as a modern would put it their actions are purely reflex. Ray sets out the four arguments by which this contention is supported; that God can produce mechanisms which act as if they were possessed of conscious awareness;[4] that men can and often do act automatically and without any attention, that is, on the brute level; that brutes, although differing widely from one another in the degree of their response to stimuli, yet neither speak nor give other signs of intelligence; that if brutes possess conscious awareness and are intelligent, they must possess immortal souls, which is absurd in view of worms and sponges. Ray comments that the question is not what God can produce but what He does produce; that if both brutes and men often act automatically, they also both act in ways that involve attention and perception; that though brutes do not employ what is strictly called speech, they reveal conscious awareness in other ways; that what happens to the souls of brutes is not within our knowledge—perhaps they persist and attain new bodies. Then he goes on to admit that in the main brutes act without intelligent control of their actions, but under impulse and to ends which they do not consciously purpose or understand: their actions if rational would involve an intelligence richer than man's reason. Yet many of their actions plainly show a measure of conscious awareness, as he proves from the behaviour of dogs, quoting first from Johann Faber[5] and then from his own observations: how can we explain the fact that a dog running in front waits for his master when he comes to a fork in the path, or at a stream jumps across if it is narrow enough, but does not attempt to do so if it is

1 *Synopsis Quadrupedum*, pp. 1–13.

2 Descartes's first work on physiology appeared in his *Discourse on Method*, published in French in 1637 and in Latin in 1644. He maintained that animals were pure automata, while man possessed a soul whose situation was the pineal body in the brain; and had discussed the matter with More in 1648–9: cf. *More's Works* (1712 edition), pp. 64–6, 70–2.   3 Cf. pp. 42–4.

4 This seems the nearest rendering of 'sensus et perceptio': for 'perceptio' in Cartesian usage is almost synonymous with cognition; cf. A. D. Lindsay in *A Discourse on Method*, pp. 239–41.

5 Of Bamberg, but living at Rome, botanist and anatomist.

too wide, unless he has some appreciation of distance and therefore some conscious awareness? Further, if brutes were automata, how explain their suffering and our disgust at cruelty to them? 'If it is argued that this is mere prejudice unworthy of a philosopher, then I shall stand by that prejudice: put it down to my stupidity or the weakness of your arguments as you like: the torture of animals is no part of philosophy.' Thirdly, why suppose that God has mocked us by filling the world with puppets and automata? the world does not exist for man's sake only. Finally, if you argue that even if brutes have conscious awareness this is a property of matter, why not apply the same argument to rational intelligence? The two are distinct, as we see in dreams: there is a principle of consciousness as well as of reason, and the brutes possess it.

His second paper deals with the generation of animals and with the three problems that were specially exercising the minds of his contemporaries. The first is that of spontaneous generation. This was a matter on which Ray had long ago taken a strong stand. In 1671, writing to Oldenburg,[1] he had said: 'Whether there be any Spontaneous or Anomalous Generation of Animals, as hath been the constant opinion of naturalists heretofore, I think there is good reason to question. It seems to me at present most probable that there is no such thing: but that even insects are all the natural issue of parents of the same species with themselves.' He had gone on to admit that since Redi[2] had gone far to explode the idea of generation from putrefaction there remained two points, the insects bred from galls and those bred in the bodies of other animals. It was a subject that he had constantly discussed with his friends, especially with Lister, who on this point shared his convictions. He had pondered over the parasites of caterpillars, the toads found in rocks, the intestinal worms in human beings; and when he wrote the *Wisdom of God* had roundly asserted that 'there is no such thing in nature as Aequivocal or Spontaneous Generation'. For this he had been attacked by an anonymous correspondent: and so next year in the second edition had expanded his statement and devoted twenty new pages to the evidence.[3] In the *Synopsis*[4] he surveys the whole case systematically.[5]

He begins with an explicit avowal: 'I entirely agree with those who think that no animal at all is born spontaneously and that there is no such thing as equivocal generation.' His reasons are (1) the production of an animal from matter is a work of Creation: God created in the beginning,

1 *F.C.* p. 56, printed in *Phil. Trans.* VI, no. 74, p. 2219.
2 In his famous work *Generazione degl' Insetti*, Florence, 1668: reviewed in *Phil. Trans.* V, no. 57, pp. 1175–6.
3 *Wisdom*, 2nd ed. Part II, pp. 74–94.   4 Pp. 14–22.
5 For T. Burrell's letter on this part of the *Synopsis* cf. *Corr.* pp. 284–5 and for Ray's reply, *Early Science in Cambridge*, p. 354.

but finished His work on the sixth day; (2) if animals are produced spontaneously, why all this apparatus of procreation, the two sexes, these superfluous organs? (3) I think it absurd to suppose that the lower can produce the higher, or the lifeless the living; (4) experiment refutes such generation, as Redi proved for insects and Malpighi for plants; (5) if an insect can be produced spontaneously, why not an elephant—or a man? (6) why are they not new species, but always the same as those born by natural propagation? (7) the opinions of the experts confirm the view, Redi, Malpighi, Swammerdam, Lister, Leeuwenhoek. Finally, he quotes and briefly repeats the answers to objections which he had set out in the *Wisdom of God*.

The second question is whether individual animals were each created in the beginning or are constantly produced by fresh generation—a problem which directly raises the issue between the tradition and the newer outlook. Here he begins not with a statement of his own opinion but with quotations from J. C. Peyer[1] who, repudiating his previous view, now affirms that in every egg from which a female is born the *idea realis* or essential principle of every subsequent birth is contained; for that God is the sole Creator and made all things in the beginning so that it is impious to ascribe to living creatures the power of forming and procreating descendants, or to regard it as due to a fortuitous combination of atoms, or to introduce any plastic or formative principle. In contrast with this he sets forward the opinion of Cudworth[2] that in the work of forming or generating animals God employs the agency of Plastic Nature—or as we should say a Life-force; and this for the reasons (1) that to assume otherwise is to burden the deity with trivial cares and (2) to diminish the sense of his majesty; (3) that the slow progress from imperfect to more perfect observable in generation would seem to be superfluous and otiose if the agent is omnipotent, and (4) that the errors and lapses such as monstrosities, due to the obduracy of matter, are inconsistent with belief in an irresistible agent. He insists that it is absurd to accuse of folly or blasphemy one who like Cudworth was eminent both for piety and for learning, however unconventional his theory may be. Ray himself, as on the whole an ovarian,[3] inclines to the belief that, since each living female contains all the ova that she can produce and becomes sterile when these are exhausted, so the probability is that all ova or *ideae* were created at the beginning; or at least that they are not created by the parent. But instead of developing

1 *Merycologia*, I, ch. v: he received a copy of this book early in 1692 from Robinson: 'I do own him to be an ingenious and careful writer: but yet in some few things I must needs differ from him, they being contrary to my opinions and observations': *Corr.* p. 246.

2 *True Intellectual System of the Universe*, ch. III, § xxxvii: in Harrison's edition, I, pp. 222–6.

3 See below for this controversy.

his own views he examines the two objections to Peyer put forward by J. C. Brunner.[1] The first, that on the mere ground of their bulk it is absurd to suppose that Eve could have contained the germs of all future generations, he rejects with long quotations from Boyle and references to Leeuwenhoek and Hooke. The second, that if all the germs, containing future animals, existed in the ova, monstrosities could not occur, he regards with sympathy: Peyer's answer that such are the work of the devil he dismisses —'why bring in a demon ex machina: these cases are due to natural causes': for himself he regards More's doctrine of Plastic Nature as more satisfactory and supports this by a long quotation from J. J. Wepfer.[2] This paper, though difficult and inconclusive, is of interest as illustrating the problems of this transitional period when the old belief in the traditional cosmogony and theology was being challenged by the new knowledge and outlook, but when the content of that knowledge was still too uncertain and incomplete to supply a firm alternative. Ray, like his contemporaries, had to maintain a precarious footing in two worlds. He was wise enough to reject blind allegiance to the old, or uncritical enthusiasm for the new. But the result reveals rather the difficulty of his position than any great success in overcoming it.

The third question arises directly out of one section of the second. 'Are the animalcules which are increased and perfected by generation situated in the ovum of the female or the semen of the male?'[3] This problem, raised by Leeuwenhoek's discovery of spermatozoa[4] and his hypothesis that the spermatozoon is the true origin of life, revived a very ancient controversy in a new form. Readers of the *Eumenides* will remember that, in the trial of Orestes for the murder of his mother, Apollo secures the defendant's acquittal by the argument that the mother is only the nurse of the seed and therefore that the father is alone entitled to the full status of parenthood. But in Ray's time the weight of opinion among naturalists was on the other side. Malpighi had proved that the nucleus existed in the ovum before fertilisation;[5] and this was regarded as the germ of the future organism which would be stimulated into activity by coition and probably by a volatile salt or effluence of the semen. Ray opens the discussion by stating Leeuwenhoek's contention; says that he does not dispute the discovery; but inclines to the other view, and sets out a number of points which tell

1 *De Pancreate*, a reply to Peyer, in his *Experimenta c. Pancreas*, Amsterdam, 1683, pp. 165–6.

2 He quotes a paper by him in *Ephem. Germ.* III, observ. 129.

3 *S.Q.* p. 36.

4 This was reported to Brouncker, President of the Royal Society, in Nov. 1677 and published in the *Phil. Trans.* in 1683 (XIII, no. 152, pp. 347–55): but thanks to his lack of scientific training and his secretive character was regarded with some suspicion.

5 In his work, described in London in 1672 (*Phil. Trans.* VII, no. 87, pp. 5079–80), on the formation of the chick in the egg.

against the new claim. Harvey,[1] de Graaf,[2] Dionis,[3] Duverney,[4] Nuck,[5] Peyer,[6] Malpighi,[7] and several contributors to *Ephemera Germanica* and to *Philosophical Transactions*, are cited in support of the various cases, observations and experiments which seem to prove the ovarian thesis. Then briefly examining the arguments of Ludwig Hannemann[8] and other spermatists, he concludes with that of those who maintain that the spermatozoon fuses with the ovum and that this fusion constitutes impregnation: this he admits is very hard to disprove, though he raises several difficulties. Finally he says: 'To be frank many doubtful points can be mooted which I confess myself unable to solve, not because they do not have definite natural causes but because I am ignorant of them.'[9]

As a summary of the evidence of doctors, physiologists and anatomists and of his own observations and reflections the paper gives a useful survey of contemporary opinion. Plainly Ray was not himself familiar with the range of enquiry which the improved microscopes of the Dutch scientist had disclosed. The 'glass microscope' which John Aubrey sent him in 1692 and with which his 'wife and young daughters' were 'indeed much pleased'[10] was quite insufficient; and he failed to appreciate the minuteness and mobility of the 'insects' which Leeuwenhoek had found in semen. Moreover, he was sceptical of Leeuwenhoek's accuracy because he had found him mistaken in regard to the seeds of plants: Leeuwenhoek had said that radishes, turnips and others had four leaves (the cotyledons), 'whereas they have but two only, with a notch or crena at the top, but that not very deep so as to make any show of two leaves'.[11] But although the new evidence conflicted with his settled convictions and was perhaps specially hard for him as a botanist to accept, he is not prepared to ignore it or diminish its significance; and in the final pages, where he admits the probability of a fusion of spermatozoon and ovum, comes near to a correct solution of the problem.

1 Most of the names here noted occur also in the second edition of the *Wisdom of God*, published in the previous year: see below, pp. 472–4. Harvey wrote *De Generatione Animalium*, 1651.

2 Died aged 32 in 1673: author of *De Organis Generationi Inservientibus*, Leyden, 1668.

3 Professor of Anatomy at Jardin des Plantes, Paris 1673–1718.

4 Professor at Jardin Royal, Paris, noted for dissection of rare animals.

5 Of Leyden, published *Adenographia*, 1691.

6 Of Basle, published *Merycologia*, 1685.

7 Published *De Ovo* in London, 1675, as Supplement to his *Anatome Plantarum*.

8 Presumably Johann Ludwig Hannemann (1640–1724), author of a tract on the blackness of the descendants of Ham.

9 *S.Q.* p. 46.

10 *F.C.* p. 175: an account of these early lenses and microscopes is given by Meyer, *Rise of Embryology*, pp. 247–56.

11 Letter of 3 March 1692 to Robinson: *Corr.* p. 246.

So he turns to his main subject; and first discusses and dismisses the easy distinction of animals into viviparous and oviparous. This is inadequate because in the strict sense all animals come from eggs; and though such eggs are of two sorts, those which contain only nucleus and a little fluid and so need a supply of further nutriment from the womb, and those which contain sufficient nutriment to complete their growth up to birth, even so the distinction is not valid because the cartilagineous fishes and certain snakes like the viper are viviparous. Rather a true distinction must be found in the respiratory system. The two main classes, Sanguineae and Exanguia, may be preserved although all organisms have a vital fluid—it corresponds roughly of course to the subsequent division into Vertebrates and Invertebrates. The former class is subdivided according to its method of breathing by lungs or by gills. Animals with lungs are again divided into those with a double ventricle, hot-blooded creatures, Quadrupeds, Cetaceans and Birds, and those with a single ventricle, Batrachians, Lizards and Snakes. Animals with gills include the Fishes, except the Cetaceans and those invertebrates often loosely called fishes. The latter class, Exanguia, are conveniently divided by Aristotle into large and small; and the large into 'Molluscs,'[1] Crustaceans and Testaceans; for these Ray prefers a different classification, Pedata, Cephalopods and Crustaceans, and Apoda, Snails and Shell-fish and Slugs: the small are the Insects. Though this is in his judgment the most accurate and natural classification, Ray does not reject the conventional Land-animals, Birds, Fishes, Insects, but definitely condemns that into Land-animals, Water-animals and Amphibians: each of these unite what ought to be separated and separate what obviously belong together. He concludes this section with a note: 'To maintain agreement with common opinion and to avoid the appearance of an affected novelty we will number the Cetaceans among Fishes although they evidently agree with viviparous Quadrupeds in everything except hair, feet and the element in which they live.'[2] He was in fact loyal to his earlier arrangement, although in his Table of Classification he explicitly rejects it and groups the Cetacea with the other Mammals.

Turning to the classification of viviparous, hairy quadrupeds (which he explains as including the 'two-footed' Manati), he rejects Aristotle's classification, until then unchallenged, and propounds a system of his own. Instead of the three classes, those with solid hoofs, with cleft hoofs and with many toes, he suggests a simple division into the Ungulates and Unguiculates,[3] those in which the toes are covered with horny hoofs and

1 So D'Arcy Thompson translates it (*Hist. Animalium*, IV, 523): it includes cuttle-fish and octopus.
2 P. 55.
3 'Give me leave', he adds, 'to invent these names for the sake of brevity and exposition' (p. 56).

those in which they are bare but carry nails. Ungulates he divides into solid-hoofed (Horse, Ass, Zebra), cleft-hoofed, and four-hoofed (Rhinoceros[1] and Hippopotamus), explaining that he knows, of course, that the cleft-hoofed have two more toes behind and above: the cleft-hoofed he divides into the Non-ruminant, the Pigs, and the Ruminants, having horns either permanent like Cattle, Sheep and Goats, or annually changed like Deer. The Unguiculates he divides into headings which are differently arranged in the essay and in the table appended to it. In the essay he begins with two main classes, wide-nailed and narrow, the former containing only the Anthropomorpha, Apes and Monkeys. The narrow-nailed are divided into the cleft-footed (Camel) and the many-toed; and these according to their teeth into a main group, the normally toothed Analoga, either with more than two front teeth or with two (Rodents), and the abnormally toothed Anomala. Finally the Analoga, other than Rodents, are divided into a larger section, either short-nosed (Cats) or long-nosed (Dogs), and a smaller, the Weasels; and the Anomala into a toothed group, Hedgehog, Armadillo, Mole and Shrew, and a toothless, the Ant-bear; the Bat and the Sloth being attached to the same division. The Elephant is omitted in the essay, but in the table appears in a section between the Camel (cleft-footed) and the many-toed, as having all its toes united but the foot equipped with blunt nails. Animals that breathe with lungs and have hearts with one ventricle he divides into three groups; the Batrachians and Tortoises; the Lizards from the Crocodile to the Chamaeleon; and the Snakes.

The book is full of interest, quite apart from its pioneering value for classification; for although as Ray admits he has not described carefully all the animals that he has seen—'for never even in my dreams did I expect to produce a history of quadrupeds'[2]—and does not now intend to make a complete list, yet his book reveals the extent and limits of contemporary knowledge; shows how far exploration had gone; and proves that scientific study of anatomy was being vigorously pursued—especially in Paris, where the collections of the Jardin Royal were at the disposal of Dionis and Duverney.

The first point to note is that here as in the *Ornithology* he rejects ruthlessly creatures that he regards as legendary. The 'Asinus indicus monoceros' or Unicorn is dismissed in a single sentence. Of the Yale, 'the deer of the ancients with its horns twisted in different directions', in spite of Pliny, he writes, 'what beast it is and whether anything of the kind exists, is deservedly doubted'.[3] But he accepts the Buselaphus,[4] 'midway be-

1 Rhinoceros has in fact three not four toes.
2 *S.Q.* p. 153.
3 *S.Q.* p. 83: there is a picture of the 'Eale' and several of the unicorn in J. Johnstone, *Historia Naturalis*, Tab. x and xxiv, Frankfort, 1652.   4 *S.Q.* p. 82.

tween a deer and a cow' (and probably a Hartebeest rather than a Gnu, although he describes it as 'playful and a mighty jumper and runner'), and states that it is probably a kind of antelope; the Tapir, which he puts next to the Hippopotamus; the Opossum, which is grouped between the Raccoon and the Badger, although he calls special attention to the pouch ('marsupium') or 'open womb' which 'distinguishes it from all other animals'[1] and which should make possible a fuller knowledge of foetal development; and the Manati, placed after the Walrus and the Seal, of which he remarks that 'if Diogenes had known of it he would not have had to pluck a fowl in order to ridicule Plato's definition of man as a featherless biped; for the Manati is a featherless biped'.[2]

Going through the book in detail we find a number of points of interest. In the Horse tribe he knows the Onager (*Equus onager*) and the chagrin made from its hide, and describes the Zebra as African and Indian:[3] in the Oxen the Aurochs (*Bison europaeus*) is only represented by two quotations, from Caesar (*De Bello Gallico*, VI, 5) and Mentzel, and a request to the German Academy to describe and depict, and the Bison (*B. americanus*) by a specimen in the royal menagerie at Westminster, which Ray assumes to have come from Florida: in the Sheep are the Fat-tail, which he had apparently seen exhibited, and the Musimon or Mouflon (*Ovis musimon*), of whose present existence in Sardinia and Corsica he is doubtful: of the African Sheep kept at Westminster and having hair instead of wool he remarks that this is a proper change since in Ethiopia the men have wool instead of hair: the Cretan Sheep which he describes from Belon appears to be the Ibex. The Goat tribe includes the Antelopes and begins with the Ibex (*Capra ibex*), whose horns he saw in Glarus, though it was only found in Valais and Salzburg, and the Chamois (*Rupicapra tragus*), of which Willughby had left a short description; he rejects as fabulous the story that it hangs from a rock by its horns. He knows the Indian Antelope (*Antilope cervicapra*) from a skin in the Royal Society's museum and one of the Africans from a live specimen in Westminster; and a Syrian Goat 'I saw in the year 1668 in the street called the Strand: it had once-twisted horns and long hanging ears, and was eating hay and barley'.[4] He regards the Red Deer (*Cervus elaphus*) and the Canadian (*C. canadensis*)—usually called Wapiti—as the same species in spite of the difference in size and horns: of the Fallow Deer (*C. dama*) he says that it is very abundant in parks in England, where there are more of such places than in all the rest of Europe; Willughby had described various forms of it in the royal menagerie in St James's Park: the Elk (*Alces machlis*), whose skin he saw stuffed in the Grand Duke's museum at Florence, he records as found in

---

1 S.Q. p. 184.
3 The Zebras are solely African.

2 S.Q. p. 194.
4 S.Q. p. 81.

'Lithuania, Muscovy and Scandinavia'; the Moose (*A. americanus*), 'a creature of the same kind', huge horns of which are to be seen at Lewes in Dr Holney's[1] house, comes from New England and other parts of North America; the Reindeer (*Rangifer tarandus*), about which the authorities differ widely, is used by the Lapps and Finns as a beast of burden; he mentions the Axis deer (*Cervus axis*) and the Roe (*Capreolus capraea*), whose 'horns are used for knife-handles'. Rather tentatively he includes the Giraffe, presumably *Giraffa camelopardalis*, with the Deer tribe on the ground of its hairy horns. In the Pigs he gives a long account of his own dissection of the domestic sow with Benjamin Allen at Braintree, and a still longer one of Tyson's description[2] of the Tajacu or Mexican Musk-boar, now called the Peccary (*Dicotyles tajacu*). The Wild Boar (*Sus scrofa*), which he saw alive in a menagerie at Florence and dead in the market at Rome, and the Babyroussa, whose teeth, long and curved though they be, are yet teeth and not as Grew supposed horns, are given only a few lines apiece. The Ungulates are completed by the Rhinoceros, of which he records that a specimen was carried round England on exhibition in 1684–5,[3] Hippopotamus of whose existence he has been fully convinced since he wrote his *Observations*,[4] and Tapir (*Tapirus terrestris*), to which are somewhat curiously attached the Capybara (*Hydrochoerus capybara*), whose teeth he describes as like those of a rodent but with twenty-four back teeth in each jaw, and the 'Animal moschiferum', plainly the Musk-deer (*Moschus moschiferus*), of which he gives a long and accurate description from Grew, commenting upon the vast number killed to supply the world with musk.

Of the Unguiculates he deals first with the Elephant, presumably the Indian species,[5] whose skeleton he had seen at Florence. He gives a long and elaborate account of its anatomy on the ground that this has been described only in English by 'Dr Moulins as I suppose'[6] and that he is now

1 Probably 'John Holney of Lewes physician', whose daughter was married at Lewes in 1691 and who died there in 1706: cf. J. Comber, *Sussex Genealogies*, I, p. 177.

2 This occupies pp. 97–120 and seems out of proportion to its importance. It is translated from *Phil. Trans.* XIII, no. 153, pp. 359–85. Cf. above, p. 354.

3 He says that it is found in the African deserts and in Bengal, but describes *Rhinoceros indicus*.

4 *Obs.* p. 27, and above p. 132.

5 There is the tail of an African Elephant among the remains of Tradescant's museum now in the Old Ashmolean, but no evidence that the species were then separated.

6 The various bearers of this name (Molines, Moleyns, or Mullins) have not been very clearly differentiated: cf. *D.N.B.* XXXVIII, p. 126. The tract here translated is *An Anatomical Account of the Elephant Accidentally Burnt in Dublin on Fryday June 17 1681*, by A.M., Med.B. of Trinity Coll. Dublin—that is, by Allan Molines, F.R.S.: cf. *Brit. Mus.* (*Nat. Hist.*) *Supplement to Catalogue of Books*, VII, J–O, p. 856. Ray refers to him also in *Wisdom of God*, 2nd ed. Part II, pp. 105–6.

making it available to all scholars for the first time. With the Camel (*Camelus dromedarius*) and the Bactrian or two-humped Camel he joins the Llama and the Pacos, insisting that they are not sheep as has been commonly supposed. Of the Apes in the strict sense he is evidently aware; for he divides the tribe into tailless and tailed: but the Baboon, probably *Cynocephalus hamadryas*, is the only large species that he mentions. Gesner, as he says, thought this to be the Hyaena of the ancients; but the legends ascribed to the Hyaena make the existence of such an animal doubtful, and for himself he believes it to be the Badger. Out of many species that he had seen he describes only some ten Monkeys of the old world, one, 'the bearded monkey called Exquima by the people of the Congo',[1] probably *Cercopithecus diana*, another seen by Robinson in the Strand, possibly *Semnopithecus entellus*, a third, probably *Nycticebus tardigradus*, in Charleton's Museum; and four of the New, one of the Howlers (*Mycetes*), a *Cebus* and probably *Midas leoninus*. Of the Cat tribe, Lion, Tiger, Leopard and Lynx are first described, the last said to live in Italy and Germany and with richer colours in Asia; then the Jaguar (*Felis onca*) and two supposed American species, one of which may be the Puma (*F. concolor*); the 'Cat a Mountain' and the domestic Cat; and finally the Bear, which is said 'to be found in the Alps, in Germany, Poland, Lithuania and Norway, to vary in colour, both black and white being known, and to reach its largest size in Nova Zembla'.[2] Of the Dog tribe Wolf and Jackal come first, then the Dog, of which he mentions ten varieties reared in England,[3] the Fox, the Civet Cat (*Viverra civetta*), which by its teeth and the shape of its head is a dog and yields 'Zibethum' or civet; the Raccoon (probably *Procyon cancrivorus*), which he saw and described in Mr Middleton's house at Lewes;[4] the Coati (*Nasua rufa*); the Grison (*Grisonia vittata*), whose smell he notes; the 'Possum', possibly *Didelphys marsupialis*, and another Marsupial; the Badger, which he describes evidently at first hand and has seen in Essex, Sussex and Warwickshire; the Otter; and the Seal (*Phoca vitulina*), which he knew from his visits to Cornwall; the Walrus (*Trichechus rosmarus*) and the Sea-cow (*Manatus americanus*), of which he seems to have formed a fairly accurate notion, comparing the nails on its front limbs to those of the Elephant, and contrasting its vegetable diet with the fish-eating of the Seal.

1 *S.Q.* p. 159, and perhaps also p. 157.

2 P. 173: he evidently had not seen any of the other species except *Ursus arctos*.

3 These are: Mastiff, Greyhound, Irish Greyhound, Hound with Beagle and Bloodhound and Gazehound, Land Spaniel, Water Spaniel, Tumbler, Cur or House Dog, Lap-dog, Shock.

4 This may perhaps be Robert Middleton, Rector of Braintree 1676–90, who became Vicar of Cuckfield in 1690 and of whom Ray speaks as a friend in letters to Burrell: cf. Gunther, *Early Science in Cambridge*, pp. 351, 354: but probably it is someone whom he met there while staying with Courthope.

The next two groups, the Weasels and Rodents, contain much of interest. The Common Weasel (*Putorius vulgaris*) he had evidently dissected and describes fully. Then follow the Indian Ferret, 'the deadly enemy of the snake tribe';[1] the Mungo (or Mongoose), listed separately from the Leyden Catalogue (these two should of course belong to the Civets); the Ferret, said by Willughby to come originally from Africa; the Polecat (*P. foetidus*), whose chief quality he illustrates by the proverb 'it stinks like a Polecat'; the Ermine or Stoat (*P. erminea*), which he knows in its white form; the Beech and Pine Martens, distinguished by white or yellow throat, but interbreeding;[2] 'Mustela zibellina', the Sable, whose pelt is the most precious of all and which 'Tancred Robinson saw in the chambers of Charleton in the Middle Temple'; the Genet (*Genetta vulgaris*), found in Spain and according to Belon kept tame in Constantinople; and the Ichneumon (*Herpestes ichneumon*), which lives on anything alive, birds or snakes as well as slugs or lizards, and is kept as a household pet in Egypt. Of Rodents, after the Hare and Rabbit and two relatives from Brazil, he deals with the European Porcupine, which 'we saw for sale among the bird-fanciers at Rome, caught in the neighbouring hills and bought cheap',[3] and with the 'Cuanda' or Tree Porcupine (*Cercolabes prehensilis*) of South America, whose four-toed hind feet and arboreal habits he notes—though he dismisses as erroneous the story accepted by Marcgraf and the Parisian doctors that it shoots its quills like arrows. Of the Beaver (*Castor fiber*) he quotes a full description by Wepfer of a specimen caught in the Rhine, and appends to this a long discussion of the Beaver's claim to have been a native of Britain. He quotes Edward Lhwyd, 'my close friend',[4] and the evidence of Giraldus Cambrensis, and the prevalence of the Welsh name 'Afange' in the names of lakes, e.g. Llyn yr Afange. Then mentioning that Lhwyd regarded Roedeer and Bears as formerly living in Wales he quotes ancient laws and rules of the chase which mention Bears and Boars, and concludes by saying that Roedeer are said to survive still in the north of Scotland. Of Squirrels he mentions five, and gives a concise account of the habits of the Red, mentioning that it is sometimes found black, and in Poland and Russia grey: he knows the Grey and the Flying Squirrel (*Sciuropterus volucella*) of New Spain and Virginia. The Rat he describes too briefly to determine its species, 'color ei cinereus obscurus seu fuscus'; it must be *Mus rattus*, not *M. decumanus*:[5]

1 *S.Q.* p. 197.
2 *S.Q.* p. 200: this distinction has been applied until recently to the Martens found in Britain: but the true white-throated *Mustela foina* is not found in this country.
3 *S.Q.* p. 206: cf. *Obs.* p. 363.   4 Cf. *F.C.* p. 233.
5 *M. decumanus* seems to have reached Britain only in 1728: so Pennant, *Synopsis of Quadrupeds*, p. 300, who says it had already (in 1771) 'destroyed the Black Rat in most places'.

the Water-rat (*Microtus amphibius*) and Musk Rat from Russia,[1] the various Mice including the two Dormice (*Myoxus glis* and *Muscardinus avellanarius*) and (erroneously) the Field Vole (*Microtus agrestis*)—'the Field Mouse with large head and short tail';[2] 'Cricetus', the Hamster, 'animal iracundum et mordax'; the Marmot (*Arctomys marmotta*) seen at Glarus,[3] the Cavia Cobaya or Guinea Pig (*Cavia porcellus*) dissected by Willughby and the Agouti (*Dasyprocta aguti*). Of the Lemming (*Myodes lemmus*) he gives Worm's account, and dismisses with scorn the notion that its mysterious hordes are due to spontaneous generation in decaying matter or in the clouds:

> I reject all spontaneous generation not only of quadrupeds but of insects... but I cannot pass over the fact that recently eaten grass has been found by dissection in the bodies of these creatures rained down from the sky: so grass as well as mice are born in the clouds: curious that the sky does not ever seem to rain down hay. You remind me of the records of showers of wheat? So herbs not only are generated in the clouds but grow there and set seed.[4]

The Hedgehog (*Erinaceus europaeus*) Willughby had dissected in May 1672: the four species of Armadillo are taken from Marcgraf: the habit of the Three-banded or Apara (*Tolypeutes tricinctus*) of rolling itself up is well described: the Mole (*Talpa europaea*) Petiver had sent him from Warwickshire: the Shrew (*Sorex vulgaris*) and its points of difference from the mouse he draws from Willughby. He describes two 'Ant-bears', *Myrmecophaga jubata* and *Tamandua tetradactyla*; and ends this part of the book with the Bat, of which he gives only one species, though saying that in Guinea they are as big as hens, and the Sloth or Ai (*Bradypus tridactylus*), for the account of which he quotes six authorities: 'it moves very slowly and hanging down in the trees; on the ground it spends a whole day in covering fifty yards'.[5]

He then turns to the 'Animals with blood, breathing by lungs, with a heart having a single ventricle, oviparous' and begins with 'the Frog or Frosh', *Rana temporaria*. He does not refer to Nidd's observations,[6] but quotes from Swammerdam on the structure of their eggs and tadpoles ('gyrini'), from Richard Waller, for a short time Secretary of the Royal Society, and from his friend Walter Needham: the method of impregnation was still in dispute and Ray's discussion of it, though he rejects speculations, is not very conclusive. He adds that from his own dissections frogs eat small land-snails as well as insects. The Tree Frog (*Hyla arborea*) he has

1 The account is from Clusius: if Russia is correct, it cannot be the American Musquash, though he identifies the two in the Preface.
2 *S.Q.* p. 218.   3 Cf. *Obs.* p. 419.   4 *S.Q.* p. 229.
5 *S.Q.* p. 246.
6 Cf. *Wisdom of God*, Part II, p. 84.

seen and heard, first at Baden[1] and commonly in Germany and Italy: he comments upon its torpor in winter; and uses the evidence of the vitality of Frogs to reject the theory of spontaneous generation in the case of the Toads (*Bufo vulgaris*)—an instance of one embedded in a rock having given him much interest.[2] Of Tortoises he deals first with the Common Land Tortoise which he had first seen at Vienna[3] (probably *Testudo graeca*), the Freshwater Tortoise (*Emys lutaria*), distinguished by its long tail and the 'Common Sea Tortoise' (probably *Thalassochelys caretta*— the Loggerhead Turtle), which he had seen caught in the Mediterranean.[4] He recounts much evidence for Turtles of vast size, big enough to feed a hundred men, or carry fourteen on their backs, or to make boats or houses with their shells: but this he does not endorse. He gives brief descriptions of a number of species, including the Brazilian *Pelomedusa expansa*, some of them taken from Grew's *Musaeum Regalis Societatis*.

Of Lizards he says that, like Serpents, though all produce eggs, in some these are hatched before birth. He begins his account of them with the Crocodile, following Worm's description, giving the Nile, Niger and Ganges[5] as its habitat, but not separating it specifically from 'the Jacare of Brazil and Cayman of the Congo';[6] he explicitly rejects on anatomical grounds the belief of Aristotle and almost all his successors that the Crocodile moves its upper not its lower jaw. There follows a medley of species: the first, distinguished by its round and scaly tail and from his description apparently *Uromastix acanthinurus*, he claims to have seen at Montpellier; the second is from Hernandez, a tame creature which if irritated shoots out blood from its eyes; the third, 'the Common Eft or Swift', varies and includes five forms, the first being probably and the fifth certainly a Newt. Then comes the Green Lizard, *Lacerta viridis*, —'very common in Italy and found in Ireland: is it Tradescant's Irish Lizard?'—the 'Tarantola' (*Tarentola mauritanica*), found in old buildings in Rome and Naples and as Ray confesses a fearsome object though as he knows harmless; then a variety of species with outlandish names, of which *Iguana tuberculata*[7] and *Tupinambis punctatus*[8] are recognisable from Marcgraf and César de Rochefort; and finally the Seps, 'more like a footed snake', and evidently from the description *Chalcides lineatus*, which he

---

1 *Obs.* p. 101.

2 Reported to him by Lhwyd from Dr R. Richardson of North Bierley, Yorkshire, *Corr.* p. 241: Ray's reply is *F.C.* p. 227: for this cf. Gadow, *Amphibia and Reptiles*, pp. 174–5. 3 *Obs.* p. 141.

4 *Obs.* p. 290: Ray lists 'Testudo caretta', but from his description of it as much smaller and the source of tortoiseshell this must be the Hawksbill (*Chelone imbricata*).

5 The species differ and Ray no doubt had seen specimens both of *Crocodilus porosus* and of *C. niloticus*.

6 The Caimans are solely American: the Congo species is *Crocodilus cataphractus*.

7 *S.Q.* p. 265.   8 *S.Q.* pp. 265–6: he notes the bifid tongue.

found on the shore at Leghorn; the Salamander (*Salamandra maculosa*), which he found in the Styrian Alps and which does not live in a fire; the Water-Salamander (Triton or Newt), of which Grew describes three species;[1] the Flying Lizard (*Draco ?* sp.), of which William Charleton had a specimen in spirits in the Middle Temple; and the Chamaeleon (*Chamaeleo vulgaris*), which he had seen in Rome and describes at length partly from his own observation of its shape and habits (especially its respiration), partly from Panarolus and the Parisians: he accepts the view that it changes its colour towards red or green owing to changes in its feelings or temperature, not from neighbouring objects.

Finally, he treats of Serpents and suggests a division into those which have fangs in the upper mandible by which they instil a drop of poison, and those which have no such fangs and are harmless. But though he begins with the Viper or Adder (*Vipera berus*), which he describes from life and from his own dissections,[2] and ends with 'Natrix torquata, the Common Snake' (*Tropidonotus natrix*), of which he obtained and dissected a specimen in October 1692 and gives a very exact account, his knowledge of the other species is not enough to enable him to place them—even the Blind-worm (*Anguis fragilis*) appearing high up on the list, although he recognises that its shape and length of tail separate it from the rest of the tribe.[3] The chapter is indeed little more than a series of names and brief notes, fifteen of them, including the Cobra (*Naja tripudians*) and the Indian Python (probably *Python molurus*), supplied by Tancred Robinson from specimens in the Museum at Leyden. Several, *Cerastes cornutus*, *Vipera ammodytes*, *Boa constrictor*,[4] *Elaps corallinus*, and perhaps *Lachesis gramineus*,[5] are identifiable. Only one, the Rattlesnake (*Crotalus horridus*), occupies more than a paragraph: but for this he inserts a very full anatomical description by Edward Tyson covering pp. 291–325 and taken from the *Philosophical Transactions*.[6] One of the rest, the 'Anguis Aesculapii', as he calls it, now *Coluber longissimus*, had been studied by him personally, Philip Skippon having bought a specimen of it when they were together in Rome in 1664.[7] Ray describes it as 'very tame and innoxious', but like others of its species it was very active:[8] 'after a while it escaped out of the box wherein we kept it and hid itself that we could find it no more'.

1 The short descriptions indicate differences of season and sex.
2 He had found three mice in one: *S.Q.* p. 286 and cf. *Corr.* p. 252.
3 The three British species he had described as 'plentiful in my own fields', *Corr.* p. 231: the Blind-worm is of course a lizard.
4 *S.Q.* p. 335: from Piso and Marcgraf. He notes the spurs on each side of the vent.
5 *S.Q.* p. 331, Ray's 'Serpens Indicus gracilis viridis'.
6 XIII, no. 144, pp. 25–45.
7 Cf. letter to Robinson of March 1691: *Corr.* p. 231.
8 Cf. Gadow, *Amphibia and Reptiles*, p. 617.

CHAPTER XV

THE *HISTORY OF INSECTS*

The Phalaenae are so numerous that I despair of coming to an end of them, much less of discovering the several changes they go through from the egg to the papilio, and describing the erucae and aureliae of each.

JOHN RAY to WILLIAM DERHAM, 6 September 1704, *Correspondence*, p. 455.

One last department of the great undertaking to which Ray had dedicated himself on leaving Cambridge and to which his friend Dr Tancred Robinson was constantly exhorting him still remained unfulfilled. He had published the Synopsis of British Plants in 1690 and the supplementary Sylloge of European Plants in 1694. The Synopsis of Animals and Reptiles had appeared in 1693 and that of Birds and Fishes had been sent to Dr Robinson on 29 February 1694.[1] There remained the Insects—a tribe including, in those days, everything from an amoeba to an earthworm;[2] and to these he turned with an energy amazing but characteristic.

It is indeed an almost heroic achievement. He was living in considerable poverty at Black Notley, remote from any libraries or collections and from contact with friends and fellow-workers. He was maintaining his other interests—in theology, in botany, and the general field of natural studies. He was constantly pressed to fresh labours by the importunity of his friends, and in 1693 edited his *Collection of Curious Travels and Voyages* at the request of Charles Hatton and Hans Sloane.[3] Moreover, he was in poor health, suffering continuous pain and sleeplessness from ulceration of the legs and unable not only to visit London and the Royal Society but even to get out into the fields.[4] Any man under such circumstances might be allowed to regard his work as finished, or at least be excused from launching out into fresh adventures.

He had indeed, as we saw when considering his first book, been early attracted to enquiry into the insects which his studies of botany brought to his notice. The remarkable insight and accuracy of his notes at that period prove that he had not only read the little available literature[5] but

---

1 So Derham, *Memorials of Ray*, p. 46. These were mislaid and not published till 1713: see above, p. 330.

2 In Nov. 1674, writing of his plans to Courthope, Ray had said: 'The History of Insects I reserve for the last' (*F.C.* p. 69). Cf. also letter to Lister in July 1676: 'It will be long before W.'s History of Insects is fitted for the press' (*Corr.* p. 126).

3 Cf. above, pp. 274–6.

4 So Smith, *Memorials of Ray*, p. 82.

5 Cf. a brief account of all the writers from Aristotle to Mouffet in a letter to Lister May 1668 (*Corr.* p. 24).

given thought and interest to the problems of entomology. During the years of his association with Willughby this interest was maintained, as is evident from references in his *History* and letters; and his friendship for Lister, who was specially studying Spiders, filled much of his correspondence with entomology. In 1668 he noted and described the Bee-like Mountain Fly: 'I found it on a high mountain Hincklehaugh near Settle: it was very importunate and troublesome'.[1] There are notes of insects seen on other journeys—of Spiders seen in March at Lincoln;[2] of Stag-beetles (*Lucanus cervus*) at 'Branson' (Bryanston) near Blandford in Dorset, 'where they are called Branson Bucks', and near Wrekin in Salop;[3] of a small black Caterpillar feeding on watercress and sea rocket at Penzance and St Ives in Cornwall;[4] of the Swallow-tail Butterfly seen in Etruria and at Montpellier, 'feeding on a sort of fennel';[5] of a Sea Worm, a yard long and living in holes in rocks, 'a peculiar creature unusual and almost absurd in appearance', found by him in the sea off Leghorn,[6] and of a tame Flea bought by Willughby on their travels in Venice, kept warm in a box lined with wool, and fed once a day: 'it lived for three months by sucking blood from Willughby's hand but died of cold in the winter'.[7] He made a few records while at Middleton, for example, one dated 1670 of a Caterpillar black with reddish hairs found hibernating in February,[8] another of a Hover-fly caught and described on 17 April 1671,[9] and another no doubt of the same period of a Caterpillar living underground and brought to him by Thomas Willisel;[10] and at least one, a note of the finding of a Puss Moth Caterpillar, refers to his stay at Sutton Coldfield, that is to 1676 or 1677.[11] But from that date his concern with other tasks seems to have left him no room for further observations: nothing that can be dated is assignable to the years between 1677 and 1690.

In this subject indeed Willughby had been particularly interested. In the *Cambridge Catalogue*, though his observations are not very searching,

---

1 *Hist. Ins.* p. 273. This must have been on the solitary expedition on which he studied the Cloudberry on the same Hincklehaugh in the end of July: cf. *Cat. Angl.* pp. 68–9. He refers to it in a letter to Lister dated 10 Sept. 1668 (*Corr.* p. 28), in which along with other observations is an interesting allusion to the account of Bees in C. Butler, *The Feminine Monarchie*.

2 *H.I.* p. 41.   3 *H.I.* p. 75.

4 *H.I.* p. 352. He was at St Ives on 30 June and Penzance on 2 July 1662.

5 *H.I.* pp. 110–11, *P. machaon*: cf. also *Corr.* p. 62 for the same caterpillar at Middleton.

6 *H.I.* p. 46, probably a Phyllodocid. His visits were in April and July 1664: cf. *Obs.* pp. 263–5 and 322.

7 *H.I.* p. 8: they visited Venice in Oct. 1663; cf. *Obs.* pp. 149–204.

8 *H.I.* p. 365—probably of *Arctia caia*: cf. the record of the same date of another larva, possibly *Triphaena pronuba*, in *H.I.* p. 358.

9 *H.I.* p. 273.   10 *H.I.* p. 364.   11 *H.I.* p. 154.

there is evidence that he was already gathering material.[1] A letter of his
from Middleton read to the Royal Society on 10 February 1670 concerning
'the worms wrapped up in leaves in a rotten willow' stated that he had
only met with them ten years before in oak-leaves, but added some sound
comments upon the hibernation of various larvae.[2] Other letters to the
Society and the material printed below his initials in the *Historia Insectorum*
bear out the claim made during his last illness that though he did not think
his studies of animals and birds worthy of publication yet he had made
some nice observations of insects. He had certainly collected and noted a
considerable amount of material dealing with the less known orders,
hemiptera, neuroptera and diptera, and was attracted by the subtleties of
instinctive behaviour. But unfortunately the naming of species in these
orders had not been established, and his notes though sometimes interesting are mere jottings, describing isolated points in an unmapped area.
Moreover, as is evident from Ray's letters to Robinson in 1685[3] and to
Sloane in 1703,[4] Ray had no access to his collections or manuscripts after
he left Sutton Coldfield. He had left insects to Willughby in the days of
their partnership; he had not touched them while he was doing the birds
and fishes; when in 1690 he turned to entomology he had to begin alone.
It was not until the last year of his life that he was lent Willughby's notes:
he used them in full and even in preference to his own similar material,[5]
but they were, as he would put it, 'huddled together in haste': and when
he died with his task incomplete could represent only stray comments upon
species hardly identifiable. He himself had worked mainly at lepidoptera,
a group which Willughby seems to have hardly touched. This is in fact
the part of the *Historia* that is of most value; and in it he got no help from
his friend.

Nor did any fresh literature on the subject make his task easier. Apart
from the work of Johannes Swammerdam,[6] whose father's collections had
given him a mass of material and whose interests ranged over insects as
well as other fields, and of his fellow-countryman Johan Goedart, an en-
graver whose notes and plates had been translated and 'methodised' by
Ray's friend Martin Lister,[7] Mouffet's *Theatrum* remained the one treatise

1 E.g. in regard to caterpillars: cf. *C.C.* p. 136.
2 Birch, *History of the R.S.* II, p. 422: on this occasion a discussion ensued about the
grubs found in willows, apparently the larvae of *Trochilium crabroniformis*.
3 *Corr.* p. 165.   4 *Corr.* p. 416.
5 Thus in *H.I.* p. 341 he prints a short account by F. W. of the larva of *Penthina
gentianana* instead of his own fuller record in *C.C.* p. 45.
6 His *History of Insects* is declared by Ray to be 'the best book ever written on the
subject': *F.C.* p. 227.
7 Presented to the Royal Society on 19 Nov. 1684 (Birch, *History of the R.S.* IV,
p. 331): the pictures of caterpillar and imago are generally accurate, but the book is of
little scientific value.

on the subject. Considering the date of its composition[1] and the miscellaneous character of its sources,[2] it is a work of real merit; and such development of interest in insects as we can trace during the next century largely derives from it. But both the descriptions and the woodcuts that accompany them are primitive and inaccurate; the classification is rudimentary; and the information as to habits fanciful and misleading. Thus in Book I the best chapter[3] is that on butterflies; and this begins with the division into diurnal and nocturnal and the statement that 'as great tyrants devour the nobles of lesser races, so these night-fliers lash with their wings and kill the day-flies lurking under the leaves'. The moths are divided according to size—nineteen large, including the Death's head, Emperor, Privet, Eyed Hawk, Tiger, Eggar, Drinker, Leopard and Elephant Hawk (*Acherontia atropos, Saturnia pavonia, Sphinx ligustri, Smerinthus ocellatus, Arctia caia, Lasiocampa quercus, Cosmotriche potatoria, Zeuzera pyrina, Chaerocampa elpenor*); seventeen medium, the Magpie, Buff-ermine and Silk-worm (*Abraxas grossulariata, Spilosoma lubricipeda* and *Bombyx mori*) being clearly recognisable; and six small, of which the first figured is certainly a beetle and the third the chrysalis of a butterfly, and including the Clothes-moth, the Cinnabar, Burnet and Forester (*Hipocrita jacobaeae, Zygaena filipendulae* and *Ino statices*). Similarly the butterflies are also grouped: sixteen large, two Swallowtails, Peacock, Red Admiral, Clouded Yellow and Dark Green Fritillary (*Papilio podalirius* and *machaon, Vanessa io, Pyrameis atalanta, Colias edusa* and *Argynnis aglaia*) being tolerably accurate though the cuts are very poor; thirteen medium, including the Brimstone (*Gonepteryx rhamni*), and three obvious moths; and ten small, all unidentifiable, and no. 8 being the same as no. 11 in the medium series. Book II deals with wingless insects of which the first class is Erucae—'the English call them by the common term "Catterpillars": but northerners call the hairy ones "Oubutts", and southerners call them "Palmerwormes".'[4] They are divided according to colour, green, yellow, the Puss-moth (*Dicranura vinula*), grey and various; *Sphinx ligustri* and *Metopsilus porcellus* are named, and the latter is said to feed on the leaves of the marsh trefoil, two pictures being given of it, one apparently *Deilephila galii*, the other *Chaerocampa elpenor*. The account ends with a discussion of the nature of a chrysalis, whether it is a separate animal or the egg of the caterpillar. Entomology, though it recognised the fact of metamorphosis, had not yet begun to study it.

1 He died in 1604 and his book was published by the efforts of Sir Theodore Mayerne in 1634.

2 It was largely derived from the work of Edward Wotton, Conrad Gesner and Mouffet's friend, Thomas Penny.

3 Ch. XIV, pp. 87, 108.   4 P. 179.

Swammerdam's *General History of Insects*,[1] which Ray used in the French version,[2] is no doubt as he maintained 'the best book that ever was written on that subject',[3] indeed in some sense it marked an epoch. The Dutch naturalist was not only a keen observer, but he used a microscope and made excellent enlarged pictures of his subjects. He fastened upon the fact of metamorphosis, observed it in insects of a large variety of types, and in his book used it as the basis for a classification. It is interesting to notice that in his survey of previous opinions on the subject he not only criticises Mouffet for having failed to observe the true nature of the 'nymph' state; but regrets that his failure has affected 'the careful and very penetrating Englishmen who have written a book on the plants growing near Cambridge'.[4] He alludes here and in a later reference to Ray's discussion of the parasites on the Cabbage-butterfly;[5] and is perhaps unfairly severe in his complaints. If so, Ray bore him no grudge; recognised the great value of his treatment of the whole life-cycle; accepted his classification as 'the best of all';[6] and followed its outline with many references to its author's work in his own *Methodus*.[7] But Swammerdam, though he mentions the number of species under each heading—alluding apparently to those in his collections—makes no attempt to describe them; and Ray does not allude to him in reference to the single species of moth (*Orgyia antiqua*) and butterfly (*Pieris brassicae*) which he figures.[8] From neither of these sources could he get any serious help or encouragement for his task.

In addition to these Ray had seen in Sicily a substantial manuscript on Insects in two volumes written and illustrated by Pietro Castelli, prefect of the botanical garden at Messina. This he had tried unsuccessfully to purchase for Willughby, and evidently thought valuable.[9] But if he learnt anything from it, it was at a time when he had no special concern with its subject.

Nevertheless in 1690[10] or thereabouts he began to collect material in

1 Published at Utrecht in 1669 'in the Belgick tongue'—Dutch; and reviewed in *Phil. Trans.* pp. 2078–80 for Oct. 1670.

2 Cf. *Corr.* p. 401: the French version was issued at Utrecht in 1682: he quotes it in the original in *Wisdom of God*, 2nd ed. p. 76.

3 So to Lhwyd in 1692 (*F.C.* p. 227) and to Sloane (*Corr.* p. 364). Books were however few—Aldrovandi, Mouffet, Goedart, Redi, and the Classics.

4 *Hist. Gen. Insect.* p. 30 (French edition 1682).

5 *C.C.* pp. 134–8.   6 *Corr.* p. 435; *H.I.* p. vii.

7 Swammerdam's classification is very slight: four classes: Lice; Crickets; Moths and Bees; Flies. He admits inability to place Glow-worms, Beetles, Scorpion, etc.

8 L.c. Tables X and XIII.

9 Cf. Skippon, *Journey*, p. 614; *Phil. Letters*, p. 361; *Corr.* p. 24.

10 On 7 July he writes to Lhwyd of 'reviewing my notes concerning Insects' and discusses the possible divisions for their classification (*F.C.* p. 208): so in *H.I.* p. 116 he gives this year as that of an early observation. He was certainly collecting hard in 1691, and reported having got 200 species in his immediate neighbourhood; cf. *Wisdom of God*, 2nd ed. p. 8.

earnest. His friend Lister had done some work on Spiders and Beetles; Willughby had left material upon the Worms, Leeches and other 'insects without metamorphosis', which were regarded as the less developed tribes of the order; Ray turned his attention to Lepidoptera. The first dated record in the new series of observations is apparently that of a caterpillar dug up in a hop-yard on 22 March 1691,[1] and then that of a large moth which 'I found sitting on the outside of the bedroom window' on 3 September 1691: this is carefully described and was evidently a Red-underwing (*Catocala nupta*). In his book Ray adds the note: 'at last in 1702 in the middle of May the caterpillar from which it emerges was brought to me'.[2] During the next four years notes and collections increase; and he got some from Tilleman Bobart,[3] who with his brother Jacob Bobart the younger[4] worked in the 'physick garden' at Oxford and sent Ray his collection of insects;[5] from the learned Saxon doctor David Krieg,[6] who afterwards went on a botanical expedition to Maryland; from his friend William Vernon of Peterhouse, Cambridge,[7] Krieg's companion in America; and above all from his good neighbour the Braintree physician Samuel Dale.[8] In 1695 he began intensive work, replacing the annotated diary or Kalendar which he had been keeping by an elaborately numbered series of detailed descriptions, both of moths and caterpillars, preparatory to the full account of each species obtainable by breeding it from the larva to the perfect insect. These notes continue for the next four years and form the latter part of his book, from p. 276 to p. 375. During these years he had further help from the other Braintree doctor, Benjamin Allen,[9] from Mansell Courtman, minister at Castle Hedingham,[10] and John Morton,[11] who visited him in 1695, from William Derham,[12] his literary executor and biographer, and from James Petiver, the well-known apothecary, botanist and collector, who visited him with the Rev. Adam Buddle,[13] the botanist, in the summer of 1699 and received from him on 11 July some insects which he

1 *H.I.* p. 355.   2 *H.I.* p. 153.
3 *H.I.* pp. 128, 148, 153, 248, 271.
4 Jacob junior succeeded his father Jacob senior on his death in 1680.
5 Cf. *F.C.* pp. 223, 233 and Gunther, *Life of Lhwyd*, p. 205: from this we learn that Tilleman Bobart was appointed to the garden at Hampton Court in 1693.
6 *H.I.* pp. 128, 214, 328.
7 *H.I.* pp. 113, 317: of Vernon the story was told that 'he followed a butterfly nine miles before he could catch him': cf. *Letters by Eminent Persons*, II, p. 101.
8 *H.I.* pp. 142, 'about the year 1692', 150, 223, 'at the beginning of June 1703'— the last dated note—etc.
9 He had come to the town in 1689: *F.C.* p. 293. Mentioned in connection with glow-worms, p. 80; 'Caterpillars seen at Mr Allen's', p. 283.
10 *H.I.* pp. 127, 337.   11 *H.I.* p. 127.   12 *H.I.* p. 211.
13 Buddle was described by Vernon as 'the top of all the moss-croppers', *Corr. of R. Richardson*, p. 73. His *Synopsis Plant. Brit.* was never published but his herbarium passed to Sloane and is preserved.

had not seen before,[1] including a Purple Hairstreak (*Zephyrus quercus*).[2]

But it was in his own home and by the efforts of his family and a working man, Thomas Simson[3] or Simpson, who may probably have been in his employ, that the bulk of the material was procured. From scattered hints jotted on the margin of the rough notes incorporated in the *History* and from a few allusions in the text we get an attractive picture of his activity during these last years of his life. We see him with his chip-boxes[4] (did not one of his captives—'apis valde mordax'—gnaw its way out?[5]), his breeding-cage,[6] and his store-boxes, the Large Box with its rows of numbered specimens, entered as P.M. in his notes,[7] the 'Pyx. select.', reserved for choicer species, the 'Pyx. rar.' and 'Mr Dale's box'.[8] Like other collectors he had his troubles; 'colours have faded with age';[9] 'by a sad mishap we destroyed it';[10] 'this is kept in the box of rarities and by an accident one wing was broken off: it was moved and put back into this year's box next to number 27';[11] 'two caterpillars were put in the same box and one devoured the other'.[12] He has his beating stick, and records one or two successful captures by it;[13] and presumably the family was equipped with nets: 'at the beginning of summer my little daughters caught many of this species flying at dusk in our garden' he writes[14] under a moth that must have been Burnished Brass (*Plusia chrysitis*). The exploits of the children are frequently recorded. Of the Merveille du jour (*Agriopis aprilina*) he notes 'my daughter Margaret found it and brought it to me: I have never seen any other specimen of the kind',[15] and she is mentioned three times in the notes for 1698.[16] Mary has a number of records, most if not all of them relating to 1696.[17] Though Catharine is only twice mentioned as adding to the collection,[18] three caterpillars are given her name: 'Katherine's Eruca',[19] which though briefly described is evidently the Lilac Beauty (*Hygrochroa syringaria*) and was reared in 1695; 'Katherine's Oak-Geometer' and 'Kath. Ash-Geometer',[20] neither of them certainly

1 So *Musei Petiveriani Cent.* IV, p. 330.
2 Cf. Petiver, *Mus.* no. 319 and *Gazophylac.* Dec. I, p. 18.
3 His baptism at Black Notley is registered on 13 Aug. 1663: he may well be the man-servant referred to in a letter of 1691 (*F.C.* p. 174). He seems to have worked in the garden, *H.I.* p. 294, and lived next door, cf. Court Rolls quoted above, p. 180.
4 *H.I.* p. 286.
5 *H.I.* p. 242: for the curious, chip-box is in Latin *pixidula abiegna*.
6 *H.I.* p. 177.
7 It held at least 150: cf. *H.I.* pp. 171, 214, etc.
8 Cf. e.g. p. 223.   9 *H.I.* p. 208.
10 *H.I.* p. 373.   11 *H.I.* p. 324.   12 *H.I.* p. 342.
13 *H.I.* pp. 144, 231, 283.   14 *H.I.* p. 182.
15 *H.I.* p. 158.   16 *H.I.* pp. 333, 336, 337.
17 *H.I.* pp. 167, 316–18, 320, 328, 373: she died in January 1698.
18 *H.I.* pp. 281, 319.   19 *H.I.* p. 287.   20 *H.I.* pp. 290–1.

identifiable though the first is said by Werneburg to be *Selenia lunaria* and the other may well be the Brindled Beauty (*Lycia hirtaria*). Jane, the youngest, is even more frequently commemorated: she seems to have made her first capture in July 1692 when she was four and a half;[1] in the autumn of 1696 she found a caterpillar[2] and caught a moth;[3] in 1697 there is another capture;[4] and she is still collecting and caught *Acidalia ornata* in 1703:[5] a geometer caterpillar feeding on dock and found in May 1695 (the Blood-vein, *Timandra amata*) is named Jane's,[6] and in 1697 are two named after her, 'Jane's Chickweed Caterpillar' and 'Jane's, found by Jo. Kelhog'.[7]

His wife is only twice mentioned, but one of these is in a very notable observation. Describing a moth, which is evidently *Pachys betularia*, under the date 29 May 1693, he writes:[8] 'It emerged out of a stick-shaped geometer caterpillar: it was a female and came out from its chrysalis shut up in my cage: the windows were open in the room or closet where it was kept, and two male moths flying round were caught by my wife who by a lucky chance went into the room in the night: they were attracted, as it seems to me, by the scent of the female and came in from outside.' This must surely be the first known record of 'assembling' and Ray has hit upon what is almost certainly the correct explanation of it.

There is something peculiarly attractive in the picture of the 'most celebrated and worthy author', as Petiver calls him, almost bedridden, confined to the little house that he had built in his native village, surrounded by his collection and galvanising the neighbourhood into an interest in adding to it. It must have been a glorious month, that July of 1695, when within a few days were brought to him the first specimens ever described of the Purple Emperor and the White Admiral, both caught in his native county.[9] From Mr James Coker of Plumtrees, his friend and neighbour in whose 'labyrinth' one of his best specimens, a male *Oeonestis quadra*, was caught[10] by Mary, to Thomas Simson, whose name occurs constantly as the capturer of moths and larvae,[11] they were proud to help him. No doubt they thought him queer when he nicknamed his moths after his daughters and wrote in his delicate Latin: 'The caterpillar which produces this moth on account of the huge size of its head I call "Grout-

1 *H.I.* p. 227: it was *Miltochrista miniata* and taken in Ray's orchard.
2 *H.I.* p. 293.   3 *H.I.* p. 236.   4 *H.I.* p. 327.
5 *H.I.* p. 228.   6 *H.I.* p. 286.
7 Both on *H.I.* p. 308; for Kelhog cf. also p. 358. The name, Kellog, occurs in the marriage register of the neighbouring village of Great Leighs in 1633, 1642 and 1646 (*Essex Parish Registers*, IV, pp. 113–14).
8 *H.I.* p. 177: the other reference to Mrs Ray is on p. 200.
9 *H.I.* pp. 126–7.
10 *H.I.* p. 219: for the house cf. *Syn. Brit.* p. 95.
11 *H.I.* pp. 280–1, 293–4, 314, 334, 355, 357.

head, id est Capito".¹ But the blacksmith's son who could carry his massive learning so lightly and live so vividly when his nights were full of pain and his work discharged under cramping conditions far from all the appurtenances of scholarship was worthy of their affection.

What we naturally miss in Ray's later work is that element of first-hand observation of the living creature in its natural environment which characterised his early study of the subject. We have quoted a few of these records of behaviour already; several others of real interest are preserved in the *History*. Here for example is one² that anticipates the *Souvenirs Entomologiques*.

On the 22nd June 1667³ I saw a Wasp, one of the largest of this tribe—I do not now recollect the species⁴—dragging a green caterpillar three times larger than itself....Before my very eyes it carried it almost the full length of a measuring rod, that is some fifteen and a half feet; and then deposited it at the mouth of a burrow which it had previously dug for itself. Then it removed a ball of earth with which it had sealed up the entrance; went down itself into the hole; after a brief stay there came up again; and seizing the caterpillar which it had left near the opening carried it down with it into the burrow. Soon, leaving it behind there, it returned alone; gathered pellets of earth and rolled them one by one into the burrow; and at intervals scratching with its fore feet, as rabbits and dogs do, flung dust backwards into the hole. It kept repeating the same operation with dust and pellets alternately until the burrow was completely covered up; sometimes it descended in order, as it seemed to me, to press down and solidify the soil; once and again it flew off to a fir tree nearby perhaps to look for resin to stick the soil together and consolidate the work. When the opening was filled and levelled with the surface of the ground so that the approach to it was no longer visible, it picked up two pine-needles lying near and laid them by the burrow's mouth, to mark, as is probable, the exact spot. Who would not wonder in amazement at this? Who could ascribe work of this kind to a mere machine?

There is another excellent description that must come from the same period, signed F.W. but in fact an observation originally made by them both and in this form obviously written by Ray, of a Leaf-cutter Bee (*Megachile willughbiella*) building its nest.⁵

---

1 *H.I.* p. 144.   2 *H.I.* p. 254.

3 He was then at Middleton with Willughby: they left three days later on their second journey to Cornwall. Willughby sent a record of this observation, as 'made by Mr Wray in company with another ingenious neighbour', to the Royal Society: it was printed in *Phil. Trans.* 22 Oct. 1671, VI, no. 76, pp. 2280–1 (*F.C.* pp. 57–8). It is interesting to note that in it he spells Ray's name in the way that he abandoned in 1670.

4 Obviously *Ammophila sabulosa*. The description is singularly exact, save for the suggestion that the spot was marked by pine-needles.

5 *H.I.* p. 245. A differently told and much less exact record dated 2 Sept. 1670 was sent by Willughby to the Royal Society (*Phil. Trans.* V, no. 65, pp. 2100–1, and Birch, *History of the R.S.* II, p. 449). He describes how Mr Snell showed him these

These bees fashion sections of rose-leaves carefully rolled up and stuck together into cylindrical chambers: they might be called 'Cartrages' [*sic*] in English from the exact resemblance they bear to the paper wrappings filled with gunpowder used for the larger guns [in Latin 'cum papyraceis involucris pulvere pyris repletis pro bombardis majoribus']: in the trunks of willows soft and decaying they dig cylindrical burrows exactly the size of these capsules: these burrows run up or down following the grain or fibre of the wood, never across it: the capsule is placed at the bottom of the burrow, or at the top if the burrow runs upwards, and exactly fills the space, the round end of the capsule touching the bottom of the burrow: at the concave end of the capsule the round end of another capsule is tightly fixed; and so five, six or seven capsules, one upon another, are found in a single burrow.

Such field-observations had perforce to be given up when the old naturalist returned to entomology. He could get out into his garden and orchard: but it was no Harmas and he too old and ill to be a Fabre. He had to rely on the activity of others for his specimens and spend his time with his collections and boxes of caterpillars. Here and there an instance of behaviour comes his way: we have recorded his note on the assembling of male moths:[1] and he managed to breed the species whose 'smooth black larva ate the leather bindings of his books'[2] (*Aglossa pinguinalis*). There is an observation of a beetle larva and its metamorphosis, with a note on its parasites, dated June 1694;[3] and learned accounts of the Field cricket[4] and the Glow-worm,[5] in which the old flair for correct interpretation of nicely noted facts is evident. But his health kept him indoors;[6] and instead he filled his days with the task of describing and trying to arrange the captures that his friends and family provided.

The letters of this period bear out our picture of his industry and show us more of his mind with regard both to the work and to the superhuman courage with which he undertook it. The earliest reference is in a letter to Lhwyd on 7 July 1690;[7] and this is of importance from its reference to the problem of classification which was already exercising his mind. He writes:

Reviewing my notes concerning Insects and considering the things themselves I find it a thing of infinite difficulty to draw up any tolerable Epitome of

'cartrages' at Astrop near King's Sutton, Northants, and told the place and how he and Ray visited it: Ray reports it in a letter to Lister (*Corr.* p. 29) and in *Wisdom of God* (p. 124). The original discovery was made by Dr King and reported to the Royal Society in April 1670 (Birch, l.c. p. 435).

1 *H.I.* p. 177.　　　2 *H.I.* p. 171.　　　3 *H.I.* p. 85.
4 *H.I.* pp. 63–4. In this he refers to the chemist (pharmacopoeus) James Garret.
5 *H.I.* pp. 78–81, in which he quotes Waller's paper in *Phil. Trans.* XIII, no. 167, p. 841. There is a letter to Lhwyd on the subject 18 July 1692: *F.C.* pp. 228–9.
6 From 1692 to 1699 there was a series of cold and sunless summers which made collecting difficult and exacerbated Ray's illness.
7 *F.C.* pp. 208–9.

the History of such as are found with us: they being almost innumerable. The
two great heads or Summa genera that I would divide them into are 1. Quae
nullam subeunt mutationem. 2. Quae metamorphosin aliquam patiuntur. The
second genus I may divide either as they appear at their first hatching from the
egg, and so they will be either Polypoda, erucae; or Hexapoda; or Apoda, eulae.
Of the first sort come Butterflies; of the second Beetles; of the third Flies of all
sorts, Bees etc. Or 2. as they appear after they have undergone their last change,
and so they may be subdivided into Coleoptera, Beetles etc. or Anelytra which
are either *alis farinaceis*, Papiliones, or *alis membranaceis pellucidis*, and those
either Tetraptera, Bees, Wasps, Hornets etc. or Diptera, Flies, Gnats etc. I am
yet doubtful about the process of Locustae forficulae and Cimices sylv. which
though winged insects, yet I suspect undergo no metamorphosis. Howbeit they
are not at their first exclusion winged, but their wings grow out afterward. Can
you give me any account of the Woodseare or Cuckoo-spittle? doth not the
animal latent therein come to be a Cimex sylv.? I am doubtful concerning Ants
whether the flying may not be the males, the creeping the females; for they are
found together in the same hills. Neither am I yet fully satisfied concerning the
flying and creeping Glow-worms. The number of Erucae alone in this island is
incredible, some plants having 3 or 4 sorts feeding upon them; and if we should
make the Papiliones a distinct genus from them, as all that write the History of
Insects have done, we should double the number of species.... The Beetle-tribe
I hold to be no less numerous than they; and the Flies perchance more. So that
I know not but that the species of Insects may not equal to or exceed those of
Plants.

Next year, on 22 September in a letter to John Aubrey,[1] he expresses his
satisfaction that Mr Bobart 'hath been so diligent in observing and making
a collection of insects: he may give me much assistance in my intended
Synopsis'. Subsequent letters to Lhwyd express gratitude for Bobart's
willingness to let him see the Collection,[2] satisfaction at its safe arrival,
although only one of the Diurnal Papilios was unknown to him 'but of
the nocturnals or Phalaenae many',[3] and an account of the naming of the
new butterfly.[4] None of his chief correspondents was specially interested
in this side of his work and references to its progress are not numerous.
'I do not altogether neglect the prosecution of the History of Insects', he
writes to Sloane,[5] 'which I intend to extend no further than to take in such
as are found within two or three miles of my habitation': but he feels the
task too big: 'Insects I find so numerous, and the observation of all the
kinds of them, but Papilios, so difficult, that I think I must give it over.
It were a fitter task for a young and ingenious person, who had the perfection of all his senses and time enough before him. I do, and must, consider
that I have one foot in the grave.'[6]

Such hesitation did not deter him from going forward with his plans;

1 *F.C.* p. 170.
2 5 April 1692: *F.C.* p. 226.   3 28 Dec.: *F.C.* p. 233.
4 16 March 1693: *F.C.* p. 234: it was *Hipparchia semele*; see below, p. 413.
5 *Corr.* p. 286.   6 To Lhwyd, 8 Oct. 1695: *F.C.* pp. 261-2.

and two years later he reported to Sloane a possibility which, had it been taken, might have given him a very necessary encouragement and his book a vastly increased value.

> A young German doctor gave me a visit, known to you better than myself, for he told me that he had been with and received several things of you. He draws insects, as far as I am able to judge, exceedingly well. He stayed at Braintree two days, which time he spent in drawing several Papilios and Phalaenae, some of Mr Dale's and some of mine.... Were I but ready with my History of English Insects, he might be of great use. But alas! I have not gone through one tribe, that of butterflies nocturnal and diurnal; nor, should I live ten years longer, were I like to come to any near prospect of the end of it, should I pursue it with that diligence and application I have done now these seven years.... You would not imagine how much time it takes one to search out and to feed them, I mean the Erucae.[1]

From subsequent reference to pictures of 'six of the rarer sorts curiously drawn and sent to me by Dr Krieg when he was with you at London'[2] the identity of his visitor is made clear. He was the giver of *Argynnis lathonia* and *Hipparchia semele*[3]—the former he had apparently brought with him to England, the latter he had caught at Cambridge, probably while staying there with Vernon, with whom he travelled to Maryland in search of plants.[4] Ray was evidently attracted by him and disappointed that on his return from America he was not persuaded to stay in England or able to come again to see him.[5] If only he had had the help of such a colleague!

Nevertheless, he pressed on. In 1699 he gives a brief report:

> Though I have taken a great deal of pains, yet have I made but little progress on the History of Insects. The most that I have done is in observing the generation and transmutations of the papilionaceous tribe, of which I have at least 200 species near my own habitation, *necdum finitae*, every year bringing new ones to my knowledge. Of these as many as I could get the eggs or caterpillars of, I have fed and endeavoured to bring to their changes, though I have failed in many. This hath taken me up no small time and pains; and yet if I had not taken this course I had never seen abundance which now I have knowledge of. The other tribes of Insects I have not been so diligent and curious in observing, yet have I made collections of such as came in my way. The next tribe I intend to have fixed upon after I had despatched Papilios was Beetles, which are no less numerous and whose way of generation is the same with Papilios, and which may be as easily fed and brought to change as they. But alas, my glass is almost out, and I am so afflicted with pains, that I have no heart to proceed any further. Indeed I could do very little all last summer, and I must *alii lampada tradere*.[6]

1 19 July 1697: *Corr.* p. 328. He wrote of this visitor on the same day to Robinson: *F.C.* p. 302.

2 *Corr.* p. 450, to Petiver: he was elected F.R.S. in Jan. 1698.

3 Cf. below, pp. 410, 413.

4 This collection was sent by Sloane to Ray for the third volume of the *Historia Plantarum* in 1699: cf. *Corr.* p. 363; *F.C.* p. 279.

5 *Corr.* p. 365.   6 To Sloane, *Corr.* pp. 363–4.

Nevertheless, though he was obliged to tell his friends that 'I am sadly afflicted with pain which renders me listless and indisposed to any business, and disables me with intention to prosecute any study',[1] they did not cease to lay fresh proposals before him. Writing to Sloane in 1702[2] he tells of this and exposes all unwittingly the reason of their importunity:

Mr Dale tells me that some of my friends in London talk of imposing a new task upon me, that is, of describing such exotic insects as are found in the museums of the virtuosi about London, which, if there be no more able and better qualified person living in or near the city, I should not be much averse from, if it please God to continue me any tolerable measure of health and ease; for that I may do sitting and without much motion. But then they must be sent down to me by parcels. As for our English insects, I think I may, without vanity, say that I have taken more pains about some tribes of them than any Englishman before me. If I were to publish a History of Insects, in each tribe I would first place the English ones by themselves and then the exotics.[3] I have by me a history of our diurnal English Papilios of my own knowledge which I drew up some years since. They are in number about forty....In case you think fit to employ me in this service, I would begin with the tribe of Papilios because therein I have taken the most pains, though they be far from the first tribe.

A few months later[4] he adds details of his intention:

I did not intend to write an universal History of Insects, but only of such British ones as have or should come to my knowledge, which I do believe would scarce amount to a third of such as are natives of these islands; and such exotics only as are to be found in the hands of the virtuosi about London, especially yourself. But these separately; first the British by themselves in each tribe and then the exotics after them by themselves. I intended to begin with the Papilionaceous tribe, not because they are the first in order of nature, but because I have taken most pains in searching them out and have described most species of them. Of these the diurnal ones are not very numerous, I having not observed about forty-five sorts of them. But of the nocturnal, should I live twenty years longer, I despair of ever coming to an end, every year offering new ones; and yet I have already observed about 300 species, and this within a small compass of ground. But these I shall so methodise that it shall not be difficult for any man to find any Phalaena he shall discover in the method, if it be there described, or else to know that it is a new one and not described by me. But enough of this, it not being like to take effect.

A few weeks later he wrote to Petiver to the same effect—that he would deal only with British species and such foreigners as he could personally inspect: 'I shall not concern myself with those published by the lady

1 13 Aug. 1701, to Lhwyd; *F.C.* p. 283.   2 *Corr.* p. 410.
3 As is done in the Butterflies of the *Historia Insectorum.*
4 27 July 1703: *Corr.* p. 431.

Merian[1] as neither with those of Goedartius, Hoefnagel, Hollar, Aldrovandi.'[2]

In the autumn of 1703 Dr Robinson urged him to publish a Methodus:[3] and he agreed that this 'might be of use, especially if I should happen to die before the History be finished'.[4] In the following spring his fears were nearly fulfilled. His letters to Sloane on 8 and 17 March[5] tell a terrible tale of suffering and violently septic ulceration. But in May he could tell Petiver[6] that he was now drawing it up and on 8 June could report to Sloane,[7] 'I have drawn up a little Method of Insects which may take up two sheets. It is very lame and imperfect, especially in the tribe of Muscae.[8] I did intend to have sent it up this day, but I fear it is now too late and the carrier gone.' Having finished it[9] and at last received Willughby's papers[10] he began the larger task and in August 1704 sent for Sloane's approval a definite proposal for publication.[11]

Having by me a competent quantity of materials for a History of Insects, collected partly by myself, partly by Francis Willughby, Esq., deceased, expecting also great contributions from my friends skilful in that part of Natural History, I intend, God producing my life and granting me a tolerable measure of ease, to draw up such an history, and have already begun and made some progress in it; which, because it will not be of half the use if published without figures as it would be if illustrated therewith, and because the graving of them is a matter of greater charge than I can sustain, I am constrained to beg the assistance of ingenious gentlemen and wellwishers to this kind of learning in contributing toward the charge of the plates the moderate sum of ten shillings which shall be well husbanded and faithfully expended on the gravers and supervisors of the work. If the sum collected doth not suffice for plates for the whole work, then I must pray a further supply from the contributors to whom what is finished and wrought off shall be delivered, who thereupon may either cease or contribute further as they shall see cause.

1 Lankester prints Lady Marian; he had previously (*Corr.* p. 328) called her Sibylla Myrion: the reference is of course to Maria Sibylla Merian, artist and entomologist of Frankfort, whose first volume of pictures of insects was published at Nürnberg in 1679. Her books are noted in the Epilogue, *H.I.* p. 398. Hoefnagel and Hollar were also artists.

2 *Corr.* p. 434.   3 *F.C.* p. 306.   4 *Corr.* p. 436.
5 *Corr.* pp. 412–13.   6 *Corr.* p. 444.   7 *Corr.* p. 445.
8 He concludes the *Methodus*, p. 16, by saying: 'To tell truth we have not sufficiently studied the origins and metamorphoses of most Flies. From the observations of our friends we hope to compose shortly a more accurate classification.'
9 It was published by Smith and Walford with the date 1705: from its contents it seems likely that he revised it again after getting Willughby's papers.
10 Ray had sent to Sir T. Willughby his father's papers long ago (cf. *F.C.* p. 100): he had written asking for the loan of them but had no reply (*Corr.* p. 431).
11 *Corr.* p. 449: the prospect of publication had been already announced to his friends; for on 25 March Sherard had told Richardson of it and added 'I wish he may live to finish it' (Nichols, *Illustrations of Literature*, I, p. 346).

In the same letter he asks for 'a sight of your exotic diurnal Papilios', and on the same day sends a similar request to Petiver[1]—warning his friends in each case that there may be some danger of damage in transit. Sloane apparently visited him, bringing him a few specimens and a criticism of his proposal; for late in August he sent a fuller draft of it, elaborating the return to be made to contributors of plates, and adding a private note to the effect that he does 'not well like the cuts of Mr Petiver's *Gazophy-lacium*; they are not so elegant and polite as I could wish mine to be'.[2] Sloane warned him that the prospects of publication on the scale that he desired were poor; and Smith,[3] who had grumbled at paying him 'thirty pounds in money and twenty copies for his friends'[4] of the third volume of the Plants, and preferred producing fresh editions of the theological works to adventuring into a new field, merely relapsed into silence—a habit of the firm where Ray's scientific works were concerned.[5] With so vast a work before him, with 'not above two hours in a day to bestow upon it',[6] and with no sort of security that his labour would survive him, it is not surprising that he should write on 1 November to Sloane:

The History of Insects must rest if I continue thus ill, and I see no likelihood of amendment unless I should overlive the winter, which I have little reason to hope or expect. However, though I fail, there are many at present more able and skilful in this part of the history of animals than myself; as first of all your-self, next Mr Stonestreet,[7] then Mr Petiver, Mr Derham, Mr Morton,[8] Mr Antrobus,[9] Mr Dandridge,[10] Mr Bobart and many more

or that he should add: 'Pardon my scribbling who am scarce able to manage a pen'.[11]

Yet during these last months he must in fact have worked with much of his old energy. A hundred pages of the moths are nearly as complete as the butterflies; and the beetles, wasps and flies are much more than loosely arranged notes. Indeed in the light of the circumstances revealed by these

1 *Corr.* p. 450.   2 *Corr.* p. 454.
3 Samuel Smith and Benjamin Walford were printers to the Royal Society and publishers for Ray.
4 *Corr.* p. 456.
5 Cf. their treatment of Hotton's enquiries about Ray, *Corr.* p. 379. This time Smith had a sufficient excuse: he died in 1703.
6 *Corr.* p. 450.
7 Rev. William Stonestreet, Rector of St Stephen's, Walbrook, a great collector of shells and fossils: cf. *Letters to R. Thoresby*, p. 120.
8 Rev. John Morton, Rector of Oxendon, Northants: cf. *Corr.* pp. 369, 379: 'an ingenious person who was once here with me', *F.C.* p. 248.
9 Robert Antrobus, 'the most curious that ever collected insects', *F.C.* p. 305: he was at Peterhouse (B.A. 1701 and fellow), a pupil of Vernon (cf. Gunther, *Life of Lhwyd*, p. 469), and afterwards a master at Eton (*Antrobus Pedigrees*, pp. 15–16).
10 Joseph Dandridge, famous for his collection of birds' eggs.
11 *Corr.* pp. 457–8.

letters the book takes on an almost poignant pathos. The careful references
in the margin to the 'P.M.' or 'Pyx. rar.' or notes like 'lost out of the
Great Box'—the dying man clung to the hope that some one of his friends
would furnish plates, and was at pains to supply the necessary information.
The descriptions in the diaries, 'Caterpillars taken in 1695' and the rest,
we can see him laboriously copying out in the appropriate place each under
its own imago; and thankfully writing 'transcribed' against the original
note when one more was added to the finished script. The draft scheme for
the beetles with its references for the correct insertion of the longer descriptions—how many painful hours went to the compilation of it. We can see
him buoyed up by the belief that, if none of the experts would complete
the work, at least an editor would see his instructions and carry them out;
and resolved to leave it, however unfinished, at least in a state which would
make editorship easy. For those who want to know the man, and have the
imagination to admire and love him, it is perhaps better to have 'Kath.
Ash-Geometer' or 'Jane's, found by Jo. Kelhog' than a donnish reference
to '*Mus. Pet.*' or '*Gazoph.*' The book in its present condition is a memorial
to his quality, to his courage and care and faith, that an editor might easily
have destroyed. But it is not easy to think gently of the people who let his
work lie forgotten and then could publish it without a word even of
explanation and apology.[1]

So, with the book mummified for us as it left his hand, the end came.
A few days before his death the collections were sent to Samuel Dale in
Braintree, and after it strong pressure was put by Sloane upon Dale to
undertake its editing. His reply, though a refusal, does him honour: but
this was a tragedy; for instead the script was taken over by Derham, who
had a real interest in the subject but no special knowledge of Ray's work.
Until the summer of 1708 he kept it; and the bundle of papers gathered
from the dead man's study may well have suffered further derangement in
his keeping. Then having done nothing he sent it to the Royal Society[2]
where, as Robinson wrote to Lister, 'they lie in Dr Sloane's hands and
when they will see a resurrection I cannot divine'.[3] In July 1709 Aston
moved that the book be printed 'without endeavouring to make it more
perfect or staying till figures could be got'; and on 6 August Robinson
wrote that it was being sent to the press. It was published early in the next
year by A. and J. Churchill, of the Black Swan in Paternoster Row, but
without any sort of preface or explanatory notice or any editing. Fortunately Ray's handwriting was always 'peculiarly fair and elegant' and his

---

1 Or of L. C. Miall, who sneers at Ray for the muddle of the book and dismisses his records and observations as 'of no great moment', *The Early Naturalists*,
p. 113.
2 So Lhwyd to Richardson on 8 Oct.: Nichols, *Illustrations of Literature*, I, p. 321.
3 *F.C.* p. 309.

Latin remained lucid and intelligible. But the book is a medley, an un-edited collection of material, such as any author undertaking a piece of research amasses before the final drafting of his work. The 'methodus' or scheme of classification, published as a sixteen-page pamphlet in 1705, was reprinted as an introduction: but in the book itself this arrangement is very loosely followed. The first ten pages deal with worms, leeches, bugs, fleas and ticks and contain only notes signed F. W. or condensed from Mouffet. Then follow thirty pages on spiders prefixed by the statement that as Willughby had died before he had completed his observations[1] and Ray had himself left this field to others,[2] he proposes to borrow the account of them from Lister's treatise.[3] Then on p. 41 begin more notes by F. W. on wood-lice and millepedes, on p. 47 a list of dragonflies with short descriptions, and on p. 53 a similar list of Hemiptera. Then follow grasshoppers, crickets, mole-crickets and the rose-gall fly; cockroach; the ant or pismire, which Ray in his *Methodus*[4] had rightly placed with the bees and wasps; then a list of Tipulae; and so to the beetles. Here there are first descriptive notes on thirty-seven beetles beginning with the Dor, thirteen lady-birds and three or four smaller ones, notes typical of Ray and certainly his own;[5] and then on pp. 89–108 fourteen sections each devoted to a genus of beetle, the genera being distinguished by the form of its antennae: this looks like an attempt by Ray to produce a classification; for in the first section is the sentence 'here belongs in my opinion my first beetle the Dor', and similar notes occur in other sections. On p. 109 he begins the butterflies and the book at once becomes more complete; classified ac-counts, in many cases both of larva and imago, of many British and a few other species follow until p. 139. Then a page giving a classified list of dragonflies is interpolated. On pages 141–240 the moths are treated in the same fashion as the butterflies—the descriptions are not always suf-ficiently clear to make identification certain, but on the whole he fastens very skilfully upon the characteristic points. Bees, wasps and ichneumons fill pages 240–62, the descriptions and the notes on habits being often full of interest. Then is inserted a long letter to Ray from Derham dated 24 June 1702, Upminster,[6] and dealing with gnats—a letter which must

---

1 Long discussions about the projection and character of webs had taken place between Ray and Lister in 1668–9 and resulted in a letter to *Phil. Trans.* v, no. 65, p. 2101; and Lister promised Willughby to join him in work on spiders and put his results at the service of the *History of Insects* (*Corr.* p. 51). Lister's first paper and list of species appeared in *Phil. Trans.* June 1671, VI, no. 72, pp. 2170–5.

2 He confessed in a letter to Lister (*Corr.* p. 29) that from childhood spiders had filled him with disgust.

3 Published 1678 with three other papers.   4 P. 11.

5 Only one is dated—June 1694: two others have notes, 'I had it from Mr Robert' (can this be in both places a misprint for Bobart?) and one is 'sent by Mr Derham'.

6 Ray had asked Derham for this: cf. *Corr.* p. 399: his reply is printed in Derham's *Phil. Letters*, pp. 316–19 as 'answer to the Editor's Letter of June 24, 1702 printed in Ray's *Hist. Insect.* p. 262': cf. *Corr.* p. 401.

have been preserved among Ray's manuscript, but is out of place here. Then two-winged and four-winged flies fill eleven pages, 266–76; at this point the book ought presumably to end. Instead there is a heading: 'Caterpillars observed and described in the year 1695', and we have a hundred pages, quite unclassified, of paragraphs giving accounts, short or long, of all the Lepidoptera[1] that Ray handled during the next four years. Many of them are marked 'transcribed' or 'entred' [*sic*] and can be found in the earlier and classified part of the book. Like the previous section on moths, the work is generally clear and careful. Appended is a list of beetles written by Dr Lister, and presented to the Ashmolean by him.

It is unnecessary to give a detailed examination of the contents, because these are set out in so far as the Lepidoptera are concerned in the first volume of A. Werneburg's *Beiträge zur Schmetterlingskunde*, and generally in F. S. Bodenheimer's massive volumes, *Materialien zur Geschichte der Entomologie bis Linné*. In Bodenheimer's first volume[2] there is a brief note of Ray's life, an analytical table of his system of classification and a series of specimens of his records. It is unfortunate that this account both by its position and its proportion does less than justice to Ray's work. He is placed not with his contemporaries, Malpighi and Swammerdam, nor in his chronological order as the first serious entomologist, but with and after the men of the eighteenth century, Réaumur, Frisch, Roesel and Bonnet: and whereas their work is described at great length, his is given a cursory glance and summarised in a few ill-chosen extracts. In the second volume,[3] however, are tables giving the identification of the insects mentioned by him; and here Bodenheimer's work, being merely condensed from that of Werneburg,[4] is of some value. Werneburg is on the whole judicious: his ignorance of the fauna of Essex occasionally leads him astray;[5] and even among the butterflies there are a few cases of carelessness:[6] but in the main he has deciphered Ray's descriptions with remarkable skill. Of the species described in the main part of the book he lists 268, of which 59 belong to the Butterflies. Of the 209 moths only 12 are marked as unidentifiable;[7] and most of the names given seem to be correct. In addition he has identified most of the two hundred odd records contained in the diaries at the end of the book. To these Bodenheimer adds

1 Sawfly and other larvae are also included: cf. pp. 286, 372–3.
2 Pp. 486–94. 3 Pp. 412–27.
4 Published at Erfurt 1864 his first vol. devoted pp. 56–86 to identifying Ray's species.
5 Thus he names a moth that is clearly *Plusia iota* (p. 154) as *P. chryson*, and *Catocala nupta* (p. 152) as *C. promissa*, and *Boarmia gemmaria* (p. 175) as *B. roboraria*.
6 Noted below: of moths his worst cases are *Dianthoecia cucubali* (p. 165, no. 20) for what is really *Apamea secalis*; *Larentia ferrugata* (p. 220, no. 85) for *Camptogramma bilineata*; and *Leucania comma* (p. 220, no. 64) for *Cucullia umbratica*.
7 Among them are p. 218, no. 106, which is certainly *Trichiura crataegi*, and p. 143, no. 4, probably *Notodonta trepida*.

a large number of the Beetles and a few of the Hymenoptera, deriving his lists from H. Kuntzen and H. Bischoff.

This testimony to the skill with which Ray picked out the salient points and to his accuracy in the use of terms to describe the size and shape of the wings and the colour and position of the markings is really remarkable. Where he gives (as he very frequently does) descriptions of the larva and pupa, food-plant and time of emergence, his species can almost always be named with certainty: where he has only the imago, and this often a species with obscure, intricate or variable markings, the task is more difficult: where he has only one in a genus of closely similar forms, it is not always possible to define precisely. But in general a careful reading of his words gives a clue.

That he obtained so many in a place not particularly favourable, and with little apparatus or opportunity, is proof both of his enthusiasm and of his gifts of accurate observation and discrimination. He is short among the Noctuids, the night-flying moths which the modern collectors secure at their sugar or in their light traps: many of his best species he only obtained by beating out and breeding their larvae. Yet he got and described a big proportion of the moths of the neighbourhood, and some which are definitely scarce; and though his arrangement is very defective there are signs that he was beginning to group the families together on structural grounds.

To conclude our account it may be of interest to look at the section in which the nearest approach to a final treatment is attained, that dealing with butterflies, which had been fully revised in the last few months of his life.[1]

The section begins with a note on the Butterfly. 'Papilio,[2] as Vossius the etymologist declares, is so called because it *papet*, that is eats and sucks honey from plants. Rather the word is derived from the Greek ἠπίολος by repetition of the letter "p." Aristotle, VIII, 27,[3] speaks of ἠπιόλους flying round the lamp which Pliny, XI, 19, translates as papilio.' Ray had always delighted in rare words, and was indeed a fore-runner of Skeat as much as of Darwin. Here is a typical instance of his ingenuity. ἠπίαλος (the rough breathing is perhaps a printer's error which escaped Dr Derham) is a rare word used chiefly in medical writers for the shivering of a malarial patient: Aristophanes uses it of nightmare: the passage in Aristotle is quoted by Liddell and Scott under the heading ἠπίολος as meaning 'moth': is it not the original meaning of the word—fly-by-nights having then as now a sinister character? This derivation at least proves that Ray was acquainted with the *Etymologicon* of Vossius. But we digress. He continues:

1 So *Corr.* p. 451, to Petiver.
2 Papilio to Ray and all his contemporaries included moths.
3 I.e. in *Historia Animalium*.

You ask what is the use of butterflies? I reply to adorn the world and delight the eyes of men: to brighten the countryside like so many golden jewels. To contemplate their exquisite beauty and variety is to experience the truest pleasure. To gaze enquiringly at such elegance of colour and form devised by the ingenuity of nature and painted by her artist's pencil, is to acknowledge and adore the imprint of the art of God.

Then he turns to his science with the heading: 'Butterflies flying by day whose sure and characteristic mark is their clubbed antennae.' So he fastened upon a distinction that has since been universally accepted.[1] He subdivides into those with shorter antennae and wings held erect when at rest, and those with longer antennae and wings open—the latter class including, as we shall see, certain day-flying moths. The former class is then split 'for convenience of recognition' into groups according to the colour of the wings.

I. 'Butterflies with yellow or yellow and black wings':

1. 'Butterfly with wide expanse of wing, beautifully varied with yellow and black, the hind wings tailed.' He gives references to Mouffet, *Theat. Insect.* p. 99, and to Petiver, *Mus.* 328 [2]—certainly *Papilio machaon*. This cannot be confused with any other British species. As localities he gives Leghorn and Montpellier, Sussex and Essex; and for the caterpillar Sussex and Montpellier. He gives a full description of the caterpillar and draws special attention to 'its two saffron coloured horns [the osmateria] almost a finger's breadth long, drawn in and out at will like a slug's'. 'In France it fed on Foeniculum tortuosum [3] and in Sussex on Burnet-Saxifrage' (*Pimpinella saxifraga*), an unusual food-plant in this country, if correctly stated here.[4]

2. A similar species from Virginia, larger but hardly separate. M. Probably *Papilio ajax*.

3. 'Butterfly with wide wings, pale yellow, marked with transverse black bars.' M. Certainly *Papilio podalirius*. 'We found it at Leghorn and also, unless my memory is at fault, in Britain.' In view of his saving clause

1 He had mentioned it as characteristic of Diurnal Butterflies in a letter to Lhwyd of 5 April 1692: *F.C.* p. 227.

2 Petiver, no doubt in compliment to the Revolution, calls it the Royal William. Hereafter M., P., and G. (Goedart) will denote references to Ray's chief authorities.

3 This plant is listed in his *Catal. stirpium in exteris observ.* p. 53. It is *Seseli tortuosum*, Linn.

4 Writing to Lister in July 1670 from Middleton, he says: 'This summer we found here the same horned Eruca, which you and I observed about Montpellier feeding on Foeniculum tortuosum. Here it was found on common Fennel' (*Corr.* p. 62). Machaon, which was formerly widespread in England, will eat many Umbelliferous plants, though with us its usual food is *Peucedanum palustre*. This plant was apparently unknown to Ray and may have been confused with the very variable *P. saxifraga*.

Ray can hardly be quoted in support of the long-held belief that *podalirius* was once a British insect.

4. 'Butterfly, early, sulphur or yellow-green.' M.P. Certainly *Gonepteryx rhamni*. 'I have often seen it in early spring and also after midsummer.' 'The male is paler, a greenish white: Petiver says he found the two—pale and dark—paired. Here the pale one is rarer and I was uncertain whether the difference was specific or sexual till I read Petiver.' Here as often he assigns the sexes wrongly.

5. A similar species from Jamaica, 'sent by my great friend Dr Hans Sloane'. Possibly *Catopsilia eubule*.

6. 'Butterfly, middle-sized, upper wings deep yellow with black edges.' M.P. Certainly *Colias edusa*, not, as Werneburg says, *C. hyale*. He describes it fully, noting the red-rimmed spot on the underwings. 'In the female white takes the place of yellow': so he had seen and recognised var. *helice*. 'I found it in a flax field at Bocking in Essex. Mr Vernon sent it to me from Cambridgeshire. Willughby noticed it frequently in Styria: so it is evident that many British butterflies are common to the rest of Europe.'

II. 'Butterflies white or white varied with black or orange':

1. 'Butterfly, the larger Cabbage white very common.' P.G. Now *Pieris brassicae*. He notes the distinction of the spots in the sexes and describes the caterpillar and chrysalis with an account of its parasites—a record condensed from that in the *Cambridge Catalogue*.[1]

2. 'Butterfly middle-sized white with two black spots on fore wings, hind yellow below.' P.G. Now *Pieris rapae*. 'This and the following first shown me by Mr Dale.' The caterpillar is carefully described.

3. The same with a faint black spot. 'This probably differs only in sex from the preceding.' It is of course the male.

4. 'Butterfly, middle-sized Cabbage, white with greenish black lines on the veins of the under wings.' P. Now *Pieris napi*. He notes that some have a double spot: 'these are female.'

5. 'Butterfly white with black veins, of the size and shape of the Greater Cabbage.' P. Now *Aporia crataegi*. 'This species is readily distinguishable.'

6. 'Butterfly small with white upper wings marked with bright orange.' 'The white marbled Male Butterfly.' Now *Euchloë cardamines*.

7. The same but with no orange: 'The white Marbled Female Butterfly.' M.P. 'The female of the preceding.' Description of imago and caterpillar marked F.W. It feeds on Thlaspi biscutatum.[2]

---

1 *C.C.* pp. 124–8; the parasites are further described under Vespa Ichneumon, *H.I.* pp. 254, 260.

2 This appears to be *Alliaria officinalis*, for which Ray gives Thlaspi cornutum as a synonym (*C.C.* p. 6).

8. 'Butterfly, small, white with a black mark in the corner of the upper fore wings: faintly marbled below.' P. 'Small white Butterfly.' Now *Leptosia sinapis.* 'Dark spots hardly seen in the other sex.'

9. 'Butterfly of middle size with wings beautifully variegated with white and black.' P. 'The Half-mourner.' Now *Melanargia galatea.* 'In June about St John the Baptist's day [24th] I saw it first flying this year, 1690, especially in marshy places: it was a very cold spring.' 'I have not yet seen the caterpillar.' 'It is very common with us round Braintree, and not rarer, I fancy, elsewhere in England.'

10. 'Butterfly of middle size, with upper wings white with fewer black spots, under wings varied with white and green.' P. 'The greenish marbled Half-mourner.' Now *Pontia daplidice.* A careful description follows; and the note: 'I had it from Mr Vernon who took it in Cambridgeshire. Mr Jezreel Jones observed it round Lisbon according to Mr Petiver.'[1] This, the rare and casual Bath White, was known for some time as 'Vernon's Half-mourner'.

11. 'Butterfly coming from a large gregarious caterpillar, dingy white with black veins.' The description of the imago and of the larva ('it feeds on Hawthorn, spins webs and lives gregariously returning after feeding to its companions') is marked F.W. It clearly applies to *Aporia crataegi,* i.e. to no. 5, and should have been inserted there.[2]

III. 'Butterflies with wings red or tawny, varied with black, white, yellow, blue, all or some':

1. 'Butterfly of the nettle very common.' M.P.G. 'The lesser Tortoise-shell Butterfly.' Now *Vanessa urticae.* Short description of imago, caterpillar and chrysalis.

2. 'Butterfly like that of the nettle but larger: we call it the elm-butterfly.' 'The greater Tortoiseshell Butterfly.' P. Now *Vanessa poly-chloros.* It is well described with the note: 'the caterpillar is rather like the preceding; I found many this year (1695) feeding on sallow.'

3. 'Butterfly like the elm-butterfly but smaller with jagged wings marked beneath with a curved white dash.' P. Now *Polygonia c-album.* A good description of the imago—'a white dash almost the shape of the letter c'—and of the caterpillar which he had certainly found at Black Notley[3]—'feeding on Hop with a long white patch on the back'—and a note: 'Mr Petiver puts forward another species, surely rather a variety with more jagged wings.'

---

1 Cf. *Mus. Pet.* p. 304: 'The only one I have seen in England Mr Will Vernon caught in Cambridgeshire. Mr J. Jones', etc.

2 Presumably this note was added after he had received Willughby's records.

3 Cf. *H.I.* p. 349, where the original description is printed.

4. 'Butterfly, large, with tawny black-marked wings, with transverse silver bars below.' 'The greater Silver-stroaked Fritillary.' P. Now *Argynnis paphia*. 'A different species from the next. Mr Dale showed me both sexes, the male having larger and blacker spots'—another transfer of sex.

5. 'Butterfly, large, tawny winged with many black spots above, and elegantly marked with silver below.' M.P. 'The greater Silver-spotted Fritillary.' Either *Argynnis aglaia* or *A. adippe*.[1] The black spots are arranged like a chess-board (tabula schaccaria). 'I have not yet been able to see the caterpillar.'

6. 'Butterfly from Riga, small, golden, beautifully marked with silver spots below.' P. Probably *Argynnis lathonia*. 'A beautiful and quite distinct species.' 'Petiver had it first from Mr David Krieg from Riga: afterwards it was found around Cambridge by Mr Vernon, Mr Antrobus and others.' In view of the following species and Ray's claim that this is quite distinct, this can hardly be any other than *lathonia*: the next early record was from Gamlingay; and on occasion a number of specimens have been taken together.[2]

7. 'Butterfly, Fritillary, large, tawny, black-spotted.' P. 'The April Fritillary.' Clearly *Argynnis euphrosyne*; not as Werneburg says *A. selene*: for 'on the lower side of the hind wings are two spots of silver and the wing is edged with silver lunules'. 'Mr Dale first showed it to me.'

8. 'Butterfly, middle-sized, the upper wings rufous marked with thin black lines down the veins and other wider transverse ones.' P. 'The May Fritillary.' The description of the imago—'the characteristic of the upper wings is a black transverse band followed by a wider rufous area, cut up into oblongs and on the hind wings marked with black spots which on the underside are larger and rimmed with yellow'—and of the caterpillar—'black with pencils of hair and two white points on each segment: it feeds on scabious'—must refer to *Melitaea aurinia*. But a note follows: 'this is smaller than the preceding and has more silver spots. Mr Petiver thinks it only a sex difference. Mr Dandridge has observed that it comes out nearly a month later': and this must refer to *Argynnis selene*. Probably the note should belong to the heading which fits *selene*; and the description of *aurinia* has been put in here by mistake—possibly belonging to no. 10 or no. 11.

9. 'Butterfly, Fritillary, from Lincoln with pale bands on the under-side.' P. with a reference to *Gaʒoph*. Tab. 18, F. 10, which reads: 'The

1 Petiver in *Pap. Brit. Icones*, Tab. III, 5–8, gives plates of both: in the notes he refers to Ray and his own *Museum* under *adippe*. Linnaeus, *Syst. Nat.* I, v, quotes Ray under *aglaia*, but does not list *adippe*.

2 Cf. Newman, *British Butterflies and Moths*, pp. 34–5.

Lincolnshire Fritillary first observed there and given me by Madam Glanville. The curious Mr Dandridge hath lately caught him not far from London.'¹ From this reference the species has been regularly identified with *Melitaea cinxia*, the 'Glanville Fritillary': but *cinxia* feeds on *Plantago maritima* and is now confined to the south cliffs of the Isle of Wight, though formerly it was found on the Kentish coast. The localities suit *Melitaea athalia* much better; for this had formerly a wide distribution, especially round London. Ray's description is long and the term 'rufous' which he applies to both upper and under wings is more appropriate to *athalia* than to *cinxia*; but his description of the black spots in the rufous bar on the upper hind wings and of the white and black-spotted tips to the under fore wings applies only to *cinxia*. We may therefore vindicate tradition and agree that *cinxia* was once common at Lincoln. Ray adds a note: 'Mr Dale gave me abundance of it: I have perhaps described it too lengthily.'

10. 'Butterfly, Fritillary, marked like a chess-board.' P. This seems to be a specimen of Petiver's. The brief description says 'it differs from the preceding because the base of the upper wings is black and the orange spots are square and below it has an edging of double lunules, pale and divided by a black line'. It is probably what Petiver later called the Straw May Fritillary, that is *Melitaea athalia* var. *tessellata*; though it may be *M. aurinia*.²

11. Another of Petiver's and either *aurinia* or a variety of the preceding.

12. 'Butterfly, Fritillary, small.' P. 'Mr Vernon's small Fritillary.' Certainly *Nemeobius lucina*, though the description is very short: 'smaller by a half than the smallest of the others: on the lower hind wings a double series of whitish spots and at their margin a row of eyes with black pupils'. 'This was first observed by Mr Vernon about Cambridge, afterwards in Hornsey-wood near London by Mr Handley and by Mr Dandridge at Boxhill, and is pretty common about Dulwich' (this note is in English).

13. 'Butterfly, large, beautiful, varied with black, red and white.' P.G. 'The Painted Lady.' Now *Pyrameis cardui*: well described and noted as 'common with us round Braintree and elsewhere'.

14. 'Butterfly, very elegant, with each wing marked with an eye-like spot.' 'The Peacock's Eye.' M.P.G. Now *Vanessa io*; a good description of imago, caterpillar and chrysalis.

15. 'Butterfly, middle-sized, with wings varied with tawny or rufous and black, with an eye in the corners of the outer wings.' P. Now

¹ Petiver, *Gazoph.* p. 28: the plate is not very exact.
² In Petiver, *Pap. Brit.* l.c., reference to this species in Ray is accompanied by a plate of *aurinia*: but Petiver's references to Ray are often careless.

Pararge megaera: described and noted as 'not uncommon after mid-summer'.

16. 'Butterfly, middle-sized, with dark brown upper wings with an eye in the corner.' P. Now *Epinephele jurtina*; 'in meadows before hay-making'.

17. The female, or as Ray says, 'the male as I suppose' of the preceding. P.

18. 'Butterfly, middle-sized, tawny and dark brown with a black eye having a double white pupil in the corner of the front wings.' P. Now *Epinephele tithonus*.

19. 'Butterfly, small, tawny and dark brown.' P. 'The Small Heath Butterfly.' Now *Coenonympha pamphilus*; 'very abundant in meadows all the summer'.

20. 'Butterfly, small, with wings red and shining like silk, margined and spotted with black.' P. Now *Chrysophanus phlaeas*; 'I found it in hedge-rows at the end of July'.

21. 'Butterfly, very small, with wings tawny above and below.' Either *Adopaea flava* or *A. lineola*. 'In the middle of the front wings is a rather wide oblique black line.' 'Early July in meadows.' No reference is given and the species is one of Ray's first descriptions.

22. The same without the oblique line; 'perhaps only sexually distinct'; the female.

IV. 'Butterflies with dusky or black wings with yellow or yellowish white or red eyes or spots'; 'including those of which neither sex shows any tawny or rufous on the upper wings':

1. 'Butterfly, large, blackish with wings beautifully marked with red and white.' M.P.G. 'The Admiral.' Now *Pyrameis atalanta*. 'I saw it frequently at Middleton in autumn among the pears: it is not less frequent here at Braintree.'

2. 'Butterfly, large, black or dusky with wings marked with white.' Certainly *Apatura iris*. There follows a very full and exact description, and the note: 'captured in July near Heveningham Castle [Castle Hedingham] in Essex in 1695 by Mr Courtman.' There is no reference to any authority, and this is presumably the first published description of the Purple Emperor.

3. 'Butterfly, of middle size, very elegant; the upper wings black with a white transverse bar, the under variously coloured.' Certainly *Limenitis sibylla*. Another excellent description and the note: 'captured in Essex not far from Tolesbury by Mr Morton and brought to me July 11th 1695.' Again there is no reference: this is also a first description of the White Admiral.

4. 'Butterfly, large, with wide wings; the outer ones dark brown; beautifully variegated with white and red spots and streaks.' 'Goedart,

Sect. 1, 5ta. edit. Lister.' A short description follows: but this does not easily suit any known species.[1] It ends with the words: 'I noticed one eye exceeding the rest with a black pupil and a light ring round it.' Hence Ray had seen the insect or Goedart's picture. A reference to Goedart's engraving and to the accompanying text[2] clears up the problem: the text speaks of a nettle-feeding caterpillar producing 'a very faire Butterfly Peacock-like Eyed'; the plate combines a larva of *Vanessa io* with the underside of *Pyrameis atalanta*. Goedart's plate 4 is, as Ray had noted, a Red Admiral with wings spread; plate 6, as he also noted, is a Painted Lady: on plate 5 is depicted the Red Admiral with wings closed: as each plate professes to be a separate species and as Lister evidently regarded plate 5 as distinct, Ray must have accepted it as such: it is not a very good picture, and Ray was always ready to believe the best of his friends. Nevertheless it is not easy to find an eye on the picture!

5. 'Butterfly, middle-sized, dusky, the wings with yellow spots above, variegated with yellowish white and black.' Certainly *Pararge egeria*. There is a clear description of the eye-like spots and a note: 'about mid-June I observed them frequent in our orchard and elsewhere, indeed during the whole summer'. Again there are no references and the species is here first described.

6. 'Butterfly, rather large, the wings dusky, with a double yellow spot and a double black eye on the outer part.' P. Now *Hipparchia semele*. There follow two descriptions, the first followed by the note: 'sent to me by Mr Tilleman Bobart',[3] the second preceded by 'in the year 1697 a similar specimen was given me by Dr David Krieg of Annaburg in Saxony, found on the Gogmagog hills in Cambridgeshire'.

7. 'Butterfly, middle-sized, all dusky, the underside of the wings marked with eyes.' P. Now *Aphantopus hyperanthus*. Very short description and no locality.

8. 'Butterfly, small, above all blackish, below bluish-green with white transverse line.' P. Now *Zephyrus quercus*.[4] Good description of male: 'in the other sex the upper fore wings are suffused with purple shining through the brown in the inner half of the wing.' 'On July 8th 1692 found paired on nettles.' 'In 1696 and 1697 we found its caterpillar, somewhat like a woodlouse.' There follows a description of caterpillar and chrysalis.[5]

1 Werneburg, *Beit. zur Schmetterlingskunde*, p. 59, and Bodenheimer, *Gesch. der Entomologie*, II, p. 415, identify it with *Pyrameis cardui*: but Ray knew *cardui* well, cf. *H.I.* p. 422; and Goedart's picture of it is no. 6 not no. 5.

2 Goedart, *Of Insects*, Pl. 5 and p. 7.

3 In Dec. 1692: this gift is described in letters to Lhwyd (*F.C.* pp. 232–4).

4 Werneburg gives *Thecla w-album*, clearly a mistake in view of the colour of the other sex and the colour and food of the larva.

5 This is transcribed from the description on p. 277, where details of food—oak—are given.

9. 'Butterfly, small to middle-sized, with broad wings, varied above with black and blue, with white fringes on the black edge.' P. There follows a long description of the underside. 'Emerged from chrysalis about the end of June: it really differs from the preceding though resembling it.' He then notes some small differences in the larva. It is plainly *Zephyrus quercus*, a bred specimen, but identical with no. 8.[1]

10. 'Butterfly, small, the upper wings blackish with a wide curved tawny mark in the middle.' P. Certainly *Zephyrus betulae*. He gives a full description, and adds: 'there is much more yellow on the under wings in the male than the female: in the latter the upper surface is all blackish'— a sound observation but with sexes reversed.

11. 'Butterfly, small, the wings purple-blue above, grey and black-spotted below.' 'The most common small blue Butterfly.' M.P. Now *Lycaena icarus*, 'common in fields in June and all summer'.

12. The female of the preceding—'very common in the same places'.

13. 'Small butterfly, dusky with golden margins.' P. Perhaps *Lycaena aegon*, as he calls the wings 'purple-blue in the middle'.[2] This is derived from Petiver—both name and note.

14. 'Butterfly, small, bluish with black edges.' P. From description 'larger than common argus: wings palely blue', evidently *Lycaena corydon*: 'Petiver found it on the Banstead hills near Epsom: Mr Dale saw it lately at Newport in Essex.'

15. Female of the preceding. P.

16. 'Butterfly, small, blue wings, the upper ones black above at the outer corner.' P. Now *Celastrina argiolus*—female. 'Mr Petiver found it in Mr Wattes' garden at Enfield.'

17. 'Butterfly, small, upper wings purple-blue, lower eyed.' M. From this very short description it seems to be *Lycaena aegon*—Ray's own specimen of no. 13 'captured by Mr Dale and shown to me'.

18. 'Butterfly, small, fuscous and marbled, of Hampstead': this and the brief description, 'yellow with a tinge of red (tan)' both from Petiver[3] suggest *Augiades sylvanus*, though Petiver's plate[4] is apparently *Nisoniades tages*.

1 Linnaeus, *Syst. Nat.*, enters it as 'P. pruni ?', 8 being 'quercus'.

2 Werneburg says *Papilio circe*; Bodenheimer *Chrysophanus dorilis*.

3 This has presumably no connection with the most famous of all Petiver's specimens, the Papilio oculatus Hampstediensis ex aureo fuscus (*Pap. Brit. Icones*, Tab. v, 2 and p. 2), which as Albin's Hampstead Eye appeared in all early butterfly books, even down to the 1895 edition of that by F. O. Morris. Certainly Ray makes no mention of eyes. The Hampstead eye as pictured by Petiver looks an aberrant *Pararge megaera*.

4 *Gazoph.* Tab. 36, 3, so too in his later work *Pap. Brit. Icones*: but there he pictures *A. sylvanus* (Tab. vi, 16, 17) and describes it 'mixed yellow and dusky and as common at Hampstead'.

19. 'Butterfly, small, all dusky with a line of white dots edging each wing.' Certainly *N. tages*: well described and with note: 'I found it in the fields on May 9th and it was not uncommon all the month.' No references, so this is a first description.

20. 'Little butterfly, fuscous with rather few white spots.' P. Probably *Hesperia malvae*, as he adds 'fewer but larger spots than that next to be described: Mr Petiver thinks it different in sex'.

21. 'Butterfly, very small, dusky, all the wings marked with spots and points pale or of whitish yellow.' P. Certainly *Hesperia malvae*. Description and note, 'I found it on May 29th in a marshy place in the fields', follow, and a statement that 'it resembles the first butterfly of the fourth class, as arranged by me, in colour but is much smaller'—this is inexplicable, the butterfly referred to being the Admiral: some error has arisen.

22. 'Butterfly small, the upper wings dusky, the under green.' Now *Callophrys rubi*. 'I saw it at Mr Dale's.' A first description.

There follows the second division of day-flying butterflies, those with long antennae which hold the wings open at rest. This includes only three. First, a 'very swift insect with short wings and heavy body that makes a humming in flight'—*Macroglossa stellatarum*, the Humming-bird Hawkmoth. This he describes in detail, noting its habit of perpetual motion, of hovering in front of flowers with its long proboscis exserted, and speculating on its refusal to settle while it feeds. 'The female', he states, 'is duller coloured: those of this sex that I obtained at Notley made no humming and I never saw them flying before the evening.' The second is one of the Burnet Moths; 'it has five blood red spots on the fore wings, three towards the edge and a double one at the base'; the caterpillar, briefly described from a note by Willughby, 'feeds on Dropwort and other herbs', and changes into a black pupa encased in a shining yellow cocoon. It is probably *Zygaena trifolii* or *lonicerae*, though the mention on Willughby's authority of 'Filipendula' as the food-plant may have led Linnaeus to bestow the specific name *Z. filipendulae* upon the commonest British species.[1] Finally, there is a small insect, like the Burnet but differently coloured, evidently one of the Foresters, and probably *Ino statices*; for Ray comments upon the fact that some specimens are of a green and others of a blue colour, which is true of *statices* though not as he suggests a mark of sex.

There follow a number of exotic butterflies. The descriptions are not classified and in some cases are mere headings. Of the nineteen species one,

---

1 Linnaeus, *Syst. Nat.* I, v, 2390, quotes Ray under this species, which has not five spots on the fore-wings. All the Burnets feed normally on trefoil.

an Indian, is described twice[1] in almost identical terms. Only three of
them are of interest. One is 'a very large black butterfly with all its wings
edged with a broad white border, found on the mountains near Geneva
and brought to Mr Willughby',[2] obviously the Camberwell Beauty,
Vanessa antiopa. Another is 'a large Alpine butterfly, the wings whitish,
the fore wings marked with black spots, the hind with eyes having a red
iris'; it is carefully described, evidently by Ray's own hand; and notes are
added: 'I found it on the very top of Mont Salève near Geneva' and 'this
species is mentioned by Mr Petiver and was brought to him by Mr Richard
Wheeler from Norway'. It is of course *Doritis apollo*.[3] The third, 'a large
butterfly from Jamaica', recorded also by Petiver from Carolina, is
certainly *Anosia plexippus*.[4]

From this perhaps over-detailed account of one section of his book it is
possible to realise how good a use Ray made of his last years. To have
described recognisably forty-seven British butterflies, at least six[5] new to
science, is no small feat; and this impression would be much strengthened
if our examination were carried on to the moths. Here Ray evidently felt
that he was breaking fresh ground; for in fact, apart from a few large and
brightly coloured species, they were then almost entirely unknown. No one
hitherto had even collected them seriously: no one had ever realised the
importance of studying their metamorphoses and working out full descriptions of each stage. He seems to have grasped what lepidopterists for
a hundred and fifty years after him were slow to appreciate that, if a true
understanding of these insects or a correct classification were to be obtained, it was not enough to amass collections of the imagines. From
Linnaeus down at least to Haworth the 'perfect insect alone was regarded
as furnishing characters for the division and subdivision of groups',[6] and
study of the earlier stages was virtually ignored. Here as elsewhere Ray
reveals a scientific endowment, a sense of the wholeness and continuity of
life and a flair for the right method of approach, which was centuries ahead
of his time. Where he has completed his task and brought together descriptions of larva, pupa and imago, his records can easily be identified;
if he had been able to carry forward his work or even to arrange the
material actually gathered, he might have saved entomology from a vast
amount of superficial and mistaken classifications. He would have done

---

1 On pp. 134 and 137, evidently a case of a note inserted again without noticing
that it has already been copied and included.

2 *H.I.* p. 137.

3 A drawing of this identical specimen was made by his friend Allen in his Commonplace Book, no. 122: it is described as 'taken on Alps by Mr Ray'. See *Essex Naturalist*,
XVI, p. 167.   4 *H.I.* p. 138.

5 It must not be forgotten that in addition he gave to Petiver several new to
him, including *Zephyrus quercus*, *Mus.* p. 330.

6 E. Newman, *Butterflies and Moths*, p. 18.

for insects what he did for plants—start future students upon the study of the organism as a whole, open up and provide material for a thorough appreciation of taxonomy, and save us from being afflicted with arbitrary and fanciful systems of classification and nomenclature.

In some respects his work on moths is better than on butterflies. He no longer has to bother with the records of Mouffet or the chance collections of Petiver. He has to pick his own way, formulate his own descriptions and work out for himself the distinctions between one species and another. It is a delight to read his clear and discriminating pictures, even if sometimes the lack of knowledge of kindred forms makes him omit just the point which would determine the exact identification of his specimen. If only the Royal Society or some of his wealthy friends had realised the mass of material that he had assembled, and recovered from Dr Dale the boxes of insects to which his notes so carefully refer, and had plates made from them so that each description might have its appropriate illustration, a century of pioneering could have been saved and Ray's own inimitable work been made available. As it is, the *History* remains a locked museum, known indeed and studied by Linnaeus,[1] but replaced in public esteem and general utility by Petiver's very inferior junk-shops,[2] and Eleazar Albin's picture-books.[3]

NOTE. *Ray and the case of Lady Glanville*

Moses Harris, author of a quaint picture-book of butterflies and moths called *The Aurelian* and published at his own charges in London in 1766, in his comment upon the Glanville Fritillary states:

This fly took its name from the ingenious Lady Glanvil, whose memory had like to have suffered for her curiosity. Some relations that were disappointed by her will attempted to set it aside by Acts of Lunacy, for they suggested that none but those who were deprived of their senses would go in pursuit of butterflies. Her relations and legatees subpoenaed Dr Sloane and Mr Ray to support her character. The last gentleman went to Exeter and on the tryal satisfied the judge

1 So Pulteney, *Sketches of Botany*, I, p. 270: study of the *Systema* makes this evident.
2 Petiver's earlier publications were simply catalogues of his plates and collections; the *Museum*, published in 'centuries' between 1695 and 1703, contains in Cent. iv a list of English butterflies, pp. 301–30, containing twenty-six species, including Forester and Burnet: the *Gazophylacium* also appeared in parts, a medley of plants, shells, insects, birds, fishes, etc. in any sequence with brief notes to the plates. In the advertisement affixed in 1703 an offer is made to depositors of 10s. of receiving 5s. worth of his published works; and a price list is given 'My eight *Centuries* to subscribers 2/6, to others 3/6; *Gazophyl.* Dec. I, 4/- and 5/-; the same mostly painted £1. 1. 6 and £1. 5. 0; the *Catalogue* separate 6d. and 1/-'—an interesting side-light upon the man and his times. For a vivid and characteristic picture of him cf. Uffenbach's account of a visit to Aldersgate on 19 July 1712 (Mayor, *Cambridge under Queen Anne*, pp. 370–1).
3 *A Natural History of English Insects*, London, 1720: his plates were very similar to those of Maria Sibylla Merian.

and jury of the lady's laudable enquiry into the wonderful works of the creation
and established her will. She not only made the study of insects part of her
amusement, but was as curious in her garden and raised an Iris from the seed
that is known to this day by Miss Glanvil's Flaming Iris.[1]

The late Sir Albert Seward in his sketch of Ray relates this legend
as if it were history,[2] deriving it no doubt from Boulger's article in the
Dictionary of National Biography.[3]

Dr Malcolm Burr in *The Insect Legion*, in the chapter dealing with the
history of entomology, states[4] that the lady appears to have been Miss
Winifred Bourchier, who married Sir John Glanville of Broad Hinton,
Speaker of the House of Commons, who died in 1661. He says that no
record of the case is extant; and suggests that it was before 1690, as after
this date 'Ray retired to his native village to nurse his ill-health'. He does
not seem to have seen Harris's original statement, as he erroneously
charges him with calling Sloane 'Sir Hans' and then points out that this
honour was not bestowed until 1716. Burr's account of Ray is unfortunately very slipshod; but there is certainly no record of the case in
W. U. Glanville-Richards's *Records of the House of Glanville*; and I have
failed to find any references in law reports.

There is, however, plenty of other evidence which makes any connection of Ray with the case impossible. Petiver in his *Gazophylacium*, commenting on Table 18, Fig. 10, the plate of the 'Lincolnshire Fritillary',
says: 'first observed there and given me by Madam Glanville'.[5] He mentions Madam Glanville several times, notably as the donor of three species
of 'western moths' on Table 7 and as 'a person extremely curious in the
knowledge of English insects'.[6] He speaks of her as alive at this time, 1703,
so that she cannot possibly be the Lady Glanville who married Sir John in
or about 1615. Further, in the Sloane MSS. in the British Museum is an
undated letter from Petiver to Madam Glanville[7] and a reply dated 1702
from Bristol from 'E. Glanvile' to Petiver.[8]

In his own description of the Lincolnshire Fritillary, Ray, as we have
seen, refers to the *Gazophylacium* plate but states that his own specimens
of it came from Dale and makes no reference there or anywhere else to the
lady.[9] It is obvious that as she was certainly alive in 1702 her will cannot
have been disputed at a time when Ray could have given evidence about it.

It may well be that the rest of the story is true; that Sloane gave evidence
and used Ray's *Wisdom of God* as proof that devotion to natural history
was a sign rather of learning and piety than of madness.

1 *Aurelian*, p. 34.   2 *John Ray*, p. 28.   3 Vol. XLVII, p. 341.
4 Cf. pp. 283–4.   5 *Gazoph.* p. 28.
6 Figs. 7, 8, and 11: *Gazoph.* pp. 12–13.   7 Sloane 4067, f. 83.
8 Sloane 4063, f. 180: this may be Elizabeth, the youngest and unmarried daughter
of Sir John Glanville—she made a will, seemingly reasonable, in 1707; or her niece,
daughter of Julius: cf. *Records of Glanville*, p. 161.   9 Cf. *H.I.* p. 121.

CHAPTER XVI

OF FOSSILS AND GEOLOGY

Nor did his artful labours only shew
Those plants which on the earth's wide surface grew,
But piercing ev'n her darkest entrails through
All that was wise, all that was great he knew
And nature's inmost gloom made clear to common view.

Ray's Epitaph translated, *General Dictionary*, VIII, p. 695.

Ray's work as a botanist and zoologist was mainly concerned with description and classification; and as such lay outside the area of speculative and controversial science. He found the study of nature in a primitive and inchoate state, and was content to devote himself to the primary business of observing, discriminating, defining and arranging the flora and fauna to which he could get access. The need for such work and the magnitude of his contribution to it are obvious: there could be no scientific biology until it had been undertaken. In perceiving and insisting that specific distinctions must be based not upon size or colour, habitat or habits, but upon structure, Ray vindicated his claim to be something more than a maker of collections and catalogues, and repudiated in advance Linnaeus's criticism of him. If his researches into plant and animal anatomy seldom led him beyond the study of taxonomy into the problems of form and function, still less into those of evolution or genetics, he very effectively laid the foundation for such studies and, as we have seen, pointed the way towards them. He had neither the opportunity nor the equipment to follow up his hinted doubts as to the fixity of species or to proceed from anatomy to physiology; and to blame him for accepting as his own a limited and preliminary task is to misjudge the contemporary position.

But from the standpoint of the modern scientist such work must necessarily seem elementary: and his devotion to it gives little proof of scientific ability as we understand it. Granted that the 'new philosophy' depended upon accurate knowledge of nature and must proceed from observation to experiment and so by induction to the formulation of hypotheses and the discovery of principles, the work of naming and classifying affected only the foundations of the study. It did not in itself or inevitably raise issues in which the new came into conflict with the old nor force its exponents to recognise that they were creating a fresh interpretation of the universe. Ray, like Linnaeus or Gilbert White, could go on with his natural history without consciously challenging the historicity of Genesis, or the traditional dating and method of the creation. It was only in his incursions into geology and the study of 'formed stones'[1] that he entered the arena of conflict.

We have already seen that towards the end of his career as a botanist he recognised the need to carry his studies beyond the living plants to the trees and ferns embedded in the rocks. That he was prepared to do so and to insist that these fossils had once been alive is, as we have noted, a remarkable indication of his courage and insight. It was a conclusion which he did not reach lightly: and though he never proceeded to a book on prehistoric organisms, the many places in his writings and letters in which he discusses the problem are among the most important of his contributions to science. For he not only took from the first and strove to maintain an enlightened position but in so doing he found himself opposed to two of his best friends, to Martin Lister in the earlier years and to Edward Lhwyd in the later. His modesty, and his preoccupation with other studies, prevented him from attaching much weight to his own views or from publishing any book on the subject except his *Miscellaneous Discourses*: but enough evidence can be found in his writings to show that in fact his outlook was far in advance of all his British contemporaries and that, without posing as an expert, he amended Hooke, the ablest of them all, rejected Lister, corrected Lhwyd and condemned without compromise the speculations of Burnet, Whiston and Woodward. Too little justice has been done to him by writers on the history of geology. Lyell, in the historical chapter of his *Principles of Geology*, dealt only with Ray's *Discourses* published in 1692 and ignored his earlier work:[2] in this he was followed by Geikie, who gives a fuller but hardly more satisfactory account of him:[3] and their attitude has been generally adopted.[4]

In fact Ray was one of the very first pioneers. Modern study is rightly dated from the work of Nicolaus Steno, whom Ray met at Montpellier and whose book *De Solido intra Solidum naturaliter contento*, quoted also as the *Prodromus*, was published in 1669. This book was translated into English by Henry Oldenburg, Secretary of the Royal Society, in 1671, the year in which Ray's first paper on the subject was written. But in 1661 he had already begun to make notes on the subject, reporting 'the serpent-stones called by naturalists in Latin cornua ammonis—we were somewhat puzzled to get them entire out of their matrices, the usual way of heating them in the fire very hot and then quenching them in the water, not always succeeding', and also 'plenty of the lapides belemnites or thunder-stones' at

1 The term 'fossil' was applied in Ray's time to all kinds of geological specimens, not as now exclusively to organic remains.

2 L.c. I, pp. 45–6 (10th ed. 1867).

3 *Founders of Geology*, pp. 73–6 (2nd ed. 1905).

4 Von Zittel, *History of Geology and Palaeontology* (English edition), pp. 17, 19, 31, merely condenses Lyell: Adams, *The Birth and Development of the Geological Sciences*, hardly mentions Ray and then borrows from Geikie.

Whitby;[1] and in 1662 he visited Alderley in Gloucestershire, 'where we found plenty of the cockle-shells and scallop stones but no cornua ammonis',[2] and Keynsham, where he saw 'the serpent-stones and star-stones found here'[3].

Up to this time there is no evidence of more than a collector's interest. Evidently he went out of his way to visit quarries noted for their fossils, and on his continental tour made extensive observations in the Low Countries, Germany, Italy and Malta. But the literature he seems only to have studied on his return. In 1667, in the second of his letters to Lister[4] describing how he bestowed his spare hours 'in reading over such books of Natural Philosophy as came out since my being abroad', he specially mentions Hooke's *Micrographia* and Kircher's *Mundus subterraneus*.[5] He certainly read Steno's epoch-making treatise as soon as it appeared and much other relevant literature. The result is seen in the references to geology in the *Observations*, where the comments are masterly.

Thus at Bruges in 1663 he described the buried forest found 'in places which 500 years ago were sea',[6] and referred to the beds of sand and cockle-shells said by Varenius[7] to have been found near a hundred foot deep when sinking a well in Amsterdam. This gives him occasion to suggest in the first case 'that many years ago before all records of antiquity these places were part of the firm land and covered with wood; afterwards being overwhelmed by the violence of the sea they continued so long under water till the rivers brought down earth and mud enough to cover the trees, fill up these shallows and restore them to firm land again'; and in the second 'that of old time the bottom of the sea lay so deep and that that hundred foot thickness of earth arose from the sediments of those great rivers which there emptied themselves into the sea...which yet is a strange thing considering the novity of the world, the age whereof, according to the usual account, is not yet 5600 years'.[8] He adds:

That the rain doth continually wash down earth from the mountains and atterrate or add part of the sea to the firm land is manifest from the *Lagune* or

1 *Mem.* p. 147. 2 *Mem.* p. 179.
3 *Mem.* p. 180. 4 *F.C.* p. 112.
5 In the next letter, epitomised by Derham, he condemns this as 'a poor book'. Athanasius Kircher, a Jesuit who travelled widely before settling in Rome, argued that there was a *vis lapidifica* inherent in the earth and a *spiritus plasticus* which shaped crystals, stalactites and fossils. 6 *Obs.* p. 6.
7 Bernhard Varenius published his *Geographia Generalis* at Amsterdam in 1650: the edition by Isaac Newton was published at Cambridge in 1672.
8 James Ussher, Archbishop of Armagh, had fixed the date of the Creation as 4004 B.C. in his *Annals*, the first part of which was published in 1650. But a date similar to this had been commonly accepted by Christian scholars, long before his time. Ray's figure is probably derived from John Pearson's *Exposition of the Creed*, first published 1659, who writing of 'the novity of the world' says 'The vulgar accounts exhibit about five thousand six hundred years' (ed. Nichols, p. 95).

flats about Venice; the '*Camarg*' or isle of the river Rhone about Aix in Provence in which we were told that the watch-tower had, in the memory of some men, been moved forward three times, so much had there been gained from the sea;[1] and many places in our land; only it is a received tradition, and may perhaps be true, that what the sea loses in one place it gets in another. That the height of the mountains, at least those which consist not of firm rocks, doth continually diminish is I think very likely, not to say certain.[2]

The most interesting of his early notes is that which deals with the fossils of Malta. The island he considers 'to have been in the most ancient times nothing else but a great rock wholly overwhelmed and covered with the sea; especially if we consider the multitude of sea shells of all sorts, sharks' teeth, vertebres of thornbacks and other fish bones petrified, found all over the island'.[3] Here follows a passage which anticipates almost verbally the remarks of Hooke quoted by Lyell[4] and Geikie.[5] Ray writes 'that these were formed by some plastic power in the stone-quarries, being nothing else but the effects or productions of Nature sporting herself in imitation of the parts and shells of these animals I can hardly be induced to believe; Nature (which is indeed nothing else but the ordinary power of God) not being so wanton and toyish as to form such elegant figures without further end or design than her own pastime and diversion'.[6] Discussing how so many sharks came there, he quotes 'Mr Steno', with whom he had probably debated the question at Montpellier.

The fullest treatment of the subject is found in the digression inserted in *Observations* as a comment upon the petrified shells that he collected at Altdorf in September 1663.[7] This paper was evidently written in the autumn of 1671; for he describes Willisel's discovery of the fossils called St Cuthbert's beads 'in the chinks of the stones in the bottom of the channel of the river Tees' as made 'this last summer (1671)': and also refers to Hooke's *Micrographia*, published in 1665, and to a contribution of Lister's to *Philosophical Transactions*, no. 76,[8] dated 25 August 1671,[9] and containing a criticism of Steno's work, Lister having already begun the collection of fossil molluscs which he finally described and depicted in his *Historia Conchyliorum*: Ray's note on the Maltese deposits is clearly a criticism of Lister's views.

1 Cf. Lyell, *Principles of Geology*, I, pp. 427–9, where this and other evidence is collected.

2 *Obs*. p. 8.   3 *Obs*. p. 294.

4 L.c. p. 42.   5 L.c. p. 69.

6 Cf. Hooke, *Posthumous Works*, p. 318, a probable case of plagiarism on Hooke's part; for though Hooke gives the date at which this paper was finished as 15 Sept. 1668, Waller says that it lay by him for many years and was not used until 1687, and it was certainly much enlarged and revised then. Except for the brief sections in his *Micrographia* Hooke published nothing on the subject till his papers to the Royal Society, 8 Dec. 1686.

7 *Obs*. pp. 113–31.   8 Vol. v, pp. 2281–4.

9 Read on 2 Nov.: Birch, *History of the R.S.* II, p. 487.

He begins by giving a full record of the places in which petrified shells are found in England, drawing partly from Camden's *Britannia*, but chiefly from his own and his friends' observations. After Whitby and Huntley Nab, north of it, Alderley and Keynsham, for which he quotes Camden, he adds Farnham, mentioned by Merret, and Richmond in Yorkshire, by Camden; then Lyme in Dorset, Adderbury in Oxfordshire, Brixworth and near Daventry in Northamptonshire, 'Verulam' in Hertfordshire, Shuckborough (Shuckburgh) in Warwickshire; and finally the Peak in Derbyshire 'from Mr Eyre' and Adderton in Yorkshire, Wansford bridge in Northamptonshire, and Gunthorpe and Beauvoir Castle both in Nottinghamshire (Belvoir actually on the Leicestershire-Lincolnshire boundary) from Lister. He goes on to give localities for 'Star-stones by some called Astroites' Shuckburgh and 'Cassington[1] near Gloucester where we also gathered them' and adds 'we have had them also sent to us out of Yorkshire but remember not where they were gathered': and for 'those they call St Cuthbert's beads found on the western shore of Holy Island'. He goes on to say that they are round, not angular like the star-stones: 'both seem to be the Spinae dorsales or tail bones of fishes petrified,[2] consisting for the most part of several plates or pieces sticking together like so many *vertebrae*, though I confess the particular pieces are shorter or thinner than the vertebres of any fish I have as yet observed.' St Cuthbert's beads are in fact remains of Crinoidea, stalked Echinoderms, while the Star-stones are various Asteroidea. Next he mentions 'Echinites or Brontiae [Echinoids] found scattered all over England of several magnitudes and shapes: I have not heard of any bed of them or great numbers found at one place': he adds that he has found them overseas not only at Altdorf but at Brescia in Lombardy 'on the sides of a hill adjoining to the city', upon the banks of the Tanaro in Piedmont 'four miles below Aste', and most plentifully in Malta. In Malta 'we also saw great store of Glossopetrae or sharks' teeth petrified of all sorts and sizes'; and he quotes 'Boetius de Boot'[3] for their discovery at Deventer and the alum-mines at Lunenburg, and Goropius Becanus[4] for them in the ditches about Antwerp and abundantly in a hill near Aachen in Germany. Finally, he gives a long list of localities for 'petrified cockles and other sea-shells' on the Continent, quoting Geo. Agricola[5] for many places in Germany and Italy; Pausanias[6] for the quarries of Megara and elsewhere in Greece; Steno for Volterra;

1 Probably Gossington, between Gloucester and Alderley.
2 He modified this view in the Preface of the *Observations* in consequence of Lister's views reported in *Phil. Trans.* VIII, no. 100, pp. 6181–90.
3 Anselm de Boodt of Bruges (1550–1632), physician to Rudolph II, who wrote *Gemmarum et Lapidum Historia* in 1609.
4 *Origines Antwerpianae*, III, p. 240 (Antwerp, 1569).
5 I.e. Georg Bauer, who died at Chemnitz in 1555, wrote *De Natura Fossilium* and books on metals.
6 *Descr.* I, 44, 9.

Jean Bauhin[1] for 'the slate-stone out of the fountain of Boll'; Scaliger[2] and Bernard Palissy[3] for various places in France; Jan de Laet,[4] who cites Morisot for Dijon and Jacobus Salmasius for Sauvignac; his own informants for other Italian localities and Goropius for many in the Low Countries. He adds 'that they have not been discovered in other parts of Europe and in Asia and Africa is certainly to be attributed to the negligence and rudeness of the people who mind nothing that is curious or to the want of learned writers who should communicate them to the world'. Considering the date at which this essay was written, the list shows diligence in his own researches and remarkable acquaintance with the relevant literature.

Having thus summarised the data he proceeds to discuss 'the opinions of the best authors concerning the original and production of these stones'; and at once sets out his own views. 'The first and to me most probable opinion is that they were originally the shells or bones of living fishes and other animals bred in the sea.' He claims that this is the general belief of the ancients,[5] and is now 'embraced by divers learned and ingenious philosophers as in the precedent age by Fracastorius[6] and in the present by Nicolaus Steno and Mr Robert Hooke after whom I need name no more to give it countenance'. He then cites from the *Micrographia*, Observ. 17,[7] the four conclusions formed from examination of Keynsham specimens which led Hooke to conclude that their figuration was due not to 'any kind of plastick virtue inherent in the earth'[8] but to actual living organisms: 'for it seems to me quite contrary to the infinite prudence of Nature...whence it has long been a general observation and maxim that Nature doth nothing in vain.'[9] He quotes Steno as being 'still more positive: for, saith he, that these shells were once the parts of animals living in a fluid, the very view of the shell evinceth as may be evident by the instance of the bivalve cockle-shells': the places in which they were

1 *Historia Fontis Bollensis*, Stuttgart, 1598.

2 *Exercit.* 196.

3 In 1580 he described and discussed fossil fishes and molluscs as remains of living organisms in his *Discours Admirable de la Nature des Eaux*, etc.

4 Of Antwerp: his *De Gemmis et Lapidibus* was published at Leyden in 1647.

5 He quotes Steno to this effect: but in fact the origin of fossils is hardly discussed except by Theophrastus, who speaks of fossil fishes and ascribes them either to the petrifaction of living forms or to a plastic virtue.

6 Girolamo Fracastoro in 1517 stated his convictions about fossils (Zittel, *History of Geology*, p. 14).

7 *Micrographia* (published in 1665) has only three pages (110–12) on the subject.

8 The phrase, very common among early speculators, derives from Theophrastus and Avicenna.

9 Hooke delivered his first discourse on earthquakes to the Royal Society in June 1667: it and its successors, the last dated 1694, are printed in *Hooke's Works*, pp. 279–450.

found, notably Volterra, must therefore have 'of old time been covered by the sea'.

He then sets out 'two very considerable objections'. The first is that if these are organic remains then 'all the earth was once covered by the sea', for they are found in the middle of Germany and upon the highest mountains of Europe. This is hard to believe and hard to reconcile with Scripture. He refuses to accept the belief that they were brought in by the deluge in the time of Noah; for this was due to rain and would have carried shells down to the sea not up from it; or if it was due, as some argued,[1] to an overflowing of the sea this would have scattered such shells instead of laying them in great beds in particular places—'such beds seem to have been the effect of those animals breeding there for some considerable time'. The argument that the mountains were raised subsequently by earthquakes he does not entirely dismiss; for he admits that this has happened: but he argues that in fact 'since the most ancient times recorded in history, the face of the earth hath suffered little change', and therefore 'if the mountains were not from the beginning, either the world is a great deal older than is imagined, there being an incredible space of time required to work such changes...or in the primitive times the creation of the earth suffered far more concussions and mutations in its superficial part than afterwards'. Secondly, many if not all of these petrified shells are not to be found alive anywhere. 'If it be said that these species be lost out of the world, that is a supposition which philosophers hitherto have been unwilling to admit.' If some few have disappeared, it is difficult to believe that whole genera consisting of many species should be utterly extinct: 'such are for example the Serpentine stones or Cornua Ammonis supposed originally to have been Nautili,[2] of which I myself have seen five or six distinct species, and doubtless there are yet many more. Add hereunto the greatness of some of these shells whereof there are found of about a foot diameter, far exceeding the bulk of any shell-fish now living in our seas.'

Having thus stated his difficulties he turns to the second opinion and cites Lister's discourse in *Philosophical Transactions*,[3] in which criticising Steno he maintains that, in England at least, all the 'quarry shells' are 'lapides sui generis and never any part of an animal'—and this for three reasons, that these shells are of the same substance as the surrounding stone, that quarries of different stone yield different sorts of shells, and that all these shells differ from species now living. He also cites Goropius as holding the same opinion. Ray adds an argument of his own that these

1 From Genesis vii, 11: 'The fountains of the great deep were broken up'.

2 Lhwyd, writing to Ray in 1692 (*Corr.* p. 254), says that Olaus Worm in 1655 was the first to identify Ammonites with Nautili. He had himself ascribed it to Boccone, 1674, and Ray had argued for Hooke, in *Micrographia*, 1665.

3 Vol. VI, no. 76, pp. 2282–3.

figured stones often resemble no part of an animal or plant—'such are Lapides Lyncurii, Belemnites [he uses the word], Lapides Judaici, Trochites, Asteriae and others'—and therefore if they are organic must belong to species lost out of the world.

He then propounds a third opinion that some are really shells petrified and others stones of their own kind shaped by some plastic power. 'He that asserts this delivers himself from the trouble of answering the arguments which urge the contrary opinion: but yet methinks this is but a shift and refuge, there not being sufficient ground to found such a distinction.'

In conclusion he writes: 'For my own part, I confess, I propend to the first opinion as being more consonant to the nature of the thing, and could wish that all external arguments and objections against it were rationally and solidly answered.'

We have summarised this essay fully because being buried in his travel-diary it has never received attention, because considering the date of its publication it is certainly the fullest and most enlightened treatment of the subject by an Englishman, and because it reveals clearly the two difficulties which beset all early geologists, their conviction of the 'novity' of the world, and their concurrent disbelief in the total extinction of large masses of prehistoric organisms.[1] Ray's readiness to question both these almost universal tenets distinguishes him from the vast majority of his contemporaries. But neither he nor they had any clear concept of geologic time. It is important to note this first considered opinion of Ray that in spite of the difficulties all 'formed stones' are organic in origin. Later on, as we shall see, the opinions of his friend Lhwyd, whom he rightly regarded as an expert, made him hesitate; and the general interest in Noah's flood which obsessed the whole development of geological studies distracted him. But in spite of occasional concessions it is clear that his own mind was always convinced of the position set out in this first essay. To it he remains constant, even if he fails to realise the consequences of it, and is therefore unable to answer objections.

During the next ten years Ray seems to have written nothing on the subject except a note on the Astroites found by Lister at Bugthorpe some ten miles east of York, and reported in *Philosophical Transactions* in January 1674.[2] On the more general question he had expressed his objections to Lister's theory and in their correspondence there is no reference to it. Only in 1684 in a long letter of 22 October to Tancred Robinson[3] does he return to it: and then on the ground that 'much contro-

1 These convictions were not confined to theologians: Hobbes himself fully accepted belief in an instantaneous and complete creation at a date about 6000 years before: cf. Masson, *Life of Milton*, VI, pp. 285–6.

2 *Phil. Trans.* VIII, no. 100, pp. 6181–90: Ray's comments are in *F.C.* p. 65; *Phil. Trans.* X, no. 112, pp. 278–9.   3 *Corr.* pp. 151–6.

versy concerning the origin of these shell-like stones found in the earth' had arisen. The Royal Society had in fact had some discussion of it, Hooke and Lister being the protagonists and Plot, whose *History of Oxfordshire* published in 1677 contained views similar to Lister's, being at that time one of the Secretaries.[1] Robinson, who had just come back from abroad and was proposed for election on 12 November, may well have asked Ray for his views.

Ray begins by repeating his conviction that they are 'the very shells of some sea-fish or got this figure by being cast in some animal mould', and refers to his essay in *Observations*. He then adds many other points: (1) An argument based upon Mentzel's account of beds of shells un-petrified found high up on a mountain near Bologna;[2] this observation 'frustrates one main objection against our opinion, viz. because no account can be given how the very shells should be brought to the tops of the mountains; when we see the thing done, it is vain to dispute against it from the unlikelihood of the doing it'. (2) 'Other bodies found in the earth are so manifestly the very things they are thought to resemble that it seems to me great weakness in any man to deny it'; such are the Glossopetrae in Malta, which are obviously sharks' teeth; if these are plainly organic remains, surely the shell-like stones found with them are also organic. (3) 'It seems strange, if these bodies were formed after the manner of the crystallising of salts, that two shells should be so adapted together at the heel and the upper and nether valve be of different figure as in natural shells'; and this is the case in multitudes of them. (4) 'Why should not nature imitate other natural bodies or their parts?' (5) 'Were these bodies produced by a concretion of salts it seems strange to me that there should be such great variety of them'; concretions of slate never appear in such forms. (6) Similarly, experiments in crystallisation have never produced anything in the least like these shells.

He then admits that he does not see any way as yet to answer some of the objections 'but by granting that many sorts of shells are wholly lost, or at least out of our seas'. But he disposes of some of Plot's arguments: (1) Plot, who was a notably credulous person,[3] had quoted Camden[4] that 'the Ophiomorphites (Ammonites) of Keynsham had some of them

1 Cf. Birch, *History of the R.S.* IV, pp. 237–8, at meeting of 12 Dec. 1683: Lister had just come to London and was very assiduous at the Royal Society.

2 Christian Mentzel published his *Lapis Bononiensis* in 1675.

3 Cf. the jests of the Staffordshire squires as to the ways in which they had 'befooled old Plot' when he was collecting material for his *History of Staffordshire*.

4 *Britannia* (ed. Gibson), c. 72: 'I have seen a stone brought from thence winded round like a serpent the head whereof, though but imperfect, jutted out in the circumference. But most of them want the head.' Gibson corrects this, c. 82: 'all our naturalists now agree that such stones are formed in nautili shells and that there are no heads belonging to them.'

heads'. Ray states that he and Willughby had enquired diligently after such stones which the common people affirmed to be found: had at last discovered a man who was said to have one; saw it and discovered that it 'had some kind of knob not at all resembling the head of any animal'. He adds that by something of this sort Camden was no doubt deceived.[1] (2) Plot argued that 'the Brontiae cannot be petrified shells of Echini Spatagi' by quoting Aristotle and Rondelet, the former stating that they are ocean dwellers and rare, the latter that spines are few and thin-set: neither of these is true: 'at Llandhwyn in the Isle of Anglesey they are more plentiful than the common Echini anywhere with us' and are thickly spined.

There is not a sign here that Ray felt able to make any concession to the belief in a plastic virtue. Indeed he is prepared to accept the fact that species have become extinct. This view was powerfully supported by Hooke in his treatise on earthquakes written in 1688 and printed in the volume of his *Works* published by Richard Waller in 1705. So far, he has set geological studies on a true course.

It was not until May 1690 that Lhwyd, whose earlier letters had been wholly botanical, asked him about 'formed stones'. Lhwyd had already begun to observe and collect these and was in correspondence with Lister about them.[2] He was at this time uncertain in his own view but inclined to adopt Lister's opinion that they were *lapides sui generis*[3]—an opinion which Lister with his usual conservatism refused to modify, but which Lhwyd admitted that Ray had overthrown in his *Discourses*. After enquiring as to Ray's collection and receiving a list and specimens of it ('I have no great variety.... Of Cornua Ammonis I have two or three varieties. I have also some resembling cockles and mussels and Belemnites and Brontiae: but few different species. Those I have were all gathered up in England except one Cornu Am. with a spine and ribs found at Altdorf in Germany. I have great variety of Glossopetrae dug up in Malta...of Oculi Cati and some snagged stones which they call baculi S. Pauli'[4]) he sent Ray a box of his own finds and asked his opinion as to the problem of their origin. Ray was at this time busied with the second and much enlarged edition of his *Collection of English Words*, and evidently had no

1 For an account and figure of an Ammonite supplied with a faked head cf. Richardson's *Introduction to Geology* (ed. Wright, 1855), p. 24; and for further reference W. D. Lang, *Proc. Dorset Nat. Hist. Arch. Soc.* LXI, pp. 98–9 (1940).

2 Several of his letters are quoted by Gunther, *F.C.* pp. 188, 194, 206, 212: but the correspondence in full is in Gunther, *Life of Lhwyd*. He dedicated to Lister his *Lithophylacium* in 1699.

3 Cf. Gunther, *Life of Lhwyd*, pp. 79, 80.

4 *F.C.* p. 207: Lhwyd's answer is *Corr.* p. 224. In a later letter Ray explains that he had contributed his specimens to Willughby's collection to which in his exploring days he had ready access (*F.C.* p. 211).

will to be drawn into controversy. He replied by saying that he had long
fluctuated and was constrained to settle in a middle opinion.[1] 'Some of the
Bodies you sent me are to my apprehension real shells'; he adds:

> I never saw nor hope to see any representations of men, women, beasts or
> birds...there may be casual figures that by the help of fancy may be imagined
> to be like land animals....I have seen some of the famed figures of this kind in
> the Popish churches but never saw any near resemblances as were pretended:
> and for Father Kircher I count him a credulous person. And truly what you
> wrote to me in a former letter of the exact figures of plants formed in stones
> seems to me to savour of the Romance, and I should be glad to see such re-
> semblances.[2]

This last request Lhwyd quickly supplied and in January 1691 Ray wrote:
'I was much pleased with the slate which hath the figures resembling the
leaves and branches upon it: for that I do not remember ever to have seen
the like.'[3] As we have seen in the Preface to the second edition of his
Catalogue of British Plants, having then received many more specimens
from coal measures in the Forest of Dean, he boldly claimed that these
fossils must in future be taken into account by botanists as forming a new
department for their studies.

The letters of 1691 and indeed the bulk of his later correspondence with
Lhwyd deal with the identification of fossils, and occasionally with the
problems of their origin. He approves Lhwyd's plan for a Catalogue of
formed stones, urges him not to limit it to the neighbourhood of Oxford
and to include 'a general discourse about the original of these stones' and
advises him 'first to draw up a method yourself without suffering your
fancy to be biassed or inclined by another man's thoughts: and then I may
send you mine'.[4] He constantly urges that 'what you have now found out
should a little stagger and unsettle you in the opinion that they are original
productions in imitation of the shells and bones of fishes'.[5] And he does
his best to prevent the problem being confused by arguments from a stone
in the belly of a calf,[6] a toad found in a rock,[7] or pearls in oysters.[8] But
now as always he avoids controversy, overrates the learning of others, and
hesitates to dogmatise in a subject not his own.

Alongside of the interest in formed stones problems of cosmogony had
in the last twenty years of the century begun to attract attention, and here

---

1 In justice to Ray it is to be noted that 'formed stones' included stalactites and
crystals, and that the line between these and organic remains was not yet drawn. Thus
Lhwyd's *Lithophylacium* includes crystals, pyrites, talc, corals, plants, shells and bones.

2 *F.C.* p. 210. 3 *F.C.* p. 213.

4 *F.C.* p. 216: cf. pp. 220, 226. 5 *F.C.* p. 223: cf. p. 231.

6 *F.C.* p. 216. 7 *F.C.* p. 227.

8 *F.C.* p. 254: 'I take them to be morbose concretions like the stone in the bladder
or the bezoar in the stomach of some animals.'

the conflict between the old philosophy and the new was most acute.
Theoretically astronomy ought to have been the battlefield. The Copernican revolution, which to us is manifestly critical, actually caused far less
stir than we should have expected. Bruno had been martyred in 1600 and
Galileo forced to recant before his death in 1642. But the difference between the cosmology of Dante and that of Milton affects only the map and
not the spirit of their picture: in *Paradise Lost* heaven and hell are further
off and more vaguely located: but the description of creation has not
changed and the tradition of Genesis stands unquestioned. Indeed no one
seems to have seen any reason to challenge it. It was only when geology
became a serious subject of study that the difficulty of maintaining the
current beliefs as to the age of the world, the duration, universality and
effects of the Deluge began to be felt. Well-meaning theologians, equipped
with a smattering of the new knowledge, strove then as now to fit the new
data and theories to the Procrustean frame of orthodoxy; and in those days
when 'the conflict between religion and science' had not arisen the
scientists joined in with equal enthusiasm.

In 1681 Thomas Burnet, fellow of Christ's College 1657–78, Master of
Charterhouse 1685, had published his *Telluris Theoria Sacra*. Though Ray
said that he had not seen it and by what he had heard did not think that it
'needed great confutation'[1] it attained great popularity, was issued in an
English version dedicated to the King in 1684, and gave rise both to criticism[2] and to rivalry. The interest in the subject led Ray to produce his
own treatise upon it. Of this he wrote to Lhwyd on 7 August 1691:
'Because my last Tract [*The Wisdom of God*] has found some acceptance,
I am at the instance of my bookseller revising and preparing for the press
a short Discourse concerning the Dissolution of the World', and on
25 October: 'The Discourse is finished and under the press....The body
or skeleton of it is a sermon I preached at St Mary's church in Cambridge
upon II Peter 3. 11—"all these things shall be dissolved".' He then outlines the heads of his treatment as they appear in the first edition of his
book, and adds:

Besides, there are two large digressions, one concerning the general deluge
in the days of Noah, the other concerning the primitive chaos and creation of the
world. In the former of these at the instance and importunity of some friends

1 *F.C.* p. 218.
2 John Beaumont (F.R.S. 1685) published *Considerations on a Book called The
Theory of the Earth* in 1693, of which Ray said: 'I think he hath fundamentally overthrown it...but a great deal of stuff he hath about the mystical and allegorical Physiology of the Ancients which I understand not, nor I believe, himself neither' (*F.C.* p. 242)
—'his book in many places smells rank of the Enthusiast' (*F.C.* p. 245). John Keill of
Oxford, the disciple of Newton, published a more adequate *Examination of Dr Burnet's
Theory* in 1698.

I have inserted something concerning formed stones as an effect of the deluge, I mean their dispersion all over the earth. Therefore you will find all I have to say in opposition to their opinion who hold them to be primitive productions of nature in imitation of shells. I intended to have reserved them for your work, but they extorted them from me, upon pretence that my Discourse would be imperfect without them; and that no man who hath written heretofore concerning the deluge hath made any mention of them; and therefore such an addition, for the novity of the matter, would be acceptable to the curious and give my book advantage of sale.[1]

According to Derham[2] he had given a similar account of the origin of the book to Tancred Robinson on 24 July, but this letter has only been preserved in Derham's epitome.[3] Derham adds that he has seen the first draft of the original sermon 'but not nearly so much enlarged as in the printed tracts'.

The book entitled *Miscellaneous Discourses concerning the Dissolution and Changes of the World* was dedicated to Ray's friend, John Tillotson, lately become Archbishop of Canterbury,[4] and received the imprimatur of the Royal Society, Robert Southwell being President, on 6 January 1692. On 18 January Ray wrote to Lhwyd: 'Wednesday last I had advice from Mr Smith that the printer Mr Mott had promised him perfect books of it on that day or the next: if he was so good as his word (as he seldom is) you may expect a copy of it the next week or the following.'[5] Evidently the book was produced rapidly; and for this Ray apologises in the Preface. After excusing himself for the two long digressions, and for writing so much ('He that writes much, let him write never so well, shall experience that his last books, though nothing inferior to his first, will not find equal acceptance.[6] But for my own part, though in general I may be thought to have written too much, yet it is but little that I have written relating to Divinity'), he adds an excuse

for being too hasty in huddling up and tumbling out books. Herein I confess I cannot acquit myself wholly from blame. I know well that the longer a book lies by me, the perfecter it becomes. Something occurs every day in reading or thinking, either to add, or to correct and alter for the better: but should I defer the edition till the work were absolutely perfect, I might wait all my life-time, and leave it to be published by my executors.... Perchance did the reader know

1 *F.C.* pp. 222–3: cf. letter to Robinson of 9 Oct.: epitomised *F.C.* p. 295.
2 *Mem.* p. 10: he dates the year 1690: but the letter is evidently 1691.
3 *F.C.* p. 294.
4 On 18 Feb. Lhwyd wrote to Lister that 'they talk much here of Mr Ray's dedication of his new book to the Archbishop and say he has a mind of some ecclesiastic preferment' (Gunther, *Life of Lhwyd*, p. 156).
5 *F.C.* p. 224: on 18 Feb. Lhwyd said that he had not received it from Smith: before the end of February he sent a long letter commenting on it to Ray (Gunther, *Life of Lhwyd*, pp. 156–60).
6 In fact the *Wisdom of God* and the *Discourses* were very successful.

my reasons for this speed, which I think it not fit now to lay open,[1] he would judge them sufficient to excuse me. However hasty and precipitate I am in writing, my books are but small, so that if they be worthless, the purchase is not great, nor the expense of time wasted in the perusal of them very considerable.

The fact that a few months later when a second edition was required he recast the arrangement and very greatly changed the contents, justifies the word miscellaneous in its title and the admission that the book was 'huddled together'. It is in fact a sermon amplified by later notes and interrupted by the long section 'Of formed stones' with which we are now concerned.[2]

This essay does little more than expand and arrange the arguments already stated in his first paper and his letter to Robinson. The case for an organic origin is set out with a long quotation from Fabio Colonna[3] in addition to Mentzel: Dr Plot's contentions are examined more fully, and a letter received in March 1685[4] from William Cole, a customs-officer of the Butts, Bristol, who had gathered a considerable collection[5] of zoo-logical and geological specimens and had worked the banks of the Severn for Ammonites, is extensively quoted. Cole claimed that his specimens, one being 'not so thick as a half-crown piece' and another 'as big as the fore-wheel of a chariot', proved that 'there are many varieties of naturally-formed stones which never were either animals or vegetables or any parts of them, not only because no such shell-fishes were ever found...but because they were never capable of being living creatures'. Ray seems not to have taken this evidence very seriously,[6] until Lhwyd's letter reporting on Cole's collection mentioned the very thin specimen. Then, though still quoting Hooke's opinion of them with respect, he can only conclude 'these Ophiomorphous stones do more puzzle and confound me, than any other of the formed stones whatsoever...they seem to have been or to owe their original to shells: and yet there is nothing like them at this day

1 He had nearly died of pneumonia in the previous March: cf. *F.C.* pp. 204–5.
2 Pp. 102–32. 3 *Dissertatio de Glossopetra.*
4 Printed in full in *Corr.* pp. 181–2: he sent a more detailed letter to Plot in Feb.: cf. Gunther, *Early Science in Oxford*, pp. 259–61.
5 This was visited and described by Lhwyd in July 1690 (*Corr.* pp. 225–6 and Gunther, *Life of Lhwyd*, p. 104). Lhwyd did his best to secure this collection for Oxford, but failed: it was bought by a Bristol doctor (Gunther, l.c. p. 469).
6 He mentions Cole's letter in writing to Robinson (*F.C.* p. 144) but without comment: in 1689 he writes to Lhwyd (*F.C.* p. 204) commending Cole but intimating that he had reported 'stories of stones engendered in clouds' of which he, J. R., was incredulous. When Cole protested that in the *Miscellaneous Discourses* Lhwyd had been given credit for discoveries due really to him, Ray expressed regret for a slip of the memory but said to Lhwyd that 'Mr Cole is a little infected with an immoderate desire of honour and apt to think well, it may be too well, of his own abilities' (*F.C.* pp. 236–7, 240).

in our or any other seas, as far as I have heard or read'.[1] From the whole
of his treatment of the problem it is clear that though anxious to set out
the different arguments fairly his own inclination is towards the organic
origin of all formed stones that resemble shells, bones or other parts of
living creatures, but that he does not regard himself as an expert and is
anxious to get the opinion of Lhwyd, 'who I hope will ere long gratify the
curious by publishing a general Catalogue of all the formed stones found
in England and his remarks upon them'.[2]

In fact a second edition of the *Discourses* was demanded almost as soon
as the first was published. In July of the same year, 1692, he wrote exhorting Lhwyd to 'carry on with vigour your history of the formed stones
and all fossils'[3] and paying an unusually warm tribute to his character.
Ray so seldom speaks his heart to his friends that the sentences are worth
quoting:

I must to your just commendation acknowledge to you in all the intercourse
I have had with you that I have discovered less of affectedness, conceitedness,
pride or vain-glory than in almost any man of my acquaintance.[4] This is all I
have to say at present and perhaps I have said more than your modesty can well
bear; however I have given you my true sense, and I have no reason to flatter,
unless to extort reciprocal commendations from you which I am conscious to
myself is no motive to me.

On 7 November in his next letter he writes: 'I hope it will not now be
long before I have a small present for you, the second edition of my
Discourses concerning the Chaos, Deluge and Dissolution of the World';[5]
and on 28 December: 'I ordered my bookseller to send you a copy of the
2nd edition of my *Physico-theological Discourses*, for now I so entitle them,
which I hope is come to your hands.'[6] The change of title, justified by the
complete rearrangement of the book and by large additions to it, indicates
that he had done much further work upon it; of this we shall speak later.
The essay on Formed Stones, Sea-shells and Marine-like Bodies, now
Chapter IV of the Second Discourse, occupies pp. 127–162 instead of
pp. 102–132. There are additions, notes from Lhwyd on fossil oysters
which Lister himself admitted to be real shells, and from Doody on a lump
of petrified fishes; a quotation from Steno to strengthen the argument that
natural salts do not crystallise in such shapes, from Robinson suggesting
that some shells may have been scattered inland by encampments of armies
or the inhabitants of towns that no longer remain,[7] and again from Steno
bearing upon the 'clay cockles' at Glympton, near Woodstock in Oxford-

1 *Misc. Disc.* p. 125.　　2 *Misc. Disc.* p. 132.　　3 *F.C.* p. 230.
4 A similar tribute is paid to him by Thomas Hearne, usually the most caustic of
critics: cf. *Collections*, I, p. 244 and Gunther, *Life of Lhwyd*, p. 48.
5 *F.C.* p. 232.　　　　　　　　　　　　6 *F.C.* p. 234.
7 A suggestion incapable as Ray recognises of solving the problem, but interesting
as foreshadowing the discovery of 'kitchen middens'.

shire. There are also three Plates with pictures of fossil shells and fishes, and an essay of eight pages prefixed to them and arguing the religious and philosophical importance of the problem of their origin. Though stating the difficulties Ray writes strongly in support of the organic origin of formed stones.[1]

This second edition was not Ray's last version of the *Discourses*. In 1703, as Derham, his executor, informs us, the booksellers were 'hasty for a Third Edition, the former impression being sold off and the book much called for'.[2] In April 1704 substantial additions were sent up to Smith; but his ill-health, retirement and death, which played havoc with several of Ray's later books, delayed publication until William Innys took over the business. Then in 1713 the third edition appeared, with a list compiled by Derham of the new material.

During the last ten years of Ray's life speculation as to cosmogony and the problems of fossils had been increased by the books of John Woodward, *Essay towards a Natural History of the Earth*, and William Whiston, *New Theory of the Earth*. Woodward, founder of the Woodwardian collection and professorship at Cambridge, had done much field-work: but though in 1694, after receiving a letter from him to Lhwyd, Ray expressed the hope that he would do great things, in 1695 when the essay was printed he wrote: 'As for Dr Woodward's hypothesis, if he had modestly propounded it as a plausible conjecture it might have passed for such: but to go about so magisterially to impose it upon our belief is too arrogant and usurping';[3] Robinson, whom he quotes in the same letter, was of the same opinion. The theory, briefly summarised, was that the Deluge had churned up and dissolved the world, and the shells and marine creatures sinking into the compost had been deposited in layers of a depth corresponding to their weight. Ray was very caustic, unusually so, about it, declaring that he 'had no proof that those bodies must be thus lodged and disposed but the negative one that they could not possibly be so otherwise' and that the facts did not bear out this conclusion. Whiston, Newton's successor as Lucasian Professor at Cambridge, was in Ray's opinion more modest;[4]

1 The slip on pp. 125–6, whereby two successive arguments are numbered 'thirdly', reappears in both second and third editions.
2 To the Reader, 3rd ed. pp. xvii–xviii: cf. *Mem.* p. 45. In Nov. 1703 Lhwyd, writing to Morton, says that having had no letter from Ray for some years he has now heard that he is preparing a new edition of the *Discourses* (Gunther, *Life of Lhwyd*, p. 494).
3 *F.C.* p. 256: for his violence cf. his expulsion from the council of the Royal Society in 1710 for insulting Sloane. He was supposed by Sloane to have written in 1700 the scurrilous *Transactioneer*, an attack on the Society and its *Transactions*, to which Ray alludes *Corr.* p. 371: cf. Weld, *History of the R.S.* I, pp. 352–5. Lhwyd's letters contain many instances of his insolence to Lister, Robinson and Ray: cf. Gunther, *Life of Lhwyd*, pp. 302–3, 305.   4 *F.C.* p. 301.

but he wrote to Lhwyd: 'the new theory seems to me pretty odd and extravagant and is borrowed of Mr Newton in great part'.[1]

His theory, influenced by the appearance of a remarkable comet in 1680, was that a comet had been responsible for the Flood, dragging the waters over the earth by attraction and increasing them by the condensation of its own tail. He also adapted from Woodward the notion that shells sank by gravity into the resulting sediment.

In addition to these two books Lhwyd had been doing much work on the subject. In 1697 he invited subscriptions in order to enable him to travel the country collecting antiquarian and geological specimens. Ray discussed the project in the spring of 1696;[2] and an interesting relic of it is in the cabinet of fossils gathered and labelled by Lhwyd for Richard Dyer of Oriel, one of his subscribers, which is now in the Old Ashmolean.[3] In 1698 Lhwyd published his *Lithophylacium*[4] or Classified List of Fossils, dating the preface 1 November at Montgomery and dedicating it to Martin Lister. Unfortunately the University Press refused to print it. He was travelling and unable to supervise the proof-reading; the book was full of errors and misprints; there were only nine subscribers to defray the cost and only 120 copies were printed; the booksellers seem to have lost heavily on it.[5] But it was a piece of pioneering and deserves to be remembered with honour.[6]

As an Appendix to it Lhwyd printed six letters dealing with geological subjects; the first to Rivinus[7] commenting on some specimens sent from Germany; the second to William Nicolson, Archdeacon of Carlisle (who sent Ray a list of Cumberland plants),[8] acknowledging and naming certain fossils; the third a summary of letters from Richard Richardson of North Bierley (who had found the toad in a stone and was a correspondent of many naturalists)[9] about Yorkshire 'lithophytes'; the fourth to John

1 *F.C.* p. 277: Whiston was deprived of his professorship on a charge of Arianism in 1710 when Nicholas Sanderson, the blind man, was elected in his place.

2 *F.C.* p. 263.

3 Dyer was fellow and dean of Oriel 1676–1724: his collection was preserved there until its origin was discovered by Dr R. W. T. Gunther, when the College transferred it to the Museum.

4 The title is *Lithophylacii Britannici Ichnographia*: for the unhappy circumstances of its publication cf. Gunther, *Life of Lhwyd*, pp. 262–3.

5 So Ray to Robinson, *F.C.* p. 304.

6 Lyell, in his historical chapter, *Principles of Geology*, ignores it: Geikie, in *Founders of Geology*, p. 78, dismisses it in a single sentence. The first edition is rare: it was reprinted and emended by Huddesford at Oxford in 1760.

7 With whom Ray had the controversy over the classification of plants.

8 Cf. *F.C.* p. 215 for his interest in North-country history; and Gibson's edition of *Camden*, cc. 765, 767, for his notes on fossils.

9 Cf. *F.C.* p. 227 for the toad. Many letters of Richardson are printed in Nichols, *Illustrations of Literature*, I.

Archer on 'star-stones' and belemnites; the fifth to Tancred Robinson and the sixth to John Ray,[1] these two being much longer and more important.

The letter to Robinson answers a number of questions, most of which deal with matters of fact. That to Ray deals with the problem of the origin of marine bodies and mineral leaves—are they due to the Deluge, or to a plastic force in nature? After explaining his perplexity over the subject and acknowledging that new evidence is constantly forcing him to reconsider it, and after raising a number of good objections to the theory of dispersion by the Flood, Lhwyd expounds his own new theory, a brief version of which, in English, is contained in a note appended to one of Ray's letters to him.[2] He rejects Plot's belief in a plastic force of salts, claiming that Ray has sufficiently refuted it in his *Discourses*.[3] Instead he conjectures that 'the mists from the sea are impregnated with the seeds of shell-fish and other fishes and crustaceans which are carried by the action of rain deep into the earth and there germinate and produce complete or mutilated specimens in stone'; and similarly that the fossil plants, mostly ferns or mosses whose seeds are minute, are formed from seeds that have found a matrix or place of growth in the rocks. He admits that this hypothesis may be open to criticism, but defends it by various, and ingenious, arguments: that it explains (1) the absence of birds and mammals of which no fossils are found; (2) the gradation between living organisms and mere stones, since these earth-born seeds degenerate; (3) the presence of Nautili and other exotic shells, since the vapours, and seeds, may come from remote parts of the world; (4) the varying and often monstrous size, since they are produced out of decaying matter;[4] (5) their position, numbers and presence not only in the earth but in the internal organs of animals.[5]

Lhwyd himself realised that his theory would have to meet criticism, and set out frankly such objections as he foresaw, declaring that those who follow nothing else but truth do not need to fear for the credit of their opinions, 'and for myself the more my love for Natural History the less is my concern with hypotheses'. The chief questions are: Can seed-stuff penetrate the pores of the rocks? Can inanimate bodies in an unnatural place grow? Can they force their growth in the hardest stone? Should we find them in heaps sticking together, as happens in the sea? Can particles of seed produce incomplete parts of the whole creature? How is it that shells thus produced show signs of nacre, and teeth of roots and of wear? Is not such a growth contrary to the course of nature, since it seems

1 Reprinted in Gunther, *Life of Lhwyd*, pp. 381–96.
2 Cf. *F.C.* pp. 272–3.   3 He refers to p. 141.
4 The logic is obscure: he assumes that these earth-nurtured births are more liable to monstrous abnormality and quotes Georg Agricola to that effect.
5 The stones in the bladder and the shell said to have been found in the intestine of a calf were thought to be identical with fossils: cf. *Misc. Disc.* p. 132.

useless and purposeless? The first he answers by reference to Dr Richardson's toad; the second on the analogy of internal growths and gall-stones; the third from observation of the roots of plants; the fourth by claiming that special places suit them and that they grow and dissolve there; the fifth by reference to monstrosities; the sixth by raising doubt as to the identification of these with real teeth; the last by acknowledging that nature does nothing in vain, but by denying that we are competent judges of her final purposes—possibly they contribute to the fertility of the soil or are useful as drugs. Even if wrong-headed, he was at least honest and ingenious.

Ray's letters during this period reflect the deepening interest. We have already cited his views upon Woodward and Whiston. In the letter of April 1695 in which he deals with the former he propounds certain enquiries about fossil plants, following up his insistence in the *Discourses* that the matter requires further investigation. In reply Lhwyd sent him a box of specimens, and he replied on 8 October that by the inspection of them he was 'inclined to think that they are the vestigia or impressions of plants themselves rather than *lusus naturae*'.[1] He quotes Woodward's assertion that he had seen 'the substantial plants themselves', and also a letter from Doody, who agreed that the plants 'had once grown'. His own verdict is:

Such a diversity as we find of figures in one leaf of Fern and so circumscribed in exact similitude to the plants themselves, I can hardly think to proceed from any shooting of salts or the like....Yet on the other side there follows such a train of consequences as seem to shock the Scripture-history of the novity of the world; at least they overthrow the opinion generally received, and not without good reason, among Divines and Philosophers that since the first creation there have been no species of animals or vegetables lost, no new ones produced.

On 3 February 1696 he reasserts his conviction as to the delineations of plants upon slate or other stones:

I did once embrace a middle way to which I see you are now somewhat inclined, but now I think that if nature may form any for her disport she may as well form all, and I see no reason for making a difference between them. And if real shells and fish-bones be found at great distances from the sea, it may rationally be thought that the formed stones which are there owe their original to them.[2]

Having thus rejected compromise and accepted the delineations as genuinely organic remains he inserted the section on palaeobotany in the second edition of his *Synopsis Britannicarum*.[3] That he should have done so in the subject on which he was best qualified to speak, at a time when the whole

---

1 *F.C.* p. 259.   2 *F.C.* p. 265.
3 Cf. above, pp. 255–6.

matter was so controversial, and when he foresaw the theological consequences of his decision, is strong testimony to his scientific insight and
integrity.

His next letter, of 8 June, begins with a reference to the account of
Agostino Scilla's work, *La Vana Speculatione* (published in 1670), which
had just appeared in *Philosophical Transactions*.[1] Ray was much impressed
by his evidence that beds of fossil shells had been 'lodged in such places as
they are found in'[2] by inundations of the sea; and he quoted Scilla somewhat fully in the third edition of the *Discourses*.[3] He recognised that this
theory must meet the fact that such beds were found 'in the upper parts of
the high mountains' as he had himself seen in the Peak country of Derbyshire. He therefore puts forward his own view. First he rejects the belief
that the shells were brought in by the Deluge, both on account of its
brevity and because as he had formerly argued a flood was 'more likely
to carry shells down to the sea than to bring any up'. Secondly, he
suggests:

> Most consonant therefore to the Scripture and to reason it seems to me that
> at first the earth was covered with water; that the land was raised up by sub-
> terraneous fires at the Divine command and that gradually first where animals
> and men were created, and then further and further, the waters being driven
> back. Afterward when the greatest part of the earth was thus raised, the skirts
> were alternated by the sediments of rivers and floods, whence and from the
> several inundations of the sea came the several beds or layers of earth. The
> finding of a bed of shells upon sinking a well at Amsterdam [the record of which
> he had noted in *Observations*[4]] at above a hundred foot depth seems to me an
> evident proof that there was then the bottom of the sea and that all the earth
> above it was but the sediment of great floods.

Here again, even if his theory is incomplete, he has clearly broken with
tradition and read the evidence of stratification correctly.

Then on 12 March 1698 Ray wrote his comment upon Lhwyd's letter,
the first draft of which had evidently just been sent to him.[5] 'Your opinion
concerning the origin of marine shells and coal-pit plants', he writes,[6]
'I must needs acknowledge to be very ingenious, and probable, nor indeed
could it otherwise find any entertainment or approbation with a person of
your sagacity and judgment. But you have such considerable objections
against it as are very hard to answer. I leave them to your further consideration'—those whose business it is to criticise the ill-founded speculations of promising pupils will applaud the tact of this condemnation. He
continues:

1 Vol. XIX, no. 219, pp. 181–201.   2 *F.C.* p. 266.
3 Pp. 138–45.   4 P. 6. Cf. above, p. 421.
   5 During the next months Lhwyd sent this letter and Ray's comments on it to
Lister and Robinson for their comments: cf. *F.C.* pp. 273–4.
   6 *F.C.* pp. 268–9.

I am of opinion with you that the several parts of animate bodies are made of the several parts of the seed.[1] I can also easily allow that these parts may be separated and act separately from each other, as appears by monsters wanting some parts and having others in undue places or enormously great. Only we see that the seeds of these bodies...grow by separating parts from a fluid and continually applying them to themselves. Now it is hard to conceive how this can be done in stones. The instance of those sorts of Fern-leaves observed in the ice of urine etc. answers not to the formation or rather the growth of an animate body from the seed, that [i.e. crystallisation] being done by one simple projection at once.

He then discusses the formation of beds of fossils, 'some of them being as it were calcined'; argues 'that such beds were sometimes the sea-shore, it requiring a very long time so to calcine these shells'; quotes the bed of shells at Harwich, lately found by his friend Samuel Dale;[2] and concludes: 'I cannot persuade myself but that these and the Glossopetrae and Raiarum vertebrae and some fossil conchae as the oysters you informed me of,[3] and many echini shells I have seen, were the very spoils of the sea and did once belong to living fishes.' He admits that 'other such figured bodies found in stone and spar must needs have a different original, as your arguments evince; and of this your hypothesis gives a plausible account. But after all we may be as much at a loss to find by what means the former were brought to the places where they are found lodged, as we now are for all.' He ends: 'The argument from the histories of conchae found in the glands of animals doth not weigh much with me, because they are but few and may be accounted for some other way.'[4]

In February 1699 he wrote to explain that Lhwyd's book had not yet reached him—the presentation copy having been sent to Vernon for him and passed on to his neighbour Edward Bullock. Apparently there had also been a failure to send him a proof-copy for correction; for he writes that he would gladly have served in this way, but supposes that the persons responsible had felt it unnecessary to ask his help in correcting what Lhwyd had written—a sentiment which he fully shares. Lhwyd had been in Wales all the year[5]—at Caldey Island in March; at Scotch Borough near Tenby in April; at Pembroke and Haverfordwest in May; at Cardigan and 'Waterfall in Radnorshire' (Rhayader) in July; at Hay in Brecon in September; at Newton, Montgomeryshire, in October; and at Dolgelly

1 He had shown this in his researches into the seeds of plants.

2 He gave a fuller account of this find in his letter, undated, in *F.C.* pp. 276–8. This must surely have been written in the winter of 1697–8 and earlier than *F.C.* pp. 267–76.

3 Lhwyd found these at 'Rangewell Hill at a village called Hedley about three miles south of Epsom' and reported his find to Lister 14 Oct. 1691 (Gunther, *Life of Lhwyd*, p. 149). It is discussed in *Discourses* (2nd ed.): cf. *Corr.* p. 257.

4 Lhwyd's comments upon this letter are printed in *F.C.* pp. 272–3.

5 Cf. Gunther, *Life of Lhwyd*, pp. 329–32.

at the date of this letter. Correspondence had been difficult; proof-reading impossible. The book suffered badly from lack of correction.

In 1701 and again in 1703 he alluded to the pleasure with which he regarded the inscribing to him of the Sixth Letter in the book and how much he approved of Lhwyd's work. But he reiterates: 'I can hardly shake off my former opinion [about the oyster-shells in Kent and Surrey] that these were originally beds of living Oysters breeding and feeding in the places where they are now found, which were anciently the bottom of the sea.'[1] He adds: 'I am also to seek about the original of beds or strata of several kinds observed in broken mountains. I cannot imagine whence they should proceed but from the sediments of land-floods or inundation of the sea.'

In the third edition of the *Discourses* the additions made in the winter 1703–4 and noted by Derham fall into two categories. There are long insertions of new evidence taken from Agostino Scilla,[2] from Woodward,[3] from an account of beds of Oyster-shells at Beckenham supplied by his friend Peter Burrell, 'merchant of London',[4] from a communication on a similar bed by James Brewer in *Philosophical Transactions*,[5] and from Lhwyd's book,[6] from which he gives an English version of the Sixth Letter. There are also certain brief comments of his own. First as to fossil-beds in mountains: 'This to me, I confess, is at present unaccountable':[7] then as to Ammonites after Cole's letter: 'Upon further consideration I find reason to agree with Dr Hooke and other naturalists, that these Cornua Ammonis are of the same genus with Nautili and differ only in species. But yet these species are subaltern genera, each having divers species under it':[8] thirdly, a note on the distribution of Ammonites: 'though altogether strangers to our seas they might as well be brought hither by force of winds or stress of weather, much more than by the general Deluge...especially if we consider that several East-India fruits have been brought over the vast Ocean and cast upon the Western Isles of Scotland':[9] fourthly, referring to his own theory of the gradual raising of the land: 'This conjecture hath no sufficient ground to support it...but truly if it had, I see not any better account of all the Phenomena':[10] and finally, a note appended to Lhwyd's letter, in which he maintains that his learned friend has sufficiently proved that these fossil-shells were not brought in by the Deluge, and adds that Lhwyd's theory is highly probable

1 *F.C.* p. 284.   2 Pp. 138–9, 140, 143–5.
3 Pp. 126, 165–7.
4 Pp. 129–30: he was father of Sir William Burrell the antiquary.
5 Pp. 131–2 from *Phil. Trans.* XXII, no. 261, pp. 484–6.
6 Pp. 175–203.
7 P. 149: cf. his last letter to Lhwyd.   8 P. 155.
9 P. 156—surely an error for West-India: the evidence for this came from Sloane, who in 1696 reported that he had received three sorts of Jamaican beans thrown up on the north-west islands of Scotland (*Corr.* p. 306).   10 P. 172.

since it certainly explains those found in the *viscera* of animals 'and if these, why not those found in the earth? I shall say no more but that those who are not satisfied with his proofs I wish they would but answer them.' He then admits that he has been 'brought over to the contrary opinion' chiefly by the beds of Oyster-shells and bones of fish: but that because this would 'infer the like original of those beds of Cornua Ammonis at Keynsham and elsewhere' he will 'allow them to have been the effects of the like principle with their fellows'.[1] It is a recantation obviously at variance with all the other additions to the book and with his own true and hitherto uncompromising judgment. He had in fact been absorbed in his work on insects and the third volume of the *Historia Plantarum* and was 'on the pit's edge', old and in constant pain. His natural tendency to accept expert opinion and to overrate the abilities of his friends combined with his ill-health and distractions to wring from him a concession that contradicted all his own work on the subject.[2]

Apart from its long and lengthening section on fossils—originally a note upon the consequences of the Deluge—the *Discourses* is a book of considerable interest. Its character, emphasised by the substitution of *Physico-Theological* for *Miscellaneous* in the title of the second edition, has perplexed modern students, who think it 'curious to meet with so many citations from the Christian fathers and prophets in essays on physical science'. Lyell, from whom this comment is quoted,[3] naively assumes that Ray, whose 'reputation placed him high above the temptation of courting popularity by pandering to the physico-theological taste of his age', ought to have adopted the modern distinction between science and religion—he is of course reading back his own difficulties into the seventeenth century.[4] But in fact to Ray, as to Milton before him and Newton after, there was no sense of inconsistency. Brought up to regard the Scriptural tradition as a primary datum for philosophy, and to insist that the God of the Bible and the Church was also the author and sustainer of nature, they did not realise the impossibility of reconciling the old beliefs with new knowledge: if, with Ray, they questioned points like 'the novity of the earth' or the effects of the Flood, they were still prepared to draw their evidence from authority or from observation almost indiscriminately and to regard the field of knowledge as a whole.

Hence to Ray's mind there was nothing incongruous in elaborating a sermon on the doctrine of cosmology into a treatise on geology; in

---

1 Pp. 203-4.
2 That Ray changed his mind at this time is indicated by Vernon (letter of Jan. 1702); cf. *Correspondence of R. Richardson*, p. 38.
3 *Principles of Geology* (10th ed.), I, p. 46.
4 Lyell narrowly and as some would say disingenuously avoided the odium which fell upon Darwin and was provoked by Huxley.

quoting Isaiah and St Matthew, Irenaeus and Cyril of Jerusalem, Heracleitus and Seneca, Julius Caesar and Ovid along with Steno and Kircher, Tancred Robinson and Edward Lhwyd; and in hesitating to reject the sages of the past in favour of their speculative and as yet very ill-equipped successors. Indeed it is not difficult to see from the authorities quoted in different parts of the book how the different strata of its composition were laid down.

Originally it was a sermon delivered not earlier than 1653 when Henry Hammond's *Paraphrase and Annotations on the New Testament*[1] was published, or possibly than 1659 when Henry More published his book on the *Immortality of the Soul*.[2] George Hakewill, whom he quotes several times,[3] had published his *Apologie of the Power and Providence of God* in 1627; Edward Brerewood his *Enquiries*[4] posthumously in 1614: Whichcote, whom he quotes twice, was preaching at this date in Cambridge. The first and last portions, Chapters I–IV and VI–XII, all of them short and based upon biblical and traditional authorities, may well have been little altered since its delivery in the University pulpit;[5] and the general plan of the sermon was no doubt that set out on pp. 3–4: as so planned it was simply a scholarly and enlightened exposition of the traditional teaching concerning the dissolution of the world. In such a thesis both the Deluge and the Creation are incidental to the main theme, arising solely out of the discussion of the possible means of dissolution, by an irruption of water or by volcanic activity.

His interest in the scientific study of cosmic and geological problems appears to have begun with his travels abroad, and to have been developed not only by his observations of the physical features, mountains, volcanoes, rivers and alluvial deposits, fossils and organic remains, but by the works of Kircher,[6] Steno,[7] Fabio Colonna[8] and Christian Mentzel,[9] Child-

---

1 Quoted in *Misc. Disc.* pp. 6, 8, 11, etc.: for Hammond (1605–60) cf. his life by Bishop Fell and *D.N.B.* Ray also quotes his *Practical Catechism*, published in 1644, on p. 229.

2 Quoted l.c. p. 32. It may well be an addition subsequent to the delivery of the sermon.

3 L.c. pp. 23, 41, 173, 175, 181, 191, 193, 198, the chief point being his refutation of the idea that the world was growing old and decaying.

4 L.c. pp. 72 (his theory of a 'vast continent towards the southern pole') and 161: Brerewood's works were of some influence: John Nidd left to Trinity a copy of the 1640 edition of his *Tractatus Ethici*.

5 Only two references to writings later than 1660 occur in these chapters—one to Wilkins's *Universal Character* on p. 203, the other to William Owtram's *De Sacrificiis* (published in 1677) on p. 204.

6 He had read the *Mundus subterraneus* in 1667 (cf. above, p. 143) and quotes the *Arca Noae*, pp. 65–6.   7 Cf. above, p. 420.

8 *Osservazioni sugli animali aquat. e terr.* 1616, quoted pp. 110–14.

9 Cf. above, p. 427.

rey,[1] Wittie[2] and Edmund Halley[3] as well as those of Hooke[4] and Plot[5] and the observations of his friends Lister,[6] Tancred Robinson[7] and Richard Waller,[8] as well as of Cole and Lhwyd. This equipment enabled him to form a clear concept of some at least of the principal factors of physical geography, to elaborate his views upon the origin and nature of springs and mountains and upon the changes which had taken place in the distribution of land and water, and to amplify very largely the parts of his discourse dealing with these subjects. In addition to expanding the contents of his original Chapter v, he introduced two long Digressions, the former on the Deluge, the latter on the Primitive Chaos and the Creation, and attached to each two Discourses, the first pair on the origin of fountains and on the nature and original of formed stones, the second (both shorter) on the equality of sea and land and on the use of mountains. The new material occupies 130 pages, or just half the book.[9]

In this long section occurs a mass of original observation and interesting discussion. He has noted the sinking of land in Derbyshire at Creech[10] (Crich), the silting up of the 'laguna' at Venice and the Camargue at the mouth of the Rhone[11]—and suggests a similar origin for the fenland 'running through the isle of Ely, Holland in Lincolnshire and Marshland in Norfolk'[12]—the erosion of the coast in Suffolk, where 'almost the whole town of Donewick [Dunwich] hath been undermined and devoured by the sea',[13] and the piling up of sand-dunes in Norfolk and in Cornwall, where 'I observed a fair Church that of the parish of Lalant [Lelant] which is the mother church to St Ives and above two miles distant from the sea almost covered with the sand'.[14] Most of these points had already been recorded by him in *Observations*.[15]

Then in discussing the cause of the Deluge and the Scriptural phrase 'the breaking up of the fountains of the deep' he criticises two theories

1 Joshua Childrey published *Britannia Baconica* in 1660. Ray quotes a passage on Childrey's native county Kent, on pp. 80–1.

2 Robert Wittie, *A Description of the Nature and Vertues of the Scarborough Spaw*, 1660, quoted pp. 53, 78–9.

3 The astronomer who published 'An Account of the Circulation of the Watery Vapours of the Sea and of the Cause of Springs' in *Phil. Trans.* XVII, no. 192, pp. 468–73.

4 Quoted pp. 105, 124–5.

5 He quotes the *History of Staffordshire*, published 1686, on pp. 44, 70–1, 121.

6 Cf. p. 120.

7 Quoting his observations on ancient walls at Rome in 1683 (p. 51).

8 Quoting a letter dated 4 Feb. 1687 to himself, p. 121: a similar letter to Hooke is in *Hooke's Works*, p. 286.

9 Pp. 39–170, the total being 259.

10 P. 44.   11 P. 46.

12 P. 47.   13 Pp. 49–50.

14 P. 54.   15 Pp. 6, 7: cf. above, pp. 421–2.

of the origin of springs. The first is the curious hypothesis of a regular circulation of water, rivers being fed from the sea by underground passages, into which the sea discharges and from which fountains rise. Ray does not deny subterranean waters, but asks how these can be forced upwards to the tops of mountains, and concludes that the theory is unproven. In contrast he introduces his own hypothesis that rain is the normal cause of all ordinary springs and adduces evidence from the brook at Black Notley, arguing from it by analogy to the great rivers.[1] In this connection he dismisses the contention that rain cannot sink more than ten feet into the ground on evidence gathered by himself in 'Pool-hole in the Peak'[2] and supplied by Sir Thomas Willughby from his own coal-mines. This leads him to deal with the second theory, that of Halley the astronomer, who had tried to prove that 'springs and rivers owe their original to vapours condensed on the sides of mountains rather than to rains'.[3] He admits that this may be true of tropical regions, but denies it of Europe, on the ground that in the Alps for six months while they are under snow there is no vapour or condensation, and yet the rivers run all winter—'nay even in summer the vapours are but rarely raised so high in a liquid form but are frozen into snow before they reach that height'; that the rivers overflow when the snow melts; and that they are obviously fed by snow, as is proved both by their colour and by 'the bronchocele or gutturine tumour [goitre], an endemial disease of the natives of those parts which physicians attribute to the water they drink, which consists of melted snow'.[4] He concludes with an observation of his own—'made in the last great frost, the biggest that was ever known in the memory of man'.[5] In the notes added 'upon a review of the precedent discourse'[6] he returns to Halley's theory and discusses it at length, illustrating his points by observations of a mist in December 1691[7] and conceding that vapour condensed on trees contributes to springs although rain is their chief constituent. In the text he quotes a letter of 12 November 1691 from Tancred Robinson on the same subject.[8] Yet in spite of this argument he allows that the Deluge was due to an unusual pressure upon the surface of the oceans forcing up the subterranean waters; and instances 'an extraordinary tide during the last year upon our coasts wherein the water overflowed all the seabanks, drowned multitudes of cattle, and filled the lower rooms of the houses of many villages that stood near the sea'.[9]

1 Pp. 74–5: he had first discussed the origin of springs (very briefly) in *Obs.* pp. 296–300.
2 P. 76: he visited it on his first tour in 1658: *Mem.* p. 124.
3 P. 82.   4 P. 89.
5 P. 91. This frost lasted from 28 Nov. 1683 till 5 Feb. 1684 and will be familiar to readers of *Lorna Doone*, ch. 42.   6 Pp. 242–55.
7 Mentioned in letter of 14 Dec. to Robinson: *F.C.* p. 295.
8 Pp. 92–4.   9 P. 100.

After the section on formed stones he deals with the earth's central fire which, while disavowing Descartes's theory of the earth as a cooled star caught up into the solar system, he accepts as analogous to the fire under 'the burning mountains or Vulcanos, Aetna, Vesuvius, Stromboli and Hecla'.[1] After discussing the nature of fire he passes to his second digression on Chaos and Creation; and suggests that the action of earthquakes and 'subterranean fires or flatus' was instrumental in raising the land above its original covering of water and throwing up mountains: this hypothesis he bases upon the raising of a hill near Pozzuoli on 29 September 1538, which he had seen and 'judged not near so great' as was reported, and from the earthquake in the Andes in 1646 mentioned by Kircher.[2] There follows a discussion on the uses of mountains, aimed at those who regard them as evidence of confusion in nature and therefore incompatible with belief in a Creator who always acts geometrically.[3] He defends hilly country as 'very beautiful and pleasant...far more grateful to behold than a perfectly level country...as any one that hath but seen the Isle of Ely must needs acknowledge'; and closes the digression with an apology for saying nothing about the creation of living organisms.

In the second edition the whole structure of the book was changed. The original sermon was transferred to the end, and the two digressions, expanded by new material, were 'made substantial parts of the book and disposed as is most natural according to their priority and posteriority in order of time'. The first discourse on Chaos is the shortest of the three: but the former digression is amplified by a discussion of the creation of animals based upon the introductory chapters of his *Synopsis Quadrupedum* which he was writing at this time, and containing a full account of the views of Peyer in his *Merycologia*[4] and of Brunner his critic[5] and a note of his own objection to Leeuwenhoek's views of spermatozoa. In the second, dealing with 'the fountains of the great deep', he inserts into the text the long note that he had printed in the Appendix and also some further details about the straits of Gibraltar,[6] a matter on which he had been in correspondence with Robinson.[7] Discussing Halley's theory of vapour and the nature of hail-stones he adds an interesting section on the buoyancy of

---

1 P. 139. 2 Pp. 155–6.

3 He quotes the Platonic dictum θεὸς ἀεὶ γεωμετρεῖ: p. 165. Ray is one of the earliest writers to appreciate the beauty of mountains, cf. Willey, *Eighteenth-Century Background*, p. 37.

4 Ray alludes to this in his letter of 3 March 1692 to Robinson, who had lent him the book: *Corr.* p. 246.

5 Author of a treatise *De Pancreate*. 6 Pp. 81–3.

7 *F.C.* p. 296. Letters of 24 and 29 Feb. 1692. Robinson had warned him that his previous statement was regarded as mistaken by Lister, Aston, Halley, Flamsteed and all seafaring men: *Corr.* p. 239.

birds in the upper air—a section touching upon the recently discovered principles of gravity[1] and upon problems of flight.

That the superior air doth support heavy bodies better than the inferior, the flight of birds seems to be a clear demonstration. For, when they are to make great flights they soar aloft....So I have often seen a flock of Wild-Geese mounted so high that though their flight be swift they seemed to make but little way in a long time...by reason of their distance. And yet one would think this were contrary to reason that the lighter air, such as is the superior, should better support a weighty body than the heavier, that is, the inferior. Some imagine that this comes to pass by reason of the wind constantly moving in the upper air which supports any body that moves contrary to it. So we see those paper kites[2] that boys make....In like manner the birds flying contrary to the wind, it supports and keeps them up. But if this were the only reason, methinks it would be very laborious for birds to fly against the wind....And therefore possibly they may be nearer the right who suppose that the gravity of bodies decreases proportionately to their distance from the earth; and that a body may be advanced so high as quite to lose its gravity and inclination to the centre. Of which I do not see how it is possible to make experiment. For, to what is said by some to have been tried, that a bullet shot perpendicularly upward out of a great gun never descended again, I give no credit at all.[3]

After the chapter on formed stones there follows a chapter on 'the great changes made in the superficial part of the earth', and their causes, which is almost entirely new. It is a collection from records of many times and countries of the submersion and separation of lands and of the raising of islands and silting up of seas. The most interesting part of it is that dealing with the effects of earthquake; for he gives a full description not only of that in Jamaica on 7 June 1692[4] but of that in England on 8 September,[5] and compares this latter with an earthquake at Oxford in 1683[6] and one which he himself observed at Sutton Coldfield in 1677. In this connection he discusses the cause of earthquakes and suggests that it is often the explosion of fire-damp in the subterranean caverns of the earth—this on the common

1 Newton's *Principia* had been published in 1687, by the initiative of Halley.
2 The earliest reference to such kites given by Murray is to Butler's *Hudibras* in 1664.
3 Pp. 103–5.
4 Pp. 186–208. He quotes the letters from the Minister on board the *Granada* in Port Royal, published by authority.
5 Pp. 209–16. He gives an account of the weather at the time, of the extent of the shock (quoting a letter from Robinson), of its motion and effects (quoting Allen, the Braintree doctor), of its noise, and of its local incidence (houses within a mile of Dewlands being shaken though neither he nor his family 'though they were above-stairs' were aware of any impression of it). There is a shorter account of it in Calamy, *My Life*, p. 236, where it is stated that the earthquake in Jamaica almost destroyed Port Royal and killed 1500 persons.
6 For this cf. Thoresby, *Diary*, I, p. 169: and *An Historical Account of Earthquakes*, published anonymously by Z. Grey (Cambridge, 1750), pp. 59–60.

but of course mistaken analogy between earthquake and thunder. Fire-damp had been brought to his notice first in 1668 by his friend Francis Jessop, who had also sent to Lister a letter on the subject published in *Philosophical Transactions* ;[1] and he records instances of explosions from it in various coal-mines, which being ill-ventilated and lighted by open candles must have been veritable death-traps. He also associates them with volcanoes, whose violence in eruption he connects with subterranean water: 'for the fire suddenly dissolving the water into vapour, expands it to a vast dimension and by the help thereof throws up earth, sand, stones and whatever it meets. How great the force of water converted into vapour is I have sometimes experimented by inadvertently casting a bullet in a wet mould, the melted lead being no sooner poured in but it was cast out again with violence.'

The third discourse is much less altered; indeed is a reprint of the main portion of the first edition. In the discussion of the erosion of mountains he quotes a letter from Lhwyd[2] describing the fall of masses of rock on Snowdon and in the Nantfrancon Pass; and on the general subject of the action of water some pages from Giuseppe Biancani (Blancanus),[3] whose book, *De Mundi Fabrica*, he had seen quoted by Hakewill but had been unable to procure till after his first edition was printed. The only other alteration seems to be a redrafting of the passage on the eternity of punishment, in which he omits two references to Origen and inserts a reference to and an extract from Whichcote's *Sermons*.

In the third edition the alterations are much less radical. Most of them are mere notes, giving corroboratory evidence and further references—some from Ovid, Josephus or the Septuagint (evidently he still read the Classics) and some from recent writers and correspondents, John Greaves,[4] Hooke,[5] Francis Maximilian Misson[6] on the earthquake in Naples on 5 and 6 June 1688 and Paolo Bonone[7] (surely a misprint for Boccone) on those in Sicily on 9 and 11 January 1693. Two of these touch on points of interest; the first, dealing with a vacuum, states that this is 'intermixed with the minute parts of all bodies; those that have more of it interspersed among

1 X, no. 117, pp. 391–5: he gives a good account of the behaviour of methane here and in a later contribution, no. 119, pp. 450–4.

2 Pp. 285–7. Ray refers to this letter as containing other portions 'omitted because I was not willing to raise a "diell" I could not lay': *F.C.* p. 245.

3 Jesuit and Professor of Mathematics at Parma, author also of *Sphaera Mundi*, 1635. Ray prints a translation of it here on pp. 296–305.

4 P. 84: he quotes the *Pyramidographia* published in 1646 when Greaves was Savilian professor and sub-warden of Merton College, Oxford.

5 P. 89: his theory of springs that the fresh water in subterranean caverns was forced up by the heavier salt water.

6 P. 292. His *Nouveau Voyage d'Italie*, published at the Hague in 1691, was translated into English in 1695.

7 P. 292: cf. *Phil. Trans.* XVII, no. 202, pp. 827–38.

their parts are more rare or thin, those that have less more dense or thick; the rarer bodies are lighter according to the proportion of matter they contain':[1] the second adducing a proof that Britain was once united to France from the fact that 'wolves and foxes and bears too were anciently in this island; it is not likely that they should venture to swim over a channel 24 miles broad...nor that men should transport such noisome, mischievous creatures by shipping'.[2] Three of the insertions,[3] in addition to those in the chapter on formed stones, are criticisms of Woodward, whose theory he roundly condemned:[4] 'that phenomenon, for the solving whereof I suspect he invented this hypothesis, is not generally true.'[5]

The two chief additions are both in Chapter v of the second discourse —the chapter inserted in the second edition. The former deals with further instances of 'atterration'—the silting up of land at the mouths of rivers— one communicated by Ralph Thoresby to the Royal Society[6] of the discovery of jetties, old boats, ox-horns, etc., ten feet underground on the old course of the Welland at Spalding; the second the finding of the old ground-level, 'a solid gravelly and strong soil', at a still greater depth 'at the laying of the new sluice or goat (as they call it) at the end of Hamore beck as it falls into Boston haven';[7] the third an account by Ramazzini[8] of ruins buried deep near Modena. The latter and longer deals with the matter of subterraneous woods and fossil trees—a subject which he had already noticed in *Observations*.[9] He begins with the passage from 'Boetius de Boot' and the reference to Dunwich, there recorded; then adduces a letter from Richardson of North Bierley[10] on the buried trees found at 'Youle' on the Humber—'some are so large that they are used for timber in building houses which are said to be more durable than oak itself; others are cut into long chips and tied up in bundles, and sent to the market towns several miles off to light tobacco[11]...they are really fir trees'.[12] Then he quotes a

---

1 P. 71.   2 P. 208.   3 Pp. 126, 211–12, 218–20.

4 Woodward had spoken violently of him and had quarrelled with Lhwyd. Ray only alludes to the controversy or to the book by John Harris in two letters, *Corr.* p. 332 and *F.C.* p. 270: but the letters of Lhwyd, Nicolson and others show how divided and embittered was the world of science by it. Cf. note at end of chapter.

5 P. 167.

6 *Phil. Trans.* XXII, no. 279, p. 1158. Thoresby became an F.R.S. in 1697 and published his *Ducatus Leodiensis* in 1715.

7 P. 222.

8 Pp. 223–5: 'this relation I transcribed out of the history of the *Works of the Learned*.' Bernardino Ramazzini, professor of medicine at Modena, wrote *De Fontium Mutinensium scaturigine*, in 1691.

9 Pp. 6–9.

10 *Phil. Trans.* XIX, no. 228, p. 526: he describes Youle as 'about twelve miles below York upon the river Humber—where the Don runs into it': the modern Goole—then a marsh not marked on the maps.

11 Is this the first mention of spills?   12 Pp. 230–1.

similar record of trees from a bog 'called the curragh in Kirk-Christ Lezayre' in the Isle of Man, noting that they lie prostrated along the direction of the prevailing wind. Next he gives evidence of buried trees 'burnt or cut down by the labour of men' and found 'in most of the great morasses, mosses, fens and bogs in Somersetshire, Cheshire, Lancashire, Westmorland, Yorkshire, Staffordshire, Lincolnshire and other counties: the wood of them is usually called Moss-wood and is black as ebony'. He quotes De la Pryme's letter to Sloane[1] for evidence that they were cut down with axes or burnt through and for the supposition that this was done by the Romans in order to deprive the Britons of their fastnesses. Next he refutes an opinion in Charles Leigh's *Natural History of Cheshire*[2] that these trees were brought by the Deluge, proving that they grew where they are found and that the fir is in fact an ancient and indigenous British tree. He concludes: 'it is a vain thing to dispute by argument against clear matter of fact'[3]—a maxim which in his geological studies he follows faithfully.

As compared with the accuracy and insight of Ray's other work this first essay in physics and physical geography seems crude and primitive. It must be remembered that the subject was in its infancy; that the whole Copernican hypothesis was by no means generally accepted; and that wild and fanciful speculations were characteristic of almost all books dealing with the structure of the earth. The remarkable feature of the *Discourses* is not their ignorance of physical facts, but the evidence of Ray's wide reading outside his own proper sphere and the acumen with which he marshals the data, grapples with problems hitherto unfamiliar, and uses his own observations as pointers towards reasonable hypotheses. It must not be forgotten that during the years covered by these editions he was producing his final volumes on the classification and history of plants, was devoting himself to the fresh study of insects, was issuing new versions of the *Wisdom of God*, and was so constantly in pain that he could often get only two hours a day for writing.

NOTE. *The controversy about Woodward's book*

It is unnecessary to give a full account of the conflicts which broke out after the publication of John Woodward's *Essay toward a Natural History of the Earth* in 1695: but Ray and his friends were so far involved in them

1 *Phil. Trans.* XXII, no. 275, p. 980. Abraham De la Pryme was born at Hatfield Chase in 1672 and graduated from St John's College, Cambridge, in 1694, afterwards living in Yorkshire and studying its archaeology.

2 Published at Oxford 1700: Leigh had become an F.R.S. in 1685, but his book is of little value.

3 P. 241.

that some at least of his letters are unintelligible without some knowledge of the facts.

The *Essay* combined an admirably clear conviction of the organic origin of vegetable and animal fossils, based upon wide experience and energetic collecting, with a fantastic theory of the character and effects of the Flood, which was supposed to have liquefied the earth so that when the waters subsided shells and plants sank into the slime to a depth proportionate to their specific weight. Woodward was a man of dogmatic temper who early in his *Essay* grouped together Colonna, Steno, Boccone, Grandius and Ray and proposed to show 'what they had already done, wherein they failed and what remains still to be done',[1] and later dismissed in two sentences the theories of streams put out by Halley and Ray respectively and insisted that they arise from the waters of the abyss:[2] he was quite incapable of recognising or admitting that the two parts of his hypothesis were of widely different value. Moreover, he was violent and bitter of speech, and regarded every comment as an insult: as such he invited criticism and provoked enmity.

Several pamphlets and letters appeared attacking his thesis; for its subject was in the air and its treatment controversial. Among these were two essays from Oxford signed L.P.,[3] but strongly asserted to be by Tancred Robinson: Robinson when challenged by William Wotton, formerly the boy prodigy,[4] whom Nicolson invited to mediate between him and Woodward,[5] denied authorship but apparently admitted having had a hand in providing material for them. Then came two letters published in *Miscellaneous Letters*, one dated Dublin 18 December 1695, signed S.G.A.[6] and dealing with the essays, the other discussing petrifactions and dated February 1696. Finally Thomas Robinson, Rector of Ousby in Cumberland, 1672–1719, produced *New Observations on the Natural History of this World of Matter and this World of Life*: according to Nicolson,[7] Robinson, who was his neighbour, told him that the 'additional remarks were printed without his order, consent or knowledge'…'being in his opinion not

1 *Essay*, p. 39: he never fulfilled this threat.
2 L.c. pp. 120–1: Ray complained of his 'haughtiness': cf. *F.C.* p. 301.
3 It is tempting but speculative to suggest that L.P. stands for Lister and Plot the protagonists of the case against the organic origin of fossils. Lhwyd, writing to Lister about the Essays (Gunther, *Life of Lhwyd*, p. 276), says 'I am sure by some words in it you can guess at it and am certain no one of Oxford could write it', but Lister did not reply!
4 Cf. above, p. 53.   5 *Letters*, I, p. 88.
6 The initials are those of St George Ashe, F.R.S. 1686, Secretary of the Phil. Soc. of Dublin and in 1695 Bishop of Cloyne: Harris declared the initials to be false, as their proper owner was not in Dublin at the date specified: Lhwyd spoke of the letter as by Ashe (Gunther, *Life of Lhwyd*, p. 305).
7 *Letters*, I, p. 95.

only too severe upon Dr Woodward and Mr Whiston, but also malicious': he appears also to have said that he had had no intention of making any of his papers public.

In 1697 John Harris, fellow of St John's College, Cambridge, in 1691 and of the Royal Society in 1696, published *Some Remarks on some late Papers relating to the Deluge,* in which he dealt almost savagely[1] with these three criticisms of Woodward. Ray, referring to this, wrote to Lhwyd: ' 'Tis true I have been much sleighted by Dr Woodward in his late invective against my worthy friend Dr Robinson subscribed by Mr Harris'[2]—from which it appears that Woodward was suspected of having written the *Remarks* himself. In any case it is, as Ray calls it, 'a scurrilous piece wherein the author hath discovered a great deal of pride, scornfulness and ill-nature': he adds, 'as for his treatment of me, though it be not very civil, yet it is not so vilely rude and contemptuous as might have been expected'.[3] In fact Harris had only spoken of Ray in answering the charge that Woodward had borrowed from him: he expressed respect for Ray; stated that his work is 'collected chiefly out of other writers';[4] and criticised the four reasons against the reality of fossil plants set out in the second edition of the *Synopsis Britannicarum.*[5]

---

1 The attack on Robinson—a good friend of the miners of Cumberland and a reputable naturalist—is as cruel as it is clever.
2 *F.C.* p. 270.   3 *Corr.* p. 332.
4 Tancred Robinson in April 1696 had reported to Lhwyd that Woodward 'is intolerable....He calls Mr Ray a mercenary scribbler, a mere copyist' (Gunther, *Life of Lhwyd,* p. 305 note).
5 *Remarks,* pp. 253–6.

CHAPTER XVII

THE *WISDOM OF GOD*

A character from whose penetrating genius and persevering industry not Botany
alone but Zoology may date a new aera: in these branches of natural history he
became without the patronage of an Alexander the Aristotle of England and the
Linnaeus of the time.

RICHARD PULTENEY of JOHN RAY, *Sketches of Botany*, I, p. 188.

In considering Ray's work on geology it has been necessary to give an
account of the *Discourses*; for, as we have seen, these are his fullest treatment of the subject and the only contribution that has received any attention from later writers. But in fact the *Discourses* both in their intention
and in their gradual enlargement follow closely the book which he published in the previous year, 1691.

The Wisdom of God manifested in the Works of the Creation is certainly
his most popular and influential achievement. Published as a slim octavo
volume of 249 pages, in a first edition of 500 copies,[1] it was reprinted in a
second edition of 382 pages in 1692,[2] in a third of 414 pages in 1701 and
in a fourth of 464 pages in 1704. It was reissued many times during the
next century;[3] it formed the basis of Derham's Boyle Lectures in 1711–12;[4]
it supplied the background for the thought of Gilbert White and indeed
for the naturalists of three generations; it was imitated, and extensively
plagiarised, by Paley in his famous *Natural Theology*;[5] and more than any
other single book it initiated the true adventure of modern science, and is
the ancestor of the *Origin of Species* or of *L'Évolution Créatrice*.

The novelty of the book consists in the fact that in it Ray turns from
the preliminary task of identifying, describing and classifying to that of

---

1 So Derham, *Mem.* p. 44. Derham also declares that the MS. was sent to Robinson
on 3 March 1691 and that Robinson sent it to the printers: the letters to Robinson
epitomised by Derham (*F.C.* p. 294) hardly bear this out: but it was Ray's regular
practice to get Robinson to read his scripts before publication: cf. Preface to the
Synopsis Britannicarum.

2 After the first edition it was divided into two parts—in the second of 206 and
176 pp. respectively.

3 Thus the tenth edition appeared in 1735. It appeared in Dove's English
Classics in 1827.

4 *Physico-Theology: or a Demonstration of the Being and Attributes of God from His
Works of Creation*: Derham acknowledges fully his obligation 'to my Friend the late
great Mr Ray'.

5 Paley seldom quotes it by name, but repeatedly borrows from it without acknowledgment: indeed almost its whole contents are found rewritten but easily recognisable
in his pages.

interpreting the significance of physical and physiological processes, studying the problems of form and function and of adaptation to environment, and observing behaviour and recording the achievements of instinct. It was inevitable, as we have urged, that taxonomy should be his first concern. Until Adam's duty of 'naming the animals' had been undertaken, the study of nature could not be more than guess-work and fancy. He found this study congested with legendary and symbolic ideas, with catalogues of plants and animals in which specific distinctions were vague and confused, and with classifications that were arbitrary or alphabetical. Order had to be brought into the chaos; and by patient collection of synonyms, exact descriptions, and a rare skill in fastening upon salient points he made possible the *Systema Naturae*.

But he never fell into the error of his great successor. Linnaeus was always content to 'suppose that the highest and only worthy task of a naturalist was to know all his species exactly by name' and in consequence 'never made a single important discovery throwing light on the nature of the vegetable or animal world'.[1] Ray had already in the introductory essays of his *Historia Plantarum* and *Synopsis Quadrupedum* discussed the structure and functions of the living organism, proceeding from the study of comparative anatomy to that of physiology and behaviour. But he had not gathered together the results of his studies or attempted to give a scientific and philosophical exposition of organic life. In the *Wisdom of God* he made a beginning in this field; and with each edition enlarged the record with fresh and often original observations and inductions. The book is packed with references to and discussions of problems, ranging from the nature of atoms or the influence of the moon upon tides, to those of the shape of bees' cells, the movements of birds and fishes, the structure of the eye and the growth of the foetus in the womb. Always he attempts to relate the significance of particular adjustments to the life of the whole, to correlate and compare a mass of detailed observations so as to disclose the conditions or 'laws' which they illustrate, to expound an integrated view of the universe in terms of its plan and purpose. He brings to the task qualifications unique in his own day; an exact knowledge of a large range of phenomena—he had studied sciences from astronomy and geology to botany and zoology at first hand; a power of judging evidence disciplined by a lifetime of effort—the disentangling of fact from fiction, of reliable from legendary records, had been his concern in every department of study; a refusal to accept irrational or supernatural interpretations—the discoveries and excitement of the times had given him confidence in the experimental method and also a readiness to admit ignorance and refuse appeals to a *deus ex machina*; an interest in process and machinery which enabled him to appreciate the importance of microscope and scalpel and

1 Cf. Sachs, *History of Botany* (ed. Garnsey and Balfour), pp. 84–5.

the value of the Cartesian philosophy; an experience of life, of its progress and of its imperfection, which refused to be content with the crudely mechanistic categories that satisfied his contemporaries Hooke and Grew;[1] a religion which if hampered by a too literal acceptance of Scripture and tradition had been formed by the most enlightened theologians of the day and was closer to that of the great thinkers of Christian Platonism than to the bibliolatry of the Puritans or the conservatism of the Churchmen.

Hampered he certainly is. So much has been obvious from his geological studies. The frame of reference built up on the book of Genesis, strengthened by centuries of contented usage, and accepted by the vast majority as divinely designed, had provided a house for humanity which had survived the intellectual upheavals of the Renaissance and the Reformation. Protestantism, refusing to accept tenancy of it on papal authority, had increased rather than weakened respect for its Scriptural basis. The 'novity' of the world, a postulate far older than Archbishop Ussher's chronology, had been given precise definition; the universality of the Deluge, unchallenged even by those who could not reconcile it with the evidence of fossil remains, gave to the fixity and permanence of species the force of a dogma; witchcraft and demonology, supported by biblical tradition and accredited by statutes and law-courts, militated against any concept of natural law and provided an easy solution for the perplexities of scientists. The ablest minds of the time, Descartes and Malpighi not less than Locke and Newton, hesitated to move out of their ancient intellectual home and constantly insisted that they had no intention of doing so. Nor was their hesitation solely due to fear that Church and State, regarding all questioning of the established order as heresy and treason, would combine to punish innovations with damnation and death. It was in fact impossible even for the most independent intellect to emancipate itself from the postulates of contemporary thought or to realise the scope of the changes for which mankind was being prepared. If Hobbes the iconoclast, Hobbes whose coarsely virile mind had no prejudice in favour of orthodoxy, yet accepted a cosmology practically identical with that of Ray and while agreeing with Copernicus yet had no glimpse of creation as anything but a sudden and complete act nor any consciousness of the age and evolution of the earth,[2] we can at least understand how inevitable was the traditional *Weltanschauung*.

Ray himself was singularly free from the limitations which the old order imposed. He had learnt from his own studies the lesson which his teachers had outlined, that reason, strictly disciplined and honestly followed, was

---

1 To both of them the divine activity in creation was limited to that of a First Cause —a watchmaker: cf. an interesting article on the *Tercentenary of Grew* by Dr Agnes Arber in *Nature*, vol. 147, p. 630, in which this point is clearly made.

2 Cf. Masson, *Life of Milton*, VI, 285–6.

the supreme instrument in science and religion. If he did not identify the spiritual with the rational, he was convinced that every tenet of theology or philosophy must be assayed by the test of its reasonableness; that loyalty to truth was loyalty to God; and that man derived his status from his capacity to share in and respond to divine wisdom. From Plato and Clement of Alexandria, from Whichcote and Cudworth and Worthington, he had received a mental and religious training which invested his scientific researches with the dignity of a sacred calling, and being debarred from other exercise of his ministry he had accepted as his vocation the task of 'thinking God's thoughts after Him'. There was for him nothing incongruous in seeing the objects of his study, the order of the universe, the life of plants and animals, the structure and functioning of nature, as the manifestation of the Mind of God. Indeed the wonder with which he regarded the works of creation, and the thrill which accompanied his growing insight into the processes of their growth and function, were to him, as to mankind in general, essentially religious. He found in this new approach to the physical world the awe and reverence, the release and inspiration which psalmists, poets, thinkers and explorers have always found; and though it was difficult to reconcile his discoveries with the formulae of Christian tradition it was impossible not to find in them a profound religious and indeed Christian significance.

That he expresses this conviction in terms of a simple and too anthropomorphic teleology, that his faith in God is often naive and sometimes conventional, that though he refuses to believe that nature has no value in itself[1] he cannot escape the contemporary confidence that its primary purpose is its utility to man, is of course obvious. His book would be scorned by those moderns for whom teleology is a forbidden word and who have forgotten how generously Darwin acknowledged his own indebtedness to Paley.[2] Condemning the over-simplification of Cicero's argument borrowed by Grew and Paley of the watch and the watchmaker,[3] they will reject Ray's similar claim for the human body 'to imagine that such a machine composed of so many parts, to the right form, order and motion whereof such an infinite number of intentions are required, could be made without the contrivance of some wise Agent, must needs be irrational in the highest degree'.[4] Yet it is at least arguable that the belief that Darwin disposed of teleology or that the evolution of a highly complex object like the eye or of an interdependent series of events like the

---

1 He explicitly denies 'the generally received opinion that man is the end of the creation as if there were no other end of any creature but some way or other to be serviceable to man': *Wisdom*, pp. 127–31.
2 Cf. *Life and Letters*, II, p. 219.
3 In *De Natura Deorum*, II, 34, quoted in *Wisdom*, p. 50.
4 *Wisdom*, p. 167.

parasitism of the cuckoo, where the efficacy of the whole depends upon the simultaneous co-ordination of a multitude of factors, can be ascribed to natural selection operating upon random variation, still deserves to be called irrational.[1] Ray, unlike the majority of scientists, was trained to think as well as to observe and experiment.

But in any case Ray does not adopt the simplified teleology of Cicero or Grew. He is too well aware of the difficulties, the evidence of incompleteness and monstrosity, of continuous creativity and variation, to be content to assign the creation solely to the direct and initial act of God.[2] The anthropomorphism which regards an omnipotent deity as both designer and craftsman seems to him not only unsound but unworthy. So he deliberately accepts Cudworth's belief in a 'Plastic Nature or Vital Principle',[3] a life force or creative spirit, embodying, so to say, the laws of nature and expressing itself in the unconscious processes of organisation and growth. The belief in such a principle is of course ancient and widespread, though it takes many forms—the soul of the universe in Plato's *Timaeus*, the directive element of Stoicism, the world-soul of Plotinus, or the seminal word of Clement or Augustine. In the seventeenth century the Cambridge Platonists adopted it partly from the classical sources but partly from the kindred notion of 'archei' in Paracelsus and Van Helmont, which gave rise also to the monads of Leibnitz. More and Cudworth, Ray's chief authorities, both employ it, though in a slightly different sense. In More it is akin to the subconscious or subliminal self: he uses it to explain sympathetic cures, maternal impressions and such phenomena as 'cannot be effected by the simple mechanical powers of matter'.[4] In Cudworth the concept is much more carefully worked out and defined;[5] 'plastic nature' is closer to the Bergsonian *Élan vital* or Life-force: it is employed to explain the evidences of frustration and failure in the natural order, of difficulty and slowness in the process of growth, and of the recalcitrancy of matter. Ray, as we have seen, follows Cudworth, endorsing his grounds for refusing to regard creation as the work of unlimited omnipotence and adding to them his own conviction of an indwelling vitality whereby, for

1 As has been argued by many students of biology from J. J. Murphy and Herbert Spencer onwards who had the advantage, denied to Darwin, of some training in philosophy.

2 He criticises Boyle, *Free Enquiry into the vulgar notion of Nature*, on this ground. Cf. *Wisdom*, pp. 32–3.

3 *Wisdom*, pp. 32, 33–4, quoting *True Intellectual System*, I, pp. 222–3 (Harrison's edition).

4 *Immortality of the Soul*, III, 12. He employed it to explain the action of Boyle's air-pump and was criticised by Boyle for so doing: cf. F. I. Mackinnon, *Philosophical Writings of H. More*, p. 305.

5 *True Intellectual System*, ch. 3 (Harrison, I, pp. 217–74): for exposition of his view cf. Tulloch, *Rational Theology in England in the Seventeenth Century*, II, pp. 269–73.

example, 'the segments and cuttings of some plants, nay the very chips and smallest fragments of their body, branches or roots, will grow and become perfect plants themselves'[1]—an argument which suggests the debates over preformation or epigenesis and the entelechies of Driesch.

This is not the place to enter into the controversies to which the doctrine has given rise or to discuss its value as a contribution to biological philosophy. But at least it proves that Ray was not blind to the weakness of the traditional belief in a direct and omnipotent creator or content with the crude teleology of Paley. He worked, here as elsewhere, within the framework of contemporary Christian thought, but with a loyalty to experiment and observation and a faith in the unity and rationality of nature which contributed powerfully to the abandonment of that framework and stimulated the quest for a truer and more scientific interpretation of the data of physical studies. It is as absurd to suggest that he foreshadowed the concept of evolution or adopted the categories of a modern scientist as it is to blame him for not doing so. But in drawing attention to the unity of nature, to the problems of form and function, to adaptation, and to a great number of strange and in some cases still unexplained phenomena he gave the first strong impulse to the scientific movement of to-day. A good case could be made for assigning to the successive editions of the *Wisdom of God* a primary place in the development of modern science.

The book itself was dedicated to Francis Willughby's sister, the Lady Lettice Wendy, widow of Sir Thomas Wendy[2] of Haslingfield, with whom he stayed in 1668. It contained, as he declared in the Preface to the first edition,[3] 'the substance of some Common Places (so in the University of Cambridge they call their Morning Divinity Exercises) delivered in Trinity College Chapel when I was Fellow of that Society which I have enlarged with the addition of some collections out of what hath since been written by the forementioned authors upon my subject'—that is 'by the most learned men of our time, Dr More, Dr Cudworth, Dr Stillingfleet, now Bishop of Worcester,[4] Dr Parker, late of Oxon,[5] and to name no more The Honourable Robert Boyle Esquire'. He describes one section

---

1 *Wisdom*, p. 35.

2 He had been Member of Parliament and Knight of the Shire and died in January 1674 (*Diary of S. Newton*, p. 71): Ray's prayer on the occasion of his death is in *Mem.* pp. 58–60.

3 The words are omitted in later editions.

4 Edward Stillingfleet, fellow of St John's College, Cambridge, 1653, had been made Bishop in 1689.

5 Samuel Parker, made Bishop of Oxford in 1686, had died two years later: he had been at Wadham under Wilkins, 1656–9; and Ray may have had glowing reports of him: but his reputation for learning and character when he died was not high; and he is quite out of place in this company.

at least as 'written above thirty years since';[1] and in another complains that 'I am sorry to see so little account made of real Experimental Philosophy in this University and that those ingenious sciences of the Mathematics are so much neglected by us'[2]—evidence that parts at least of his Cambridge script are contained in the work as printed. Most of the books quoted belong to a later period. But if, as is probable, his discourses were delivered in 1659–60, it is likely that the frequent references to More's *Antidote against Atheism* belong to them. Indeed it is probable that they were originally an exposition of and comment upon the theme of the second book of More's treatise.

There is unfortunately no evidence that the two men were ever acquainted: indeed beyond the fact that they had many mutual friends and common interests there is nothing to suggest personal contact. More was thirteen years older than Ray, had been educated at Eton and Christ's College, where he became a fellow and tutor and spent the whole of his placid, studious and saintly life.[3] As an undergraduate he had been fascinated by philosophy, the classical philosophy of the schools. Then he had discovered Plotinus, Marsilius Ficinus, and the *Theologia Germanica*;[4] and set himself to the study of mysticism and the life of disciplined contemplation. But he kept a strong interest in nature and the new scientific pursuits, was a fervent admirer and in 1648–9 correspondent of Descartes, whom he defended in his letter to 'V.C.'[5] in 1662 but afterwards learnt to criticise;[6] and became a member of the Royal Society in 1664. After a volume of *Philosophical Poems* in 1647 and several other smaller publications he produced his first considerable book, *The Antidote against Atheism*, in 1652 and with it took a leading place among the 'latitude-men' or Christian Platonists of Cambridge. The book is so definitely the starting-point of Ray's *Wisdom of God* and differs from it so markedly, partly by reason of their respective dates, partly from divergence of temper and outlook, that a brief account of it is inevitable.

1 *Wisdom*, p. 71. This date and the Commonplaces at Cambridge are mentioned in his letter to Robinson in Nov. 1690 (*F.C.* p. 294).
2 *Wisdom*, pp. 125–6.
3 Unfortunately little is known of More's training or friends: Ward's *Life* is edifying but tells nothing: the autobiographical preface to the Latin edition of his *Works*, published in 1679, represents him as passing rapidly from natural philosophy to mystical theology.
4 Possibly under the influence of his tutor Robert Gell, a learned man with a strongly mystical outlook and an interest in astrology: cf. Gell's *Remains* (2 vols.), published 1676.
5 Stated by F. I. Mackinnon, *Philosophical Writings of H. More*, Introd. p. xv, to be William Coward, author of *The Grand Essay* etc., but in 1662 Coward was a baby—born 1656 or 1657: Mackinnon's facts are often inaccurate.
6 In the preface to his *Divine Dialogues* (1667): cf. Tulloch, *Rational Theology*, II, pp. 368–77.

The *Antidote* is in three books. The first deals with the philosophical arguments for the existence of God and contains a valuable exposition of More's own characteristic ideas.[1] The second, which supplied Ray with a number of quotations and the outline of his discourse, deals with the argument from the phenomena of nature. The third is concerned with the supernatural, with a large collection of records of witchcraft, including satyr-dances, werewolves and fairy rings,[2] magic such as the Pied Piper of Hamelin,[3] and the exploits of ghosts and poltergeists.[4] This last book, with its grotesque records and aggressive credulity, following upon the wisdom and rationalism of its two predecessors, comes as a shock to the modern reader and is a valuable reminder of the transitional character of the period. More was a great scholar, a fine and singularly attractive personality, the friend and colleague of the most liberal and scientific students of his day. But he accepts and describes in detail the crudest superstitions with the same care and confidence as he devotes to the records of physics and biology. He rejects with scorn the suggestion that these phenomena are the outcome of 'melancholy' or morbid imagination; refuses even to ascribe them to the agency of the devil; insists that 'these hags' supply valid testimony to an objective supernaturalism and assimilates their evidence to that of the miracles of the New Testament as a decisive argument against atheism. We have to remember that Sir Thomas Browne, author of *Religio Medici*, and Joseph Glanvill, champion of the Royal Society, shared More's conviction of the reality of witchcraft, and that Ray's friends Andrew Paschall and John Aubrey were exchanging letters about the prodigies at Barnstaple,[5] if we are to understand the outlook of his contemporaries.[6] It is no small proof of the outstanding greatness of the *Wisdom of God* that it entirely ignores the whole of this third book. Indeed in all his writings and letters Ray never gives a sign that he regarded magic as true or legitimate, or accepted for a moment the belief that such phenomena had any objective value: on the contrary he explicitly condemns the practice of the Black Arts and only mentions supernatural interpretations in order to reject them.

The second book of the *Antidote* supplies the sequence and underlies the

---

1 It is reprinted together with a schematised exposition of his philosophy and some interesting comments and notes by Mackinnon, *l.c.*

2 More admits that two of his authorities, Johann Weier and Remigius, disallow these, but he himself, with Bodin, the early economist, accepts them.

3 Derived from Weier, *De Praestig. Daemon.* I, 16.

4 More spent much time on these phenomena in association with Joseph Glanvill: cf. *Saducismus Triumphatus.*

5 Cf. Aubrey, *Miscellanies*, pp. 149–57.

6 It is to be noted that John Smith of Queens' College, whose *Select Discourses* were edited by Ray's friend John Worthington, has a sermon against witchcraft which is wholly free from More's credulity and superstition: cf. *Select Disc.* pp. 443–68.

contents of a large section of the *Wisdom*. It begins with a chapter upon
the 'external phenomena of Nature' regarded as matter and motion and
argues that since matter is admittedly uniform it could not if self-moved
produce 'variety of appearance' and if inert must be actuated by some
spiritual substance; if it is not uniform, yet chaotically differentiated movement cannot produce 'so wise a contrivance as is discernible'; in any case
motion must proceed from 'a knowing principle': 'therefore that I contend for is this that, be the matter moved how it will, the appearances of
things are such as do manifestly intimate that they are either appointed all
of them, or at least approved, by the Universal Principle of Wisdom and
Council'. He inclines to the belief that God both originates the general
operations of natural law and also 'where such laws are not to His purpose
rectifies and compleats them'.[1] Then turning to details he considers the
motion of the earth and the position of its axis, gravity and the *fuga vacui*,[2]
and other physical problems; the use of mountains and metals; the sea and
navigation: these chapters are not very satisfactory, for physical science
and geology were in their infancy. Next he deals with living organisms in
whose beauty and structure there is evidence of an intellectual principle;
with plants, their seeds, 'signatures' and usefulness to man; with the usefulness of animals, the distinction of the sexes, the structure of beasts,
birds and fishes and their special adaptation to their way of life; with birds
and especially the Bird of Paradise; and finally with the anatomy and
passions of mankind, concluding with an eloquent eulogy of his scientific
achievements.

It is unfortunate that we know almost nothing of More's own training
in the subjects of which he is discoursing[3] beyond his admission that his
final studies at Cambridge were of Aristotle, Cardano and Julius Scaliger;[4]
and that he gives very few references to authorities. There seems no
evidence that he had himself studied anatomy or dissection; and little of
the occasion and extent of his contact with Descartes, who certainly dissected and vivisected, but whose books on this subject, *L'Homme* and the
Traité de la Formation du Fœtus, were only published posthumously in
1664,[5] or with the British scientists whose atheism shocked him into
writing the *Antidote*. But he had been for his first seven years at Christ's

1 *Ant.* (1712 edition), pp. 37–40.
2 More here quotes Boyle: Boyle replied rejecting the *fuga vacui* as an 'unphilo-
sophical principle': cf. *Free Enquiry*.
3 P. R. Anderson's study of More, entitled *Science in defence of liberal Religion* (New
York, 1933), has no evidence except references to the influence of Descartes.
4 Julius Scaliger (1484–1558) was a doctor, and commented upon Aristotle's History
of Animals and Theophrastus on Plants: his son Joseph Justus was the great classical
scholar.
5 The general outline of these works had been sketched out thirty years earlier.

in touch with Joseph Mead,[1] who was described by Worthington[2] as 'an accurate philosopher, a skilful mathematician, an excellent anatomist' and was a man of originality and influence. He cites Boyle's experiments with the air-pump; Girolamo Cardano's evidence on the Bird of Paradise,[3] the Camel and the Mole; and Erasmus and Johnstone for the antipathy of the ape to the snail.[4] The list of plants with which he illustrates his sections on signatures is presumably from Dioscorides; his human anatomy probably from Vesalius.[5] Otherwise Pliny and Aristotle are his masters, and he cites a few other details from the classics. His one allusion to the method of creation refers to the 'teeming' of the wombs of the earth in the manner of Milton.[6] Most of his arguments deal with the usefulness of nature to man;[7] in this he sees God's purpose most clearly: but many are genuine, if sometimes in fact fabulous, examples of the adjustment of form to function. Much the longest of them deals with the Bird of Paradise; and there against better evidence he accepts Cardano's story of its footlessness and life in the air, of the making of its nest in the hollow of the male's back, and of the binding together of male and female during incubation by the streamers, the two wiry tail-feathers, of the male.[8] The impression which the book gives is of a diligent and well-read but somewhat credulous amateur, a man with a great love of nature but without first-hand scientific training or much understanding of physical or physiological principles.

It is not difficult to conjecture the original form of the Commonplaces which Ray delivered in Trinity College Chapel on this theme, or indeed to infer which passages have been added or rewritten. His text is the great utterance of the Psalmist: 'How manifold are thy works O Lord: in wisdom hast thou made them all.'[9] He begins with a brief illustration of the first clause—a section which has been revised, though perhaps not very

---

1 Died 1638: has strong claim to be considered the first of the Cambridge Platonists.

2 Life prefixed to his Works.

3 *De Subtilitate* x, refers also to Pighafetta, Scaliger, Hernandez, Eusebius of Nürnberg, Aldrovandi, and De l'Ecluse.

4 Citing Johnstone, *Hist. Nat. de Quadruped.* III, 2 (published in 1652): Ray regarded Johnstone as 'a mere plagiary and compiler of other men's labours': cf. *F.C.* pp. 101, 160.

5 More knows of the valves in the veins, but does not refer to Harvey.

6 Cf. *Ant.* p. 68 and *Paradise Lost*, VII, ll. 453–98. Milton left Christ's College in the year after More entered it: but his epic was not published till 1667.

7 Yet he insists that 'the creatures are made to enjoy themselves as well as to serve us': *Ant.* p. 73, quoted by Ray, *Wisdom*, p. 128.

8 Johnstone, *De Avibus*, p. 171, had said that the bird tied itself to a tree by these and hung down from them when it wanted a rest. The species is probably *Paradisea minor*.

9 Ps. civ, 24.

recently.[1] Insisting that the universe is not, as Descartes had urged, limitless, he yet admits that the stars are almost infinite in number, myriads remaining to be discovered by improved telescopes, and argues that each fixed star is the centre of a planetary system. Beasts he places at 150 species, birds at 500, insects at more than 10,000;[2] and he adds a comment upon the greatness of these numbers and the amazing variety and adaptation to their environment of the several types. Then he turns to his main theme, and first examines and rejects the theories of Aristotle, Epicurus and Descartes, quoting Tillotson and Wilkins as against the first, Cicero, *De Finibus* and *De Natura Deorum*, Cudworth and Stillingfleet against the second, and Boyle, Cudworth, Harvey and Lower against the third. His comment is not unsympathetic and ends with the argument for a Plastic Nature to which we have already referred, and with a confutation of the Cartesian theory that animals are mere automata similar to that in his *Synopsis Quadrupedum*.[3] Next he surveys, somewhat in More's manner, the various divisions of the natural order, prefacing his account with an interesting section upon the 'Atomic hypothesis', in which rejecting the uniformity and the great multiplicity of atoms he concludes: 'I do not believe that the indivisible particles are exceeding numerous [i.e. of many different sorts]: but possibly the immediate component particles of the bodies of the plants and animals may be themselves compounded.'[4]

Treating first the heavenly bodies he accepts enthusiastically 'the new hypothesis'— of Copernicus and Galileo—and predicts that it will explain the problems of stellar and solar physics when its true method has been fully discovered. He ridicules the notion of the sun whirling round the earth; treats the moon as 'somewhat like the earth' and 'in all likelihood' inhabited and notes its influence upon tides and possibly on the weather; and argues that the courses of the planets, the movements of the fixed stars and the eclipses of the sun and moon display, as Cicero insisted, 'an excellent and divine reason'.[5] The four 'elements' of traditional usage, which he next discusses, he describes as 'four great aggregates of bodies of the same species' rather than 'principles or component ingredients'. Of fire he has little to say except of its use—the ignorance of any true concept of its physical character leads him to identify it with 'the vital flame residing in the blood'. Of air he notes its value as the fuel of the vital flame, essential to fishes which die in a sealed vessel, as Rondelet had proved, to insects breathing through 'many orifices on each side their bodies', the

1 He takes the number of plants from Bauhin's *Pinax* as 6000 and after criticising Bauhin for including many twice over surmises that there must be double this number or perhaps more. He makes no mention of his own *Historia*, in which over 6000 species had been described apart from those in the Appendix.

2 He refuses to do 'as most naturalists have done' and count caterpillars and perfect insects distinct species.
                                    3 *Wisdom*, pp. 38–40; *S.Q.* pp. 2–13.

4 *Wisdom*, p. 45.   5 *Wisdom*, pp. 45–50.

trachea, so that if these are stopped with oil or honey the insect dies, or if partially stopped is convulsed and locally paralysed,[1] to plants which 'have a kind of respiration', as was discovered by Malpighi, and to the foetus in the womb.[2] Here he has a long discussion of the placenta and other organs of embryonic life, with the remark that 'the foetus doth live as it were the life of a fish', and with discussion of the means by which the diaphragm and muscles of respiration are set in motion. Of water and earth he speaks very briefly. The next section, on Meteors,[3] deals with rain and wind, rain being defined as 'nothing else but water by the heat of the sun divided into very small invisible parts, ascending in the air, till encountering the cold it be by degrees condensed in clouds and descends in drops'. He notes the distillation of salt water in rain and the failure of science to distil sea-water artificially. Of wind he says nothing, except its obvious uses. Ray was not an expert physicist; indeed in his time physics and chemistry hardly existed: the chief merit of these chapters is their avoidance of the cruder fancies of the tradition. With this part of his subject he dealt more fully in the *Discourses* of the following year.

When he proceeds to the examination of geological, botanical and zoological phenomena his record increases in value. Of 'stones, metals, minerals and salts' he mentions briefly the varieties of structure and form, including a note of warning as to 'those formed in imitation of shells...if these be not shells themselves petrified as we have sometimes thought'.[4] He then discusses the 'loadstone' 'endued with an electrical or attractive virtue' and admits that 'the hypotheses invented to give an account of its admirable phenomena seem to me lame and unsatisfactory'...'its attractive power of iron was known to the ancients, its verticity and direction to the poles of the earth is of later invention'. Rejecting the idea of the transmutation of metals, he adds a note to say that this was written above thirty years ago 'when I thought I had reason to distrust whatever had been reported or written to affirm it'. Thus he, unlike Boyle and Newton, had then repudiated the whole business of alchemy in searching for a means of turning baser metals into gold; and there is nothing in the remarks that follow this note to suggest that he had changed his mind.

It is naturally with 'vegetables or plants' that his thought becomes rich and suggestive. His purpose, to show that nature is neither accidental nor arbitrary, expresses itself first in a series of questions. Why does each species produce the same leaf, flower, fruit and seed 'though you translate

1 This seems to have been an experiment of his own, though he quotes Pliny for the fact.

2 'Which I have often dissected': *Wisdom*, p. 57—probably with his friend Walter Needham in 1654: cf. above, p. 46.

3 Meteor, like fossil, is a word which had not at this time gained its specialised meaning.   4 *Wisdom*, p. 68.

it into a soil which naturally puts forth no such kind'? Why does each maintain its proper growth and size so that 'you can by no culture or art extend a Fennel stalk to the stature and bigness of an Oak'? Why are some long-lived, others 'annual or biennial'? What determines the situation of the leaves 'by pairs or alternately or circling the stalk', or of the flowers 'singly or in company to come forth on the bosoms of the leaves or on the top of the branches', the shape of leaves and petals, 'the figure and number of the *stamina* and their *apices*, the figure of the *Style* and seed-vessel and the number of cells into which it is divided'? 'There seems to be necessary some intelligent Plastic Nature which may understand and regulate the whole economy of the plant.'

Then follows a brief account of plant physiology, seeds and seed-leaves, roots, fibres, 'vessels to contain the proper and specific juice, and others to carry air for respiration'; a section closely following More on their beauty and symmetry; a further treatment of 'the masculine or prolific seed in the apices of the stamina' which he had admitted in the *Historia Plantarum*, of the provision for the protection and distribution of the female seed (he mentions 'the pulp of the fruit or pericarpium' and 'the pappose plumage growing upon the tops of some'), and of the minuteness of the seeds of mosses that grow upon roofs ('when shaken out they appear like vapour and ascend...so that we need not imagine they sprung up spontaneously there'). Finally, after a paragraph on adaptation—the various means of propagation, the climbing and thorn-bearing habits—he dismisses the doctrine of signatures as 'rather fancied by men than designed by nature to point out any such virtues as they would make us believe', and referring presumably to the *Cambridge Catalogue* says: 'I have elsewhere, I think upon good grounds, rejected them; and finding as yet no reason to alter my opinion I shall not further insist on them.'

Still more interesting is the final section of this part of his book, that on Animals. Continuing his argument against the accidental or arbitrary in Nature, he draws attention first to the provision for the maintenance of the species, to the facts of sex and the sexual instinct, of gestation and the nurture of the young. Apropos of this he quotes 'two or three considerable observations'; first, as to the value of the yolk as enclosed in the belly of the newly hatched chick and supplying it with food while it adapts itself to eating and digesting other diet; secondly, as to the feeding of young birds in the nest when the parents 'though in all likelihood they have no ability of counting...yet do not omit one of them'; thirdly, 'the marvellous speedy growth of birds hatched in nests and fed there till they be fledged...at which perfection they arrive within the short term of about one fortnight'; and finally, the skill with which the nests are placed and constructed (he describes the hanging nest of an Indian Weaver-bird, probably *Ploceus philippinus*), the ardour with which the young are brooded and fed, and the courage with which the parents 'fly in the face of a man that shall molest their young'—a thing 'contrary to any motions of sense or instinct of self-preservation'.

This naturally leads him to notice 'various strange instincts...directed to ends unknown to them'; that the calf pushes with his horns even before they shoot; that poultry and the partridge 'at the first sight know birds of prey and make sign of it by a peculiar note of their voice to their young who presently thereupon hide themselves'; that young mammals find and suck the paps; that 'ducklings though hatched and led by a hen leave her and go into the water though they never saw any such thing done before and though the hen clucks and calls and does all she can to keep them out'; that 'birds of the same kind make their nests of the same materials, laid in the same order and exactly of the same figure so that by the sight of the nest one may certainly know what bird it belongs to: and this they do though they never saw nor could see any nest made, that is though taken out of the nest and brought up by hand'. Finally, he gives a full account of the comb and cells of the hive-bee; and his own observation of the 'Tree-bee' (*Megachile willughbiella*) as already recorded;[1] and of the ant, which 'as all naturalists agree hoards up grains of corn for the winter, and is reported by some [Pliny being mentioned] to bite off the germen of them lest they sprout: which I look upon as a mere fiction; neither should I credit the former relation, were it not for the authority of Scripture, because I could never observe any such storing up of grain by our country-ants'. He concludes this section by reference to the squirrel and its 'hoards of nuts', to Boyle's account of the beaver[2] and to a miscellaneous list of traditional illustrations borrowed from More.

Next he discusses the 'care that is taken for the provision of the weak' —their burrows or spines, their speed and acuteness of sight and hearing, and their fecundity as compared with the rapacious. This leads to the general subject of the adaptation of form to function. After quoting, with comments, More's account of the Mole he adds notes on the 'Tamandua or Ant-bear' from Marcgraf and Piso,[3] on Woodpeckers who 'have a tongue which they can shoot forth to a very great length[4]...and at pleasure thrust deep into the holes of trees to draw out cossi or other insects, as also into anthills'[5]—he comments also on other points in their

1 *Phil. Trans.* v, no. 65, p. 2100 and *H. I.* p. 245; cf. above pp. 396–7.

2 In *S.Q.* p. 213 he notes its building of houses but without reference to Boyle.

3 Cf. *S.Q.* p. 241, where its long tongue and habit of eating ants is more briefly noted.

4 Cf. *Orn.* p. 136 for an accurate account of its anatomy; *F. C.* p. 115; above p. 315.

5 A good observation: the Green Woodpecker feeds its young entirely upon ants, gathered by its tongue and packed into its crop.

special structure, and on the Chameleon.[1] There follow paragraphs on the
flight of birds with special reference to the function of the tail in steering
and maintaining balance, and on the swimming of fishes, the air-bladder,
the fins and tail, and the muscular system. Finally, after a few more general
observations he decisively rejects as fabulous the belief that the Bird of
Paradise has no legs.

Dealing next with the usefulness to man of the animal creation he starts
with the objection 'that these uses were not designed by nature in the
formation of things: but that the things were by the wit of man accommo-
dated to those uses'—a very proper argument. First quoting More's
Appendix, wherein the matter is briefly treated, he urges that 'we only
find, not make these useful dependencies'[2]—that is, that things are thus
constituted whether we use them or no; and adds his own conviction that
'since we find materials so fit to serve all the necessities and employ the
wit of an intelligent being... which without them had been left helpless...
to them that acknowledge the being of a Deity it is little less than a
demonstration that they were created intentionally, I do not say only, for
those uses'—that is, that if we accept belief in God on other grounds it is
legitimate to pass from the fact that things are so constituted as to be useful
to the belief that they are so constituted in order to be useful: it is notable
that he posits a prior belief in God and that he does not limit the purpose
of creation to its utility.

He then breaks into an interpretation of the divine intention, which in
the manner of the book of Job he presents as a speech of God—a long and,
for Ray, unusually eloquent passage in which he sets out the way in which
man's life and character is conditioned and educated by his environment.
The passage begins: 'Methinks by all this provision for the use and service
of man the Almighty interpretatively speaks to him in this manner, I have
placed thee in a spacious and well-furnished world'; and four pages later:

I persuade myself that the bountiful and gracious Author of man's being and
faculties and all things else, delights in the beauty of his creation and is well
pleased with the industry of man in adorning the earth with beautiful cities...
if a country thus planted and adorned be not preferred before a barbarous and
inhospitable Scythia... then surely the brute beasts' condition and manner of
living is to be esteemed better than man's, and wit and reason was in vain bestowed on him.

We have drawn attention to this passage because, quaint as it may appear to a modern scientist, it represents an attitude towards Nature which
is new and vastly important. This delight in the worth of the world as

---

1 Quoting Panarolus, whom he quotes also, along with others, in the long account
in *S.Q.* pp. 276–83.
2 In ch. XI, par. 2 of the Appendix to *Antidote* (p. 174).

aesthetically satisfying, intellectually educative and spiritually significant reflects the best Hebrew, Greek and Christian thought, but is in strong and striking contrast to the philosophy and religion both of the Catholic and the Protestant tradition. The antithesis between the natural and the supernatural had led Christendom to a scorn and depreciation of Nature. From the third to the sixteenth century the note that Ray strikes had been almost entirely mute. To Augustine as to Luther nature belonged to a plane irrelevant if not actually hostile to religion. Its beauty was a temptation, its study a waste of time, its meaning so distorted that there is a radical difference between Nature and grace. The best Puritans, though their theology was Calvinistic, yet derived from their profound consciousness of the presence of God a valuation of Nature incompatible with their doctrinal position; and no one would suggest that the Elizabethans or Milton failed to appreciate or find religious value in the natural order. But the direct insistence upon the essential unity of natural and revealed, as alike proceeding from and integrated by the divine purpose, had not found clear and well-informed expression until Ray's book was published. It gave to the development of science a status and sanction of the highest value; and, so far as Britain is concerned, freed it from the conflict with ecclesiastical authority until it had become an established element in the national character.

After a section, obviously of later date and rather clumsily interpolated here, on the 'incredible smallness' of the organisms lately disclosed by Hooke and Leeuwenhoek, Ray concludes this first portion of his book with a defence of the value of scientific studies and a plea for their extension. It is addressed to his hearers in Cambridge and probably formed part of the original 'Common Places'.[1] The language extends and enriches that of the final paragraph of the preface to the *Cambridge Catalogue*;[2] and contains some of his best-known utterances:

It may be (for aught I know and as some Divines have thought) part of our business and employment in eternity to contemplate the Works of God....I am sure it is part of the business of a Sabbath-day....Let it not suffice to be book-learned, to read what others have written and to take upon trust more falsehood than truth, but let us ourselves examine things as we have opportunity, and converse with Nature as well as with books....Let us not think that the bounds of science are fixed like Hercules his pillars and inscribed with a *Ne plus ultra*.... The treasures of Nature are inexhaustible....I know that a new study at first seems very vast, intricate and difficult: but after a little resolution and progress, after a man becomes a little acquainted, as I may so say, with it, his understanding

1 But some phrases about the study of language and oratory occur also in a letter to Robinson of 15 Dec. 1690 (*Corr.* p. 229: cf. *Wisdom*, p. 123)—that is, while he was preparing *Wisdom* for the press.
2 See above, p. 83.

is wonderfully cleared up and enlarged, the difficulties vanish, and the thing grows easy and familiar....Some reproach methinks it is to learned men that there should be so many animals still in the world whose outward shape is not yet taken notice of or described, much less their way of generation, food, manners, uses, observed. If man ought to reflect upon his Creator the glory of all his works, then ought he to take notice of them all and not to think anything unworthy of his cognizance.[1]

After this Ray turns to the two 'particular pieces of Creation...the whole body of the earth and the body of man' which he wishes to consider in greater detail—an arrangement which reflects More's order and no doubt goes back to his original discourses.

The most interesting part of his account of the earth is his spirited advocacy of the 'new hypothesis'—of Copernicus and Galileo—as against the old, which he allows to be at first sight in accordance with common sense and tradition and with Scripture. He shows the absurdity of geocentric ideas and their inability to explain the facts of planetary motion 'so that whosoever doth clearly understand both hypotheses cannot, I persuade myself, adhere to the old and reject the new without doing some violence to his faculties'. He argues that the senses are often mistaken, 'the sun or moon appear no bigger at most than a cart-wheel and of a flat figure...and to give a parallel instance, when the clouds pass nimbly under the moon, the moon itself seems to move the contrary way'. As to Scripture, he says, 'when speaking of these things it accommodates itself to the common and received opinions and employs the usual phrases and forms of speech (as all wise men do though in strictness they be of a different opinion) without intention of delivering anything doctrinally concerning these points or confuting the contrary...howbeit because some pious persons may be offended at such an opinion, I shall not positively assert it, only propose it as an hypothesis not altogether improbable'— a pronouncement unusually enlightened for a Churchman of Ray's date, and typical of him in its desire to avoid controversy.

As to the human body he first produces eight general arguments to show its perfect adaptation to its functions: then deals with certain organs separately: and devotes fullest attention to the eye, giving a very detailed account of its structure, describing the various parts and discussing their correlation. This, though a few phrases like the comparison of the blackened inner surface of the *Tunica uvea* to a tennis-court[2] are borrowed from More, is evidently based upon considerable study. Pierre Herigone, 'that learned mathematician',[3] whose *Optica* is quoted, is the only

1 *Wisdom*, pp. 124–30.
2 *Wisdom*, p. 174 from *Ant.* II, xii, 3 (p. 80).
3 His *Cursus Mathematicus* was published in Paris in 1644.

authority¹ cited by name; and the source of Ray's knowledge (or indeed of More's before him) is unknown.²

After closing his survey of the human body with an account of the sex-organs and the mechanism of embryonic life Ray adds a note: 'My observation and affirmation is that there is no such thing in nature as aequivocal or spontaneous generation, but that all animals, as well small as great, not excluding the vilest and most contemptible insect, are generated by animal parents of the same species with themselves.' This is, as we have seen, one of Ray's strongest convictions, but he seldom expresses it so forcibly. It would be interesting to know whether his words belong to the earliest stratum of his book. Presumably they do not; for in support of them he quotes Redi's famous research,³ published in Florence in 1668, and Wilkins's statement that his experiments had been repeated by members of the Royal Society; and describes Malpighi's experiment with earth, which refuted belief in the spontaneous appearance of plants. He reiterates and expands this argument in the *Synopsis Quadrupedum*⁴ and in later editions of the *Wisdom of God*.

The remaining section of the book is given to the religious consequences of the conclusions reached—the joy of thankfulness, the duty of right conduct, and the value of faith. It is redeemed from formalism and platitude by his transparent sincerity and robust practicality.

Ray had begun work on this book in the autumn of 1690.⁵ He mentions it—'a small tract which I have drawn up'—as 'come abroad' on 27 May 1691, being published by Samuel Smith, who had now become his regular agent.⁶ By August he can speak of it as having 'found some acceptance';⁷ and in April 1692 'the second edition with large additions' was ready for publication.⁸ During these months he had effected a thorough revision, dividing the book into two parts, providing a detailed synopsis of its contents, and inserting a large amount of new material, ranging from passages of a sentence or two to sections of several and in one case of

1 Boyle, *On Final Causes*, is quoted for the use of the nictitating membrane in birds and frogs. Cf. *Works* (ed. 1772) v, p. 435.

2 The earliest book in English seems to be Richard Banister's 'breviary of the eyes', *Treatise of* 113 *diseases of the eyes*, a translation of Jacques Guillemeau's *Traité*, published in 1662. William Briggs, fellow of Corpus Christi College, Cambridge, 1668, published his *Ophthalmographia* in 1676.

3 *Generazione degl' Insetti*, reviewed in *Phil. Trans.* v, no. 57, pp. 1175–6.

4 He quotes the *Wisdom of God* in S.Q. p. 20.

5 So letter to Robinson, *F.C.* p. 294.

6 To Lhwyd, *F.C.* p. 219: he seems to have sent a copy to Robinson in March; *F.C.* p. 294.

7 *F.C.* p. 221.

8 *F.C.* p. 226: he adds 'I have no copies of it for myself, and so cannot present my friends with any; which I am sorry for'.

seventy[1] pages, and containing both quotations from other writers and a large amount of personal observations. Ray in compiling his books on plants and animals must have used loose sheets of paper and much scissors and paste; and he obviously employed similar methods in the preparation of new editions of this book. But the interpolations, though occasionally they damage the proportion and arrangement, are skilfully inserted and can only be discovered by careful collation.

The personal observations range over a wide field, and many of them are full of interest. Appended to his discussion of embryonic life is a section[2] dealing with the permeation and circulation of air, in which he refers to the fishes found in subterranean rivers and to the 'fossil fish', the Misgurn that he had seen at Nürnberg.[3] A few pages later is an account of his observations on the character of the sea-floor and of the effect of the tidal movements of water upon the structure of marine plants; and he expresses his conviction that 'in the great depths of the sea there grow no plants...and are no fish'.[4] Then he has an interesting section on the number of eggs laid by birds and on the instincts which lead butterflies to lay their eggs on the food-plants appropriate to their caterpillars, and bees to store up 'Erithace which we English call Bee-bread' for their eulae;[5] and shortly after certain notes on birds—the habit of swallowing 'little pebble-stones or other hard bodies' and 'they first try them in their bills to feel whether they be rough or angular for their turns'—the presence in certain species of digestive fluid in the glandular surface of the gullet—the removal of faeces from the nest by the parents or their ejection outside it by the young[6]—the diving of ducks when pursued by a dog and their habit of only raising the bill above water till the danger has passed.[7] In dealing with structure he adds notes of the adaptation of the Swallow for catching insects in flight and 'the sign of rain when this bird flies low in which there may be some truth because the insects may when the superior air is charged with vapours have a sense of it and descend near the earth'; of the plumage, flattened leg-bones and backward feet of 'Colymbi or Loons'; of the hollow bones and air-sacs of birds, of the plumules and hooks of their feathers, of the 'oily pap or liniment' in the glands on the rump, and of the use of the tail in steering;[8] of the adjustment of the tail

---

1 In Part II, pp. 73–143: a large part of it deals with spontaneous generation.

2 *Wisdom*, II, pp. 68–70.   3 Cf. *H. Pisc.* p. 118.

4 *Wisdom*, II, pp. 72–4.

5 *Wisdom*, II, pp. 109, 114–15: that on birds includes Lister's observations of the Swallow which when the daily egg was removed 'proceeded to lay nineteen in succession', already reported in *Ornithology*, Preface.

6 It is perhaps permissible to say that when I first recorded this of young Hawks I thought the observation new!

7 *Wisdom*, II, pp. 120–2, 128.   8 *Wisdom*, II, pp. 134–41.

and blubber in Cetaceans;[1] of the scaly bodies of insects; and of 'the proportioning the length of the neck to that of the legs'.[2]

In the second part of the book (in which in this edition the pagination begins afresh) there is the long insertion, the first section of which deals with Spontaneous Generation. Here he justifies his fuller treatment 'because some, I understand, have been offended at my confident denial of all Spontaneous Generation, accounting it too bold and groundless'.[3] He first sets out the evidence of Swammerdam, quoting the *History of Insects* in its French version, of Malpighi on galls, and of Lister in his edition of Goedart, and alluding to the familiar experiments of Redi. Then he deals with the generation of frogs supposed to be rained down from the clouds, and describes it in detail from the observations made by himself and 'my learned friend Mr John Nid, Fellow of Trinity College, long since deceased'. After quoting a letter from Tancred Robinson he cites his own record of larval parasites in the caterpillars of *Pieris brassicae* from the *Cambridge Catalogue* and adds that 'the fly may with the hollow and sharp tube of her womb punch and perforate the very skin of the *eruca* and cast her eggs into its body'.[4] As to toads found in stones, while asserting that he is not fully satisfied of the fact, he suggests that the creature when young may find a small hole in the stone, 'creep into it as a fit *latibulum* for the winter, grow there too big to return and so continue imprisoned'.[5] As to plants, he admits more hesitation; yet if spontaneous generation were possible 'what need of all that apparatus of vessels, preparation of seed, and as I also suppose distinction of masculine and feminine,[6] or of the seed-leaves which are found in all plants except "the grass-leaved tribe" and imply a previous seed'. The rest of the long insertion contains a large number of further observations of special structures and adaptations which reinforce his previous treatment of the subject in the earlier part of the volume.

The larger proportion of the new material consists of quotations from and references to other and contemporary scientists; and here the additions supply an impressive testimony to the extent and variety of Ray's knowledge. His previous books had proved his mastery of the work done in botany and zoology; but had not revealed his familiarity with the whole range of anatomical and medical research. These insertions contain a tolerably complete survey of the history of science down to the date of their composition. Some of them are mere lists of references such as a modern writer would relegate to footnotes. Most contain proof that he had read carefully the writers from whom he quotes and had selected with

1 *Wisdom*, II, pp. 142–3. 2 *Wisdom*, II, pp. 148–51.
3 *Wisdom*, II (2), p. 74. 4 *Wisdom*, II (2), p. 92.
5 *Wisdom*, II (2), pp. 92–3. 6 *Wisdom*, II (2), p. 94.

discrimination the best and most reliable authorities in each department. They should be studied in connection with the essays prefixed to his *Synopsis Quadrupedum*, which were being written at the same time and in two cases deal with the subjects here introduced.

Naturally his own friends, Robinson,[1] Sloane,[2] Lister,[3] Sir Thomas Browne,[4] Peter Dent,[5] and John Aubrey,[6] figure in the list. Then there are the great doctors of his own country, Boyle, to whom he refers with a peculiar veneration;[7] Francis Glisson, Regius Professor of Physic in Cambridge for forty-one years from 1636 though living in London;[8] Thomas Wharton,[9] the eminent anatomist, whose *Adenographia* was published in 1656; Thomas Willis, Sedleian Professor at Oxford, an original member of the Royal Society, whose work on the intestines[10] appeared in 1659; Richard Lower, Willis's pupil, whose Tractate on the Heart[11] was published in 1669 and whose experiments on blood-transfusion attracted the attention of the Royal Society for several years; William Croone, who had been with Ray in France and whose *De Ratione Motus Musculorum* here cited[12] appeared in 1664; and a much younger Cambridge man, Clopton Havers, from whom the Haversian canals are named and whose *Osteologia Nova* here quoted by Ray[13] was only published in 1691 and became a standard work. These were all men with whom he must have had some personal contacts.

Of other scientists he quotes several botanists not previously mentioned by him—Reinerus (Reinier) Solenander,[14] the sixteenth-century doctor, and Hendrik van Rheede van Draakenstein, 'that immortal patron of natural learning' and creator of the *Hortus Malabaricus*;[15] and as students of plant physiology, in addition to Malpighi and 'our own ingenious

1 He quotes the MS. of his travels in Italy, *Wisdom*, II, pp. 83–4, and letters to himself, pp. 194–5 and (2), pp. 83, 88–9.
2 *Wisdom*, II (2), p. 81, acknowledging a present of edible coleopterous larvae from America.
3 *Wisdom*, II, p. 109 and (2), pp. 75, 80–1.
4 *Wisdom*, II (2), p. 102, recording an experiment on a frog.
5 *Wisdom*, II (2), p. 127, on the rising and sinking of cartilaginous fishes in water.
6 *Wisdom*, II, p. 128, on the habits of the hunted hare—from papers privately lent.
7 *Wisdom*, II, p. 113: he had of course quoted Boyle in the first edition.
8 *Wisdom*, II (2), p. 64: the reference is to his work on the intestines.
9 *Wisdom*, II (2), p. 65, on the kidneys.
10 Referred to, *Wisdom*, II (2), p. 64.
11 Quoted in *Wisdom*, II (2), pp. 139–41; and apparently used by Ray in his own suppressed tract on Respiration: cf. above, p. 286.
12 *Wisdom*, II (2), p. 65: the title of the book is not mentioned.
13 *Wisdom*, II (2), p. 61: the reference is to the 'inunction and lubrification' of the joints.
14 *Wisdom*, II, p. 104: his works were collected as *Reineri Solenandri Consilia Medica*, Frankfurt, 1609.
15 *Wisdom*, II, pp. 192–3, in his account of the Coconut Tree.

countryman' Thomas Brotherton,[1] Claude Perrault, the famous architect, and Edme Mariotte. In regard to zoology he refers principally and freely to the French Academists of Paris,[2] though Joseph Duverney[3] is the only one of them to be named: this school, with its access to the rich material in the royal gardens, was doing work of supreme value both in quality and quantity. He refers also to 'Dr Moulins', that is, Allan Molines of Dublin, for the elephant,[4] and to 'Olaus Borrichius in the Danick Transactions' (the *Acta Hafniensia*) for the muscular strength of the Hedgehog.[5] But it is in human anatomy that his references are most numerous. In general he refers his readers[6] to 'the excellent figures of Spigelius and Bidloo'—Adrian van den Spieghel, whose complete works had been published in Amsterdam in 1645, and Govard Bidloo, whose *Anatomy* appeared in the same city in 1685 and who visited England[7] and became F.R.S. in 1701—to Michael Lyser's exposition of the subject,[8] and to the methods of inflating and injecting the organs developed by Jan Swammerdam, Caspar Bartholin and Anton Nuck—Bartholin's work having been collected and edited by his son Thomas in 1673 and Nuck's *Adenographia* published in 1691, both at Leyden. To Nuck's book Ray is indebted for a detailed account of the mammary glands[9] and to a paper of his recording experiments at Leyden for statements regarding the 'aqueous humor' in the eye-ball.[10] Dealing with the intestines he refers to Theodor Kerckring,[11] whose *Spicilegium anatomicum* was published at Amsterdam in 1670, and to Johann Conrad Peyer,[12] whose *Merycologia* is quoted in his *Synopsis Quadrupedum*.[13] For the kidneys he names Bellini[14] as the first discoverer of their tubular structure and adds references to De Graaf,[15]

---

1 *Wisdom*, II (2), p. 128, referring to B.'s experiments recorded in *Phil. Trans.* no. 187.

2 *Wisdom*, II (2), pp. 108–11, 126.

3 *Wisdom*, II (2), p. 39, on the structure of the ear: his *Traité de l'organe de l'ouie*, Paris, 1683, and in Latin at Nürnberg, 1684.

4 *Wisdom*, II (2), pp. 105–6. Cf. above p. 382.

5 *Wisdom*, II (2), p. 111: Borrichius was at once alchemist and doctor: he died in 1690.

6 *Wisdom*, II (2), p. 19.

7 He was physician to William III and visited Oxford: cf. Gunther, *Life of Lhwyd*, p. 320.

8 *Culter Anatomicus* (1653).

9 *Wisdom*, II, pp. 144–5: he quotes *Adenog.* ch. 2.

10 *Wisdom*, II (2), pp. 30–1.

11 Elected F.R.S. 1678.

12 *Wisdom*, II (2), p. 64.

13 Cf. above pp. 376–7.

14 Lorenzo Bellini, *Exercitatio anatomica de structura renum*, Amsterdam, 1665.

15 Regnier de Graaf, inventor of the syringe for injections, 1668: *Opera*, Leyden, 1677.

Rudbeck,[1] Bils[2] and others: for the muscles Steno[3] and Borelli:[4] and for the 'lymphaeducts' and dropsy Tulp,[5] Meekren,[6] Pechlin[7] and Blasius.[8] There are also a number of references to Leeuwenhoek[9] and to Malpighi.[10]

It is legitimate to infer from this very remarkable list not only that Ray did his best to keep up with the general progress of scientific work, but that he had a specially close connection with the great publishing activities of the Dutch schools at Amsterdam and Leyden, a fact which throws light on the publication in Holland of his own last *Methodus Plantarum Emendata*, printed as we have seen by the house of Waasberg in Amsterdam (though with London on its title-page) in 1703 through the kindness of Dr Hotton the botanist of Leyden. The history of the disposal of his *Historia Piscium* also proves the close contact between Dutch and British publishers, a contact that in fact goes back to the early part of the century and, after the Revolution, was powerfully revived. At this time the Dutch universities and medical schools were beginning to take the place hitherto held by Italy and especially Padua; and that Ray drew so largely upon them is not due solely to geographical and cultural affinities but to his appreciation of the soundness of their researches. If Malpighi and Borelli still kept up the pre-eminence of the Italians, Swammerdam and Leeuwenhoek in biology, and a number of those whom he quotes in medicine, were making contributions of the highest value.

Two more editions appeared during Ray's lifetime, and though additions were not numerous each contained new material. The third, published by Smith and Ben Walford in 1701, is slightly rearranged. The final section of Part I, dealing with the 'whole body of the earth',[11] is made the first section of Part II:[12] the pagination of the two parts is made consecutive:[13] the material annexed at the end of the second edition is distributed more

---

1 Olof Rudbeck, *Nova exercitatio anatomica exhibens ductus* etc., Leyden, 1654.

2 Lodewijk de Bils, *Observationes anatomicae*, edited by Steno, Leyden, 1680: Ray met him in Brussels in 1663, *Obs.* p. 10.

3 Nicolaus Steno, *Observationes anatomicae*, Leyden, 1662: he was the geologist.

4 Giovanni Alphonso Borelli, *De Motu Animalium*, Rome, 1680.

5 Nicolaus Tulp, *Observationes medicae*, Amsterdam, 1641.

6 Job van Meek'ren, *Observationes medico-chirurgicae* (Latin translation by Abraham Blasius, Amsterdam, 1682).

7 Johann Nicolaus Pechlin, *De purgantium Med. facultatibus*, Leyden, 1672.

8 Gerard Blasius, *Medicina generalis*, Amsterdam, 1661, and two anatomical books, 1673 and 1674.

9 *Wisdom*, II (2), pp. 69, 75, 123.

10 *Wisdom*, II (2), pp. 64, 75, 128.

11 2nd ed. pp. 175–206.

12 The title-page of Part II being altered to include it.

13 With a curious error: Part I runs from p. 17 to p. 208; Part II begins with p. 193.

THE WISDOM OF GOD   475

appropriately:[1] these are definite improvements. There are also a few additions of his own: observations on the wide geographical and climatic range of wheat;[2] on the migration of birds and of salmon with a reference to the annual return of the Gannet to the Bass Rock and the crossing of Quails from Italy to Africa 'lighting many times on ships in the midst of the sea';[3] on the various notes of birds, with observations on the calls and alarm notes of fowls, no doubt observed in his own poultry-yard at Dewlands;[4] on sleep, with a personal reference to the absence of pain and its return soon after awaking;[5] and notes on eternal life as not 'a torpid and unactive state';[6] on the importance of disproving spontaneous generation;[7] and on the occurrence, which he now admits, of toads enclosed in stones.[8]

Longer insertions are derived from the works of others, though these are far less numerous than in the former edition. He quotes twice from John Keill's *Examination of Burnet's Theory of the Earth*;[9] from John Cockburn's *Essays on the Nature of the Christian Faith*;[10] and from Edward Tyson's dissection of the 'Orang Outang or Pigmy';[11] and once from Sloane, with reference to a plant from Jamaica;[12] from Martens's *Voyage to Spitzbergen*;[13] from Boccone's *Natural Observations*;[14] from Pietro Marchetti on baldness;[15] from the famous Richard Bentley, then Master of Trinity;[16] and from César de Rochefort's *History of the Antilles*, reporting cases of longevity.[17] In addition, at the close of the book he has

1 Thus e.g. 2nd (2), pp. 123–4 = 3rd, pp. 138–9; 2nd, pp. 128–9 = 3rd, pp. 115–16; 2nd, p. 138 = 3rd, pp. 250–1; 2nd, pp. 113–14 = 3rd, pp. 301–3. The table of contents is altered to show the new division into parts, but not the new sequence, though the pages are correctly numbered.

2 *Wisdom*, III, pp. 125–6, based on a quotation from Pliny.

3 *Wisdom*, III, pp. 143–5.   4 *Wisdom*, III, pp. 178–9.

5 *Wisdom*, III, pp. 244–6.   6 *Wisdom*, III, pp. 191–2.

7 *Wisdom*, III, p. 328.   8 *Wisdom*, III, p. 329.

9 *Wisdom*, III, pp. 87–93, 208–9: published Oxford, 1698.

10 *Wisdom*, III, pp. 107–9, 252–4: *An Enquiry into the Nature, Necessity and Evidence of Christian Faith*: two parts, 1699.

11 *Wisdom*, III, pp. 232–4, 375–7: *Orang Outang sive Homo Sylvestris, or the Anatomy of a Pigmy*, London, 1699: the subject was a chimpanzee.

12 *Wisdom*, III, p. 214.   13 *Wisdom*, III, p. 210.

14 *Wisdom*, III, p. 235: published Bologna, 1684.

15 *Wisdom*, III, p. 255: *Observationum medico-chirurgicarum rariorum Sylloge*, Amsterdam, 1665; Ray worked with him at Padua in 1663–4.

16 *Wisdom*, III, pp. 367–73. Bentley delivered Boyle lectures, 'A Confutation of Atheism', of which Ray quotes from 'the structure of human bodies', in London in 1692.

17 *Wisdom*, III, pp. 205–7. There is a curious misprint on p. 205, 'Galileans' for 'Caribbeans', corrected in *Wisdom*, IV, p. 232. Ray probably quoted from the English translation by John Davies, London, 1666: the passage cited is pp. 342–3, a chapter on the age of the Caribbeans, quoting Vincent Le Blanc on Sumatra, Francis Pirard on the Brazilians, René de Laudonnière's *Voyage into Florida*, 1586, and 'Mapheus' (i.e. Vegio Maffeo, *De Educatione*, 1511). Ray associates longevity with equable temperature, and instances the bed-ridden as proof.

half-a-dozen pages dealing with reasons for the great number of species, especially among insects.

The fourth edition, 'printed by J.B. for Sam. Smith and sold by Jeffery Wale', preserves the same arrangement with correct pagination and contains relatively few additions. There is an apology to Boyle, based on a passage in his *Christian Virtuoso*, which Ray accepts as correcting his previously quoted opinion;[1] a short paragraph quoting Lister and Dr Jones for their views on pain;[2] a section of his own on food, disease and pain;[3] a quotation from Dr William Cooper on the regulation of the movement of the blood;[4] and an observation from his friend William Derham, Rector of Upminster, upon a vast number of frogs seen in Berkshire.[5]

We have examined the expansion of the book in some detail because it gives an impressive picture of the range of its author's interests and of the vigour which he preserved until the last days of his life. In 1691 the work had been an experiment. The rapid demand for new editions had given him the opportunity to develop and enrich the theme so that the book became a store-house of data covering the whole field of contemporary science and containing many points, especially of his own observing, which are startlingly modern in character. No one who reads his paragraphs about birds, for example, can fail to recognise his extraordinary flair for fastening upon issues of profound significance which we are still unable to interpret satisfactorily. How is it that the hen, recognising the hawk which she has never seen before, utters her peculiar alarm-note and that her chickens obey it and take cover? Or that the impression of a defective clutch stimulates the successive production of eggs—as in Ray's Swallow or the classic case of the Wryneck? What is the impulse which compels migration and what the compass by which its course to a particular islet is set? Ray fastens upon these and similar questions, as deserving the attention of scientists: they would be asked in almost identical language by any intelligent ornithologist to-day and we should hardly be in a better position to answer them. It is small credit to biology that in the main so little attention has been paid to them; for in fact if honestly examined they raise fundamental obstacles in the way of the orthodox doctrines loosely described as Darwinism, and perhaps of all purely mechanistic hypotheses. In any case no one who studies the mass of observations and enquiries here propounded can doubt that Ray was vastly more than a systematist, and that his work deserves the high opinion which has until lately been bestowed upon it.

1 *Wisdom*, IV, pp. 55–7: discourses published in 1690–1.
2 *Wisdom*, IV, pp. 374–5.   3 *Wisdom*, IV, pp. 281–3.
4 *Wisdom*, IV, pp. 318–21, quoting *Phil. Trans.* no. 280.
5 *Wisdom*, IV, pp. 365–7.

To ask the right questions is always an indication of real insight and often the most useful contribution to progress possible at the time. In Ray's day it is obvious that biology could achieve results only in the field of the identification and classification of species: beyond this it could but record observations of structure and behaviour and leave their interpretation to the future. Study of physiology and psychology, of organic processes and development must wait until chemistry and physics had emerged out of the twilight of alchemy and guess-work. While scientists had not investigated the chemical character even of the simplest inorganic substances or begun to explain the nature of combustion, any understanding of the functioning of organisms or of their origin and adaptation must be conjectural and premature. It was not an accident due to the dominance of Isaac Newton, nor was it wholly regrettable, that in 1719 Walter Moyle, once one of Ray's most ardent disciples, wrote "I find that there is no room in Gresham College for Natural History; Mathematics have engrossed all ":[1] biology must wait until preliminary studies had been carried forward. Nor is it surprising that in the 150 years after Ray's death the advance of science was almost solely in physics and chemistry, and that in the realm of his own achievements, botany and zoology, little was done except in enlarging the knowledge and revising the classification of flora and fauna and in describing their anatomy and habits. For intelligent solutions of the problems to which Ray had drawn attention the time was not yet ripe.

For the necessary study his work and particularly the very influential book which we have been considering provided a stimulus and a defence. He not only focussed attention upon a number of fascinating subjects, but he convinced the champions of the old outlook that such attention was a proper exercise of man's faculties, a legitimate field for Christian enquiry. His own honest and reverent mind, fearless in facing facts but slow to dogmatise prematurely or to reject established opinion until the evidence was clear, commended the new philosophy and encouraged less adventurous Churchmen to support its claims. When Bishop Butler expounded the implications of Ray's work in his *Analogy*,[2] when John Wesley made its message a part of his philosophy,[3] when Gilbert White gave it worldwide fame in one of the most popular of English prose

---

1 *Works* (ed. Sergeant), I, p. 422.

2 *The Analogy of Religion, natural and revealed, to the constitution and course of Nature*, 1736, probably the most important work in Christian philosophy of the eighteenth century.

3 In his *Survey of the Wisdom of God in the Creation* (for the composition of which cf. *Journals*, ed. N. Curnock, iv, p. 295) he not only pays a notable tribute to Ray's book but draws largely from it.

classics,[1] they demonstrated that the Church was ready to abandon its mediaeval *Weltanschauung* and reassert its faith in the worth of the works of the Lord. In consequence in Britain there was a century and a half of scientific progress undisturbed by theological controversies and fostered by the spokesmen of religion. During that time if applied science and the industrial revolution aroused occasional outbreaks of revolt against the machine, the love of birds and flowers, the desire to understand nature as well as to admire her products, and the discovery in such activity of rich recreational, educative and religious opportunities became a national characteristic. During that time there was developed a type of theology, of which *The Wisdom of God* is the first example, capable of giving appropriate expression to the Christian faith in a scientific age. This is John Ray's proper memorial.

NOTE. *Ray's scientific genius*

As an illustration of Ray's almost uncanny flair for points of real interest and as proof of the claim that he was a man of genius far ahead of his time it may be well to take one of his characteristic observations and examine it more closely.

In February 1676, when he was at Coleshill, Lister wrote to him to acknowledge his gift of the Latin *Ornithologia* and to send a number of notes about birds. Among these is the following: 'One and the same Swallow I have known by the subtracting daily of her eggs to have laid nineteen successively and then to have given over.'[2] This Ray promptly inserted in the Preface of the English version of the *Ornithology*:[3] and years later quoted in the third edition of his *Wisdom of God*,[4] making it clear that he appreciated the remarkable significance of the fact. He wrote:

Another experiment I shall add to prove that though birds have not an exact power of numbering yet have they of distinguishing many from few and knowing when they come near to a certain number: and that is that when they have laid such a number of eggs as they can conveniently cover and hatch, they give over and begin to sit: not because they are necessarily determined to such a number; for that they are not, as is clear because they have an ability to go on and lay more at their pleasure.

1 *The Natural History of Selborne*, 1789, 'has had up to now about seventy editions' (Fisher, *Birds as Animals*, p. 5).
2 *Corr.* p. 117.   3 P. 9 (unnumbered).
4 Pp. 131–2.

Then after quoting the Hen he adds: 'This holds not only in domestic and mansuete birds, for then it might be thought the effect of cicuration[1] and institution, but also in the wild; for my honoured friend Dr Martin Lister....'

The first subsequent reference seems to be that in Edward Jesse's *Gleanings in Natural History*,[2] published in 1832. He quoted Ray's record; cited examples of the same phenomenon in the case of the Lapwing, Blackbird, Lark and Long-tailed Tit; and said that it was noticed by Sir Andrew Halliday (physician to William IV and author of several valueless books) 'about twenty years ago'. This attracted no attention. But in December 1833 J. D. Salmon of Stoke Ferry, Norfolk, reported a similar remarkable egg-laying in the case of a Wryneck.[3] This was quoted at length by W. Yarrell in the first edition of his *History of British Birds*.[4] Alfred Newton, who produced the fourth edition of Yarrell, condensed this account but added several other examples to it, including the famous case in which under similar stimulus the Wryneck laid up to forty-two eggs consecutively in each of two successive years 1872 and 1873.[5] He also mentioned the fact in his *Dictionary of Birds*. The popularity of these two books gave wide publicity to the record and egg-collectors proceeded to take advantage of the habit.

Nevertheless, although a few more reports were published,[6] the physiological interest of the observation does not seem to have been fully appreciated until Dr F. H. A. Marshall drew attention to it in his Croonian Lecture in 1936.[7] In the section of this lecture dealing with 'the effects of exteroceptive stimuli in altering the succession of the physiological processes which constitute the sexual cycle' he cites Ray's original record and the others here mentioned; claims that the evidence is known to all ornithologists; and discusses its significance both in relation to sexual behaviour in general and to the particular processes by which the result is effected. It is hardly to the credit of physiology that so remarkable an observation, fully attested by a mass of evidence, should have been so long

1 I.e. taming—a good example of Ray's use of rare words: cf. *New English Dictionary*, II, p. 414.
2 Vol. I, p. 191.
3 *Magazine of Natural History*, VII, pp. 465–6.
4 Vol. II, p. 154.
5 Vol. II, p. 491; references to later cases are given in detail.
6 O. Davie, *Nests and Eggs of N. American Birds*, Columbus, 1898 (a Flicker laying 71 eggs in 73 days); E. Witschi, *Wilson Bulletin*, XLVII, p. 183 (1935), dealing with the Sparrow: neither of these seems aware of Ray's observation—or of any others than their own; H. Friedmann, *The Cowbirds*, 1929, p. 184, divides birds into determinate and indeterminate in this respect, but his evidence seems insufficient and unreliable. I owe these references to Dr Marshall.
7 *Phil. Trans.* B. 1936, vol. 226, p. 443.

ignored, and should not even yet, as Dr Marshall informs me, have found a place in the text-books. It is notorious that biology has suffered seriously by the divorce between field-studies and the laboratory—the very divorce against which Ray continually protested. But it is an irony that an observation whose importance he rightly emphasised should have received no attention for more than one century and no scientific study for more than two.

CONCLUSION

There is little more that need be said. Derham printed the confession which the Rev. William Pyke, who had succeeded Plume as Rector of Black Notley in 1686, reported as made by Ray on his death-bed. It is a plain statement of faith and of loyalty to the Church of England such as his whole life corroborates. Sloane, who sent an offer of sympathy and help to Margaret Ray, preserved a few letters from her in which she consulted him about approaching Sir Thomas Willughby for the payment of the half year's annuity, told him that the books were to be sold and were being catalogued by Dale,[1] that the insects and all the papers about them had been delivered to Dale for his (Sloane's) use, and that Willughby's papers were safe and would be returned when instruction was given. Later, when Sir Thomas had responded 'in charity' to Sloane's petition, she wrote again to express gratitude, to return the papers and to report that 'the circumstances of the family cannot but be strait when Mr Ray did not leave £40 per year among us all out of which taxes, repairs and quit-rents make a great hole'.[2] By his will dated 13 April 1704 and proved in the Commissary Court of the Bishop of London for Essex,[3] Ray had left £4 to the poor of the parish and £5 to the Library of Trinity College; Dewlands to his wife and afterwards to his daughters; and to them lands in Hockley, 'Bird's lands' at Black Notley[4] and £200 respectively. Mrs Ray was still alive in 1727 when the youngest daughter, Jane, who had married the Rev. Joshua Blower, Vicar of Bradwell juxta Coggeshall, wrote two letters to Sloane.[5] Jane's grand-daughter, apparently the last descendant of the naturalist, was still living, 'a tall quiet maiden lady, of some seventy years ripening',[6] in Essex in 1851. Of the others it appears that 'Margaret married John Thomas, of Langford, yeoman, and Catharine Thomas Beadle, of Billericay, farmer'.[7]

Having been buried, according to Dale[8] by his own choice, in the churchyard rather than the chancel, a small monument was erected over him at the initiative of Bishop Compton and by subscription from his friends; and an inscription, dignified, lengthy and more appropriate to the

1 Some 1500 volumes were sold by auction in London about two years after his death.
2 *Corr.* p. 478.
3 Under present conditions it has proved impossible to check this record of his will, which is drawn from the article by Mrs Pattisson in the *Englishwoman's Magazine*, II.
4 Cf. above p. 180.
5 Cf. *Proceedings of the Essex Field Club*, IV, pp. clxii–iii.
6 Cf. *The Cottage Gardener*, V, p. 221.
7 So *Englishwoman's Magazine*, II, p. 274.   8 *F.C.* p. 7.

taste of the age than to the character of its subject, was composed by the Rev. William Coyte. It is printed in full and in an English version in the article on Ray in the *General Dictionary*.

The story of his posthumous books, the *Historia Insectorum* and the *Synopsis Avium èt Piscium*, has already been told; and enough has been said in the Introduction of the neglect which has deprived us of any adequate record of his life.

GENERAL INDEX

[N.B. Names of recent writers are given with initials only.]

INDEX OF FLORA

Helianthemum guttatum, 250
  polifolium, 245
Helleborus foetidus, 122
Herminium monorchis, 122, 150, 260
Herniaria glabra, 144, 245, 293
Hieracium murorum, 122, 143
Hippocrepis comosa, 127
Hippuris vulgaris, 96
Hottonia palustris, 88
Humulus lupulus, 99
Hutchinsia petraea, 250, 265
Hydrocharis morsus-ranae, 91
Hydrocotyle vulgaris, 128
Hymenophyllum tunbridgense, 245 n.
  unilaterale, 245
Hypericum androsaemum, 125
  calycinum, 285
  elodes, 92
  hirsutum, 122
  montanum, 124, 127
  perforatum, 124
Hypochoeris maculata, 122

Iberis amara, 265
Illecebrum verticillatum, 128, 141
Impatiens noli-tangere, 114, 266
Indigofera anil, 303
Inula crithmoides, 125–6
Iris foetidissima, 6
  germanica, 235
  pallida, 235
  susiana, 235
  tuberosa, 235
Isatis tinctoria, 265
Isoetes lacustris, 248, 266

Juncus acutus, 126
  filiformis, 245
  maritimus, 126
Juniperus communis, 236
  sibirica, 236

Lactuca saligna, 86
Lamium amplexicaule, 88
Larix europaea, 236
Lastraea thelypteris, 96
Lathyrus hirsutus, 141
  montanus, 121
  nissolia, 122
  palustris, 96
Laurus nobilis, 239
Lavatera arborea, 38, 118, 126, 144
Lepidium campestre, 265
  latifolium, 265
  ruderale, 128, 265
  smithii, 265
Ligustrum vulgare, 102
Limosella aquatica, 122
Linaria cymbalaria, 96, 198
  elatine, 198
  repens, 129

Linum bienne, 128
  perenne, 93, 205
  usitatissimum, 234
Liparis loeselii, 93
Liquidambar styraciflua, 240
Listera cordata, 118
Lithospermum purpureo-caeruleum, 124, 128
Lloydia serotina, 265
Lobelia dortmanna, 154, 266
Luzula pilosa, 121
Lychnis alba, 234
  chalcedonica, 234
  dioica, 234
  flos-cuculi, 234
Lycopersicum esculentum, 227
Lycopodium alpinum, 113, 119
  clavatum, 112
  selago, 112
Lysimachia thyrsiflora, 246, 265
Lythrum hyssopifolia, 93, 199

Malachium aquaticum, 87, 261
Malaxis paludosa, 227, 245
Malva moschata, 122
Mammea americana, 238
Mangifera indica, 237
Manihot utilissima, 239
Matthiola sinuata, 125, 264
Meconopsis cambrica, 125
Medicago hispida, 130
  maculata, 98, 126
  minima, 92
  sylvestris, 268
Melittis melissophyllum, 129
Mentha piperita, 263
  rotundifolia, 144, 260
Mercurialis annua, 144
Mertensia maritima, 154
Metroxylon sagus, 235
Meum athamanticum, 149
Moenchia erecta, 260
Montia fontana, 122, 261
Myosurus minimus, 130
Myrica cerifera, 303
  gale, 93, 266
Myriophyllum spicatum, 89, 266
  verticillatum, 89, 93, 198, 266
Myristica fragrans, 237

Nardus stricta, 92
Narthecium ossifragum, 128
Nasturtium amphibium, 264
  officinale, 264
  sylvestre, 264
Neottia nidus-avis, 130
Nepeta cataria, 98, 101
Nerium oleander, 240
Nicotiana tabacum, 110
Nopalea coccinellifera, 237

INDEX OF FAUNA

CAMBRIDGE: PRINTED BY W. LEWIS, M.A., AT THE UNIVERSITY PRESS

Printed in the United States
By Bookmasters